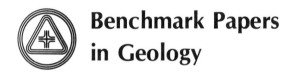 **Benchmark Papers
in Geology**

**Series Editor: Rhodes W. Fairbridge
Columbia University**

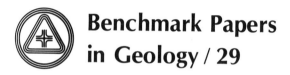

Benchmark Papers
in Geology / 29

A BENCHMARK® Books Series

METALLOGENY AND
GLOBAL TECTONICS

Edited by

Wilfred Walker

Dowden, Hutchinson & Ross, Inc.

STROUDSBURG, PENNSYLVANIA

Distributed by

HALSTED
PRESS

A Division of
John Wiley & Sons, Inc.

LIBRARY OF CONGRESS CATALOGING IN PUBLICATION DATA

Main entry under title:
Metallogeny and global tectonics
   (Benchmark papers in geology/29)
   Includes indexes and bibliography.
   1. Ore-deposits—Addresses, essays, lectures.
2. Geology, Structural—Addresses, essays, lectures.
3. Geodynamics—Addresses, essays, lectures. I. Walker,
Wilfred.
QE390.M47 1976   553′.1   76–3605
ISBN: 0–87933–208–5

Exclusive Distributor: **Halsted Press**
A Division of John Wiley & Sons, Inc.
ISBN: 0–470–15051–3

ACKNOWLEDGMENTS AND PERMISSIONS

ACKNOWLEDGMENTS

THE INSTITUTION OF MINING AND METALLURGY—*Transactions of the Institution of Mining and Metallurgy*
Evolutionary Trends in Ore Deposition

INTERNATIONAL GEOLOGICAL CONGRESS—*Report of Section 4, 24th International Geological Congress*
Time- and Stratabound Ore Deposits and the Evolution of the Earth

MCGRAW-HILL BOOK COMPANY—*The Ore Magmas*
Excerpts

PERMISSIONS

The following papers have been reprinted and/or translated with the permission of the authors and copyright holders.

AMERICAN GEOLOGICAL INSTITUTE—*International Geological Review*
Basic Problems of Investigation in the Field of Metallogeny

AMERICAN GEOPHYSICAL UNION—*Journal of Geophysical Research*
Mountain Belts and the New Global Tectonics
The Origin of Metal-Bearing Submarine Hydrothermal Solutions

AMERICAN INSTITUTE OF MINING, METALLURGICAL, AND PETROLEUM ENGINEERS, INC.—*Transactions of the Institute of Mining and Metallurgical Engineers*
Structure of Ore Districts in the Continental Framework

AMERICAN JOURNAL OF SCIENCE (YALE UNIVERSITY)—*American Journal of Science*
The Distribution of Mineral Dates in Time and Place

CANADIAN INSTITUTE OF MINING AND METALLURGY—*The Canadian Mining and Metallurgical Bulletin*
The Evolution of a Metallogenic Province at a Consuming Plate Margin: The Andes Between Latitudes 26° and 29° South
Geological Observations on the Delbridge Massive Sulphide Deposit
Nickel Sulphide Deposits—Their Classification and Genesis, with Special Emphasis on Deposits of Volcanic Association

Acknowledgments and Permissions

CROWN COPYRIGHT (CANADA)—*Symposium on the Basins and Geosynclines of the Canadian Shield*
   The Abitibi Orogenic Belt

CROWN COPYRIGHT (NEW SOUTH WALES)—*Journal of the Geological Society of Australia*
   A Plate Tectonic Model of the Palaeozoic Tectonic History of New South Wales

DOVER PUBLICATIONS—*The Origins of Continents and Oceans*
   Author's Forward

ECONOMIC GEOLOGY PUBLISHING COMPANY—*Economic Geology*
   Geological Setting of the Betts Cove Copper Deposits, Newfoundland: An Example of
   Ophiolite Sulfide Mineralization
   Geology of the Kalamazoo Orebody, San Manuel District, Arizona
   Neptunist Concepts in Ore Genesis

ELSEVIER SCIENTIFIC PUBLISHING COMPANY
   Earth and Planetary Science Letters
   Mesozoic Structural-Magmatic Pattern and Metallogeny of the Western Part of the
   Pacific Belt
   The Geochronology of Equatorial Africa
   Some Conclusions
   Geosynclines
   Figure 23

GEOLOGICAL ASSOCIATION OF CANADA—*The Geological Association of Canada, Special Paper Number 9*
   The Evolution of the Rice Lake—Gem Lake Greenstone Belt, Southeastern Manitoba

GEOLOGICAL SOCIETY OF AMERICA—*History of Ocean Basins*
   Abstract

INTERNATIONAL TIN COUNCIL and K. F. G. HOSKING—*2nd Technical Conference on Tin of the International Tin Council*
   Aspects of the Geology of the Tin-Fields of South-East Asia

INTERNATIONAL UNION OF GEOLOGICAL SCIENCES—*International Union of Geological Sciences*
   On the Depth Problem of Postmagmatic Deposits

MACMILLAN (JOURNALS) LTD.—*Nature*
   Continent and Ocean Basin Evolution by Spreading of the Sea Floor

MINING JOURNAL LTD.—*Mining Magazine*
   Further Reflections on Ore Genesis and Exploration

QUEEN'S COLLEGE—Metallogenic Provinces and Metallogenic Epochs

C. A. REITZELS FORLAG—*Meddelelser om Grønland*
   The Petrology of the Skaergaard Intrusion, Kangerdlugssuaq, East Greenland

UNIVERSITY OF CHICAGO PRESS—*The Journal of Geology*
   Island-Arc Evolution and Related Mineral Deposits

SERIES EDITOR'S PREFACE

The philosophy behind the "Benchmark Papers in Geology" is one of collection, sifting, and rediffusion. Scientific literature today is so vast, so dispersed, and, in the case of old papers, so inaccessible for readers not in the immediate neighborhood of major libraries that much valuable information has been ignored by default. It has become just so difficult, or so time consuming, to search out the key papers in any basic area of research that one can hardly blame a busy man for skimping on some of his "homework."

This series of volumes has been devised, therefore, to make a practical contribution to this critical problem. The geologist, perhaps even more than any other scientist, often suffers from twin difficulties— isolation from central library resources and immensely diffused sources of material. New colleges and industrial libraries simply cannot afford to purchase complete runs of all the world's earth science literature. Specialists simply cannot locate reprints or copies of all their principal reference materials. So it is that we are now making a concerted effort to gather into single volumes the critical material needed to reconstruct the background of any and every major topic of our discipline.

We are interpreting "geology" in its broadest sense: the fundamental science of the planet Earth, its materials, its history, and its dynamics. Because of training and experience in "earthy" materials, we also take in astrogeology, the corresponding aspect of the planetary sciences. Besides the classical core disciplines such as mineralogy, petrology, structure, geomorphology, paleontology, and stratigraphy, we embrace the newer fields of geophysics and geochemistry, applied also to oceanography, geochronology, and paleoecology. We recognize the work of the mining geologists, the petroleum geologists, the hydrologists, the engineering and environmental geologists. Each specialist needs his working library. We are endeavoring to make his task a little easier.

Each volume in the series contains an Introduction prepared by a specialist (the volume editor)—a "state of the art" opening or a summary of the object and content of the volume. The articles, usually some thirty to fifty reproduced either in their entirety or in significant extracts, are selected in an attempt to cover the field, from the key papers of the last century to fairly recent work. Where the original works are in foreign languages, we have endeavored to locate or commission translations. Geologists, because of their global subject, are often acutely aware of the oneness of our world. The selections cannot, therefore, be restricted to any one country, and whenever possible an attempt is made to scan the world literature.

To each article, or group of kindred articles, some sort of "highlight commentary" is usually supplied by the volume editor. This should serve to bring that article into historical perspective and to emphasize its particular role in the growth of the field. References, or citations, wherever possible, will be reproduced in their entirety—for by this means the observant reader can assess the background material available to that particular author, or, if he wishes, he too can double check the earlier sources.

A "benchmark," in surveyor's terminology, is an established point on the ground, recorded on our maps. It is usually anything that is a vantage point, from a modest hill to a mountain peak. From the historical viewpoint, these benchmarks are the bricks of our scientific edifice.

RHODES W. FAIRBRIDGE

CONTENTS

Contents

Contents

CONTENTS BY AUTHOR

METALLOGENY AND GLOBAL TECTONICS

INTRODUCTION

Metallogenesis: the genesis of metal—the scope of the term is awe-inspiring. One adds detail: sources of metal, magmatic rocks, depositional settings. This, though, is approaching the scope of economic geology, and we should perhaps consider what de Launay had in mind when he founded the science of metallogeny.

A contemporary, Finlayson (1910), gave full credit to de Launay for "the geographical method of studying ore deposits," and sixty years later Ramović (1968) wrote:

> By the end of the 19th and at the beginning of the 20th century Louis de Launay (1892, 1906, 1913) founded metallogeny which was gradually developed into a separate branch quite distinct from the science of economic geology. Up to that time it was considered to be a part of the latter. De Launay made the first metallogenic maps for certain European countries and introduced and explained such terms as "Metallogenic Epoch" and "Metallogenic Province." He also explained why metallogenic maps differ from maps representing mineral deposits and from topographic maps of mineral occurrences.

Epoch and province, time and place, became the prime considerations in metallogeny. The morphological character of mineral deposits was relegated to a subordinate position. The description of mineral deposits is as important today as it was to Agricola (1556) and to Elie de Beaumont, who was described by Bateman (1951) as "the father of modern thought of the formation of mineral deposits." But mineral deposit description is not the theme of metallogeny. Finlayson showed this in his metallogenic map of the British Isles: not a single deposit or even an ore field is depicted; the map shows only tectonic provinces and granitic intrusions. The mineralization related to the tectonic provinces

1

was described in the accompanying text, which was, perhaps, a little too extreme a division. Time and place in relationship to ore is the theme of this volume, with "global tectonics" giving a cohesive understanding to some of the patterns. Global tectonics: continental drift, sea-floor spreading, and plate tectonics can all be embraced by the term. Geotectonic cycles, each with initial vulcanism, then early, syn-, and late tectonic orogeny, are accepted by many geologists for the last 200 million years, by some for 600 million years. Mobile belts of a particular era like the Paleozoic may include one or more geotectonic cycles, and within the term "global tectonics" we must now consider ten eras, covering all rocks yet dated, back to almost 4 billion years ago. In this context, no one will argue that Alfred Wegener (1967; see Paper 4, this volume) was the pioneer. In the words of King in the introduction to the new translation of Wegener's fourth edition, (1929), "Alfred Wegener's theory of continental drift may well be judged the most stimulating and fertile of the major geological concepts of this century."

Bilibin (1968; see Paper 2) placed metallogeny on a sound base; Hess (1962; see Paper 5), with his concept of sea-floor spreading, did as much for global tectonics. By the end of the 1960s the two themes were ready to be joined. We are now in the blossoming stage: the two themes have honorable histories, they show every indication of being compatible, but the graft has yet to bear fruit; no one has yet claimed success in finding new ore. One has the feeling that the concept-development stage is well advanced, routine metallogenic mapping in plate-tectonics concepts is getting started, and we are starting to think of the histories, the life cycles, of metals. Concrete results cannot be far in the future.

The search for ore led me to metallogeny, metallogeny to the pattern of geology. In the outcome, as a "Precambrian" geologist, I found myself systematizing 4 billion years of global geology, and systematize it does. Bilibin recognized the relationship of metallogeny and the tectonic cycle, Gastil (1960; see Paper 11) recognized that thermal events are worldwide and have been taking place since at least 3 billion years ago, and in this volume the theme is that metallogeny and global tectonics have gone hand in hand for 4 billion years.

My own role (Walker, 1974) has been to demonstrate that the tectonic cyclicity recognized for rocks of the last 200 or 600 m.y. (the Alpine and Paleozoic eras) was also the dominant feature of older rocks. Just as the Alpine and Paleozoic eras were worldwide, so too were the eight pre-Paleozoic eras. Intervals between the eras, presumably corresponding to changes in the pattern of convecting mantle cells, were marked by dramatic changes in climate, from frigid, on the evidence of tillites, to tropical, on the evidence of kaolinite. Huronian-type intracratonic basins also mark these intervals of change. In each tectonic cycle and interval, similar patterns of mineralization appear.

A nomenclature for the cycles and for the eras and intervening per-

2

iods has yet to be accepted. The suggestions in Table 1 are from Walker (1975). As with the Paleozoic, it is recommended that the era be the period of development of a mobile belt system.

Table 1

Era	Approximate age in m.y.	Interval
Alpine	200–10	Pleistocene
Paleozoic	570–270	Permo-Triassic
Baikalian	900–600	Varangian
Grenvillian	1600–950	Roan
Karelian	1950–1650	Cuddapah
Nullaginian	2200–2000	Transvaal
Superior	2800–2350	Huronian
Pilbaran	3100–2800	Great Dyke?
Swazilandian	3500–3200	
Godthaabian	3900–3600	Ameralik

Serial, nonreversible changes of ore deposition throughout time were considered by Pereira and Dixon (1965, see Paper 13): as cumulate, the continental crust, is skimmed off; the residium, the mantle, is modified. And so linear and cyclical changes accompany each other.

This volume starts with the classical papers of both metallogeny and global tectonics, goes on to the merger, expands the theme from 200 m.y. to 4,000 m.y., develops the theme with papers illustrating metallogenesis in each stage of the tectonic cycle and in the intervals between eras, shows the additional effects of major cross-structures and metallogenic provinces, and ends with a paper on metallogenic mapping, the systematics of metallogeny.

In covering the field, one attempts to give credit to the leading worker. It is rare to find that he gives no credit to earlier workers, and one soon goes back through generations of thought to Agricola, Werner, and Elie de Beaumont. In practice, most of the leaders are well known—Wegener and Hess, de Launay and Bilibin, Holmes and Wager, Haddon King and Garlick, Lowell and Guilbert, Hosking, and Mitchell, Sawkins, and Sillitoe. Some workers, Naldrett and Goodwin for examples, give frequent reports on the developing state of the art (at times with co-workers) and the paper selected may be the original or the most embracing: I prefer the latter. The ideas of the Conzinc group, led in Australia by Haddon King, spread to Canada in the late 1950s under Moss and Edwards, but the most illustrative paper was not written for another ten years, by Boldy (1968; see Paper 20), who came late to the scene. Meanwhile, in Ireland, Pereira (1963; see Paper 26), another member of the group, was developing broader ideas on changes with time.

3

Several volumes readily available complement this one and should equally be on the bookshelf of those interested in metallogeny and global tectonics. Some of the papers in these volumes merit Benchmark status but their omission from the short list of suitable titles has allowed inclusion of papers of similar merit. Specifically, no papers are included that appear in the compilation editions of *Scientific American* and the *Journal of Geophysical Research* nor the symposium volumes of *Economic Geology* and the Geological Societies of Canada and Australia. Similarly, Krauskopf's section on source rocks in Barnes's *Geochemistry of Hydrothermal Ore Deposits* is on most bookshelves.

As volume editor, I give as background my years at Durham (which immediately followed those of Arthur Holmes, whose stature and thoughts on global geology and time were the air we breathed) under the leadership of Wager, who in those days had gone as high on Everest as any man and whose intensive work centered on differentiation. The Canadian branch of the Hunting organization provided my early training, particularly with Stu Scott, and their worldwide interests have stood me in good stead. More recently, Barringer Research has provided an atmosphere of advancing exploration technology in geophysics and geochemistry, which is a challenge every geologist involved in that organization must meet; as a consultant it is a privilege to work with them. Finally, I would thank Dave Strong for inviting me to attend the NATO Institute of Advanced Studies meeting in Newfoundland on metallogeny and plate tectonics. With Sam Sawkins, Dick Sillitoe, Andrew Mitchell, Magnus Garson, Phil Guild, Alan Clark, John Gabelman, Takeo Sato, Ei Horikoshi, Walther Petrascheck, Frank Vokes, Roly Ridler, Andy Glikson, Nick Badham, George Constantinou, Bob Hodder, and Eigen Stumpfl, we stood up to Newfoundland weather and the machinations of the brilliant at Memorial University of Newfoundland, who as reviewers as well as knowledgeable guides had the pleasure of tearing apart the best work of the supposed fundi. So if there is a sense of family about this volume, it is not surprising. Most of us do know each other.

For a short period in the mid 1950s—just a few months in fact—I was with Anglo American, where Britt Brock was doing some highly original thinking on the broad scale of geology (see Brock, 1972). Some introductory comments may help the present reader as much as they have helped me.

A MATTER OF SCALE

Brock (1972) pointed out that

> Analysis entails pulling a thing to pieces to find out what it is composed of. It is the fashionable scientific method. Both in physics and chemistry, analysis as a method of science has been spectacularly successful. In physics it has led to the splitting of the atom; in chemistry, to the

discovery of new elements and new wonders concerning molecules and their arrangement. Analysis has been so successful that there is a strong tendency to regard it as the method of science, almost to the exclusion of its counterpart, synthesis. The tendency is strongly discernible in the attitude of field geologists who, in following that precept, work only downwards from their field scale–restricted upwards by the boundaries of their area, but unrestricted downwards except for the lower limitations of microscope, spectroscope and chemical analysis. The common cry, when megascopic and microscopic methods have failed to give the answers to a tectonic problem, is for more and more field work. Wide-angle viewing, the third member of the trio has either been forgotten or is outside the terms of reference.

TABLE OF SCALES

RECONCILIATION OF TWO OPPOSED THOUGHT-PROCESSES

ASTRONOMY

APPROX. RANGE OF SCALES	STRUCTURAL STUDY	RELATED MINERAL STUDY	SCALE OF VIEWING	
1 : 35 m	GLOBAL TECTONICS	THE GLOBE AS AN ORDERED BODY WITH ORDERED STRUCTURAL AND MINERAL PATTERNS	GLOBAL VIEWING	END of SYNTHESIS — PROPER START FOR ANALYSIS
1 : 10 m	CONTINENTAL FORM & STRUCTURE	CONTINENTAL MINERAL PATTERNS	WIDE ANGLE VIEWING	NEGLECTED ZONE
1 : 5 m	SUB-CONT OR LOBE	MINERAL REGIONS	WIDE ANGLE VIEWING	
1 : 1 m / 1 : 500,000	REGION / TERRITORY	MINERAL REGIONS	WIDE ANGLE VIEWING	—ANALYSIS WRONGLY BEGINS WITH THE FIELD SCALE
1 : 200,000 / 1 : 25,000	FIELD WORK	METALLOGENIC PROVINCES	MEGASCOPIC VIEWING	SYNTHESIS ↕ ANALYSIS
1 : 2,500	WORK	MINERAL DEPOSITS	MEGASCOPIC VIEWING	
1 : 1				
50 : 1 / 500 : 1	LABORATORY	MICROSCOPY: MINERAL AND ROCK DETERMINATION PARAGENESIS ETC.	MICROSCOPIC VIEWING	
	NUCLEAR PHYSICS	ABSOLUTE AGE DETERMINATION		

The potentially commanding position of global viewing on the table of scales shows it to be the starting point for the natural tendency to analyze. Before analysis from the vantage point can begin, however, one must be sure that the synthesists have gone that high, and that their data have not become oversynthesized in the process. For synthesists seem to be a race apart, having abandoned the fashion of scientists.

Brock was breaking his own trail, and no doubt the way was hard, but Suess and Stille, Aubouin, Dewey and Bird, Tuzo Wilson, and Bilibin are all synthesists, and perhaps leaders of scientific fashion.

Editor's Comments
on Papers 1 and 2

1   DE LAUNAY
     Sur les types régionaux de gîtes métallifères
     English translation: *On Regional Types of Metalliferous Strata*

2   BILIBIN
     Excerpts from *Metallogenic Provinces and Metallogenic Epochs*

The theme of Papers 1 and 2, the origins and development of metal-logeny, has had many contributors. The role of de Launay has already been noted in the introduction, and for that note and much of the following, I am indebted to Ramović (1968).

The concepts and techniques of de Launay were accepted by Lindgren for the western United States, Miller and Knight for Ontario, Wong for China, Watanabe for Japan, Obruchev for Siberia, Fersman for Mongolo-Okhotsky, Smirnov for Vostochnogo Zabaykalya, Huttenlocher for the Alps, and others.

In 1923, Niggli described the recurrence or cyclicity of magmatic groups and quantifying was attempted by Daly (1933). Buddington (1927) related metal concentration to petrographic provinces. Stille, in several papers in the 1940s, and Schneiderhöhn in 1941 developed ideas on the simultaneous evolution of orogenic and magmatic cycles, contributing to the rapid development of metallogeny. Petrascheck and Cissarz were early followers of Stille and Schneiderhöhn. Stille went so far as to tabulate magmatic activity (initial geosynclinal, main period, late, etc.) in different parts of the world against the Phanerozoic time scale. Preparation for Bilibin was thus well advanced, and one can only blame the language barrier, German and Russian to English, for a twenty-year delay in the transmission of metallogenic concepts to the English-speaking world.

The prefaces to Bilibin are almost introduction enough, but as the work is the key to this volume, a few comments will be made. Table I is the synthesis of the work. The sections in Table I correspond, although not strictly so, to the numbering (1 to 19) on pages 10 to 24. It is helpful in using this rather difficult work to transpose these numbers to the right side of the table. Bilibin did not make specific application of his cycle to geological locations, and the most simple application is by recourse to Aubouin (1965; see Paper 3), who matched similar development to the last 200 m.y., the Alpine era; to Dewey and Bird (1970; see Paper 7), who matched it to the Alpine and Paleozoic eras; and to McCartney and Potter (1962), who matched it to the Paleozoic era. Zonenshain and co-workers (1974; see Paper 9) showed repetitions of the cycle within the Alpine era in the Far East, and Scheibner (1973; see Paper 10) showed repetitions within the Paleozoic era in New South Wales.

Reprinted from *Compt. Rend.*, 130, 743–746 (Jan.-June 1900)

SUR LES TYPES RÉGIONAUX DE GÎTES MÉTALLIFÈRES

L. de Launay

« Chaque région géographique présente un type particulier de gîtes métallifères, caractérisé, tant par la nature même des minerais que par l'allure de leurs dépôts. Un gîte de Suède ne ressemble pas à un gîte du Plateau Central, ni celui-ci à un gîte d'Algérie ; et surtout l'ensemble de la richesse minière offre des conditions très différentes dans l'un ou l'autre de ces trois pays. Mais ce contraste, qui est très frappant quand on parcourt l'Europe du nord au sud transversalement à ses grandes chaînes de plissement, disparaît, au contraire, presque complètement, si l'on se déplace de l'ouest à l'est en suivant l'une ou l'autre de ces chaînes de plissement. Le faciès, que nous avons d'abord considéré comme caractéristique d'une région géographique, l'est, en réalité, beaucoup plutôt d'une chaîne géologique, et l'expérience acquise dans une des parties de cette chaîne peut servir, jusqu'à un certain point, dans toutes les autres. Par exemple, la chaîne hercynienne constitue, d'un bout à l'autre, un ensemble très homogène, et la Meseta espagnole, le Plateau Central français, les Vosges, la Forêt-Noire et la Bohême ont une distribution de la richesse minière singulièrement analogue.

» Ce fait intéressant, qui n'avait pas encore, je crois, été mis en lumière, peut avoir plusieurs causes.

» Tout d'abord, l'idée vient aussitôt que le type régional des gisements métallifères doit correspondre au type non moins particularisé des diverses manifestations éruptives en un même point, à cet air de famille que des études récentes ont interprété par une différenciation progressive des mêmes magmas. Les analogies, que peuvent présenter, d'un bout à l'autre d'une même chaîne, les roches éruptives, aussi bien que les formes de dislocations, se retrouveraient alors dans les gisements en relation avec ces deux catégories de phénomènes. On peut même se demander si l'âge de la chaîne, avec lequel l'âge de la plupart des gisements qu'elle contient concorde approximativement, ne marque pas, dans l'évolution géologique de la croûte terrestre, une phase déterminée, ayant imprimé aux minerais comme aux roches un caractère propre et, en quelque sorte, un numéro d'ordre. Mais, tout en admettant fort bien ces explications, je crois qu'on peut leur ajouter une hypothèse générale, ayant peut-être l'avantage de coordonner, en les expliquant, un très grand nombre de faits d'observation.

[*Editor's Note:* An English abstract follows this original.]

» Cette hypothèse, c'est que le type régional des gîtes métallifères, type
en rapport, comme je viens de le dire, avec l'âge de la chaîne de plissement correspondante, résulte, avant tout, de la profondeur jusqu'à laquelle
l'érosion a pu, depuis ce plissement plus ou moins ancien, décaper les
parties hautes du gîte; autrement dit, plus un gisement serait ancien, plus,
les autres conditions restant les mêmes, nous en connaîtrions en moyenne
les parties profondes, tandis que, dans les gîtes très récents, les parties
qui nous sont accessibles sont encore celles qui, au moment de la cristal-
lisation des minerais dans les filons, touchaient à la superficie.

» S'il en était ainsi, nous nous trouverions résoudre, simplement en
considérant une série de gîtes de plus en plus anciens, une des questions les
plus intéressantes de la Géologie appliquée, celle des *variations originelles
en profondeur* des formations métallifères.

» Dans les théories professées jusqu'ici, on a toujours attribué à ces
variations un rôle considérable et qui paraît logique, en remarquant l'influence qu'avaient dû exercer sur le dépôt des eaux la diminution de
température et de pression, la volatilisation des gaz au voisinage de la surface, etc. J'ai essayé pourtant de montrer, dans une série de Communications antérieures ([1]) que la plupart des faits interprétés ainsi comme ré-
sultant de modifications originelles dans les conditions de dépôt étaient,
en réalité, en relation très nette, non avec la surface topographique du
sol au moment de ce dépôt, mais avec la surface actuelle, très différente
de la précédente, telle que l'a produite une longue érosion, et qu'il fallait
les expliquer par des altérations superficielles et secondaires, par des
remises en mouvement tout à fait récentes. Il n'en résulte pas que des modifications originelles n'aient pas dû se produire jadis, au moment où les
eaux métallisantes circulaient dans les crevasses du sol et se minéralisaient
à la rencontre des fumerolles dégagées par les roches en ignition. Mais
ces phénomènes devaient présenter une amplitude supérieure à celles sur
lesquelles peuvent porter la plupart de nos travaux de mines et une hypothèse telle que celle que je propose paraît seule permettre de les étudier.

» On peut ajouter, d'ailleurs, que dans les régions très anciennement
érodées, l'abrasion, le nivellement du sol ont été généralement poussés si
loin que le niveau hydrostatique actuel se confond presque avec la superficie, en sorte que la zone alternativement immergée par des eaux oxy-
dantes ou desséchée, sur laquelle portent surtout les altérations météo-
riques, est très restreinte, tandis qu'elle sera, au contraire, considérable

([1]) *Comptes rendus,* 22 et 29 mars, 14 juin 1897.

dans les chaînes récentes à profil encore très accidenté. Si donc on observe de tels phénomènes d'altération sur un gisement de ces chaînes anciennes, c'est, en général, dans des conditions très différentes, sur un compartiment de l'écorce qui a subi cette altération à une époque ancienne et qui, depuis ce moment, a été, par un enfoncement relatif, soustrait à l'ablation générale.

» Je ne puis songer à entrer ici dans le détail des faits (¹). Je vais seulement préciser encore par l'exemple de trois zones géographiques de plus en plus récentes et dont les deux dernières surtout sont à des stades incontestablement très différents de leur érosion.

« Si l'on part du nord de l'Europe, en Scandinavie, on trouve presque uniquement des gisements d'un type bien particulier : d'énormes amas d'oxydes de fer dans le terrain primitif; des masses et imprégnations de pyrites, pyrrhotines nickélifères et cuprifères, en relation plus ou moins directe avec des roches basiques de profondeur; des gîtes de ségrégation; puis, au sud, quelques veines à minerais du groupe stannifère en relation avec des granulites (un peu de bismuth, de molybdène et des métaux rares, zirconium, niobium, etc.). Au contraire, extrêmement peu de filons proprement dits (Kongsberg, etc.) et ne présentant pas de grandes fractures prolongées; pas de métaux volatils comme le mercure (sauf des traces à Kongsberg) et pas d'altérations superficielles donnant des carbonates de zinc, fer ou plomb aux affleurements des gîtes sulfurés. Le Canada est, de l'autre côté de l'Atlantique, exactement l'homologue de la Scandinavie.

» Dans le Plateau Central ou les autres tronçons de la chaîne hercynienne, le rôle des ségrégations pyriteuses au contact des roches basiques se restreint; on voit s'accentuer les gîtes du groupe stannifère en relation avec des roches très acides, qu'il est naturel de considérer comme ayant pu remonter plus haut dans un même plissement que les fonds de creuset basiques. Mais surtout ce qui prédomine, ce sont les grands décrochements, les longues fractures nettes, où ont cristallisé, en zones régulièrement concrétionnées, les trois sulfures presque inséparables de zinc, plomb et fer, avec du nickel, du cobalt et de l'argent.

» Enfin, si nous passons en Algérie ou sur un autre rameau secondaire des Alpes, le long des chaînes tertiaires, ce qui attire l'attention, ce sont les nombreuses fractures éparpillées, irrégulières, telles qu'il peut s'en produire dans les terrains déchiquetés, bouleversés, des saillies de plissement, fractures souvent sans continuité en profondeur, les filons-failles parallèles aux plis (dont on aurait de meilleures types en Amérique), ou les veines directement reliées avec des roches d'épanchement. Les minerais des magmas basiques profonds, tels que le nickel, le fer chromé ou oxydulé, les masses pyriteuses, ont disparu; on ne voit plus non plus de ces métaux qui sont une émanátio directe des roches acides pendant leur rochage et qui ont exigé, par suite, une haute pression avec une température élevée, comme l'étain, le bismuth, etc.; ce qui domine, c'est le plomb, le zinc, le fer, le cuivre, le mercure, en un mot, les mé-

---

(¹) Ils seront l'objet d'un article prochain dans la *Revue générale des Sciences.*

taux dont les sels ont pu rester en dissolution dans des eaux peu thermalisées et circulant à la façon de nos eaux thermales actuelles, où l'on en retrouve parfois de semblables.

» En même temps, les actions secondaires atteignent une intensité toute particulière, comme on peut le constater pour les gîtes de zinc de l'Attique, de la Sardaigne ou du sud de l'Espagne presque entièrement transformés en calamine, pour les gîtes de plomb du Taurus cilicien transformés en cérusite, pour les gîtes de fer de tant de mines de Styrie, des Alpes, des Pyrénées etc., où domine la sidérose.

» La superposition de ces trois zones, en négligeant le métamorphisme, représenterait non pas précisément les variations d'un même gîte ou d'un même faisceau de gîtes en profondeur, — car il faudrait encore tenir compte du caractère propre aux magmas cristallins, dont ces gîtes métallifères paraissent une émanation assez directe, — mais les divers types de gisements que pourrait espérer remonter une coupe idéale, menée à travers une chaîne de plissement jusqu'à ses racines les plus profondes. On arriverait ainsi à la notion : formations de profondeur, formations voisines de la surface, formation d'épanchement pour les minerais, comme on y est déjà arrivé pour les roches. »

11

1

ON REGIONAL TYPES OF METALLIFEROUS STRATA

L. de Launay

This English abstract was prepared expressly for this Benchmark volume by Dowden, Hutchinson, & Ross, Inc., Stroudsburg, Pennsylvania, from "Sur les types régionaux de gîtes métallifères," Compt. Rend., **130**, 743–746 (Jan.–June 1900)

Each geographic region is characterized by a particular type of metalliferous deposit—identified as much by the nature of the ore found there as by the general appearance of the deposit.

As one travels from northern to southern Europe, the contrast between strata is very evident. However, if one follows a particular flexure from west to east, it becomes clear that a given facies, until now considered characteristic only of a particular geographic region, is actually characteristic of an entire geologic chain; thus, any knowledge gained in one part of this chain will be applicable, to a certain extent, to all other parts.

It is hypothesized here that the regional type of metalliferous deposit is related to the age of the corresponding fold belt, and so results from the depth to which erosion has removed the highest formations; in other words, the older the deposit, the better we can know the deeper parts, whereas in very recently formed deposits, the parts accessible to us are those which were near the surface at the moment of the crystallization of ores in the veins. If this hypothesis is true, we can resolve the question of original variation in the depth of metalliferous formations merely by looking at a series of successively older deposits.

Until now, geological theory has attributed an important role to these original variations. However, the majority of events interpreted hitherto as resulting from original modifications in the condition of deposition are, in reality, directly related not to the topographic surface of the ground at the moment of deposition, but to the present surface; and these original modifications must be explained by superficial and secondary alterations, that is, by movement of very recent origin.

2

Reprinted from Y. A. Bilibin, *Metallogenic Provinces and Metallogenic Epochs*,
E. A. Alexandrov, trans., Queen's College Press, Flushing, N. Y., 1968,
prefaces and pp. 1–35

METALLOGENIC PROVINCES AND METALLOGENIC EPOCHS

Y. A. Bilibin

PREFACE TO THE ENGLISH TRANSLATION

Professor Yu. A. Bilibin was one of the most famous geologists of the Soviet Union. He was born in 1901 into the family of an army officer and was graduated in 1926 from the Department of Geology of the Leningrad Institute of Mines. Following his education he did geological field work in remote areas of the Far East, Siberia, Kazakhstan, and Soviet Turkestan. Making geological trips under difficult conditions over the Siberian taiga, polar tundra, and hot deserts of Turkestan, Bilibin showed courage and endurance. He earned the reputation of an experienced traveler and perceptive and analytical naturalist. Mountains, rivers, and settlements were named after him. The best known of these is the Bilibino Gold Mine. His early scientific interests were connected with petrology of abyssal magmatic rocks and geology of endogenetic ore deposits, predominantly those of gold. In 1947 he made public for the first time the results of his studies in the field of endogenetic metallogeny of geosynclines. The rest of his life up to 1952, when he died, was dedicated to the study of this problem.

The present book represents the posthumous publication of his lectures in endogenetic metallogeny of geosynclines, presented in 1950-1951 at the Leningrad State University. This book reflects the basic ideas of Bilibin on the problem of endogenetic metallogeny. His main scientific principle consists in interpretation of the processes of regional mineralization from the point of view of their historical geology. Thus he finds a close connection among tectonic evolution, sedimentation, and magmatism. Using this approach, Bilibin investigated the metallogeny of geosynclines most thoroughly. The geosynclines are considered as mobile zones of the earth's crust which evolve through a series of very definite stages into stable folded belts. Each such stage features specific formations of magmatic rocks and associated endogenetic ore deposits. The consecutive creation of these formations during the evolution of orogenic regions results in a regular change of endogenetic mineralization in time and in space, providing distributions known as metallogenic epochs and metallogenic provinces.

At the beginning, Bilibin's concept was received with caution. But later, its profound scientific logic, supported by numerous case histories and their great practical value as a guide in prospecting for mineral deposits, won the approval of most of the geologists in the Soviet Union. Bilibin's ideas are still very influential and will continue for a long time to affect the research and the thinking of Soviet geologists. This is why Bilibin's book, published over ten years ago, remains important today.

Certainly, Bilibin's ideas did not remain stagnant and have been further developed by his followers. The modernization has proceeded in two main directions. The first direction developed because Bilibin considered the transformation of a geosyncline into a folded belt during a single cycle of evolution. This required elaboration of his points of view with respect to the polycyclic orogenic regions. The second direction is connected with the analysis of distribution features of magmatic formations and mineralization forming consecutively during the evolution of the geosyncline, within its main tectonic zones. However, all these modifications have not damaged but have rather strengthened the scientific credo of Bilibin.

Up to the present time, Bilibin's scientific ideas have been successfully used by the geologists of the Soviet Union. The translation of his book into English will make it possible to check the practical value of his principles in the United States and other countries with specific geological structures and specific mineralization of their territories. Let us hope that regardless of the result of such verification, the publication of this translation will contribute to the progress of geology as a whole and its applied branch of the study of mineral deposits.

Professor Academician V.I. Smirnov.

Moscow, December 17, 1966.

PREFACE TO THE RUSSIAN EDITION

The book by Yu. A. Bilibin on Metallogenic Provinces and Metallogenic Epochs is derived from lectures given by the author at the Leningrad Zhdanov State University in 1950-1951. It reflects most fully the author's ideas concerning this problem, which are the product of his fruitful, practical activities and scientific endeavour. A method of regional metallogenic analysis providing an efficient way of prospecting for endogenetic mineral deposits is suggested here by the author. This method was used with success in various regions of the Soviet Union.

In his work Bilibin generalized the results of the time-consuming efforts of large groups of scientists and field geologists whom he directed. The existing metallogenic theories proposed by Russian geologists and by authors outside the Soviet Union were critically examined and further trends in metallogenic science were delineated. This book completes the scientific activity by Bilibin and reflects a new original trend in a century-old practice of prospecting, which has resulted in and continues to result in discoveries of various mineral deposits as well as of entire ore-bearing regions.

Bilibin's work is especially important at the present time, when large groups of Soviet geologists are preparing metallogenic maps and are analyzing the distribution features of mineral deposits in the earth's crust to increase the mineral resources as fast as possible.

Translator's Preface

Bilibin's book about metallogenic provinces and metallogenic epochs laid the foundation of metallogeny—a new branch of the geology of mineral deposits. Bilibin had predecessors, but his approach to this problem is not only broader than theirs but original. This book was selected for translation and publication as the first issue of the Geological Bulletin of Queens College because it can serve as an introduction to the field of metallogeny and because of its potential value to economic geologists, students of historical geology, petrologists, and geochemists, interested in the evolution of the earth's crust and the distribution of mineralization in time and space.

The translation of this book was made with the intention of keeping it as close as possible to the original Russian text. A map of metallogenic provinces by V.I.

Smirnov (Fig. 1) and the diagram by N.M. Strakhov (Fig. 2) depicting the cycles of sedimentation during the Postalgonkian history of continents were added to the English translation of Bilibin's book. R.M. Finks, D. Thurber and L. Cogan helped with the editing and proofreading of the English manuscript.

Bilibin's criticism of Western categories appears too categorical. His conclusions are made with great authority, but he does not claim his scheme to be universal, if applied outside the Soviet Union. He designed it mainly for the territory of his country. However Bilibin's principal idea remains inspiring and should be used as a method in studying the metallogenic epochs and provinces from the theoretical, practical and economic points of view.

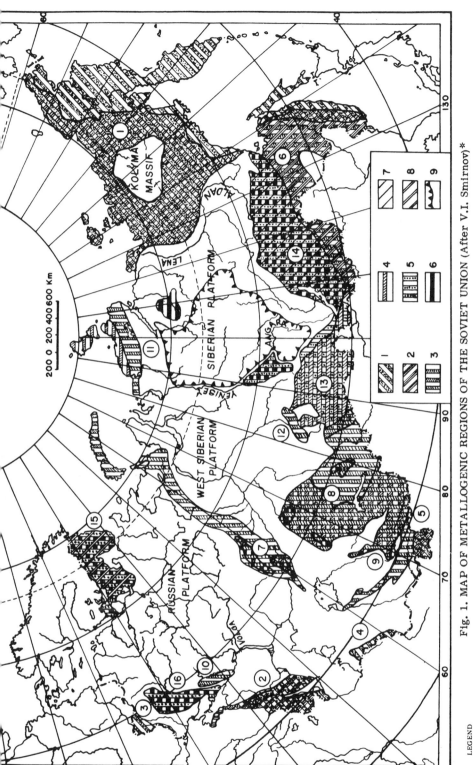

LEGEND

Distribution of metallogenic epochs of geosynclinal cycles	Ore-bearing provinces (figures in circles)

Fig. 1. MAP OF METALLOGENIC REGIONS OF THE SOVIET UNION (After V.I. Smirnov)*

1. Alpine (Mesozoic-Cenozoic**)   3. Hercynian (Upper Paleozoic)   5. Proterozoic
2. Kimmerian (Triassic-Jurassic).   4. Caledonian (Lower Paleozoic)   6. Archean

Alpine. 1. Far Northeast.   2. Caucasus.   3. Carpathians.   4. Kopet Dag.   5. Pamirs.
Kimmerian.   6. Transbaikal-Maritime Region.   7. Urals.   8. Kazakhstan.   9. Turkestan.   10. Donets Basin.
   11. Taimyr.   12. Tom-Kolyvan Zone.
Caledonian. 13. Altai-Sayan Zone.
Proterozoic. 14. Southern part of Siberian Platform.   15. Baltic Shield.   16. Ukrainian Shield.

*) Essays of Metallogeny, Moscow, 1963.
**) Translator's notes—in parentheses.

15

I. INTRODUCTION

The concept of metallogenic provinces and metallogenic epochs. If the distribution of various types of ore deposits over the globe is examined, it will be evident that the deposits of various metals are distributed quite irregularly. In some parts of the world, copper deposits occur most commonly while deposits of other metals are less frequent in these areas and deposits of a third group of metals are absent entirely. At the same time, deposits of gold, lead, zinc, tin, and other metals predominantly occur in various other parts of the world. It is difficult to indicate two large regions on the earth which contain entirely similar associations of ore deposits. Each of these regions is characterized by its own specific features, a characteristic type of association of ore deposits and various magnitudes in quantitative evolution of the various genetic types of deposits.

This fact drew the attention of scientists for many years and resulted in the development of the notion of metallogenic provinces. It is quite clear how important this notion is from the practical and theoretical points of view. From the practical point of view it is useful in the evaluation of economic prospects of large regions or provinces according to the ores of given metals and thus to the determination of the most efficient line of prospecting. From the theoretical point of view the notion of metallogenic provinces is useful in understanding the distribution features of various types of ore deposits and thus in determining the causes and conditions of their formation.

Despite the fact that the notion of metallogenic provinces has existed for quite a long time, it has had a purely empirical meaning and is still interpreted in empirical terms even today. It is commonly understood that the metallogenic provinces are geologically defined and sometimes even geographically defined provinces, which are characterized by the development within their boundaries of a definite type or types of ore deposits, and acquire on that account metallogenic features. With such an interpretation, the concept of metallogenic provinces is merely of value as a record, and confirms only the established empirical features with respect to the geographical distribution of one or another kind of ore deposits.

However, the notion of metallogenic provinces as it was formulated earlier does not mean anything with respect to their origin, the causes of differences between the provinces or their features. It is true that Lindgren attempted to include into the definition of metallogenic provinces some ideas about the conditions of formation of ore deposits. But this attempt remained only a verbal exercise. Lindgren defined the metallogenic (metallogenetic) or minerogenic (minerogenetic) provinces as

considerable areas at the surface and underground where favorable conditions existed for deposition of certain mineral deposits. However, it makes no difference whether we say that definite mineral deposits occur in a given region or that in the region the conditions were favorable for the formation of mineral deposits. None of these definitions, which differ little in their meaning indicate ways and means for further investigations, nor do they bring us closer to solving the problem of the origin of metallogenic provinces.

In each metallogenic province the ore deposits did not form during some single stage of mineralization but during several successive stages. However, the idea of metallogenic provinces sometimes refers only to one definite stage of mineralization in a given province. For example, references are made to the platinum-bearing province of the Urals (fig 1), to the mercury-antimony-bearing province of Soviet Turkestan and to others. In other cases the notion of metallogenic provinces refers to the entire association of mineral deposits of a given province which formed during all stages of mineralization, such as the provinces of the Urals, of Soviet Turkestan, of the Eastern Transbaikal region, and others including in this notion the entire association of ore deposits in each of the provinces. Thus far there is no agreement about the understanding of the meaning of the term "metallogenic province."

Intervals of geological time favorable for the deposition of a definite mineral may be designated as metallogenic or minerogenic epochs. At the same time it is considered that the metallogenic or minerogenic epochs cover not some single metallogenic province, but a whole group of provinces, sometimes even separated from each other by quite great distances or by a considerable part of the globe. For example, it is known that the Late Mesozoic metallogenic epoch of lead-zinc mineralization covered North America, the eastern part of Asia some regions of Europe, and other areas. Not of lesser extent was the Tertiary metallogenic epoch of the epithermal gold-silver mineralization which occurs in the entire inner zone of the Pacific belt (the Andes, Cordillera, Japan, Philippines, Indonesia, and New Zealand) and to some extent in Europe (Hungary, Transylvania) The individual stages of mineralization which develop only within the boundaries of some metallogenic province are usually not considered as metallogenic epochs. A purely empirical definition of metallogenic or minerogenic epochs presented earlier, as well as that of minerogenic provinces, does not indicate practical ways for further investigations and for solution of the problem.

The problem of interpreting the causes, and the conditions of formation, of minerogenic provinces and minerogenic epochs of exogenetic mineralization, has been treated more satisfactorily because the processes of exogenetic mineralization take place at the earth's surface in an environment which can be easily observed. Changes of this environment in the course of the history of the earth have been covered with sufficient success by historical geology. In most cases, historical geology supplies clear answers to the question about the causes and conditions of formation of minerogenic provinces and epochs of exogenetic mineralization, giving answers which delineate clearly the directions for further study of the problem.

It is commonly known that geologists studying the problems connected with the history of the geological evolution of the earth's crust are persistently unwilling to deal with the investigation of such an "obscure" part of the history as the evolution of magmatism and endogenetic mineralization. This is to the detriment and in violation of one of the basic principles of dialectical materialism* —the principle of universal approach in research. At the same time it is quite obvious that this is one of the main tasks of historical geology. Deprived of any substantial assistance, the geologists who study the problems of endogenetic mineralization have to help themselves by their own means. Therefore, it is not surprising that the study of the problem of the evolution of endogenetic mineralization in the earth's crust in general, and the causes and conditions of formation of minerogenetic epochs and provinces in particular are lagging very much behind as compared with the study of the same problems in the field of exogenetic mineralization.

The present work is concerned mainly with the endogenetic minerogenic provinces and epochs, as a problem not sufficiently elucidated and which has not been carefully studied. This appears to be more practical because Soviet geologists have come to some interesting conclusions in the study of endogenetic minerogenic provinces and epochs, making substantial revisions of the above mentioned concepts of metallogenic provinces, which, however, have so far not been reflected in geological literature.

Attempts to interpret the minerogenic provinces and epochs. Attempts to explain from the geological point of view the causes and conditions of formation of the endogenetic minerogenic provinces and epochs, and thus to attempt to interpret the observed empirical regularity in spatial distribution of various types of endogenous mineral deposits, have been made many times and continue to be made up to the present time. The well known batholithic concept by Emmons, which has been quite recently widely accepted outside the Soviet Union, is one of the factors which has impaired this solution of the problem of metallogenic provinces.

According to the batholithic concept developed by Emmons, the entire endogenetic postmagmatic mineralization is connected with granitoid batholiths. The batholiths are characterized as a universal source of mineralizing solutions, i.e., their postmagmatic solutions contain all (possible) ore-forming elements. Thus, the capacity of all granitoid batholiths to form ore deposits is more or less the same. The main differences are observed only as the result of the difference in levels reached by the erosion of batholiths. An attempt to explain all features and observed differences between

the metallogenic provinces by the depth reached by erosion, appeared from the point of view of Emmons to be entirely natural. Despite an obvious failure of such a point of view, it was quite popular and 15 to 20 years ago * quite widely accepted, even in the Soviet Union. It is enough to say that the geological exploratory organizations of the Soviet Union were advised to solve the problem of prospecting for new deposits of some minerals, for new ore-bearing regions, and for new provinces, from the point of view of depth reached by erosion. However, geologists following this approach were unable to discover any new ore-bearing regions or ore deposits and could not give a more or less satisfactory explanation of the distribution of already known deposits in the territory of the Soviet Union. There was no unanimity among the partisans of this point of view even in the interpretation of the depth reached by erosion.

It is surprising that even at the time when Emmons' hypothesis had failed completely, one of the textbooks on ore deposits published in 1950 contained an assertion that all metallogenic provinces are in principle of the same nature, while the differences between them are explained only by the difference in the level reached by erosion.

The failure of attempts to explain the visible differences between the metallogenic provinces solely by the depth reached by erosion forced the partisans of this point of view to call for an additional factor in the form of a concept of metallogenic epochs. It was assumed that endogenetic mineralization of a definite type developed as the result of specific conditions during definite geological epochs in a group of metallogenic provinces, sometimes quite remote from each other. This concept explained deviations from Emmons' hypothesis. In this interpretation some deposits in the Soviet Union turned out to be derivatives not of a definite metallogenic province, but of a definite metallogenic epoch, since a more or less similar mineralization developed in a group of other provinces—in North America, Western Europe, and others.

However, this additional factor could not explain the entire diversity of metallogenic provinces either. A conclusion appeared inevitable that in different metallogenic provinces apparently similar intrusions produce different types of mineralization. In one case they produced gold mineralization; in another, tungsten; and in a third case, lead-zinc mineralization; and so on. An idea of specialized intrusions took the place of the Emmons' concept of batholiths. This was an undoubtedly progressive idea which was inevitably followed by consideration of the problem of the succession of one generation of intrusions by another and the reasons for such a succession. In other words, the problem became a search for some definitely regular trend in the evolution of intrusive activity and endogenetic mineralization. This possibly explains why this trend did not develop further in the countries outside the Soviet Union, where it was represented by the formalistic attempts of Buddington, Grout, and others, to correlate the nature of mineralization with definite petrographic types of intrusive rocks. This trend turned out to be without prospects.

An attempt to find a solution to the problem created by Emmons' ideas was made by some geologists in a

---

* Philosophy advocating the priority of matter over mind (translator's note).

* In 1935–1940 (translator's note).

quite different direction. They especially stressed the inheritance of many features of the endogenetic mineralization in each separate metallogenic province and indicated that there are specific gold-bearing provinces in which quite different metallogenic epochs produce mainly gold mineralization. On the other hand there are copper-bearing provinces, in which copper mineralization is predominant, and so forth. These differences between the metallogenic provinces were explained by the initial, irregular distribution of various metals during the period of formation of the earth's primary crust. These points of view are reflected quite strongly by papers presented at the XVII Session of the International Geological Congress in Moscow in 1937. But these ideas failed to explain the appearance in the same metallogenic provinces of endogenetic mineralization of quite different kinds. It appeared that the orthodox theory of magmatic endogenetic mineralization had reached a dead end and a large number of geologists outside the Soviet Union began to abandon it. This tendency became so strong that it captivated some Soviet geologists.

The abandonment of the theory of endogenetic mineralization of magmatic origin proceeded in two directions. The first group of geologists recognized deep regional faults as the most important factor determining the nature of mineralization. The second group designated the host rocks as the most important factor. The first tendency was represented by the well known hypothesis of the so-called ultratectonists who believed that the endogenetic mineralization is not at all connected with the magmatic activity and that its source is in the submagmatic abyssal depth (this is approximately a restoration of the views by Descartes of the middle of the XVII century). The faults reaching abyssal depths create outlets for the ore-bearing emanations and the development of mineralization of one or another type is entirely controlled by the "powerful tectonic machine." According to the ultratectonists it is of no use to try to learn about the causes of endogenetic mineralization, since the primary source of the endogenetic mineralization is at depths inaccessible to our knowledge. Naturally, such an extreme trend did not develop further and did not earn any immediate recognition.

Another trend which deviated from Emmons' position was an attempt to find the source of endogenetic mineralization in host rocks. However, there were quite considerable differences in the interpretation of the nature of connections between endogenetic mineralization and the host rocks. Some of the geologists returned to the old hypothesis of lateral secretion, the advocates of which attempted during a period of one and a half centuries to earn some serious recognition. They believed that hot waters of magmatic origin or even those of surface origin strongly heated as the result of circulation at great depth, increased their useful load by leaching of the host rocks. Therefore, the composition of endogenetic mineral deposits depended on the type of rocks which the waters permeated on their way from the deep sources. Thus, according to these geologists, one should look for the explanation of differences and specific features of metallogenic provinces not in the specific nature of the evolution of magmatic processes but in the particular features in the evolution of sedimentary rocks. Other investigators, who consider the host rocks as the source of endogenetic mineralization, present a more complex picture. They believe that magma acquires definite ore-bearing features only in case of assimilation of one or another

kind of host rock. In some cases under these conditions there is a metallogenic enrichment of magma by specific ore components, while in other cases magma is enriched in volatile components. These volatile components contribute to the capacity of magma to separate the metals in hydrothermal solutions. Finally, in a third case, magma changes its composition as the result of assimilation and acquires the capacity to keep one kind of ore component and to supply others to the hydrothermal solution. But in all these variants the specific ore content in magma depends on the "powerful process of assimilation." The character and all features of a given metallogenic province depend on the evolution of the processes of assimilation and the character of rocks being assimilated.

According to P.N. Kropotkin, basic magmas are the richest in metallic components. But they are not able to produce these ore components in the form of hydrothermal solutions because of low silica content. On the other hand, in acidic magmas this capacity is developed to the highest degree, because they are rich in both silica and water. But these magmas have a low content of metallic components. The best conditions for development of endogenetic mineralization are when the early basic rocks (intrusive or extrusive) occur in places where later granitoids evolve and the processes of assimilation of basic rocks by the acidic rocks are well developed. As the result of this assimilation the granitic magma is enriched by ore components and in turn enriches the hydrothermal solutions with their ore components.

A.P. Nikol'skiy's views are concerned only with the origin of the tin-bearing granite magmas. He notes quite correctly that usually the tin-bearing granite magmas are most frequently associated with the regions featuring large shale deposits. Nikol'skiy interprets this connection in the ordinary manner. He believes that granite magma is enriched directly with tin during the assimilation of shale and becomes tin-bearing as the result of this process. However, Nikol'skiy has difficulty in giving a more definite answer to the question about the depth at which this assimilation takes place, since it turned out that in some cases shales occur only at shallow depths and are underlain by rocks of a different lithological composition. Therefore, according to Nikol'skiy's belief the tin-bearing metallogenic provinces are formed only in regions where shales are older than the granitic intrusions and make a considerable part of the host rocks.

Similar points of view were developed in more detail by Kh.M. Abdullayev. He believed that any granite magma is potentially ore-bearing but alone it is unable to mobilize its ore-bearing potential to produce economic ore deposits. This capacity to form ore deposits is acquired by magma only as the result of assimilation of host rocks. The host rocks enrich magma with volatile components, sometimes they supply magma with ore components of metallogenic importance, and finally they determine which of the ore components remain in the intrusive rock being formed and which become part of the hydrothermal solution.

Every parent magma is invariably acidic. It does not change its composition substantially as the result of assimilation of aluminosilicates. After the assimilation of carbonate or ferromagnesian rocks, the magma changes its composition toward less acidic granitoid (granodiorites, syenites, monzonites and others). Ac cording to this idea, two groups of granite intrusions are distinguished along with the affiliated ore deposits; the noncontaminated (or more correctly, granite intrusions contaminated with aluminosilicate material) and the contaminated intrusions accompanied by all known ore deposits. Abdullayev classifies all known ore deposits into these two groups. Otherwise, Abdullayev's views are not substantially different from those of Emmons. Abdullayev's granite intrusions are always represented by batholiths — this is the only possible type of ore-bearing intrusions he considers. According to him the ore-bearing capacity of batholiths is of a universal nature. That means that the batholiths may produce deposits of any metals, but it is only after the assimilation of host rocks that this capacity becomes a reality. As in Emmons' interpretation, the zoning around the batholith is of a purely spatial nature. This indicates that as the hydrothermal solutions are gradually depositing their mineral components, they are removed farther away from the batholith into the regions of lower pressures and temperatures.

Thus, one or another type of mineralization in each given metallogenic province does not depend on the nature of intrusive rocks, which are in all cases more or less the same type, but rather depend on the character of sedimentary rocks and the evolution of the processes of assimilation by granite magma. During this process the composition of the granitoids is changed and the features of their ore content are determined. Abdullayev believed that prospecting for one or another kind of mineral deposit should be directed by considering the character of sedimentary host rocks, at the same time taking into account the degree of assimilation.

A third group of geologists, recognizing the decisive contribution of the host rocks, attempt to connect the processes of endogenetic mineralization with the phenomenon of granitization. This trend was most completely expressed in Sullivan's work. He believes that the granitizing solutions remove many ore components from the host rocks creating the hydrothermal ore-bearing mineralizing solutions. Those metals, the ionic radii of which are close to those of the rock-forming elements of the granite, may become part of metasomatic granites forming during this process. The remaining metals are removed into the hydrothermal solution. Sullivan's views contain many strained interpretations and in many cases are contrary to the facts to such an extent that they resulted in several substantial objections even outside the Soviet Union.

Limiting himself to this short review of various points of view concerning the origin of metallogenic provinces and the factors controlling their nature for the time being, the present writer is not making a critical analysis of these points of view and will concentrate on such an analysis after examining the ideas of the metallogenic school created by the Allunion Geological Research Institute (VSEGEI). It should be mentioned only that not a single one of the various trends could solve entirely, or even partially, the problem of metallogenic provinces and metallogenic epochs.

Comparatively recently, systematic studies were started in this direction at the VSEGEI. A group of scientists who undertook the task of studying this problem came to several interesting conclusions concerning the causes of the development of metallogenic provinces, their features, internal structure, and so on.

Thanks to these conclusions, the general principles of regional metallogenic analysis were elaborated to a first approximation. These principles allow one to make a purely analytical approach to the evaluation of economic prospects for definite metals on the basis of information on the geological structure of a given region. This approach makes it possible to outline within this region definite structural zones which have the best prospects for each particular metal. The practical use of recommendations for the direction of prospecting made as a result of such an analysis culminated in discovery of ore deposits and ore-bearing regions. An examination should be made both of the points of departure from which this study was started and of the conclusions which resulted from these studies. The main ideas which served as the basis for this analysis are set forth below.

The processes of mineralization resulting in the formation of mineral deposits and, in particular, of ore deposits, represent one of the aspects of a unique and complex process of the geological evolution of the earth's crust. In their historical evolution they are closely connected with other aspects of the evolution, i.e., sedimentation, tectonic movements (evolution of structures), magmatic activity, and metamorphism. The processes of mineralization can and should be studied only from the point of view of their historical evolution and in closest correlation with all other aspects of the process of geological evolution of the earth's crust.

According to the two main groups of geological processes leading to the formation of mineral deposits, namely, the processes of hypogene or endogenetic origin, and the processes of supergene or exogenetic origin, it is expedient to subdivide the science of metallogeny into two major divisions—the endogenetic metallogeny and the exogenetic metallogeny. But since both the endogenetic and the exogenetic mineralizations reflect only the various aspects of a unique process of the geological evolution of the earth's crust, these two divisions of metallogeny are closely interconnected with each other.

At the present time, it is generally accepted that in the present stage of evolution of the earth's crust there are only two main types of structural elements within the continental block -- the platforms and the mobile belts, which may be in various stages of geological evolution. The processes of endogenetic mineralization continue for very long periods of time and are most intensive and diverse within the mobile belts of the earth's crust, where, because of these processes, the overwhelming majority of mineral deposits is concentrated. Within the boundaries of platforms, endogenetic mineralization was quite intensive during definite stages of their evolution. But such stages appear very sporadically, and the mineralization itself is considerably more monotonous. Therefore, elucidation of the correlation between the processes of endogenetic mineralization and the general course of geological evolution should first be studied in the endogenetic mineralization of mobile belts.

A subdivision of the territory of the Soviet Union into structural metallogenic regions was achieved to a first approximation, to determine the nature of this correlation. The territory covered by the study was subdivided into a series of large regions or provinces, inside of which were singled out smaller structural-metallogenic areas, distinguished from each other by some difference in the history of their geological evolution and the complex of endogenetic mineral deposits. The endogenetic mineralization was analyzed in 30 of these small regions. This analysis consisted of five successive stages.

1. First of all an attempt was made to group all known endogenetic mineral deposits of each region into natural associations united by a common magmatic source, common structure controlling the ore deposition, close time of formation, association with roughly one stage in the geological evolution of the region, etc. Such natural communities of ore deposits or mineral deposits were designated as ore or mineral complexes.

2. The intrusive rocks of each region, connected with each other by common origin from one parental magmatic source, and sometimes of a composition quite different from each other, are grouped into similar natural communities or intrusive complexes.

3. The establishment of a genetic connection of each mineral complex with one or another intrusive complex was not always an easy task, especially if this had to be done only according to the data from the literature. It is possible that in individual cases errors were made, but, as will be seen later, there is enough reason to think that in the overwhelming majority of cases such connections were established correctly.

4. The next problem was concerned with the establishment of the order of formation and age of intrusive complexes and of the corresponding mineral complexes. This made it possible to follow the evolution in time of the nature of endogenetic mineralization of each separate region. For convenience this succession was represented in special tables (by regions), where over 250 mineral and intrusive complexes were recorded.

5. Finally, the last problem in systematizing the factual information was the correlation of both the separate stages and the general course of the geological evolution of each region with the development of endogenetic mineralization within each region. In that way, the historical approach applied in the investigation of the problem of metallogenic provinces was sharply different from the earlier approach which consisted merely in recording the information. A comparison of tables of endogenetic mineralization for definite regions prepared according to the historical approach made it possible to make several important general conclusions. These conclusions were further used as the basis for the regional metallogenic analysis.

1) First of all it turned out that similar mineral complexes occur in different regions. These complexes include often identical and related associations of endogenetic mineral deposits, despite an often considerable distance between such regions and sometimes quite significant differences in geological age.

2) In a similar way the intrusive complexes in different regions and of different geological age turned out to be similar from the point of view of the association of intrusive rocks which made a part of these complexes.

3) Similar mineral complexes in different regions turned out to be connected with similar intrusive complexes. This extremely important circumstance speaks for the existence of a genetic connection between the endogenetic mineralization and intrusive rocks. Other-wise different regions studied independently from each other would not demonstrate such clearly developed common features. This circumstance confirms also the well known thesis by S. S. Smirnov about the existence of intrusive (or magmatic) complexes characterized by a specialized ore-bearing nature. And, finally, it indicates that the confirmation of genetic connection between mineral complexes and intrusions in different regions was made basically correctly. It is true that on the basis of revealed general features in individual cases some corrections of the originally outlined correlation had to be made when those correlations were doubtful. Nevertheless, during this comparison some subordinate differences have appeared both in the nature of the mineral deposits and in the nature of corresponding intrusive complexes in different regions. Apparently in one case these differences represent a purely spatial local feature of each individual region. In another case these differences are due to the differences in geological age. In a third case they are explained by the difference in the level reached by erosion.

4) It turned out that the intrusive complexes and the corresponding mineral complexes of different character succeed each other in time according to a more or less identical order, under conditions of different geological age and in regions quite remote from each other. The specific succession of mineral and intrusive complexes in individual regions varies to some extent deviating from this average pattern in one or another direction. Usually these deviations are demonstrated by the absence of one kind of complex while other complexes developed twice, or less frequently—several times. The normal succession of mineral and intrusive complexes of contiguous age changes less frequently. More or less identical successions of mineral and intrusive complexes made it possible to elaborate the patterns of their normal or most common succession in time.

5) During attempts to correlate, according to this pattern, the tables representing stages of endogenetic mineralization in separate regions with one or another stage in geological evolution of the corresponding mobile belts, it became evident that similar stages of endogenetic mineralization and intrusive activity correspond to similar stages in geological evolution of corresponding mobile belts, despite difference in geological age and position in space. Thus, the pattern of normal succession of mineral and intrusive complexes in time has a considerably deeper meaning and is closely connected with the general course of the geological evolution of the corresponding mobile zones. In other words, it represents not only the pattern of normal distribution of endogenetic mineralization in time, but as well a pattern of distribution during the course of geological evolution of mobile belts of the earth's crust. Thanks to this fact, it became possible, following the above mentioned pattern, to subdivide the process of endogenetic mineralization and intrusive rocks into several successive stages perfectly coinciding with those stages which may be designated according to any other geological process (sedimentation, tectogenesis, and others).

In this way, as the result of generalization about a large amount of information on endogenetic mineralization collected over a considerable part of the territory of the Soviet Union, it became possible to confirm the initial thesis about the existence of close correlation between the evolution of endogenetic mineralization magmatism, tectogenesis, and accumulation of sediments. The pattern of a normal succession of intrusive and endogenetic mineral complexes in the course of geological evolution of mobile belts of the earth's crust is represented by Table 1. Before starting on an analysis of this table, one should become familiar in very general terms with the main conclusions made by the members of the above-mentioned VSEGEI team of geologists with what concerns the distribution in space of various types of endogenetic mineral deposits.

The knowledge of the distribution in time of various types of deposits cannot be directly used as a guide to prospecting, even if it is closely linked to the general course of the evolution of mobile belts. Such guides would be useful only by analyzing the features of spatial distribution of various types of endogenetic deposits. Therefore it is necessary to make a preliminary switch from the distribution features of mineral deposits in time to the features of their distribution in space.

If the spatial distribution of various endogenetic mineral complexes is examined within the boundaries of definite metallogenic provinces, it becomes evident that no single mineral complex develops over the entire area of the given metallogenic province. On the other hand, each separate mineral complex corresponding to a definite stage in the evolution of endogenetic mineralization of a mobile belt was developed most intensively only within the boundaries of definite structural zones. At the same time this complex is absent or is represented only by a very insignificant development within the boundaries of all other structural zones of a given metallogenic province. Usually together with the deposits of a mineral complex those intrusions with which deposits of this mineral are genetically connected are most strongly developed within the same structural zones. This gives the writer reason to conclude that the formation of a structural zone as a definite structural unit must coincide closely in time with the intrusion of rocks corresponding to this structural unit and to the formation of ore deposits. In other words, the formation of structures as such, the intrusion of igneous rocks and the formation of ore deposits represent only the different aspects of a unique process of the geological evolution of a structural zone, corresponding to the late and sometimes the final stages of its formation which covers a long period of sedimentation in this structural zone. Each such structural zone also represents at the same time the structural-facies zone *, since it is characterized by definite sedimentary facies.

In such a manner, the features of sedimentation, structures, magmatism, and endogenetic mineralization distinguish each such structural zone from other structural zones of the same region. In the text to follow such zones will be called structural-metallogenic zones.

The examination of the mutual position of the successively forming structural-metallogenic zones reveals quite definite features. The earliest structural-metallogenic zones, either during the entire process of the evolution of the given mobile belt or during definite major stages of this evolution, are frequently associated with structures of an anticlinorial type. As indicated by facies analysis, these structures start to form as

---

*) Portion of the earth's crust characterized by definite tectonic evolution which, together with other physiographic factors, controlled sedimentation, magmatic activity, mineralization, and structure (Translator's note).

early as the period of sedimentation and progress in their structural evolution faster than the structures of a synclinorium type. Naturally, the earliest intrusive and mineral complexes are associated with these early structural-metallogenic zones.

During the further evolution and formation of the structural-metallogenic zones of the region, these earliest zones occupy what appears to be the axial position. They predetermine the entire plan of the distribution of later structural-metallogenic zones in a given region. The later structural-metallogenic zones form on their limbs, i.e., on both sides of the earliest axial zone. The still younger structural-metallogenic zones are displaced farther in the same direction. But sometimes the structural-metallogenic zone of subordinate importance develops also on the back side of the preceding zone, i.e., being superposed over the axial structure. The youngest structural-metallogenic zones correspond to the structures of a synclinorium type, which are the latest in completing their formation, or to the structures of a foredeep type at the contacts of folded structures of the mobile belt with the rigid massifs forming the boundaries of the belt, or with the rigid massifs adjacent to the eugeosyncline zone.

It would be an error to believe that the long process of geological evolution of each structural-metallogenic zone is terminated by a single manifestation of intrusion and endogenetic mineralization. An examination of specific structural-metallogenic zones indicates that such structural-metallogenic zones occur occasionally, but they are relatively rare. Usually the stages of tectogenesis and the corresponding endogenetic mineralization manifest themselves repeatedly within the structural-metallogenic zones. These stages of tectogenesis correspond successively to the late and latest intrusions and mineral complexes. All this complicates considerably both the structural as well as the petrologic-metallogenic character of the zone. But in most cases it is nevertheless possible to distinguish one or two stages of tectogenesis, intrusive activity, and mineralization which are strongly predominant within the structural-metallogenic zone and contribute to its quite distinct structural-metallogenic aspect.

Since sedimentation is usually discontinued during the final stages in the evolution of each mobile belt, the structural-metallogenic zones do not represent the structural-facies zone and are frequently associated with the zones of major faults. It is quite natural that the latest structural-metallogenic zones are located less conformably with respect to the earlier zones and may be superposed over these zones or intersect them.

All the abovementioned information about the mutual position of the structural-metallogenic zones is certainly a very rough outline representing only the main tendencies in the spatial distribution of the successively forming structural-metallogenic zones. Under natural conditions all this is frequently complicated and changes to some extent, but the basic meaning of these features is the fact that the mutual position of the successively forming structural-metallogenic zones is subordinated in each given region (mobile belt) to a definite structural pattern which is quite consistent over the entire area of the region. Under these circumstances the later intrusive and mineral complexes correspond to each later structural-metallogenic zone. Similar intrusive and ore complexes correspond to the structural-metallogenic zones forming at the same time, even if these zones are far apart from each other. The knowledge of the general structural pattern in the distribution of the structural-metallogenic zones within a region, and the knowledge of the association of intrusive and mineral complexes, as well as their succession in time typical of this pattern, can be used as a starting point in regional metallogenic analysis of each terrain.

The execution of regional metallogenic analysis has practical and theoretical aspects. The practical aspect of the regional metallogenic analysis consists in singling out some territory, starting with large regions and terminating with the smallest areas and local structures which show the most promise for particular minerals. This approach increases the efficiency of prospecting and exploratory work. The theoretical aspect of regional metallogenic analysis consists of the elucidation of distribution features of mineral deposits within the studied territories. At the same time the general features of formation of mineral deposits within the earth's crust, causes and conditions under which the metallogenic province formed, and so on, are also studied.

The method of making the regional metallogenic analysis including its general principles, and initial positions from the point of view of which this analysis could be achieved, remained not at all worked out till very recent times. Until recently some geologists expressed doubts about the possibility of solving this problem by analytical means. The approach which was usually applied in regional metallogenic analysis consisted of a separate study of mineral deposits, intrusive rocks and structures of the terrain or region under study and correlation of the results of such study by means of a geological map. Under these circumstances it was usually possible to determine some geological features in the distribution of mineral deposits within the territory under study. However, these were only particular (individual) regularities which contributed also to the interpretation of the already known distribution of mineral deposits. But this information contributed very little to further prognosis and forecasting of the discovery of new ore-bearing regions, areas, and structures. There could have been no question, however, about a systematic and complete metallogenic analysis of some considerable territory.

It can be considered at the present time that the general principles of regional metallogenic analysis have been worked out, though not completely, but in all cases to a degree which makes it possible to a first approximation, to make an approach to entire metallogenic provinces, and to interpret within their boundaries their structure and to designate the most promising structural-metallogenic zones. On the other hand, an approach can be made to the causes of formation of metallogenic provinces, their principal features, and differences between them.

II. PATTERN IN THE EVOLUTION OF ENDOGENETIC MINERALIZATION OF MOBILE OROGENIC BELTS

General Characteristics

It is known that the mobile belts of the earth's crust start to form and develop during the initial stages of their evolution as geosynclinal zones. It was believed in the past that the younger geosynclinal zones represent nothing but the relics of more ancient and considerably more extensive geosynclinal zones which gradually closed by reducing their width as the result of the processes of folding. At the present time among the Soviet geologists the point of view is popular that each new geosynclinal zone starts to form independently and is not the relict of an earlier geosyncline. N.M. Strakhov believes that it is possible to follow the evolution of a geosyncline only by beginning with a definite moment which is considered as the moment of the start in its development. But it remains unclear what conditions existed in a given area of the earth's crust before the geosyncline started to form.

After going through a very long and complex process of evolution, the geosynclinal zone is finally drained, or, as is said, closed, being transformed to a folded belt which continues to undergo further geological evolution until it reaches a state which is close to that of a platform. Thus, in the evolution of a mobile belt, there are two major stages--the geosynclinal stage and the stage of a folded belt. But since the final transformation of the geosyncline into a folded belt is accompanied by intensive folding, by major intrusions and by strong development of the process of endogenetic mineralization, it seemed practical to designate this comparatively short stage of the transformation of the geosyncline into a folded belt as an independent stage of evolution. In such a manner three stages were originally distinguished in the evolution of mobile belts. The early stages in the evolution of the mobile belt were represented by the evolution of the geosyncline, the intermediate stages in evolution featured the transformation of the geosyncline into a folded belt, while the late stages in evolution corresponded to the further development of a folded belt.

During subsequent investigation of the problem it became clear that the composition of magmatic rocks, and especially that of the endogenetic mineralization, changes so strongly in the middle of the late stages of the evolution of the mobile belts that under conditions of a three-member division of the process of evolution of mobile belts it was always necessary to distinguish the first half of the late stages. It became clear that a division of the late stages into late and terminal stages of evolution is more practical. Such division into four members is adopted in the present book.

A further study by N.M.Strakhov, of the early stages in the evolution of mobile belts, especially in correlation with the pattern of sedimentation, forced some change in the idea about the nature of the intermediate stages in evolution and caused reconsideration of the practical value of subdividing in turn the early stage of evolution into an initial and an early stage. However, the factual information concerning this problem has not yet been sufficiently studied. Such a subdivision appears somewhat premature at the present time, but it is possible that in the course of further investigations its introduction will be practical. To make a more accurate definition of various stages in the evolution of mobile belts and their correlation with the pattern of evolution of sedimentation according to Strakhov, it is necessary briefly to examine this pattern, the more so since it is of paramount importance in the study of exogenetic metallogenic provinces.

Evolution of sedimentation according to N. M. Strakhov. Strakhov's aim was to investigate the evolution of the process of sedimentation starting with Paleozoic time within the limits of the continental blocks of the earth's crust, i.e., excluding the oceanic areas which are so far inaccessible to our study. Besides studying the evolution of facies and thicknesses, Strakhov studied the changes of the general area of sediment accumulation within the continental block including both the mobile belts and the platforms. The latter conceal, to some extent, features related only to the mobile belts. Strakhov stresses that starting with the Paleozoic history of the earth, periods of major transgressions and regressions always coincided on platforms with those in areas of mobile belts. But because of the flatter topography of the platforms, even the insignificant vertical movements in these areas could result in extensive transgressions and regressions. At the same time, within the geosynclinal regions, characterized by considerably stronger relief, even strong vertical movements did not substantially affect the general area of sediment accumulation. Thus, transgressions and re-

gressions on the platforms emphasized those vertical movements which also took place simultaneously in the geosynclinal regions.

The well known diagram of Strakhov (Izvestiya Akademii Nauk SSSR, 1949, No. 6. Translator's note) represents changes of the areas of sedimentation within the boundaries of continental blocks of the earth's crust during the post-Precambrian period of its evolution. In this diagram the geological periods are shown on the scale of geological time along the vertical axis. The shaded areas along the horizontal axis represent areas of sedimentation during each time interval in percent of the total area of the continental block. The black circles designate the position in time of the orogenic phases of folding of various magnitudes.

Strakhov makes the following conclusions after an analysis of this diagram. Three major stages in the evolution of accumulation of sediments are clearly distinguished during the post-Precambrian history of the earth's crust. These stages are separated from each other by a strong reduction of the total area of sedimentation. The Caledonian stage comprises Cambrian, Ordovician, and Silurian periods, the Variscan stage— Devonian, Carboniferous, and Permian, and the Alpine stage—the Mesozoic and Cenozoic eras. Several cycles of sedimentation or major transgressions can be outlined within each stage when the areas of sedimentation within the continental block reach their maximum. These cycles of sedimentation are separated from each other by considerable regressions or periods of reduction of the general area of sedimentation.

Three major cycles are clearly distinguished in the Caledonian stage of sedimentation. There are also three cycles in the Variscan stage, but the third cycle is somewhat complicated and is followed by an additional small transgression. The structure of the Alpine stage of sedimentation is more complicated and Strakhov outlines here only about five cycles of transgression. It is necessary to take into consideration that during this time two major mobile belts developed—the Mediterranean and the Pacific, and the periods of regressions and transgressions in both did not coincide. This probably explains the great complexity in the structure of the Alpine stage of sedimentation. A certain complication of the Variscan stage of sedimentation, as compared with the Caledonian, depends most likely on the same causes. Strakhov stresses that periods of considerable regressions coincide well with the most important phases of folding.

This outline is empirical and reflects only the actually known areal distribution of sediments of various geological ages. Each major stage of sedimentation (Caledonian, Variscan, Alpine) characterizes the development of mobile belts of a definite age; after the completion of their development new mobile belts start forming. The evolution of each mobile belt is divided sharply according to this pattern into three subsequent cycles separated from each other by major phases of folding. Thus, the cross section of each mobile belt contains seemingly three successive sedimentary and structural levels—the lower, the middle, and the upper. The third and last major phase of folding in the evolution of the mobile belt develops after the completion of the third cycle of sedimentation. In the course of the geological evolution of the mobile belt, the sedimentation is discontinued or almost discontinued and the mobile belt is transformed from the geosynclinal state to the state

REGRESSIONS ⇄ TRANSGRESSIONS

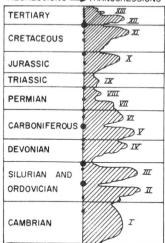

TERTIARY	XIII, XII
CRETACEOUS	XI
JURASSIC	X
TRIASSIC	IX
PERMIAN	VIII, VII
CARBONIFEROUS	VI, V
DEVONIAN	IV
SILURIAN AND ORDOVICIAN	III, II
CAMBRIAN	I

Cycles of sedimentation during the Postalgonkian history of continents (transgressions and regressions)
After N. Strakhov

Black circles — orogenic phases of various magnitude.
Shaded — areas of sedimentation in percent of the total areas of the continental block.

of a folded belt. Processes of magmatism and endo-genetic mineralization continue in this belt, but their manifestations correspond in time to sedimentation in the younger newly started mobile belt.

Thus, the third and last major phase of folding cor-responds to that which was defined earlier by the writer as the intermediate stages in the evolution of the mobile belt, i.e., to the stages of transformation of the geo-synclines to a folded belt. However, a more accurate examination of Strakhov's scheme and an analysis of the magmatism of the subsequent cycles of sedimentation make it practical to change to some extent the defini-tion of the intermediate stages. Quite naturally, the formation of each of the three structural levels of the mobile belt, corresponding to the three successive cycles in sedimentation, represents a unique process. The various aspects of this process are sedimentation, for-mation of the structure of sedimentary sequences, de-velopment of magmatism, and the endogenetic miner-alization. Not a single one of these particular processes can be separated from the unique and general process in the evolution of the structural level. Thus, if the third phase of folding together with the accompanying magmatism and endogenetic mineralization is referred to as the intermediate stage in the evolution of the mo-bile belt, the immediately preceding sedimentation also has to be referred to this stage. In other words, apart from the moment of transformation of a geosyncline to a folded belt, the entire process of formation and de-velopment of the third uppermost structural level has to be classified as the third stage of the evolution of a folded belt. This appears to be more securely founded because the nature of magmatism begins to change con-siderably with the development of sedimentation of the third cycle. The composition of magma approaches the composition of intrusive rocks of the third orogenic phase.

Each of the three cycles of sedimentation corre-sponding to the three successively forming structural levels of the mobile belt is characterized by its own rock complex. Strakhov examined these rock complexes for the entire continental block, including the platforms. If the influence of the latter is excluded, the rocks of the three cycles of sedimentation featured by the mobile belts can be characterized in the following terms. Rocks of the first cycle of sedimentation are characterized most of all by a considerable development of the geo-synclinal volcanic facies, i.e., rocks of the so-called spilite-keratophyre formation. Jaspers and slates are genetically connected with these rocks, frequently with higher than the average contents of iron and manganese. Other characteristic members of the profile at this time are the limestones, usually forming reefs, and the vol-canic-carbonate series. The contribution of terrigenous rocks (conglomerates, sandstones, and shales) varies at this time. Sometimes they are entirely absent, in other cases they are present in considerable, but nevertheless subordinated amount. The development of sedimentation of the first cycle inevitably culminates in subsequent development of the orogenic processes of terminating this cycle of sedimentation. Quite typical intrusions accompanied by endogenetic mineralization are connected with this first orogenic phase. These are intrusions of ultrabasic and occasionally basic rocks accompanied by a complex of predominantly magmatic mineral de-posits.

The accumulation of sediments during the first cycle along with the folding and intrusions concluding the cycles

complicates the structure of the mobile belt. However, these factors do not make the mobile belt too rigid and result inevitably in the development of sedimentation of the second cycle, taking place under conditions of another kind of tectonic regime. At this time volcanism of the geosynclinal type becomes weaker and despite the continuation of formation of volcanic series, their role in the formation of a stratigraphic sequence is considerably reduced, as well as the role of the ac-companying cherty sediments. The role of carbonate rocks varies in different mobile belts, but the impor-tance of terrigeneous rocks evidently increases as the result of formation of considerable elevations in relief in connection with the first phase of folding. The for-mation of the second structural level of the mobile belt is terminated by the second major orogenic phase. This creates the folded structure of sediments of the second cycle of sedimentation and complicates consider-ably the structure of the first level. Large intrusions are also connected with this second phase of folding. They correspond according to their composition to the complexes of a broad range of differentiated basic mag-mas ranging from gabbro to plagiogranites or grano-syenites. A complex of endogenetic deposits, predom-inantly of contact metasomatic and in part hydrother-mal veins, is connected with these intrusions.

The second major phase of folding complicates con-siderably the structure of the folded belt by contribut-ing to its greater rigidity and creating the most con-siderable differences in topographic elevations in the entire history of the evolution of the mobile belt. The greater rigidity of the mobile belt during this period creates a favorable ground for the manifestation of suc-cessive stages in the development of faults. These faults contribute more to the complication of the folded struc-ture and are usually accompanied by minor intrusions filling the fractures with a very diverse endogenetic mineralization. Usually the mobile belt is not suffi-ciently rigid for the termination of geosynclinal condi-tions and the mobile belt is involved in a new subsidence resulting in the evolution of the third cycle of sedimen-tation. At this time the terrigenous formations, such as sandstones and especially shales with a variable amount of conglomerates, are strongly predominant in the sequence of sedimentary strata. The contribution of carbonate rocks, as well as that of volcanic rocks, is strongly reduced. The geosynclinal volcanism is either discontinued or manifests itself very sporadically in both time and space, forming only thin flows of vol-canic formations predominantly of acidic and inter-mediate composition. The duration of the third cycle of sedimentation and the thickness of terrigenous strata formed vary greatly in different mobile belts and depend entirely on the preceding history in the evolution of the mobile belt.

The evolution of sedimentation of the third cycle cul-minates usually in a third major phase of folding, ac-companied by a complete regression of the sea from the geosyncline, and transformation of the geosyncline into a folded belt. The largest granitoid intrusions ac-companied by a very typical complex of hydrothermal deposits, pegmatites, and to some extent by contact-metasomatic deposits, are connected with this phase of folding. The intermediate stages in the evolution of mobile belts are characterized by the third cycle of sedimentation with the formation of terrigenous strata, subsequent folding, granitoid intrusions, and endogenetic mineralization.

The late and terminal stages in the evolution of the mobile belt are characterized predominantly by continental and fresh water sedimentation forming strata of insignificant thickness, in some cases covered by marine ingressions of short durations and small area. As a result of the general consolidation of the folded belt, the subsequent contribution of the folding is strongly reduced. On the other hand the importance of faults strongly increases. Intrusive activity is represented predominantly by minor intrusions and dikes accompanied by a very diverse and quite intensive, predominantly hydrothermal endogenetic mineralization. The gradual weakening of tectonic movements, resulting finally in the cessation of magmatic activity and the termination of the processes of endogenetic mineralization, transforms the mobile belt into a platform.

This is a short outline of the geological evolution of mobile belts. It shows that the accumulation of sediments, development of structures, magmatic processes, and endogenetic mineralization are closely interconnected and interdependent and reflect only different aspects of a unique process of evolution which develops according to a very regular pattern. Since a diverse complex of exogenetic mineral deposits is connected with the evolution of sedimentation, it is obvious that complexes of endogenetic and exogenetic deposits are also closely interconnected with each other. This problem will be examined at the end of this work. Now let us concentrate on a more detailed analysis of those intrusive and endogenetic mineral complexes which may be distinguished in the course of the geological evolution of mobile belts. The succession of intrusive and endogenetic mineral complexes is represented by Table 1. (See centerfold).

1. Initial and early stages of evolution

1. The spilitic-keratophyre magmatic complex is the earliest magmatic complex in the course of the geological evolution of the mobile belts of the earth's crust and at the same time it is the most typical representative of the so-called geosynclinal volcanism. This complex is represented mainly by extrusive rocks and in part by tuff formations. However, frequently a small number of intrusive formations is present in this complex. These intrusive formations are small, they formed near the surface, and they represent basically the separate portions of the same magmas which did not reach the surface and were located at shallow depths among the extrusive rocks of the same composition.

A well known conspicuous feature of lava flows of the geosynclinal period is the predominance among them of basic rocks and to some extent of rocks of a mesocratic, composition, such as diabases of various structures, diabase porphyries, porphyries, and other rocks, which being underwater flows, usually become the subject of greenstone alteration, forming spilites. Sometimes in certain zones of the mobile belt flows of predominantly diabase (basalt) composition form, while in other zones the lava flows are predominantly of porphyry (andesite) composition.

Feldspars in these rocks are represented by the basic or intermediate plagioclases or by secondary (metasomatic) albite. These rocks are connected by continuous transitions with the more acidic members of the same series—albitophyres and quartz albitophyres. The two last groups of rocks develop together with the basic rocks and rocks of intermediate composition.

Their quantitative contribution to the volcanic series of various ages and in various structural zones of the same geosynclinal zone can vary from a considerable predominance over the rocks of basic and intermediate composition to a strong subordination to the latter, or even a complete absence. In most cases there is a general trend which consists of smaller contribution of the more acidic members of this series during the early stages of volcanic eruptions and a slight increase in amount of acidic rocks toward the end of volcanic activity. A close association of acidic and basic members of this series and the existence of gradual transitions, as indicated by chemical composition and petrographic nature, is convincing evidence of the genetic connection between the two members.

The acidic members of this series are characterized by a strong predominance of sodium over potassium. This is explained mainly by the processes of secondary albitization. But the composition of intrusive rocks of the geosynclinal stage of evolution and that of some extrusive rocks which avoided albitization indicates that the predominance of sodium over potassium (although not prominent) is typical of the primary composition of more acidic rocks. Though rocks of dacite and liparite composition with normally high content of potassium feldspar occur occasionally in similar series, they are of very subordinate importance and cannot be considered characteristic in any way.

A high content of potassium cannot be considered contradictory to the nature of extrusive rocks of the early stages in evolution of mobile belts. In several regions, characterized by such extrusive rocks, individual horizons are encountered, sometimes along with whole series of extrusive rocks featuring higher alkalinity (trachyte, trachyliparites, and others), and occasionally typical alkaline (leucite) rocks with considerable predominance of potassium over sodium. To this group of rocks belong the orthophyre complex of Upper Ludlowian age on the eastern slope of the Urals, extrusive rocks of the Pambak alkaline complex in Transcaucasia, and probably the extrusive rocks of the Ishim alkaline complex in Kazakhstan. True, there is no certainty that the formation of the last complex belongs to the geosynclinal stage (early substages) in the evolution of the mobile belt. The alkaline complex of Pambak and the orthophyre complex of the Urals are followed again by lava flows typical of the early stages in the evolution of the mobile belts. The extrusive rocks of the Pambak complex are relatively thin and occur over a small area, while the orthophyre complex of the Urals covers a tremendous territory and reaches locally a thickness of 800 m. The formation of extrusive rocks of such a composition during the geosynclinal stage indicates the presence of individual particular stages of evolution or, probably, separate local structural zones in which the evolution of magma took place. This did not occur along the usual path resulting in the generation of differentiates rich in silica and sodium, but rather produced rocks rich in potassium and poor in silica or containing no quartz.

It should be stressed that the eruption of spilitic lavas is especially typical of the first cycle of the geosynclinal accumulation of sediments. They are also present, though to a lesser extent, during the second cycle of sedimentation following the first major phase of folding. This refers apparently to both the rocks of the spilite-Keratophyre series and to the rocks rich in potassium. It is possible that the importance of the latter increases to a certain extent during the second cycle of sedimentation.

Spilite-keratophyre series of a similar kind are common in the territory of the Soviet Union. To this series belong the Mesozoic-Cenozoic lavas of the Minor Caucasus, and Ordovician, Silurian, Devonian, and in part the Carboniferous series of volcanic origin of the eastern slope of Urals; the Ordovician, Silurian, and Devonian volcanic series in Kazakhstan; the Cambrian lavas of Tuva and Kuznetsk Alatau and others. All these volcanic series represent typical formations of the geosynclinal period, cover territories occupying enormous areas, and frequently represent the predominant part of tens of percent of a given stratigraphic sequence.

Minor near-surface intrusions are genetically and spatially closely connected with the extrusive rocks of the spilite-keratophyre complex. These minor intrusions represent portions of magmas which did not reach the surface and which cooled off and solidified at some depth. They are usually represented by diabases, gabbro-diabases, diorite-porphyries, quartz albitophyres, and other rocks, and form bodies of conformable or close to conformable type (intrusive sills, flattened laccoliths), or sometimes dikes. Similar intrusive bodies of near-surface type occur in many extrusive series of spilite-keratophyre formations, but sometimes they can be distinguished only with difficulty from the extrusives. In addition they are folded together with the extrusives. The diabases of Karelia which are connected with the Proterozoic spilitic formation, the Dzhusalin intrusive complex in northern Kazakhstan, similar intrusive formations of the northern part of Gornyy Altai, and others, can be classified as intrusive complexes of such kind.

An insignificant or low grade copper mineralization which commonly is of no economic interest is usually connected with the spilite-keratophyre magmatic formation, in particular with its near-surface intrusions. However, there is quite a good probability that the large copper deposits of Lake Superior in the United States (deposits of native copper with zeolites in basic lavas), which supplied about 4,000,000 tons of copper, belong to this magmatic formation. These facts indicate that under favorable conditions deposits of economic grade can form in connection with these rocks.

2. In their typical evolution the ultrabasic intrusions follow lava flows of the spilite-keratophyre formation and are most frequently connected with the first major orogenic phase which completes the accumulation of sediments of the first cycle. Thus, the formation of ultrabasic intrusions is the inevitable final stage of magmatism of the first cycle of sedimentation. In some cases the formation of ultrabasic intrusions is apparently directly connected with the processes of folding in the proper sense; in other cases they are connected with slightly later faults, representing, however, only one of the stages in the formation of folded structures of this period. This refers not only to the ultrabasic intrusions, but as well to the considerably later batholithic granitoid intrusions connected with the periods of folding. Occasionally the ultrabasic intrusions recur in the course of the geological evolution of the mobile belt, as for example, the Early Caledonian and Early Variscan ultrabasic intrusions of the Urals. Apparently, this was caused by the duplication in time of the earliest stages in the evolution of the mobile

belt. Actually, the Caledonian mobile belt of the Urals, before terminating its development, was involved in a new subsidence in connection with the formation of the enormous Variscan mobile belt comprising Mongolia, Turkestan, Kazakhstan, and the eastern slopes of the Urals. The evolution of this new mobile belt also started with outpourings of lavas of the spilitic formation and subsequent ultrabasic intrusions. Because of this, these stages were repeated in the evolution of the mobile belt of the Urals.

The ultrabasic intrusions occupy quite large areas in such regions as the Cambrian ultrabasic rocks of the Sayan mobile belt (Salair, Kuznetsk Alatau, Gornaya Shoriya, Western and Eastern Sayan, Tuva). In other regions they are associated with strictly localized structural zones (Kazakhstan, Minor Caucasus). In Central Kazakhstan it is very clearly demonstrated that the belt of ultrabasic intrusions, reaching from the Maykain region across the region of Karaganda into the Chu-Ili Mountains, divides the territory of Kazakhstan into two parts. These parts are the northwestern, which terminated its Caledonian evolution normally, and where the Variscan structures are of a superposed nature, and the southeastern part, where Variscan structures proper are developed and where also the Early Variscan ultrabasic intrusions formed. In Minor Caucasus the ultrabasic intrusions of Eocene age are clearly associated with the narrow localized Gokcha overthrust zone.

The ultrabasic intrusions are represented by peridotites, pyroxenites, dunites, serpentinites, and others, frequently accompanied by the subordinate development of related rocks of basic and less frequently of intermediate or acidic composition (gabbro, gabbro-diorites, diorites, plagiogranites). In some ultrabasic complexes the ultrabasic and more acidic rocks (gabbro, diorites, and others) are represented about equally. In some cases the ultrabasic rocks are quantitatively subordinate. To this type belong, for example, the early Caledonian gabbro peridotite complex of the eastern slope of the Urals.

Most of the deposits of platinum, chromite, and asbestos are connected with the ultrabasic intrusive complexes. Deposits of titano-magnetites may be connected with pyroxenites. The intensity of mineralization connected with the ultrabasic intrusive complexes is quite different in different metallogenic provinces, but it increases sharply in those provinces where there is a stronger contribution made by iron and other metals of its group (chromium, titanium, manganese, vanadium, and others) while the contribution by tin and nonferrous metals is smaller. Deposits of chromite, however, which are not always of economic grade, are quite common and occur in most of those provinces where ultrabasic intrusions are present. Chromite deposits of economic grade may be affiliated with the Precambrian as well as with the Paleozoic and Alpine ultrabasic intrusions. Thus, there are no substantial changes in the intensity of chromite mineralization connected with the ultrabasic intrusive complexes in the course of the post-Precambrian geological evolution of the earth's crust.

Deposits of platinum associated with the ultrabasic intrusive complexes of the mobile belt occur more locally in the Soviet Union, mainly in connection with the early Caledonian gabbro-peridotite complex of the eastern slopes of the Urals. Only insignificant amounts of platinum, and mainly platinum group metals (osmium, osmium-iridium, and others), are connected with all other ultrabasic complexes of the Soviet Union. According to a hypothesis by V.M.Sergievskiy it would be more correct to associate the economic deposits of platinum not with ultrabasic complexes, but with complexes of basic rocks in which the ultrabasic rocks form a subordinate part. Such a hypothesis appears quite probable and agrees well with the absence of economic deposits of platinum in connection with most of the ultrabasic complexes in a proper sense.

Platinum deposits of the Urals are so extensive that they naturally strongly affect the statistical data on the distribution of economic platinum deposits according to their geological age. If it is taken into consideration that in the second largest region of platinum deposit, in South Africa, platinum deposits of a platform type are of Precambrian age and that the source of platinum placers of Colombia is believed to be old, one may conclude that platinum deposits of economic importance occur predominantly in gabbro-peridotite intrusive complexes of Precambrian and Paleozoic ages.

Deposits of asbestos occupy an intermediate position between chromite and platinum in the frequency of occurrence, and develop mainly in the Caledonian provinces of the Soviet Union. Outside the Soviet Union the known economic deposits of asbestos are mainly of Precambrian and Paleozoic ages. No major asbestos deposits of Mesozoic and Tertiary ages are known.

Quite large, high-grade supergene deposits of nickel silicate ores occur in association with the ancient zone of weathering preserved on ultrabasic rocks. A small amount of cobalt is usually present in these ores. Small deposits of magnesite occur also in association with the zone of weathering of ultrabasic rocks.

In contrast to later major intrusions connected with the subsequent periods of folding, there are no minor intrusions filling the fractures with which any noticeable mineralization is connected. The intrusive bodies of ultrabasic rocks are frequently intersected by dikes of diabase or gabbro-diabase. The latter most likely represent a dike suite of the same ultrabasic intrusions and do not reveal the presence of any noticeable mineralization.

3. Complexes of basic intrusion (gabbro, norites, anorthosites) occur occasionally as the result of independent development; in other cases they are accompanied by differentiates of ultrabasic composition (the gabbro-peridotite complex of the Urals) or by intermediate or acidic differentiates up to the plagiogranites and granosyenites. The examination of the strongly differentiated intrusive associations of complex composition representing the latter follows. The position of a basic intrusive complex in the course of the geological evolution of mobile belts is not quite clear, since the well known rock complexes of this kind have been studied insufficiently within the boundaries of the Soviet Union. If one judges from the gabbro-peridotite complex of the Urals, occupying what appears to be an intermediate position, and by some similarity of the metallogeny of these complexes to the metallogeny of ultrabasic complexes with respect to age, they are analogues of ultrabasic complexes. In other words the gabbro-peridotite complexes may also be connected with the first major period of folding, but develop under slightly different tectonic conditions from the ultrabasic complexes. This problem requires further study and accumulation of factual information.

Comparatively few types of mineral deposits are connected with complexes of basic intrusions. These are mainly deposits of magnetites, titanomagnetites, apatite-magnetite ores, porphyry copper ores (of the type of the Volkovskoye deposit in the Urals), disseminated copper-nickel ores, and massive copper-nickel ore. However, the latter type of deposit is more typical of platform regions. Deposits of copper-nickel ores also occasionally contain platinum, cobalt, gold, and other metals in amounts of economic interest.

4. The gabbro-plagiogranite and gabbro-granosyenite intrusive complexes usually form the largest intrusive bodies connected with the second period of major folding in the evolution of the mobile belt. This folding follows the sedimentation of the second cycle, which separates these complexes from the ultrabasic rocks in time. Two types of differentiation or two trends in differentiation may be traced in these complexes. In one case there are differentiates ranging from gabbro through gabbro-diorites and tonalites to plagiogranites. Such complexes may be called gabbro-plagiogranitic complexes. By their composition they are quite close analogues of rocks forming a series represented by diabases, porphyries, and quartz albitophyres in the spilite keratophyre complex. In other cases there are associations of gabbro, gabbro-monzonites, monzonite-diorites, syenite-diorites, syenites, granosyenites, occasionally also granites, alkaline granites and syenites, in some cases nepheline syenites and pseudoleucite syenites. These complexes can be named gabbro-granosyenite complexes. By their composition they approach the orthophyre extrusive rocks which occasionally constitute a part of the spilite keratophyre formation. Thus, in both the extrusive and intrusive rocks of the early stages in the evolution of mobile belts there are two trends in differentiation of basic magma. One trend leads to the final differentiates rich in sodium and silica, while the other leads to the differentiates rich in potassium and to some extent depleted of silica.

There are several successive intrusive phases featuring rocks of various composition in both the gabbro-plagiogranite and in gabbro-granosyenite complexes. The early phases are represented by the basic rocks (gabbro, gabbro-norites), while the later phases are represented successively by more acidic rocks. Because of sufficiently accurate field studies, it is always possible to establish the relations between the rocks of separate intrusive phases. This indicates that the differentiation of magma took place deep below the surface and not within the intrusive body. On the other hand, the composition of the successive portions of magma indicates the general trend of the process of differentiation from basic magma to more acidic. A close connection in time and space between the more basic and more acidic members of such complexes, common serial features of mineralogical, petrochemical, and geochemical order, support the idea of the close genetic relation of both. Thus, the granitoid rocks belonging to such complexes have to be considered as the granitoid differentiate of basic magma.

In some cases the gabbro-plagiogranite and gabbro-granosyenite intrusive complexes developed separately; while in other cases, developing together, they formed substantially common intrusive complexes. As can be judged from available data, the gabbro-plagiogranite complexes occupy an equal position in the course of evolution of mobile belts, i.e., they are connected with the

Table 1.

Evolution pattern of endogenetic mineralization of mobile belts

Stages of evolution	State of structural development	Predominant magmas	Sedimentation	Fault structures	Types of magmatic bodies	Magmatic rocks	Associated Mineral deposits
Terminal	Young platform	Basalt magma.	Erosion, zone of Weathering		Extrusive bodies	Basalts. Trachybasalts.	Mercury (?)
	Shelf	Granitoid different-iates of basalt magma	Erosion. Zone of weathering. Fresh water, lagoonal, and continental sedimentation. Subaerial Volcanic series.	Tear Faults		There is no visible connection with magmatic rocks.	1. Lead-zinc (Karatau type). 2. Copper (Dzhezkazgan type). 3. Fluorite. 4. Barite. Witherite.
				Deep regional faults and tear faults connected with them	Minor Intrusions	Granites, alaskites, alkaline granites granosyenites, syenites, (gabbro-diabases) or no visible connections with magmatic rocks.	1. Silver-cobalt-nickel-bismuth-arsenic -(copper). 2. Silver-lead-zinc-copper (late group). 3. Fluorite. 4. Barite. Witherite. 5. Siderite. Magnesite. 6. Hematite. Manganese.
Late	Folded Belt			Tear faults		No visible connection with magmatic rocks.	Mercury. Antimony. Arsenic. (Tungsten)
				Tear faults. Collapse structures.	Extrusive bodies	Andesites, liparites trachytes, and others. Quartz monz-onite-porphyries, granodirite por-phyries, graniter porphyries, and others.	Gold, silver (epithermal). Copper, molybdenum, alumina. (metasomatic quartzites).[1]
		Andesite magma		Tear faults	Minor intrusions	Diorite, syenite-diorites, granodio-rites, syenites, granosyenites (ne-pheline syenites),	Contact metasomatic. 1. Magnetite, (cobalt). 2. Lead-zinc (tin).
		Granitic magma		Tear faults. Loose foliation structures.		Granite-porphyries, granodiorite-porphyries, liparites dacites.	1. Silver-lead-zinc (early group). 2. Arsenic 3. Tin (sulfide-cassiterite association). 4. Tungsten.
					Minor intrusions	Diorites, quartz-diorites, grano-diorites, granite-porphyries, lamprophyres, and others.	1. Gold. 2. Molybdenum.
Intermediate	Folded Belt	Granitic magma	Erosion	Third major folding. Faults connected with this folding.	Batholiths	Granites: alaskite type leuco-cratic, aplitelike biotite granites, pegmatites, aplites .	Hydrothermal: 1. Tin, tungsten, bismuth, molybdenum. 2. Fluorite, pegmatites, greisens; 1. Muscovite. 2. Lithium, beryllium, tin, tungsten, tantalum, columbium.
						Granodiorites, granites, (quartz-diorites)	Contact-metasomatic; 1. Tungsten. 2. Molybdenum. Hydrothermal; 1. Gold , arsenic, tungsten (scheelite) 2. Molybdenum. 3. Copper (lead-zinc).
		Andesitic magma	Erosion.	Tear faults	Minor intrusions	Quartz albite-por-phyries, berezites2), quartz diorites, and others.	Gold

Stage	Tectonic setting	Magma	Sedimentation / Erosion	Tectonic structures	Extrusive bodies / Minor intrusions	No visible connection with magmatic rocks	Mineralization
Early	Geosyncline	Granite differentiates of basalt magma	Marine terrigenous sedimentation	Tear faults. Collapse structures. Tear faults. Foliation structures.	Minor Intrusions	Andesites, liparites, trachytes, and others. Quartz monzonite porphyries, grano-diorite-porphyries, granite porphyries. Quartz albite porphyries.	Mercury, antimony, arsenic (tungsten). Gold, silver (epithermal). Copper, molybdenum, alumina (metasomatic quartzites). Sulfide deposits (partially forming veins) 1. Copper. Pyrite. 2. Silver-lead-zinc. 3. Gold-barite.
Early	Geosyncline		Erosion	Second major folding. Faults connected with this folding.	Batholiths, Minor intrusions.	1. Gabbro, diorites, tonalites, plagio-granites. 2. Gabbro, monzonites, syenites, granosyenites, granites (nepheline syenites).	Contact-metasomatic: 1. Magnetite. Copper. 2. (Cobalt, arsenic). 3. (Tungsten, molybdenum). Hydrothermal: 1. Gold. Arsenic. Tungsten (scheelite). 2. Copper. Molybdenum.
Early	Geosyncline		Marine terrigenous-volcanic series		Lava flows. Minor Intrusions.	Spilites, porphyries, keratophyres.	1. Copper 2. Iron. Manganese.
Initial	Geosyncline	Basalt magma. Ultrabasic magma.	Erosion	First major folding. Faults connected with this folding.	Batholiths. Concordant intrusions.	Gabbro, norites, anorthosites, diorites, and others.	1. Titanomagnetites. 2. Copper-nickel. 3. Copper (porphyry).
Initial	Geosyncline					Serpentinites, peridotites, dunites, pyroxenites, gabbros.	1. Platinum. Platinum group metals. 2. Chromium. 3. Titanomagnetite. 4. Asbestos. Supergene: 1. Nickel. 2. Magnesite.
Initial	Ancient platform	Basalt magma. Ultrabasic magma. Alkaline magma.	Marine volcanic and carbonate series		Lava flows. Minor Intrusions.	Spilites. Porphyries. Albitophyres. Quartz albitophyres.	1. Copper. 2. Iron. Manganese.
Initial	Ancient platform				1. Lava flows. Sills. 2. Lopoliths 3. Pipes, stocks, laccoliths.	1. Basalts, diabases (traps). 2. Norites, peridotites (granites), and others. 3. Alkaline peridotites. 4. Nepheline syenites. 5. Monzonites, syenites, and others.	1. Magnesite. Copper-nickel. 2. Platinum. Chromium. Diamonds. Apatite. Titanomagnetite. (Gold). (Lead-zinc).

1) Acidic or mesocratic igneous rocks altered by hydrothermal-metasomatic processes. According to Russian terminology the metasomatic or secondary processes consist in deposition of quartz, some sericite, alunite, pyrophyllite, kaolinite, andalusite, and diaspore.

2) Berezites — pyritized, metasomatically altered rocks of granitic composition, with quartz veins occasionally containing gold mineralization (:translator's footnote).

29

second major orogeny but they differ to some extent with respect to the tectonic environment in which they developed. The gabbro-plagiogranite formed evidently under some more unstable environment, under conditions of more subsidence and deposition of a greater amount of sediments in this area, followed by a stronger tectonic deformation of these sediments. The gabbro-syenite complexes are formed under conditions of lesser tectonic movements and somewhat stronger rigidity, smaller preceding subsidence and deposition of smaller amount of sediments, and lesser tectonic deformation of these sediments. Even within the boundaries of one region—within the limits of structural zones of various stability—differences can be observed in the composition of this complex. So, for example, within the boundaries of Rudnyy Altai, the well known Zmeinogorsk Early Variscan intrusive complex of gabbro-plagiogranite composition formed. In this area the sediments deposited before the intrusion were considerably deformed. An intrusive complex of similar compositions and of same age is developed on the other limb of the Zaisan syncline, within the structural zone of Chingiz. Here, however, the preceding sedimentation was of lesser magnitude, the rocks are less deformed, and in addition to plagiogranites, there are syenites, granosyenites, and other igneous rocks which make up a substantial part of the complex. There are also other cases of a similar nature on the territory of the Soviet Union. In addition it should be remembered that the gabbro-plagiogranite complexes occur only in mobile belts, and only during the early stages of their evolution, i.e., under conditions of greater lability. At the same time the gabbro-granosyenite complexes occur in the platforms, as well as during the final stages in the evolution of mobile belts, when the belts acquire a considerable degree of rigidity. According to this evidence the first type of differentiation of basic magma belongs to the geosynclinal type while the second type belongs to the platform type.

The quantitative relations between the basic intermediate and acidic differentiates in gabbro-plagiogranite and gabbro-granosyenite complexes vary quite strongly. Usually the rocks of intermediate and acidic composition are dominant over the basic rocks and the main differences between various complexes consist chiefly in a different volume of intermediate and acidic rocks. If the alkaline rocks (nepheline syenites and others) make up part of the complex, their quantitative importance may be quite different. The variety of rocks in different complexes may also be different. Sometimes such complexes, besides major batholiths, produce small intrusions. However, it is not clear whether they represent slightly eroded batholitic intrusions or independent minor intrusions filling the fractures.

Complexes of gabbro-plagiogranite and gabbro-syenite composition apparently differ to some extent in their ore-bearing properties, but these differences have not yet been checked in a sufficient number of cases, and their discussion is premature so far. It should be mentioned that according to the ideas of Korzhinskiy, the high chemical potential of sodium in intrusive rocks (plagiogranites) or sodium (granosyenites) affects substantially the mineral equilibria and the mineralogical composition of the skarns being formed.

The mineralization connected with gabbro-plagiogranites and gabbro-granosyenite complexes is quite common and quite diverse. Of greatest economic importance are the contact-metasomatic (skarn) deposits

of iron, mainly magnetite. Hematite deposits of this type are of lesser importance. The iron deposits of the Urals, such as Magnitnaya Mountain, Vysokaya Mountain, and Blagodat belong to this type. The deposits of Magnitnaya Mountain are connected with the early Variscan (pre-Early Carboniferous) Gumbeyka-Magnitogorsk intrusive complex in which the leucocratic rocks (granodiorites, granosyenites, and others) are predominant. The deposits of Vysokaya Mountain and Blagodat are connected with the considerably younger pre-Middle Devonian Tagil-Kushva complex represented by gabbro, syenites, and in part by nepheline syenites which were discovered relatively recently. In the Urals this repetition of complexes of a similar composition accompanied by similar mineralization reveals the duplication of the early stages of evolution of the Uralian mobile belt, which was discussed earlier in connection with the ultrabasic rocks.

The contact metasomatic copper deposits are of great economic importance, an excellent example of which are the deposits of the Tur'inskiye copper mines in the Urals. There are also occasional contact-metasomatic deposits of arsenic and, occasionally, cobalt, lead, zinc, or even gold, small amounts of scheelite and molybdenite often occur.

In connection with these complexes, besides the contact-metasomatic deposits, hydrothermal deposits forming veins also occur which on the whole, however, are of smaller economic importance than the contact metasomatic ore deposits. These are predominantly hypothermal deposits or deposits transitional from the hypothermal to the mesothermal deposits of gold, copper scheelite, and molybdenum. On the whole the association of magnetite, copper, gold, scheelite and molybdenite is very typical of the metallogeny of these complexes and can be followed through a great number of metallogenic provinces.

Besides the above-mentioned Tagil-Kushva, Tur'inskiy, and Gumbeyika-Magnitogorsk complexes of the eastern slopes of the Urals, the Kryk-Kudukian intrusive complex of a gabbro-plagiogranitic composition (northern Kazakhstan) connected with the Taconic orogenic phase also belongs to similar intrusive complexes within the boundaries of the Soviet Union. The early Variscan intrusive complex of Zmeinogorsk (Rudnyy Altai), which is also of a gabbro-plagiogranite composition; the Tannu-Ola complex in Tuva, connected with the Taconic phase of orogeny; and the Kongur-Alangez intrusive complex of Miocene age in Transcaucasia, which includes also nepheline syenites, also belong to similar intrusive complexes.

While the series of mineral deposits typical of such an intrusive complex are quite well developed within the boundaries of the eastern slope of the Urals, in Transcaucasia, Kuznetsk Alatau, and in other regions, the intensity of mineralization of the individual complexes and the role of individual members of the metallogenic association in these complexes vary appreciably. First of all, attention is drawn to the fact that the largest, most deeply eroded batholiths are characterized by the lowest ore content. The intrusive complex of Kryk-Kuduk in northern Karakhstan and the Zmeinogorsk complexes of Rudnyy Altai may be placed into this category. On the other hand the less abyssal intrusions and, in some cases, apparently even the near-surface intrusions of the Urals and Transcaucasia have a considerably more intensive mineralizing capacity.

It is necessary to stress at this point that the idea about the depth reached by erosion applies more to the complexes of the same type than to complexes of other types, since there are attempts to explain the observed differences between complexes of entirely different types by the difference in depth reached by erosion.

The role of depths reached by erosion is probably of minor importance in comparison with the differences in the group of metals characteristic of various regions and connected with intrusive complexes of this type. On the one hand, the nature of this group of metals is probably controlled by definite features of the process of abyssal differentiation of the parent basalt magma, in turn controlled mainly by the tectonic conditions of the mobile belt. On the other hand, the composition of that abyssal environment in which and from which the initial basaltic magma formed and developed at depths should be of great importance. And, finally, a definite contribution was made by the geological environment in which the ore was formed. All these factors—the depth reached by erosion, the tectonic conditions and the process of differentiation of abyssal magma controlled by these conditions, and the nature of the geological environment in which the ore deposition took place—change considerably in the course of the geological evolution of the same mobile belt.

The compositional features of the abyssal basalt magma are the most stable in time for each given mobile belt. These features probably controlled the relatively stable local variations of each metallogenic province which can be frequently followed through a whole series of ore complexes of various ages. Such differences in the composition of the products of mineralization connected with the gabbro-plagiogranitic and gabbro-granosyenitic complexes are demonstrated quite clearly. Thus, for example, in the Urals, generally characterized by copper and ferrous metals connected with these rock complexes, most intensive is copper and magnetite mineralization. In Kuznetsk Alatau gold mineralization is very strong. On the contrary, in Transcaucasia, gold is of very little importance, which is typical in general of the entire Mediterranean belt.

5. The complex of sulfide deposits includes quite diverse types of deposits. The deposits of pyrite, copper sulfide, copper-zinc, lead-zinc or polymetallic, barite-polymetallic, barite-gold, barite, and other deposits are accompanied by small veins with lead zinc mineralization, auriferous quartz of barite mineralization, which is, however, of little economic importance. In some regions the contact-metasomatic (skarn) deposits of lead zinc ore should probably be classified as belonging to the same complex. However, if the complex is considered as a whole, the sulfide and barite deposits are most characteristic of the complex and are of the greatest economic importance.

With respect to genetic aspects and their position in space all these deposits are normally connected with the minor intrusions of quartz albite-porphyries. In some cases there is no visible connection between these deposits and magmatic rocks. The sulfide deposits are frequently associated in space with extrusive series of an intermediate and acidic composition. Zavaritskiy assumed that in the Urals there is a genetic connection between these deposits and those extrusive rocks in which they are localized. This point of view cannot be applied to other regions of the Soviet Union where similar deposits occur since the ore deposits are younger

than the intrusive series and younger than faults cutting the extrusive rocks in which the ore deposits were formed.

Based on a study of Rudnyy Altai, an opinion was expressed that in the Altai Mountains there is a genetic connection between the ore-bearing intrusions of quartz albite-porphyries and the plagiogranite complexes of the Zmeinogorsk type of intrusive complex. A systematic analysis and the generalization of material from other regions of the Soviet Union where sulfide deposits occur indicate that this point of view is hardly correct. First of all the intrusions of quartz albite-porphyries and gabbro-plagiogranite complexes belong to quite different depths of formation and are connected with different types of tectonic deformations. If intrusions of quartz-albite-porphyries are formations typical of lesser depths and near-surface zones (as well as the sulfide deposits genetically connected with quartz-albite-porphyries), then the gabbro-plagiogranite complexes of the Zmeinogorsk type are formations characteristic of the batholiths of the intermediate depths of formation. The intrusions of quartz albite-porphyries are quite clearly connected with fractures cutting across the folded structure, while the gabbro-plagiogranite complexes are affiliated with folded structures. The regions of distribution of both types of intrusions and the structural zones in which they occur hardly coincide. Within the area of Rudnyy Altai the gabbro-plagiogranite complex of Zmeinogorsk and the complex of quartz albite-porphyries accompanied by sulfide mineralization developed within the same region. However, an analysis of their distribution in space indicates that the major gabbro-plagiogranite intrusions of the Zmeinogorsk complex are in the axial part of the Rudnyy Altai structure, which is in general very poor in sulfide deposits. Intrusions of quartz albite-porphyries and all principal sulfide deposits are in two clearly defined structural zones located on the northeastern and southwestern limbs of this structure. Since they are connected with distinct faults, an impression is created that the axial part of the entire structure, as the most consolidated, served as a kind of core, along the periphery of which faulted zones developed. There are minor intrusions of quartz porphyries and albite porphyries in Salair and Central Kazakhstan with which sulfide mineralization is connected, but here major intrusions of gabbro-plagiogranite composition are absent entirely and there is no reason to assume their presence at depth. On the other hand, the Caledonian gabbro-plagiogranite intrusions (the intrusive complex of Kryk-Kuduk) are very common in Kazakhstan, while the intrusions of quartz albite porphyries and sulfide mineralization are absent. In the area of this eastern slope of the Urals the sulfide deposits are within extrusive rocks metamorphosed to such an extent (transformed to quartz-sericite schists) that it is impossible to find out whether there are minor intrusions of quartz albite porphyries within this complex. The connection of sulfide mineralization with larger intrusions of a gabbro-plagiogranite composition is definitely absent in this area.

These examples are sufficient to demonstrate that the gabbro-plagiogranite intrusions and the minor intrusion of quartz-albite porphyries accompanied by sulfide mineralization develop, in general, independently of each other and besides, occur in a slightly different tectonic environment. However, in some regions, after the manifestation of gabbro-plagiogranite intrusions, a tectonic environment is created which is favorable for the formation of small intrusions of quartz-albite-porphyries accompanied by sulfide deposits. This certainly does not indicate that the intrusions of quartz-albite-porphyries have to be considered as derivatives of large gabbro-plagiogranite intrusions. Both represent entirely independent intrusive complexes formed one after another at different depths. The fact that intrusions of quartz albite-porphyries and the sulfide deposits, representing formations of very shallow depths, or even epithermal formations, occur at the same level as the quite abyssal batholiths of gabbro-plagiogranite composition, indicates that in time they are separated by a period of a fairly deep erosion.

It should be noted that the series of sulfide deposits is quite clearly divided into two groups which are frequently developed in different metallogenic provinces. To one group belong the deposits of pyrite, copper sulfide and copper-zinc deposits (Urals, the mobile belt of the Sayans). To the second group belong the polymetallic and barite-polymetallic deposits (Rudnyy Altai, Salair). In some metallogenic provinces (Transcaucasia, Central Kazakhstan) deposits of both groups were developed together, sometimes even within the boundaries of the same ore-bearing field. It is so far not clear which geological factors determine the appearance of one or another group of sulfide deposits. In any case, attention is drawn to the fact that the particular nature of sulfide deposits usually correlated well with the general iron-copper or tin-polymetallic aspect of the endogenetic mineralization of a specific province.

Transitional group of deposits

Complexes of near-surface deposits connected with small dikes and extrusive rocks or deposits not revealing a visible connection with magmatic rocks develop at the end of the early stages in the evolution of mobile belts. Deposits of metasomatic quartzites with disseminated copper deposits of the Kounrad type; deposits with high content of alumina; near-surface (epithermal) gold-silver deposits; and low temperature deposits of mercury, antimony, and arsenic (realgar, orpiment) belong to these complexes. The same series of near-surface and low temperature deposits appears in some metallogenic provinces during considerably later stages in the evolution of mobile belts, namely, at the boundary between the late and terminal stages of their evolution. Therefore, this group of deposits cannot be considered characteristic of the early stages of evolution and has to be classified as a transitional group.

It was mentioned earlier that the first major orogeny completing the sedimentation of the first cycle with which the ultrabasic intrusions are usually connected, does not make the mobile belt sufficiently rigid for the strong manifestation of minor dike intrusions. The second major orogeny completing the accumulation of sediments of the second cycle, with which major intrusions of gabbro-plagiogranites and gabbro-granosyenites are usually connected, contributes, over a considerable period of time to the rigidity of the mobile belt necessary for the formation of these intrusions. Probably in connection with this rigidity, following this particular folding and the major intrusions accompanying the folding, an entire complex of minor dikes and extrusive surface flows frequently develops, with which a substantial endogenetic mineralization is connected. A similar tectonic environment is also created after the manifestation of the third major orogeny completely transform-

ing the geosyncline into a folded belt. Minor dikes and, during some periods, lava flows are also typical of the late and terminal stages of evolution. Thus, a similar tectonic environment is favorable for the formation of a series of near-surface deposits at the end of the early as well as at the end of the late stages in the evolution of mobile belts.

The composition of these mineral deposits and their similarity during both periods may be explained most probably by the connection of this mineralization with similar magmatic complexes originating in the same abyssal zone. Actually at the end of the early stages there are manifestations of predominantly granitoid differentiates of basic magma, which are closely followed during the intermediate stages of the evolution of mobile belt by intrusion of independent granitoid magmas. Thus, the series of near-surface deposits of copper, antimony, and arsenic develop precisely during the change from one kind of magma to another. As will be seen later, this mineralization occupied exactly the same position at the end of late stages of evolution, when the independent granitoid magmas of the late stages are followed by the granitoid differentiates of basic magma of the terminal stages of evolution.

There is probably no difference between the low-temperature and near-surface deposits formed at the end of the early and at the end of the late stages of evolution. Therefore, they can be given common characteristics. The succession in time of these three complexes is not always clear. But there is a possibility that this succession varies to some extent in different metallogenic provinces and during different stages in the evolution of mobile belts.

6. The complex of metasomatic quartzite deposits with disseminated copper ore and high alumina content always develops in close spatial and genetic connection with magmatic rocks. Usually there are minor intrusions of granodiorites, granodiorite-porphyries, quartz-monzonite-porphyries, granite porphyries, and other igneous rocks formed near the surface or at moderate depth which are clearly connected with tear faults. Frequently they are developed in very close spatial and genetic connection with extrusive rocks. The disseminated copper ores are usually in the intrusive rocks which are transformed by the processes of hydrothermal alteration into metasomatic quartzites. Besides copper, a small amount of molybdenite is frequently present in these deposits and in some cases is of economic interest. A very low grade content of gold is occasionally present but the possibility of its economic utilization is not clear at this time. It is quite probable that the molybdenum ore deposit of Climax, Colorado, in the United States, the largest molybdenum deposit in the world, belongs to a similar complex, though it is purely a molybdenum deposit.

Minerals with a high content of alumina are usually developed to a certain extent in metasomatic quartzites which contain the disseminated copper ores and which formed from intrusive rocks. These minerals form a concentration of economic grade only in those quartzites which form as the result of replacement of the acidic and to some extent to the intermediate extrusive host rocks. Minerals with a high content of alumina are represented by andalusite, corundum, diaspore, dumortierite, alunite, kaolinite, and others. The processes which result in their concentration according to the current points of view vary from the high temperature contact—metasomatic (corundum–andalusite rocks of Semiz–Bugu in Central Kazakhstan) through pneumatolytic to hydrothermal and fumarole–solfatara type rocks (metasomatic quartzites with alunite and kaolin). The last two types of metasomatic quartzite indicate that the process took place under near-surface conditions.

The age relations between metasomatic quartzites and near-surface gold–silver deposits could not be established within the territory of the Soviet Union. In Turkestan, in the deposit of Ak–Tash, there is superposition of later cinnabar mineralization over the dumortierite-bearing metasomatic quartzites. In some near-surface gold deposits of the Soviet Union and the United States, there is transformation of extrusive host rocks to metasomatic alunitic and kaolinitic quartzites. These quartzites are very similar to the conventional metasomatic quartzites accompanying disseminated copper ores. This indicates the similarity of the processes of their formation.

Typical metasomatic quartzites are known in Central Kazakhstan, Soviet Turkestan, and Transcaucasia. In the first two metallogenic provinces they are of Variscan age and belong quite clearly to the late stages in the evolution of mobile belts. In particular, in Central Kazakhstan, the intrusive rocks accompanied by the formation of metasomatic quartzites contain xenoliths of acidic potassic granites which supply the rare metals mineralization and belong to the intermediate stages in the evolution of the mobile belt. In Transcaucasia the metasomatic quartzites are of Tertiary age and belong to the early stages in the evolution of the mobile belt.

Most of the known deposits of metasomatic quartzites of the world are of Tertiary age and only a very small number are of Variscan age.

7. The complex of the epithermal gold–silver deposits is especially widely developed in the internal zone of the Pacific metallogenic belt. These deposits occur in the western parts of South America, Central and North America, in Japan, the Philippines, and New Zealand. Deposits of this type are considerably less developed within the Mediterranean metallogenic belt (Hungary, Transylvania), which is in general poor in gold deposits. The age of deposits in most cases is Upper Tertiary. In individual cases it can be Lower Tertiary or Upper Mesozoic. Older deposits of this type are unknown.

In most cases the near-surface gold–silver deposits demonstrate an obvious connection, from the point of view of space, age, and origin, with the subaerial volcanic rocks which are predominantly of an andesite, and to a lesser extent of a liparite, composition. In some cases they are related to trachytes and even phonolites (Cripple Creek deposit in Colorado). The age relationship with volcanic rocks is clearly established by the ore-bearing veins cutting across lava covers and flows and by the obvious overlap of the ore veins by overlying flows or covers. There is frequently an affiliation in space of the ore-bearing veins with volcanic necks. Besides the extrusive rocks, in some deposits there are intrusive rocks of similar composition which formed near the surface (Comstock in the United States). In rare cases mineralization of this type develops without visible connection with any kind of magmatic rocks.

In some deposits of this type the intersection of

gold-bearing quartz vein by quartz–stibnite veins may indicate that the epithermal antimony mineralization is younger. However, it is quite probable that the relative succession in time of these two types of epithermal deposits is not quite the same in different metallogenic provinces.

8. The association of low temperature deposits of mercury, antimony, and arsenic is quite broadly represented in the territory of the Soviet Union. While in some metallogenic provinces these deposits undoubtedly belong to the late stages in the evolution of mobile belts (eastern Soviet Union, Kazakhstan, Turkestan), in other metallogenic provinces (Transcaucasia) they complete the early stages of evolution. There is no visible connection between these deposits and the magmatic rocks. The age of mineralization is Variscan, Mesozoic, or Tertiary. Caledonian and Precambrian deposits of this type are not known in the Soviet Union nor outside the Soviet Union. It should be stressed in describing the characteristics of the spatial distribution features of these types of deposits that these deposits form independent structural-metallogenic zones in one kind of metallogenic province (Caucasus, Turkestan), while in other provinces (Eastern Transbaikalia, and others) they are represented only as a later stage of endogenetic mineralization in structural metallogenic zones of other types.

This type of deposit includes all economic mercury deposits and most of the economic antimony deposits. The presence of finely dispersed wolframite in some antimony deposits can be mentioned as a typical detail. The economic importance of wolframite in these deposits is not clear so far. Deposits of arsenic represented by realgar and orpiment are almost of no practical importance because of very insignificant reserves. However, the ore is frequently of a sufficiently high grade.

2. The Intermediate Stages of Evolution

It was indicated earlier that sedimentation of the third and last cycle in the evolution of the mobile belt differs considerably from the accumulation of sediments during the first and second cycles by a strong predominance of terrigenous Flysch facies with a quite substantial amount of shale. Magmatism and endogenetic mineralization of the intermediate stages in the evolution of mobile belts differs to some degree from those of the earlier stages. The early stages are characterized predominantly by the basic and ultrabasic intrusions and such granitoids (acidic sodic and potassium granitoids of higher alkalinity) which according to all data are derivatives of basic magmas. The normal granitoid magmas, which do not reveal any indications of genetic connection with basic magma, are strongly predominant during the intermediate stages of evolution. The group of metals typical of the early stages of evolution, represented by platinum, chromium, titanium, iron, nickel, copper, zinc, lead, and barium, disappears almost entirely during the intermediate stages of evolution and is replaced by another complex. This complex is poorly represented during the early stages of evolution and includes tin, tungsten, molybdenum, and gold. Such a sharp change in the complex of metals during the intermediate stage of evolution confirms once more the independent nature of magmas of this period and lack of genetic connection of these magmas with basic magma. The endogenetic mineralization of the intermediate stages is connected with both the major batholithic intrusions as well as the minor pre-batholithic or post-batholithic dikes.

9. The complex of the pre-batholithic dike intrusion. In some metallogenic provinces with faults preceding the third major orogenic phase, extensive series of dikes are sometimes connected with the subordinate stocks of intrusive rocks and are accompanied by gold mineralization. According to their original composition, the intrusive rocks are represented by quartz diorites, diorite-porphyries, granodiorite porphyries, granite-porphyries, and other rocks. But in connection with intensive hydrothermal alterations they are usually transformed into rocks of quartz-albite porphyry and berezite * types. Separate series of dikes can be followed for distances of tens of kilometers and in turn can be grouped into more extensive zones reaching, with some discontinuities, hundreds of kilometers in length. The individual dikes are hundreds of meters to several kilometers long, while their thicknesses range from several meters to several tens of meters. The dikes and stocks of this complex are definitely younger than the major batholiths of potassic granitoids. These dikes are silicified in the contact zones with granitoids.

The gold deposits connected with these small intrusions are represented by quartz veins and veinlets in the dikes or, sometimes, in the host rocks. Gold deposits form ladder veins, an irregular network of small veinlets, separate irregular pockets, more or less continuous quartz veins following the central part of the dikes or the selvages of the latter.

The association of ore minerals is represented in one case by native gold, scheelite, arsenopyrite, pyrite, and, in later generations, by stibnite. Scheelite, arsenopyrite, and antimonite are in some cases of definite economic interest. In other cases the association of ore minerals is represented only by native gold and pyrite. In ancient (Precambrian) deposits of this type pyritization sometimes affects considerable areas in host rocks. The ore minerals and gangue minerals forming these deposits are affected by contact metamorphism in the contact zone of large granitic batholiths. Needles of actinolite develop in quartz; various silicates form in carbonates of the veins; pyrite is successively replaced by pyrrhotite, by magnetite, and by ferrous silicates. Ferrous silicates appear when metamorphism reaches the highest rank. As the result of this process, gold is probably dispersed in the host rocks.

Deposits of this type are known in Proterozoic, Paleozoic, and Mesozoic mobile belts, and in all cases are of major economic interest. This indicates that the given type of gold mineralization is stable with respect to time and is continuous with depth under conditions of quite different levels of erosion. The gold-pyrite association is apparently characteristic of considerably deeper zones.

10. The complex of moderately acidic granitoids represented by granodiorites and normal (biotite-hornblende and biotite type) granites corresponds to the earlier intrusive phases of major batholithic intrusions concluding the third cycle in the accumulation of sediments and is connected with a major orogenic phase draining the geosyncline and transforming it entirely into a folded

---

* Pyritized granitic rocks with quartz veins, occasionally auriferous (Translator's note).

belt. In some metallogenic provinces the granodiorite and normal granites correspond to the same phase o intrusion and are connected by mutual transitions. I other provinces they are not quite synchronous and th granodiorites are always slightly younger. However, th endogenetic mineralization connected with both, as we as the tectonic conditions under which the mineralizatio develops, are similar to such an extent that it is ex pedient to consider these rocks as a unique intrusiv complex, despite some difference in time of their forma tion. The somewhat older acidic or supersaturated po tassic granites, despite the fact that they frequentl represent the evolution of the same intrusive process are considered by the present writer as an intrusiv complex of independent age, because they are frequentl connected with other tectonic stresses and are charac terized by a quite different endogenetic mineralization.

Granodiorites and normal granites of the early in trusive phases of this period are most closely connecte with the orogenic processes. They form intrusions dur ing the orogenic phases or follow these phases. Mos frequently they form major and sometimes quite larg intrusions of batholithic type occupying areas of hundred and thousands of square kilometers. In other cases the form considerably smaller intrusive bodies of stock typ but the latter in many cases probably are only cupola of batholiths.

The mineralization connected with the complex moderately acidic granitoids of the intermediate stag of evolution is represented mainly by gold, molybdenur tungsten (scheelite); sometimes also by arsenic, coppe zinc, lead, and other metals. However, it is not alwa certain that deposits of lead and zinc belong to th group. Deposits of this group are divided into two su groups: the skarn deposits and the hydrothermal de posits. The skarn deposits are formed in connectic a carbonate medium and form quite large accumulatior of skarn-scheelite ore and less frequently those of skar molybdenum ores. Such are the skarn deposits o Turkestan, the molybdenum-tungsten deposits of th Northern Caucasus, and some others. Sometimes in suc deposits small amounts of tin, gold, and arsenic ar present. In some cases the contents of these meta reach economic concentrations—however, of relative low grades. Thus, the group of useful minerals these deposits is represented by metals typical of th independent granitoid magmas intruding during the ir termediate stages of the evolution of mobile belts. Hig contents of magnetite, copper, and cobalt are not typic of these deposits. In a few instances, when the contac metasomatic deposits of magnetite, copper, and coba were classified in this group of deposits, it most fr quently turned out to be an error. In reality these de posits are connected with gabbro-plagiogranites ar gabbro-granosyenite intrusive complexes of the ear stages of evolution.

Among the hydrothermal deposits of this comple of greatest practical interest are gold deposits whic sometimes contain small amounts of arsenic (arsen pyrite) or tungsten (scheelite). Auriferous veins Kochkar in the Urals and a group of other deposits molybdenum, tungsten (scheelite), copper (Bayan-Au district of Central Kazakhstan), and others are of con siderably smaller practical value.

The geological age of endogenetic mineralization this complex varies from Precambrian to Upper Meso zoic. The skarn deposits of scheelite and molybdenu are known predominantly in connection with the Variscan and Upper Mesozoic granitoid intrusions. These deposits are not associated with more ancient intrusions. On the other hand the gold deposits are more numerous in older, Precambrian intrusive complexes, and less common in connection with younger intrusions, in which the main contribution to gold mineralization belongs to various small intrusions.

11. The complex of acidic and supersaturated granites of the intermediate stages in the evolution of mobile belts forms directly after the complex of the moderately acidic granitoids or after a relatively very short time interval, but sometimes directly in connection with tectonic stresses of another nature. In some metallogenic provinces it is very clearly established that batholiths of acidic and supersaturated granites form in connection with the development of shear zones cutting the already formed folded structure. In some cases these shear zones dip at relatively steep angles, in other cases they slope quite gently (25° to 30°), representing in the latter case typical overthrust zones.

The penetration of granites along such shear zones, which in some cases reach a quite considerable width, makes the formation of injection gneisses at the contacts of the intrusion easier, usually accompanied by strong granitization of the host rocks. Observations indicate that movements along these shear zones, extending over a considerable time interval, took place simultaneously with the penetration of granite magma. Especially typical are the processes of migmatization and granitization of host rocks around the ancient Precambrian intrusions. But frequently their existence is confirmed also in considerably younger granite intrusions of Variscan age (Murzinka-Alabash intrusions in the Urals) and Mesozoic ages (acidic granite intrusions of the Eastern Transbaikal region). The processes of metasomatic granitization of host rocks are specially typical of this granite complex.

The spatial distribution of acidic granite intrusions within the metallogenic provinces is controlled by the same pattern as the distribution of intrusions of moderately acidic granitoids. But because of their connection with the shear zones the acidic intrusions sometimes form more clearly outlined structural zones (Eastern Transbaikal region).

This complex of rocks is represented by biotite granites, alaskites, leucocratic granites, aplitic granites, and pegmatitic granites, usually in association with a substantial generation of aplites and pegmatites. The formation of lamprophyre dikes is not typical, in contrast with the complex of moderately acidic granitoids. The apical parts of granite domes are frequently the subject of extensive processes of greisenization.

The two following most important types of deposits may be distinguished in connection with this intrusive complex; the pegmatites and the high temperature hydrothermal deposits, predominantly those of rare metals[*]. Pegmatites of acidic potassic granites represent the well known complex of deposits. These deposits are the main source of beryllium and lithium,

muscovite, minerals used in ceramics, some precious and semiprecious stones (emerald, aquamarine, topaz, and tourmaline), to some extent tantalum, columbium, and tin. The pegmatites affiliated with acidic potassium granites are quite common in association with granites ranging in age from Lower Archean to Upper Mesozoic. Pegmatites of Precambrian age are the main source of muscovite and minerals used in ceramics. The Proterozoic pegmatites are of economic importance mainly as the source from which tin placers were derived (about 7% of the world reserves). Pegmatites of quite various ages are of economic importance as sources of beryllium, lithium, and gems.

The high temperature hydrothermal deposits of this complex include the deposits of tin, tungsten, and to some extent molybdenum, bismuth, and fluorine (fluorspar). The above mentioned metals form deposits of individual metals as well as complex deposits in which tungsten is frequently associated with tin, or with molybdenum and bismuth in other cases. The combination of tin with molybdenum is considerably rare. There are usually tin and tin-tungsten types of mineralization in metallogenic provinces where complexes of acidic potassic granitoids are well developed. On the other hand, tungsten or tin-tungsten mineralization occurs in provinces with lesser development of these granites.

The intrusion of alkalic granites belongs to the concluding moment in the accumulation of sediments of the third cycle characterized mainly by terrigenous strata which are largely shales. Therefore they are the most common host rocks for intrusions of acidic potassic granites and the accompanying mineralization featuring rare metals. This circumstance is sometimes used as a reason to consider tin mineralization as the result of assimilation of large masses of shales by granites (Nikol'skiy). However, there is no sufficient evidence to make this conclusion. In those cases in which the accumulation of terrigenous strata did not reach a considerable thickness, the acidic potassic granites and the affiliated mineralization represented by rare metals occur in rocks of an older structural stage, which are frequently represented by carbonate rocks. In these cases the rare-metals' mineralization of such metals as tin and tungsten is represented by different mineral associations—the skarnlike deposits in carbonate rocks, frequently without any association with the contact of intrusive bodies.

Among other metals, arsenic (in arsenopyrite) occurs most frequently in deposits of rare metals. There is sometimes a small amount of sulfides of copper, lead, and zinc, but in general these metals are not typical of rare metals deposits. When the content of sulfides is substantial, tin deposits belong to the sulfide-cassiterite association. In this case these deposits are younger and are connected with minor intrusions, which are frequently of a less acidic composition than the acidic potassic granites. An appreciable amount of conventional sulfides occurs in some quite large tungsten deposits of the Eastern Transbaikal region. By an analogy with the sulfide-cassiterite association Shcheglov assumed the existence of a sulfide-wolframite association which is slightly younger than the quartz-wolframite association and which is connected with less acidic granitoids. The latter circumstance is very typical of some deposits of the Eastern Transbaikal region. However, this very important and interesting problem requires further study.

[*]) According to the Russian terminology, tin, tungsten, molybdenum, bismuth, antimony, mercury, vanadium, cadmium, gallium, indium, germanium, thallium, and the rare earth metals belong to this group (translator's note).

In deposits of the quartz-cassiterite and the quartz-wolframite association the content of arsenopyrite sometimes reaches considerable concentrations, while sulfides of copper, zinc, and lead, in most cases, are of no economic importance. Such a complex copper-tin deposit as Cornwall in England, which produced about 1,000,000 tons of copper, is a unique exception, as was indicated by S.S.Smirnov. Molybdenite and bismuth are quite common components in deposits of tin and especially in deposits of tungsten. Indeed, the content of molybdenite is frequently of an economic grade in complex deposits of molybdenum-tungsten ores. As a whole, deposits of molybdenum are less typical of this complex than are deposits of tin and tungsten.

Fluorite is a very common gangue mineral in many high temperature deposits of rare metals connected with acidic and supersaturated granites, especially in deposits of tin and tungsten. However, fluorite seldom produces concentrations of economic grade. Very few fluorite deposits of this complex are known to be of economic interest.

3. Late Stages of Evolution.

The major orogenic phase, with which the batholithic complexes of moderately acidic and ultraacidic potassic granites are connected, completes the accumulation of sediments of the third cycle, entirely drains the geosyncline, and transforms it into a folded orogenic belt. The accumulation of sediments within the latter is strongly reduced and sometimes is entirely discontinued. There are practically no marine facies, if one does not count marine ingressions of short duration covering very small areas in some metallogenic provinces. Typical is continental and subaerial fresh water sedimentation, sometimes with a quite strong development of volcanic and coal-bearing facies. In contrast to volcanism of the geosynclinal period, these predominantly subaerial lava flows formed during the late stages of evolution. The extrusive rocks are represented mainly by normal acidic rocks and by rocks of intermediate basicity such as liparites, dacites, and sometimes andesites. During the later periods the basicity of extrusive rocks as a whole increases. Some metallogenic provinces are characterized by the formation of a thick series of subaerial extrusive rocks directly following the batholithic intrusions of granitoids. The frequent occurrence of lava flows over an eroded surface of granite batholiths indicates that there was a quite deep erosion between the periods of their formation.

The role of folding during the late stages of evolution is strongly reduced because of a strong consolidation of the folded belt. On the other hand, the role of faulting invariably increases. As a result, the batholithic intrusions are overshadowed by dikes which are frequently consolidated under conditions near the surface or at shallow depths.

The composition of intrusive rocks and that of an endogenetic mineralization are basically inherited from the intermediate stages of evolution, acquiring at the same time some supplementary features. These features are characterized by the prevalence of normal granitoid magmas, which are, however, on the whole less acidic than their counterparts during the intermediate stages of evolution. The small size of intrusive bodies, and frequently the shallow depth at which they

were consolidated, contribute to the development of granite-porphyry textures which are very common. The acidic granites and granite-porphyries occur quite frequently. Rocks of granodiorite, granodiorite-porphyries and lamprophyre types are predominant, while the quartz-diorites and quartz diorite-porphyries are frequently present. In some intrusive complexes rocks occur with a clear tendency toward a syenitic composition (granosyenites, syenite-diorites). The alkaline granites and granite porphyries are very rare.

The leading metals in endogenetic mineralization of the later stages of evolution are gold, molybdenum, and tin (sulfide-cassiterite formation). The absence of tungsten, which is so typical of the mineralization of the intermediate stage, is not understood. The answer to this question will be clarified if A.D.Shcheglov's assumption about the existence of a sulfide-wolframite association connected with minor intrusions of the late stages of evolution is confirmed. The composition of intrusive rocks begins to show a trend toward syenites, while mineralization in some complexes contains lead, zinc, and iron (magnetite), metals which are not typical of the intermediate stages of evolution.

As a whole, the magmatism and endogenetic mineralization of the late stages are quite close to the magmatism and endogenetic mineralization of the intermediate stages of evolution, but develop in a very different tectonic environment.

12. The complex of postbatholithic gold deposits connected with minor intrusions of quartz diorites, granodiorites and their porphyry varieties is very typical of many metallogenic provinces. Intrusive bodies of this complex represent typical small intrusions in their independent development. Small stocks reminiscent to some extent of pipe-like bodies and numerous dikes are the predominant forms of intrusive bodies. The sizes of stocks are not large, usually hundreds of meters in diameter, less frequently several kilometers in diameter. In horizontal projection the stocks have round or oval outlines, sometimes quite intricate, with many juts and bays. The contacts of stocks as demonstrated by natural outcrops and in mines are vertical or almost vertical. This indicates that the stocks have the shape of pipes and not that of domes. The flowage lines in such stocks, as indicated by the orientation of hornblende crystals, are also oriented steeply, almost vertically. The diorite stocks usually occur in groups with which areas containing auriferous veins are associated. Separate stocks occur less frequently.

The dikes may occur either with stocks or without stocks. The dikes are not large, reaching several meters in thickness and tens or hundreds of meters in length. Larger dikes occur less frequently. Sometimes they form quite extensive series. More frequently they form narrow localized zones in which the spacing between the dikes is smaller and which are characterized by the presence of auriferous veins.

The petrographic composition of igneous rocks is not always the same. In one kind of metallogenic province the hornblende quartz diorites and diorite porphyries are usually typical and very common. The basic varieties of hornblende gabbro-diorite occur very infrequently. The formation of these rocks was probably caused by the processes of assimilation. Sometimes the more acidic varieties which produce the gradual transitions from quartz diorite porphyries through granodiorite porphyries to granite porphyries or quartz porphyries are also present. Lamprophyres of spessartite type frequently form gradual transitions to hornblende diorites and diorite porphyries. Such a composition of intrusive rocks is especially characteristic of the complexes accompanied by gold mineralization. Rocks of a granodiorite composition are most common in complexes in which mineralization is of a gold-molybdenum type.

All rocks usually have a strongly developed hypabyssal aspect. Porphyry structures, granite-porphyry, and porphyry-like textures are extremely common. In the case of granular texture, the shallow depth of consolidation is indicated by very elongated plagioclase and hornblende crystals, the strong idiomorphic nature of plagioclase relative to potassic feldspar and quartz, and the frequent presence of micrographic intergrowths of quartz and potassic feldspar in spaces between the strongly idiomorphic crystals of plagioclase and other minerals.

The diorite stocks and areas with relatively closely spaced dikes, in general occur independently of the granitoid batholiths and at a considerable distance from them. This distance ranges from several kilometers to tens of kilometers. Sometimes, however, the stocks are not far from the batholiths or even are in marginal zones and at the contacts of batholiths. The latter frequently serves as a reason to consider such stocks and dikes as dike facies of batholiths. The diorites are usually part of batholiths and are younger than the granitoid rocks.

The location of diorite stocks inside the granitoids, the presence of sharp chilled contacts in diorites, the penetration of diorite dikes into granitoids, the presence of granitoid xenoliths in diorites—all of this indicates beyond a doubt that the diorites are younger than the granitoids.

Rocks of this complex are frequently of a hybrid nature. This is apparently connected in many cases with the penetration of magma along the explosion vents. There are sometimes transitions of hybrid rocks to eruptive breccias. The brecciation of rocks also takes place when there is a repeated intrusion of later and more leucocratic portions of magma along the same canals. Despite the considerable size of most of the intrusive bodies, some of them are complex intrusions. This refers not only to the stocks but to the dikes as well. Many lamprophyre dikes formed as the result of intrusion of several portions of a given magma along the same fracture accompanied by brecciation of rocks formed earlier.

Gold mineralization connected with the complex of small post-batholithic intrusions is very close in its mineral composition to gold mineralization of small pre-batholithic intrusions. Quartz-pyrite and quartz-scheelite-arsenopyrite ores also occur here. The latter are also especially typical and are frequently accompanied by stibnite in low temperature generations of mineralization. In most cases the deposits are located in diorite stocks or in their immediate vicinity. When the intrusive bodies are represented only by dikes, the

gold deposits are closely associated in space with the latter or occur within the areas where the dikes are most numerous. In some cases the gold-bearing veins of this type do not reveal any visible connection with intrusive rocks and can be classified in a given genetic group by analogy with other similar deposits, according to their age relationship with the major batholithic intrusions (pre-batholithic or post-batholithic).

The alterations of host rocks are usually manifested by sericitization and carbonatization, also sometimes by chloritization and to some extent sulfidization. As a whole these rock alterations are considerably less intensive than in similar prebatholithic deposits.

The mineral composition of the gold-molybdenum deposits of the Eastern Transbaikal region is quite different. These deposits are genetically connected with the post-batholithic granodiorite dikes. The mineral composition of these deposits varies, but they are usually characterized by tourmaline, quite common both in ore bodies and in altered host rocks. The complex gold-polymetallic ores, with a considerable amount of sulfides or arsenic, copper, lead, and zinc, instead of gold-molybdenum ores, occur in separate deposits such as Darasun and Dzhalinda. This entire group of ore deposits is quite peculiar in its mineral composition, but in the course of the geological evolution of the Eastern Transbaikal mobile belt it occupies a position (following the batholith of potassic granites) which is occupied in other metallogenic provinces by the diorite intrusions accompanied by pure gold (or gold-scheelite-arsenopyrite) mineralization.

13. The complex of tin deposits of the sulfide-cassiterite association and the early group of silver-lead-zinc deposits are put together conditionally to some extent. However, in some metallogenic provinces there is an indication of a close genetic connection and even mutual transition between them. The group of deposits of the sulfide-cassiterite association is usually found only in those metallogenic provinces where the tin-tungsten mineralization is quite well represented and is connected with very acidic granites of the intermediate stages in the evolution of mobile belts. In those metallogenic provinces where tin mineralization is poorly represented or where tungsten deposits occur without tin deposits, there are no deposits of the sulfide-cassiterite formation.

The deposits of sulfide-cassiterite formation usually develop in connection with small dikes, which are sometimes closely connected with a series of acidic extrusive rocks. Less frequently the sulfide-cassiterite mineralization develops without a visible connection with magmatic rocks. Small intrusions are represented by granites, granodiorites, granite-porphyries, granodiorite-porphyries, liparites, and others. On the whole, they are characterized by a less acidic composition when compared with those acidic granites which accompany the deposits of the quartz-cassiterite formation. Frequently these minor stanniferous intrusions form later than the series of acidic extrusive rocks following the formation of granitoid batholiths and sometimes rest on the eroded surface of these batholiths. These intrusions frequently demonstrate a clear connection with tear faults, in some cases transverse to the general orientation of folded structures.

The deposits of the sulfide-cassiterite association, as well as the small intrusions, with which they are genetically connected, are both typical products of shallow depth or the near-surface zone. Phenomena of inverse pulsational zoning are sometimes clearly identifiable in these deposits.

As a rule, deposits of the sulfide-cassiterite association are not typical of the Precambrian and Caledonian mobile belts, and develop very little in mobile belts of Variscan age, but are represented by numerous occurrences in Mesozoic and Tertiary mobile belts. Tin mineralization of Tertiary age is represented mainly by the deposits of the sulfide-cassiterite association.

In many sulfide-cassiterite deposits the late generations of ore minerals are represented substantially by lead-zinc-sulfide ores. This indicates the almost continuous transitions between the deposits of the sulfide-cassiterite formation and the early group of lead-zinc deposits. Tin-polymetallic deposits occur in several metallogenic provinces. Finally, most of the lead-zinc deposits of the early group are characterized by the presence of a small amount of cassiterite.

The early and late groups of silver-lead-zinc deposits are present in Central and Western Europe. The deposits of the early group are cut across by dikes of Permian quartz porphyries, which are in turn intersected by the ore-bearing veins of the late group. The presence of a small amount of cassiterite, as well as sometimes that of scheelite or wolframite, is quite common in silver-lead-zinc deposits. These minerals are absent in deposits of the late group, but sometimes minerals of cobalt and nickel occur and in some cases there are suggestions of a transition between the late group of silver-lead-zinc deposits and the silver-cobalt-nickel deposits. In typical deposits of both groups there is some difference in gangue minerals. Quartz and carbonates are most typical of deposits of the early group. In addition to these two minerals the gangue of the late group frequently contains considerable amounts of barite and fluorite. Strontium minerals appear sometimes in the gangue of the late group.

The silver-lead-zinc deposits of the early group either do not demonstrate a visible connection with magmatic rocks or else are affiliated with small intrusions of moderately acidic rocks (granites, granodiorites, granosyenites, syenites, lamprophyres, and others). Deviations in the composition of these rocks toward syenites, granosyenites, and others, which is not characteristic of the sulfide-cassiterite association deposits, are quite common. In contrast to the lead-zinc deposits of the early stages in the evolution of mobile belts, where sulfide deposits are strongly predominant over the hydrothermal veins, the late stages in evolution are characterized most of all by replacement deposits in limestones and to some extent by contact-metasomatic deposits. Especially typical is the difference between the lead-zinc deposits of the early and the late stages with respect to the replacement of host rocks. The sulfide ore bodies of the early stages form by replacement of silicate rocks, mainly by replacement of mesocratic and acidic lavas and their tuffs. These ore deposits are usually inert with respect to carbonate rocks. On the other hand, the lead-zinc deposits of the late stages produce replacement ore bodies (or sometimes contact-metasomatic deposits) very readily in carbonate rocks,

but are entirely inert with respect to the silicate rock in which they are usually represented only by hydro thermal veins.

In some metallogenic provinces (for example, in th Eastern Transbaikal region) the earlier generations o mineralization in lead-zinc deposits may be represente by pyrite-arsenopyrite ores, sometimes of economi importance as a source of arsenic. The later and lowe temperature generation of silver-lead-zinc ores in pure sense may be connected with the arsenopyrite ore by inverse pulsational zoning. In several metallogeni provinces this group of lead-zinc deposits is charac terized by its connection with sheeted structures forme as the result of subsequent deformation of a complete folded structure.

If the preceding ore complex of gold and gold-molyb denum deposits inherited entirely the main features o its composition (gold, molybdenum) from the intermediat stages in the evolution of mobile belts, the character istic features of the composition of the intermediat (tin) as well as the final (lead-zinc) stages of evolutic are clearly blended within the complex of sulfide-cas siterite and the early group of silver-lead-zinc deposit Such blending of compositional features is also typic of some tungsten deposits, suggesting the existence a sulfide-wolframite association. If this assumption confirmed, it will probably be discovered that deposit of this association belong to the same complex. Th existence of a sulfide-wolframite association is the mo probable because: (1) deposits which suggest its exis tence (some deposits in the Eastern Transbaikal regio demonstrate a very clear inverse pulsational zoning, a in the case of the sulfide-cassiterite association and th early group of silver-lead-zinc deposits; (2) all of th most characteristic metals of mineralization shown b the intermediate stages of evolution (tin, molybdenum gold, and arsenic) produce economic-grade concentra tions in deposits of the late stages, except for tungste which is practically absent in deposits of the late stage It is most probable that deposits of the supposed sulfide wolframite association belong in fact to the late stages evolution.

14. The late group of contact-metasomatic magneti deposits has the tendency to associate itself in time several regions with the deposits of the sulfide-ca siterite association and the early group of silver-lea zinc deposits. In areas where the sulfide-cassiteri and silver-lead-zinc mineralization is absent, the la contact-metasomatic deposits of magnetite appear i dependently. In contrast with the contact-metasomat magnetite deposits of the early stages of evolution, whe copper is frequently of substantial importance, the la group of these deposits is more characterized by th presence of lead, zinc, sometimes also tin, frequent cobalt and scheelite. Thus, in this complex of deposi there is a blending of compositional features typic of the intermediate and final stages in the evolution mobile belts with, however, considerable weakening the features inherited from the intermediate stages evolution.

The same trend in composition is observed in i trusive rocks with which this group of deposits is g netically connected. Instead of normal granitoids of t intermediate and the early late stages of evolution, t gabbro-diorites, diorites, especially frequently syenite diorites, granosyenites, monzonites, and other rocks with a somewhat higher alkalinity occur here considerably more frequently. Intrusive rocks of the complex are very frequently of hybrid type and are represented in more ancient (Precambrian) metallogenic provinces by rapakivi or similar granites.

The age relations of this complex with the sulfide-cassiterite and the early group of silver-lead-zinc deposits are not always sufficiently clear. The age of the contact-metasomatic magnetite deposit in the southern Maritime region of the Far East is presumably a little older than the sulfide-cassiterite and silver-lead-zinc deposits; however, there is no direct proof of this.

15—17. Complexes of metasomatic quartzites, gold-silver deposits, deposits of antimony, mercury, and arsenic. The series of near-surface deposits, which include the metasomatic quartzites, the gold-silver deposits and deposits of antimony, mercury, and arsenic (realgar, orpiment), is later than the ore complexes characterized earlier and occupies a transitional position between the late and final stages in the evolution of mobile belts. These deposits are similar to the analogous deposits of the early stages in evolution and were described together with them (complexes 6, 7, and 8).

4. The Terminal Stages of Evolution.

The general tectonic environment of the final stages in the evolution of the mobile belts is similar to that of the late stages, but all characteristic features of the late stages are developed here considerably more strongly. The sedimentation is either discontinued or has strong features of subaerial or fresh-water sedimentation. Series of subaerial extrusive rocks of an acidic composition with a definite trend toward a slightly higher content of alkalies are quite common in some metallogenic provinces. There are almost no folded structures, but, on the other hand, tear faults are especially typical. The deepest fractures in the earth's crust form during this time. The intrusive activity subsides considerably, there are no batholithic intrusions, the intrusive bodies are represented by small intrusions or dikes consolidated near the surface. In composition these rocks are mainly granitoids with higher contents of alkalies—granosyenites, syenites, alkaline granites, alaskites, and others. Sometimes the quite large intrusive bodies consist entirely of granites with a micropegmatite texture. In some cases, together with similar granitoids, rocks of a basic composition develop—of the diabase, gabbro diabase, and olivine monzonite types.

The endogenetic mineralization is represented by hydrothermal deposits. The complex of metals changes drastically as compared with the intermediate and late stages in the evolution of mobile belts. The metals typical of the intermediate and late stages (tin, tungsten, molybdenum, and gold) disappear. Most characteristic are lead, zinc, copper, silver, cobalt, nickel, barium, to some extent iron, manganese, and others, i.e., the same metals which are characteristic of the early stages in the evolution of mobile belts. However, the metals of the early stages of evolution which occur only in deposits of a magmatic type (platinum, chromium, titanium, and vanadium) are absent during the mineralization of terminal stages.

18. The complex of hydrothermal deposits of the West European type includes a very large group of hydrothermal deposits of various mineral compositions which formed at not very great depth or are sometimes epithermal. The group of silver-cobalt-nickel deposits, usually with arsenic, copper, bismuth, and frequently with the genetically closely related group of silver-lead-zinc deposits, belongs to this type of deposits. Among the latter occur both hydrothermal and metasomatic and contact-metasomatic deposits. However, the latter two types of deposits are less typical of this group than of the early group of silver-lead-zinc deposits. There are also some differences in the mineral composition of the gangue. Quartz and carbonates (calcite and dolomite) are the most typical gangue minerals in the early group; in the later group barite and fluorite appear in considerable amounts in addition to them. Sometimes minerals of strontium are present (celestite and strontianite) which are considered to be not typical of the hydrothermal deposits. Independent deposits of barite and fluorite are frequently present in this complex, as well as occasional deposits of witherite, hydrothermal iron deposits (hematite, siderite), and sometimes metasomatic deposits (siderite), hydrothermal manganese ore, and metasomatic magnesite deposits.

This complex of ore deposits is developed quite completely only in a very small number of metallogenic provinces. To these provinces belongs the Variscan metallogenic province of central and western Europe, where all of the abovementioned types of deposits making the complex are quite well represented. However, in most of the metallogenic provinces only the individual types of deposits which are part of this complex occur. Thus, for example, the metasomatic siderite deposits of Bakal and the magnesite deposits of Satka are in the Precambrian metallogenic province of the western slope of the Urals, the fluorite deposits of the eastern Transbaikal region are represented in the metallogenic province of the eastern Transbaikal region and so on.

This entire complicated complex of deposits occurs either without a visible connection with magmatic rocks or in connection with small intrusions of peculiar granitoids characterized by higher alkalinity, such as alaskites, alkaline aegirite and riebackite granites, granosyenites, syenites, and others. Micropegmatite textures are common in the granites. Rocks of the diabase type, gabbro-diabases, and others are frequently part of these rock complexes.

19. The complex of telethermal deposits is the latest in the course of geological evolution of mobile belts and always occurs without visible connection with magmatic rocks. The telethermal lead-zinc deposits of the Mississippi-Missouri type, the cupriferous sandstones with endogenetic mineralization, some barite deposits, witherite, fluorite, as well as probably some hydrothermal hematite and manganese deposits belong to this complex. Mineralization of this type was active during quite a long geological time interval after the termination of folding and intrusive activity within the mobile belt and frequently penetrates into the rocks of the upper structural level having horizontal bedding and overlapping the earlier folded structure.

This complex of deposits is represented especially clearly in central and western Europe. Almost all authors who studied and described the metallogeny of these regions have noted that the endogenetic mineralization is peculiar to the last stages of the Variscan evolution of the region, and continues or is repeated in many regions during the Mesozoic or even early Cenozoic times without change of the main feature of its mineral composition. Thus, for example, the Tertiary and Late Variscan barite, fluorite, cinnabar, and lead-zinc deposits of the Massif Central in France are completely identical. A similar situation exists in the Bohemian massif, in the Upper Rhine region, in southern and southeastern England, etc. In addition, the Late Variscan deposits are frequently overlapped transgressively by Upper Permian sediments, and supply to the latter clearly identifiable clastic material. The Triassic cupriferous sandstones of the London basin, which are believed to be endogenetic, and telethermal lead-zinc deposits in Mesozoic carbonate rocks, of the Rhine region and eastern Silesia, are believed to be of a replacement type and should be considered as phenomena of this kind. In most of the metallogenic provinces, among the deposits of the terminal stages in the evolution of the mobile belt, only the group of silver-cobalt-nickel deposits, and possibly deposits of siderite and magnesite, do not produce the latest types of telethermal deposits. However, cobalt is present in quite substantial amounts in cupriferous sandstones of Rhodesia, which apparently belong to deposits of such type.

In the territory of the Soviet Union, the lead-zinc deposits of Karatau; barite and witherite deposits of Kopet-Dag; cupriferous sandstones of Dzhezkazgan, Atabasar, and Chidertinskiy region; small Permian deposits of fluorite, barite, and lead-zinc of the western slope of the Urals; the fluorite deposit of Amderma; and possibly the iron-manganese deposits of Atasu in Kazakhstan; and others can be classified as telethermal deposits.

5. General Remarks About the Pattern of Evolution of Endogenetic Mineralization

The distribution pattern of endogenetic mineral deposits in the course of the geological evolution of mobile belts of the earth's crust reflects quite naturally only the main trends in the evolution of endogenetic mineralization. The characterization of individual complexes is only very generalized and the local features of individual metallogenic provinces could not be completely described by these characterizations. These features are very important in the evaluation of prospects in definite regions and in one or another metallogenic province. In connection with this, some remarks have to be made with respect to the application of this pattern.

1. All main mineral complexes which occur within the boundaries of various mobile belts were examined. In such a manner, the table is a kind of a summarizing and generalizing table, including all the possible types of intrusive and mineral complexes of mobile belts in the succession in which they follow each other in time. However, if an examination is made of definite mobile belts in the Soviet Union, it turns out that in each of these belts not all of these complexes are present. Only some of them are represented, while the others are absent. Thus, each mobile belt is characterized by its own association of intrusive and mineral complexes, more or less different from a similar association in other mobile belts. It is quite natural that each

such association of intrusive and mineral complexes reflects the features of the geological evolution of the appropriate mobile belt.

Relatively more consistent among the intrusive complexes are the complexes of large batholithic or batholith-like intrusive bodies connected with major stages of folding, i.e., the complex of ultrabasic intrusions, the complex of gabbro-plagiogranite intrusions, and, finally, the complex of normal and acidic granites of the intermediate stages of evolution. However, the composition of given complexes in different mobile belts and even within different structural zones of one and the same mobile belt may vary considerably. These variations are represented by gabbro-granosyenite and gabbro-granosyenite complexes, complexes of the moderately acidic granites of the intermediate stages of evolution, the acidic and supersaturated granites, and other complexes. The ore-bearing capacity of the intrusive rock complexes in different mobile belts may vary even more.

The complexes of minor dike-like intrusions connected with faults are especially diverse in different mobile belts and frequently the metallogenic aspect of one or another province is created specifically by these small intrusions. The effect of the difference in the geological evolution of the mobile belt on the nature of intrusive rocks and mineral deposits being formed is especially strong here.

2. In some mobile belts the geological evolution is considerably complicated and accordingly the association of intrusive and mineral complexes is also complicated. So, for example, the evolution of the mobile belt of the eastern slope of the Urals during Paleozoic time was considerably complicated. This mobile belt went through the early stages of its evolution during Caledonian time. Later it became involved in the geological evolution of the newly started large Variscan mobile belt involving the western provinces of China, western Mongolia, Soviet Turkestan, and a considerable part of Kazakhstan. Together with this gigantic mobile belt, the eastern slope of the Urals once more went through its early stages of geological evolution. This explains why mineralization of the early stages was especially intensive here. The Caledonian platinum-bearing gabbro-peridotite complex and the Variscan chromite-bearing ultrabasic complex which produced the magnetite deposits of Vysokaya and Blagodat Mountains, and the Variscan Gumbeyka-Magnitogorsk granosyenite complex generating the magnetite deposit of Magnitnaya Mountain are present in this region. It is quite natural that in other mobile belts the intermediate stages of the geological evolution could have been duplicated with a double or multiple development in time of the batholithic intrusions characteristic of the intermediate stages of evolution.

3. The intensity of manifestation of various intrusive and mineral complexes even under conditions of similarity in their association may be very different in different mobile belts, since those complexes which are quite intensively developed within one mobile belt are poorly represented within another mobile belt, and vice versa. Frequently there is different intensity in the development not only of the individual intrusive and mineral complexes, but of whole groups of complexes which correspond to the major stages in geological

evolution. Thus, for example, the complexes of the early stages may be represented quite strongly, while the complexes of the intermediate and late stages are poorly developed. This strongly controls the metallogenic aspect of each separate province. So, for example, in the Urals, the complexes of the early stages of evolution, the Caledonian, as well as the Variscan, are quite strongly represented. The intrusive and mineral complexes of the intermediate stages of evolution are represented quite strongly, but in all cases less strongly than those of the early stages. The intrusive and mineral complexes of the late and terminal stages in evolution within the region of the eastern slope of the Urals are scarcely represented at all. On the other hand, the Mesozoic mobile belt of the eastern part of the Soviet Union is characterized by a weak representation of intrusive and mineral complexes of the early and terminal stages of evolution, while the intrusive and mineral complexes of the intermediate and late stages are quite intensively developed. These features control to a considerable degree the metallogenic aspects of one or another region.

4. Local features may be strongly reflected in the composition of the mineral complexes. Each mineral complex usually includes a whole series of mineral deposit types. In one kind of mobile belt the mineral complex may be represented predominantly by one kind of deposit, while in another type of mobile belt the same complex is represented by other types of deposits. Thus, for example, the rare metal mineralization connected with acidic granites of the intermediate stages is represented in one kind of mobile belt predominantly by tin deposits or tin-tungsten deposits, in other mobile belts— by tungsten and molybdenum-tungsten deposits, with tin deposits poorly developed or absent. The sulfide mineralization in one kind of mobile belt is represented predominantly by pyrite and cupriferous-pyrite ore bodies (the Urals); in other mobile belts this mineralization features polymetallic and barite-polymetallic deposits (Rudnyy Altai). The diversity of mineral deposits in different mobile belts may be even greater in the mineral complexes of the terminal stages of evolution.

Causes of such differences have not been established. It is not always possible to explain these differences by depth of erosion. The nature of the geological evolution of each separate mobile belt, as well as the composition of the abyssal substratum which produces magmatic reservoirs by its melting, is of substantial importance in creating these differences.

5. The distribution pattern of endogenetic deposits in the course of the geological evolution of mobile belts of the earth's crust is based mainly on information concerning Paleozoic and Mesozoic-Cenozoic mobile belts. This pattern reflects mainly the principal trends in evolution of these belts. The pattern may also be applied to some degree to the analysis of endogenetic mineralization in Late Proterozoic mobile belts. The entire course of geological evolution during early Proterozoic and Archean times must have been quite different. Therefore it is necessary to prepare a special outline of the distribution in time of intrusive and mineral complexes. This outline is similar to the outline examined above only in some of its most important features.

The association of intrusive and mineral complexes and certain features of their composition are probably different to some extent in mobile belts of various geological ages. The difficulty in studying this problem is in the absence of criteria for discriminating the effect of difference in geological age from the effect of difference in depth reached by erosion, since there is reason to suppose that the ancient mobile belts were affected by deeper erosion.

It is known, for example, that the anorthosite-charnockite intrusive rock complexes are typical of the Precambrian and part of the Early Paleozoic mobile belts, and are almost absent in younger mobile zones. In Precambrian mobile belts, besides anorthosites and charnockites, there are usually also gabbros, norites, monzonites, syenites, granosyenites, and other rocks. All these rocks are typical of the gabbro-granosyenite complexes of post-Precambrian mobile belts, which are apparently analogous to Precambrian anorthosite-charnockite complexes. It is most probable that the absence of anorthosites and charnockites in younger complexes is explained by a different depth reached by erosion of these complexes and not by the different course of magmatic differentiation during the Precambrian as opposed to later periods.

In a similar way the rapakivi granites are most typical of the Precambrian intrusive complexes. Minor intrusions of hybrid syenite-diorites and granosyenites of the late stages, with which the late group of contact-metasomatic magnetite deposits is connected, are their analogues in the course of the evolution of post-Precambrian mobile belts.

On the other hand, some mineral complexes typical of the younger mobile belts do not occur in earlier mobile belts. Thus, for example, the complex of metasomatic quartzites and the complex of low temperature antimony, mercury, and arsenic deposits appears for the first time only during Late Variscan time, while the complex of epithermal gold-silver deposits appears at the very end of Mesozoic time. It is possible to surmise again that this is the result of a difference in the depth reached by erosion, since all the above-mentioned complexes represent typical epithermal formations and therefore should be the first which are destroyed by erosion. Other mineral complexes formed at the same depth—for example, the sulfide deposits, the silver-cobalt-nickel deposits, and others—are also quite well developed in Precambrian formations.

Further study of this problem is needed to elucidate the associations of intrusive and mineral complexes typical of mobile belts of various geological ages.

6. The youngest mobile belts, the Mediterranean and the Pacific belts, have far from completed their geological evolution. An analysis of their intrusive and endogenetic mineral complexes and the accumulation of sediments indicates that they are just finishing the early stages in their evolution, and have started the sedimentation of the intermediate stages of evolution in separate areas. Thus, the intrusive and mineral complexes of the intermediate, late, and terminal stages of evolution have not as yet developed in these mobile belts.

7. The evolution of the individual areas of the same mobile belts are far from being synchronous. They overtake each other in development. Therefore the same or similar intrusive and mineral complexes do not develop simultaneously in different areas of a particular mobile

belt. Frequently such areas, developing during different time intervals, differ from each other by the association of intrusive and mineral complexes. The idea of metallogenic provinces is most of all applicable to these areas. The Minor Caucasus and the Black Sea can serve as examples of areas or provinces of the same mobile belt developing during different intervals of time. The Minor Caucasus is completing the early stages of its evolution at the present time, while the Black Sea basin has already progressed to the accumulation of sediments of the third stage in the evolution of a mobile belt. Similar examples can be presented also for the Paleozoic mobile belts.

8. It is necessary to make a more accurate definition of the intrusive and mineral complex. A con- siderable amount of data indicates that the intrusives and the corresponding mineral complexes do not form during a specific short time interval in the geological evolution of the mobile belt, but comprise a specific more or less long, period during which several complexes of similar composition may form consecutively. Thus, the very idea of indicated intrusive and mineral complex is considerably more complicated than was during the discussion of Table 1.

All reservations made earlier indicate clearly that this table is only an approximate outline reflecting the most basic trends in the evolution of intrusive magmatism and endogenetic mineralization of mobile belts, and far from covering the entire complexity and diversity of natural phenomena.

III. INTRUSIVE AND ENDOGENETIC MINERAL COMPLEXES OF PLATFORM REGIONS

The most substantial difference between the magmatism of mobile belts of the earth's crust and that of the platforms is the strongly subordinated role, or frequent absence, of granitoid magmas in the platforms. On the other hand, basic and ultrabasic magmas, as well as the alkaline magmas, which are accompanied frequently by a diverse group of magmatic rocks, are typical of the platforms.

One of the magmatic complexes typical of the platforms is the so-called trap formation which includes both extrusive and intrusive rocks. In contrast to the manifestations of basic magmas during the early stages in the evolution of mobile belts, the differentiation of basic magmas within the platform regions is considerably less well developed. Besides, they are of a specific nature and are characterized mainly by the enrichment of differentiates in iron due to decrease in magnesia content. The processes of such differentiation frequently proceed quite far, although the resulting differentiates belong only to the group of basic rocks. Differentiates of intermediate and acidic composition are not typical, apart from small segregation dikelets.

The endogenetic mineralization connected with trap complexes is represented by magnetite deposits and segregations of copper-nickel sulfides. Sometimes there are very small occurrences of hydrothermal lead-zinc or gold mineralization.

In some platform regions (South Africa, Canada, and in part the Kola Peninsula) more abyssal complexes occur, consisting of basic magmas and their differentiates, which frequently produce large intrusions of the lopolith type. These intrusions are a kind of abyssal analogue of the trap formation. The processes of differentiation are represented here in a more advanced form than in trap formations. But in all cases the differentiates are quantitatively considerably less well developed than the nondifferentiated basic rocks. In some cases (Bushveld complex of South Africa) differentiates of basic magmas are formed in considerable quantity; in other cases (Sudbury in Canada) granites of the alaskite type are of more substantial importance. Sometimes differentiates of a diorite composition are formed, in contrast to the geosynclinal regions, where acidic differentiates of basic magmas are represented mainly by rocks of plagiogranite composition. Potassic granitoids frequently of higher alkalinity, are of substantial importance in such a differentiated platform complex.

Copper-nickel deposits (Sudbury), as well as deposits of magnetite and titanomagnetite, are connected with similar complexes even to a greater degree than with trap formations and deposits of platinum and chromite (Bushveld complex), if the ultrabasic differentiates are present.

Complexes of alkaline rocks represent a magmatic association which is no less typical of platform regions. In contrast to the basic rocks, this formation is characterized by commonly occurring and far reaching processes of differentiation, frequently producing rock complexes of an extremely diverse petrographic composition. The differentiates range in composition from alkaline peridotites and pyroxenites to the most leucocratic rocks of an acidic and supersaturated granitic composition (alaskites), as well as alkaline and ultra-alkaline rocks (nepheline syenites, ijolites). Usually they occur together, but in one kind of rock complex the rocks of the alkaline peridotite composition are considerably more predominant than other differentiates (Siberian platform, kimberlites of South Africa). In other cases on the other hand, they are of subordinate importance (Kola Peninsula) or are entirely insignificant (Aldan platform) as compared to the leucocratic differentiates. The alkaline extrusive rocks occur very frequently together with the alkaline intrusive rocks.

Three groups among the platform-type complexes of alkaline rocks may be distinguished by their petrographic composition and character of endogenetic mineralization. Various magmatic, pegmatitic, and sometimes peculiar hydrothermal deposits containing quite diverse minor and rare elements are connected with the intrusive complexes represented by large bodies of alkaline rocks of the nepheline syenite type. To these deposits belong the apatite-nepheline rocks of the Kola Peninsula; the cryolite deposits of Greenland; the hydrothermal deposits of parisite (fluocarbonate of cerium earths); magmatic and pegmatite deposits of zirconium, tantalum, columbium, and others.

Magmatic deposits of titanomagnetite (Brazil, Arkansas in United States) are frequently connected with deeper complexes of alkaline peridotites usually accompanied by alkaline and ultra-alkaline rocks. Diamond deposits are connected with the near-surface complexes of alkaline peridotites of the kimberlite type.

The third group of alkaline rocks of platform regions consists of strongly differentiated intrusive complexes including rocks ranging from gabbro through monzonites to granosyenites and alkaline granites on one side, and nepheline and pseudoleucite rocks on the other side. By their composition these complexes remind one of the gabbro-syenite complexes of the early stages in the evolution of mobile belts, but they differ by having a more definitely alkaline character. The contact-metasomatic magnetite deposits, sometimes with small amounts of copper, scheelite and molybdenite, and the

hydrothermal deposits of lead and zinc, gold, and molybdenum, are connected with these alkaline rocks, as was the case with the gabbro-pyroxenite complexes of mobile belts.

As a whole the magmatism and metallogeny of platform regions are closest to the magmatism and metallogeny of the early stages in the evolution of mobile belts. Basic and alkaline magmas, and the mineralization affiliated with them, are strongly represented within the platforms while the granitoid differentiates of basic magmas and their metallogeny are at a minimum. It is typical that in both the platforms and the mobile belts, similar complexes of mineral deposits are connected with similar magmatic rock complexes occurring in very different structural environments.

IV. DISTRIBUTION IN TIME OF THE MAIN METALS OF ENDOGENETIC DEPOSITS

A group of scientists of the Allunion Geological Institute (VSEGEI) in Leningrad made an interesting attempt to examine the distribution of world resources of the main metals of endogenetic deposits in separate mineral complexes on the one hand, and among the deposits of different geological ages on the other hand. Both the already recovered amounts of these metals and the reserves still remaining underground were taken into account, with a very moderate amount of extrapolation in separate cases. The distribution in time of copper involved 140,000,000 tons of its world resources, 46,000 tons for gold, and so on. Let us consider now the principle results of this study.

Tin. The three following associations designated some time ago by S.S. Smirnov are of main economic importance for tin: the pegmatite and quartz-cassiterite associations connected with acidic and supersaturated granites of the intermediate stages of evolution of mobile belts and the sulfide-cassiterite association of the late stages. It is known that cassiterite also occurs in deposits of the sulfide association, in skarn deposits in connection with moderately acidic granites of the intermediate stages, and in the late group of contact-metasomatic magnetite deposits. But all these types are only of a mineralogical interest or are of no significance with respect to the world resources of tin.

The pegmatite formation of tin deposits is of substantial importance only in Precambrian mobile belts (Nigeria), mainly as the primary source from which cassiterite placers were derived. This formation makes about 10% of the world reserves of tin. Deposits of this formation occur in Caledonian, Variscan, and Mesozoic mobile belts, but their economic importance is negligible. Deposits of Caledonian age make 2% of the world reserves, while deposits of Variscan age make 24% of these reserves. Deposits of both ages are represented almost exclusively by deposits of the quartz-cassiterite association.

Deposits of Mesozoic age make up about 50% of the world reserves; 45% of the Mesozoic tin reserves are represented by quartz-cassiterite associations and 5% by sulfide-cassiterite association, the latter acquiring an economic importance for the first time. The Cenozoic tin deposits make up 14% of the world reserves of tin and are represented exclusively by deposits of the sulfide-cassiterite association. As a whole, the world re-

sources of tin ore are distributed among the separate associations in the following way: the pegmatite association, 10%; the quartz-cassiterite association, 71%; and the sulfide-cassiterite association, 19%. It is interesting that tin deposits in the most ancient mobile belts are represented by the most abyssal types of deposits, while in the youngest mobile belts they are represented by the non-abyssal types—the sulfide-cassiterite associations.

Tungsten. Three types of economic tungsten deposits can be recognized: pegmatites, skarns connected with complexes of moderately acidic granitoids of intermediate stages, and hydrothermal deposits associated with complexes of acidic and supersaturated granites. Tungsten deposits of Precambrian and Caledonian ages are of negligible importance. Variscan deposits make up about 16% of the world reserves, 10% of the Variscan world reserves of tin are hydrothermal, 5% are in skarns, and 1% in pegmatites. Most of the world reserves of tungsten (81%) are Mesozoic; 69% of these are hydrothermal and 12% in skarns. The Cenozoic hydrothermal tungsten deposits make up only 3% of the world reserves.

As a whole, the world reserves of tungsten are distributed among the separate types of deposits in the following manner: 1% in pegmatites, 17% in skarns, and 82% in hydrothermal deposits.

Molybdenum. Despite a close similarity of high temperature deposits of tin, tungsten, and molybdenum, and the fact that they frequently occur together, the distribution of molybdenum in time and among separate types of deposits is very different. The skarn deposits of molybdenum include only 1% of its world reserves; the quartz veins and stockworks associated with batholithic intrusions make 2% and, in affiliation with small post-batholithic intrusions, 4% of the world reserves. Most of the world reserves of molybdenum (93%) are represented by deposits of the metasomatic quartzite type. The distribution of molybdenum reserves according to the age of deposit is of no little significance. The Precambrian, Caledonian, Variscan, and Mesozoic ore deposits contain respectively 1, 2, 1, and 5% of the world reserves of molybdenum. The predominant amount of the reserves (91%) is concentrated in Cenozoic ore deposits. It is possible that such large concentration was caused by the formation, in the course of the geological evolution of the earth's crust as a whole, of new and the most productive type of molybdenum deposits.

Gold. Gold is very scattered with respect to the age of formation as well as among the separate complexes of early, intermediate, and late stages in the evolution of mobile belts. The difficulty of classifying the world reserves of gold according to the complexes is complicated also by the fact that it can be connected with both the small intrusions and batholithic intrusions. Most of the geologists outside the Soviet Union, according to Emmons, correlate all of the gold deposits with batholithic complexes. Though this could have been corrected to some extent in the course of the classification of the world reserves of gold according to the complexes, the importance of gold connected with batholiths appears to be somewhat exaggerated.

Precambrian deposits contain 61% of the world resources of gold. This is mainly gold connected with batholiths (31% of the world reserves), and the so-called "other groups" (27%), including the Witwatersrand gold deposit, the primary origin of which is not clear. Among the Precambrian gold deposits the pre-batholithic complexes and the sulfide deposits are of insignificant value, comprising respectively 2% and 1% of the world reserves.

During Paleozoic time the formation of gold deposits decreased sharply: the Caledonian deposits contain 6% of the world resources and the Variscan deposits represent 4% of the world reserves. Among the Caledonian deposits the pre-batholithic complexes containing 4% of the world reserves are of greatest importance, 1% of the reserves is in deposits connected with the plagiogranite and granosyenite complexes, and 1% is in all other types of rock complexes. Among the Variscan gold deposits 1% of the world reserves is represented by deposits connected with batholiths, 1% by pre-batholithic rock complexes, and 2% by sulfide ore deposits.

There was a considerably increased gold mineralization during Mesozoic time, producing 16% of the world resources of gold. In this case the decisive role was played by the pre-batholithic complexes making up 14% of the world reserves. Of lesser importance are the post-batholithic complexes of minor intrusions. During the Cenozoic era 13% of the world reserves of gold were concentrated in the near-surface deposits.

To sum up, 61% of the world resources of gold are in the Precambrian deposits, 10% in the Paleozoic deposits, and 29% in the Mesozoic and Cenozoic deposits. If the long duration of the Precambrian time is taken into consideration as compared with the Mesozoic-Cenozoic time, then the greatest intensity of gold mineralization developed primarily during the Mesozoic-Cenozoic time.

The separate complexes of all ages as a whole contributed to the world resources of gold as follows: the so-called other groups (conglomerates of Witwatersrand), 27%; plagiogranite and granosyenite complexes of the early periods, 1%; the complex of sulfide deposits, 3%; the pre-batholithic small intrusions, 21%; batholiths, 32% (probably exaggerated); post-batholithic small intrusions, 3%; and the epithermal deposits, 13%.

Copper. The distribution of copper deposits in time is characterized by weak mineralization during the Precambrian to Mesozoic time and by a strong increase of mineralization during the Cenozoic time. The Precambrian deposits produced 29% of the world resources

of copper, of which the main part (19% of the world reserves) is in telethermal deposits (cupriferous sandstones of Rhodesia and the Belgian Congo), 1% in sulfide formation, 3% in contact-metamorphic deposits, 4% in copper-nickel deposits of the liquation type, and 2% in other groups of deposits.

The Caledonian deposits contain 14% of the world resources, mainly in sulfide deposits (13% of the world resources) and in part in porphyry copper ores (1% of the world resources). However, the Caledonian age of the last group of deposits cannot be considered as being reliable. The Variscan copper deposits contain only 8% of the copper resources. Deposits of telethermal type make up 3.5% of the world resources; the porphyry copper ores, 3%; the sulfide deposits, 1%; and other groups of deposits, 0.5%. Among the Mesozoic copper deposits containing only 2% of the world reserves of copper, half of the reserves is represented by porphyry copper ores and half by the contact-metasomatic deposits.

The Cenozoic deposits contain 47% of the world reserves of copper. These are mainly the porphyry copper ores comprising 33% of the world reserves; in a considerably lesser degree sulfide ores (2.5% of the world reserves) and the contact-metasomatic deposits (3% of the world reserves); and deposits of other types containing 8.5% of the world reserves of copper.

The world resources of copper as a whole in deposits of all ages are distributed among the separate types of deposits in the following way; deposits of liquation ore deposits of liquation type, 4%; contact-metasomatic deposits (predominantly in connection with the intrusive complexes of the early stages of evolution of mobile belts), 7%; sulfide deposits, 17.5%; porphyry copper deposits, 38%; telethermal deposits of cupriferous sandstones, 22.5%; and other groups of deposits, 11%.

Lead. It was not possible to subdivide the lead deposits according to the data found in the literature into separate mineral complexes. Therefore the world reserves of lead can be examined only from the point of view of subdivision according to age. This subdivision is quite uniform: the Caledonian deposits make up 17%; the Variscan, 27%; the Mesozoic, 30%; and the Cenozoic lead deposits, 26% of the world reserves of lead.

Nickel and Cobalt. Nickel and cobalt reveal an almost identical distribution according to age: the Precambrian deposits make up 62%; the Paleozoic, 17%; the Mesozoic, 14%; and the Cenozoic, 7% of the world reserves. Their distribution among the complexes of ore deposits is different. Eighty-five percent of the world reserves of nickel are in liquation type copper-nickel deposits, 14% of the world reserves are in silicate type deposits (zone of weathering), and about 0.5% in hydrothermal nickel-cobalt deposits of the terminal stages of the evolution of mobile belts. The liquation type deposits of cobalt make 63% of the world reserves; the zone of weathering, 10%; while the contribution of hydrothermal nickel-cobalt deposits increases to 9%. About 6.5% of the world resources of cobalt are in complex iron-cobalt deposits of the early stages of evolution of mobile belts.

The importance of separate complexes in groups of deposits of various ages is not the same. The Precambrian and Mesozoic groups of nickel deposits are represented mainly by liquation type deposits, the Cenozoic are represented exclusively by the silicate types of deposits, while the Paleozoic nickel deposits are equally divided between both types. The distribution of cobalt reserves is similar to that of nickel, with the difference that in Precambrian ore deposits 5.5% belong to the hydrothermal type, among the Paleozoic ore deposits 3% of the world reserves belong to the hydrothermal and 6.5% to the complex type of iron-cobalt deposits.

Platinum and Chromites. No numerical calculations were made for platinum and chromite deposits; however, their distribution according to geological age is quite significant. The main parts of the world reserves of platinum are of Precambrian and Caledonian ages, while the younger deposits are of no significance. The chromites are distributed in time quite uniformly and produce concentrations in Precambrian, Caledonian, Variscan, Mesozoic, and Cenozoic deposits.

Conclusions. By analyzing the distribution of the world reserves of separate metals in time, it is possible to delineate approximately four different types:

1. Rare metals (gold, tungsten, molybdenum) reveal the tendency to increase in intensity of mineralization

with time and reach a maximum during the Mesozoic time (tin, tungsten), or in Cenozoic time (molybdenum).

2. Cobalt, nickel, and platinum show a decrease in intensity of mineralization in time and reach a maximum during the Precambrian and Lower Paleozoic times.

3. Gold and copper produced two maxima during the Precambrian and Mesozoic eras (gold) or the Cenozoic era (copper), separated by a period of less intensive mineralization.

4. Lead and chromites are distributed in time quite uniformly and do not reveal any strong maxima or minima.

It is typical of copper that its maximum during the Cenozoic era is connected with the same complex of deposits as the Cenozoic maximum of molybdenum mineralization. Mercury and antimony produced similar maxima during the Cenozoic time. A table showing the distribution of the world reserves of some metals in percent and in time (Table 2) follows.

TABLE 2

Eras	Tin	Tungsten	Molyb-denum	Gold	Copper	Lead	Nickel	Cobalt
Cenozoic	14	3	91	13	47	26	7	6.5
Mesozoic	50	81	5	16	2	30	14	14.5
Variscan	24	16	1	4	8	27	16.5	17
Caledonian	2	-	2	6	14	17	-	-
Precambrian	10	-	1	61	29	-	62.5	62
Total	100	100	100	100	100	100	100	100

V. GENERAL FEATURES OF THE EVOLUTION OF
MOBILE BELTS AND DISTRIBUTION OF
MAGMAS, METALS, AND SEDIMENTS WITHIN
THESE BELTS. COMPLEXES OF MAGMAS AND
THEIR METALLOGENY.

Nature of the Problem. Earlier the reader was introduced to the purely empirical distribution features of separate intrusive and mineral complexes in the course of the geological evolution of mobile belts of the earth's crust. These features were outlined according to the summary, analysis, and generalization of an enormous amount of information dealing with endogenetic mineralization in the territory of the Soviet Union. Therefore it can be assumed that these features reflect quite accurately at least the main tendencies in the evolution of intrusive magmatism and the endogenetic mineralization accompanying it. However, the problem facing the investigator who makes an analysis based on the method of dialectical materialism consists not only in the delineation of some purely empirical natural laws, but in their deep understanding and in their correct

interpretation. Only in this way can be created an appropriate foundation for further successful progress in research and correct plans drawn for its line of development and methods.

The question of the features of the evolution in time of magmatism and endogenetic mineralization was put on the agenda and started being analyzed a relatively short time ago. Therefore data available at the present time are insufficient for its complete and thorough elucidation. The accumulated material makes it possible at least to express one's observations which would help in further study of this question and which would delineate ways along which the study should proceed. Therefore, the following ideas should be considered not as final, but only as an attempt to discuss this

45

complex problem and to designate ways for its further investigation.

Distribution of Intrusive Magmatism. If the change in the composition of intrusive magmatic rocks in the course of the geological evolution of mobile belts is considered, it is not difficult to detect in the course of this evolution definite features, some of which were already indicated during the discussion of separate intrusive complexes.

The earliest manifestation of intrusive activity in mobile belts is represented by the small near-surface intrusions of basic magmas in close association with respect to space, age, and genesis, with lava flows of the spilite-keratophyre magmatic association. The mesocratic and acidic differentiates of basic magmas at this time are strongly subordinated. The next major intrusions of the first orogenic phase consist of basic and ultrabasic magmas. The mesocratic and acidic differentiates are present among the rocks in even smaller amounts than in the spilite-keratophyre association and in the accompanying small intrusions. It is not uncommon to find the more acidic differentiates present among the extrusive rocks in larger amounts than among the closely related intrusive rocks. The latter rocks are to some extent younger and are characterized by the development of more basic differentiates.

Among the extrusive rocks of the second cycle of sedimentation (when they are present) the mesocratic and acidic differentiates of basic magmas appear in considerably greater amounts. Besides quartz albitophyres, corresponding to the acidic and substantially sodic magmas, potassic magmas characterized by higher alkalinity appear, producing rocks of a trachyte type. The next major intrusions of the second orogenic phase are characterized by the same features of composition. Among them the development of basic rocks of gabbro and gabbro-norite type is strongly subordinated to other rock types. The predominant rock types are the mesocratic and acidic differentiates of basic magmas, or quartz diorite and plagiogranite types (acidic sodic granites), or of syenite-diorite and granosyenite types (potassic, of higher alkalinity). Thus, during the second cycle of sedimentation, and during the period of subsequent folding, there is a quite clear change from the predominant basic magmas and their ultrabasic differentiates to the strongly predominant mesocratic and acidic differentiates of basic magmas.

During the intermediate stages in the evolution of mobile belts there is again a clear change from granitoid differentiates of basic magmas to independent granitoid magmas which do not reveal any genetic relation to basic magma. The independent granitoid magmas, which are usually less acidic than during the intermediate stages, are preserved during the late stages of evolution. Only at the very end of late stages, sometimes together with the latter, basic magmas or their predominantly mesocratic differentiates, appear.

The terminal stages in the evolution of mobile belts are characterized by potassic granitoids of higher alkalinity, in many instances very close to those potassic granitoids which are characteristic of the second half of the early stages. Judging from all these data the granitoid differentiates of basic magmas appear again during the terminal stages in the evolution of mobile belts. The acidic sodic differentiates, so typical of geosynclinal magmatic manifestations, are not present

at this time, probably because of a considerably greater rigidity of the orogenic belt.

Thus, the general course of the evolution of magma acquires the following pattern: the magmatic activity in mobile belts begins with the manifestation of basic and ultrabasic magmas, which appear to be inherited from the ancient platforms, on which the mobile belts develop. During the further evolution of the mobile belt these magmas are followed by granitoid differentiates of basic magma and later by the independent granitoid magmas. After the transformation of the geosyncline into an orogenic belt, there is a subsequent change of magmas in an opposite direction as follows: the independent granitoid magmas of the late stages change during the terminal stages of evolution to granitoid differentiates of basic magma, while the young platforms forming as the result of prolonged evolution of the mobile belt, are again characterized by the nondifferentiated basic magmas.

The Distribution of the Principal Metals. The distribution of deposits of various metals in the course of the geological evolution of mobile belts is even more clearly expressed and is typical. Magmatic deposits of various genetic types containing platinum and platinum group metals, chromium, iron, titanium, vanadium, phosphorus, nickel, cobalt, and copper are connected predominantly with basic and ultrabasic magmas of the first half of the early stages. These are predominantly the siderophile elements and in part the lithophile elements (elements of the iron group) and the chalcophile elements (copper).

The hydrothermal and contact-metasomatic deposits predominantly those of iron, copper, lead, zinc, barite, partially cobalt, arsenic, silver, and gold are typical of the second half of the early stages in connection with the granitoid differentiates of basic magma.

The intermediate stages are characterized by the hydrothermal, partially contact-metasomatic deposits of tin, tungsten, molybdenum, and gold, as well as the pegmatite deposits of the common elements of granite pegmatites (lithium, beryllium, boron, tantalum, columbium, and others). The elements most typical of the early stages of evolution are almost absent here. This causes an extremely sharp change in the composition of products of endogenetic mineralization with the transition to the intermediate stages of evolution. This again confirms the correctness and validity of separating the independent granitoid magmas from the granitoid differentiates of basic magma.

During the late stages of evolution, the composition of mineralizing solutions is inherited primarily from the intermediate stages (tin, tungsten, molybdenum, gold). However, at the end of the late stages, together with this characteristic complex of metals, elements alien to this complex (iron, silver, lead, zinc) begin to appear. It is extremely significant, as it was mentioned earlier that a small amount of basic magmas and differentiates of these magmas is typical of the magmatic manifestations of the late stages.

Finally, a particularly strong change in the composition of endogenetic mineralization is again observed during the terminal stages in the evolution of mobile belts. Tin, tungsten, molybdenum, and gold disappear. Copper, iron, silver, lead, zinc, nickel, cobalt, as well as barite, begin to be the most important. In other words, this is the entire group of metals, which is typical of the granitoid differentiates of basic magmas of the second half of the early stages in the evolution of mobile belts.

Thus, the change in the composition of endogenetic mineralization of mobile belts proceeds in strict accordance with the change in composition of magmatic rocks.

Distribution of Sediments. The composition of the stratified series, forming during the process of sedimentation in the mobile belts, also changes quite regularly. The carbonate-volcanic series are most typical of the early stages. Iron, magnesium, calcium, and sodium are the most important elements during this stage. The sedimentary mineral deposits are represented predominantly by ores of iron, manganese, and aluminum. The importance of the carbonate and volcanic series decreases during the second half of the early stages of evolution (the third cycle in the accumulation of sediments). Silicon and aluminum are the most important elements accumulating during this stage.

During the late and terminal stages of evolution, the accumulation of sediments is strongly reduced. The coal-bearing and salt-bearing strata are usually associated with the sediments of these stages.

Complexes of Magmas, Their Metallogeny and Causes of Their Formation. After summarizing the information about the distribution of magmas and metals in the course of the geological evolution of mobile belts, it became quite clear that there are three main complexes of magmas, each accompanied by a typical group of metals. If, in each group, elements of minor importance are disregarded, then the following metallogenic characteristics can be made for each complex of magmas.

A. The complex of basic and ultrabasic magmas produces mainly magmatic deposits. The most important metals in these deposits are platinum, and platinum group metals, chromium, iron, titanium, vanadium, phosphorus, nickel, cobalt, and copper.

B. The complex of granitoid differentiates of basic magma produces contact-metasomatic and hydrothermal deposits for the most part. The most important metals in these deposits are iron, copper, lead, zinc, barium, nickel, cobalt, arsenic, and silver.

C. The complex of independent granitoid magmas generates mostly hydrothermal, in part pegmatitic and contact-metasomatic, deposits. The principal metals and elements in these deposits are tin, tungsten, molybdenum, and gold; as well as lithium, beryllium, boron, tantalum, columbium, and others.

Comparison with Platforms. If magmatic rock complexes of mobile belts are compared with those of platforms, it will be evident at once that besides the known features of similarity, there is quite a strong difference between both rock complexes. The complex of basic and ultrabasic magmas in mobile belts and platforms is substantially the same and is characterized by the same genetic types of deposits and by the same group of metals. However, the quantitative value of deposits of these metals in mobile belts is not the same as in platforms. The liquation type deposits of copper-nickel-cobalt ores are strongly predominant in platform regions and are of little importance in mobile belts. Deposits of native platinum are equally typical of the platforms and of mobile belts; however, there are more chromite deposits in mobile belts. Finally, the magmatic

deposits of titanomagnetites and apatite-magnetite ores are more typical of the mobile belts.

The complex of granitoid differentiates of basic magmas, together with the deposits and metals typical of this complex, is widely developed in the mobile belts and is very poorly represented in the platforms (the Aldan platform). In the latter case these are exclusively the potassic differentiates of higher alkalinity (granosyenite series) characterized by the absence of acidic rock differentiates of the plagiogranite series. This probably explains why no deposits of the sulfide type, inherent to the mobile belts, are ever found in the platforms.

Finally, the complex of independent granitoid magmas with a group of metals typical of this complex is not represented in the platforms at all.

In discussing the platforms one cannot avoid mentioning the complex of alkaline and ultraalkaline rocks characteristic of the platforms, with which are connected the magmatic deposits of titanomagnetite; the apatite-nepheline ores of Khibiny; deposits of alkaline pegmatites, which are genetically very close to the former; and the hydrothermal deposits of cryolite. The alkaline rocks are part of some granosyenite complexes of mobile belts, but they are usually accompanied by an insignificant mineralization. In the platforms they are characterized by a considerable independent development and therefore they have to be classified as an independent complex of platform type magmas. The alkaline ultrabasic rocks, with which deposits of diamonds, knopite, and others are connected, never occur in complexes of platform type magmas within the mobile belts.

The Diagram of Distribution of Metals. Thus, each metal or element demonstrates a predominant association with the platforms or with definite stages in the evolution of mobile belts. If the occurrence of metals of the intermediate and late stages of evolution, characterized generally by the same group of metals, is considered as a whole, then the distribution of each metal between (a) the intermediate and late, (b) early, and

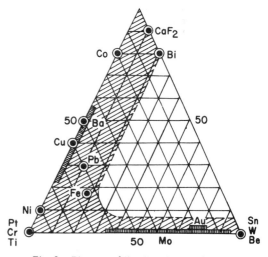

Fig. 3. Diagram of the distribution of metals.

(c) terminal stages of evolution (i.e., between the three items), can be represented by means of a triangular diagram (see Fig. 3). In this diagram the right apex of the triangle corresponds to the intermediate and late stages of evolution; the left, to the early stages; while the upper apex of the triangle corresponds to the terminal stages of evolution. The points representing such metals as tin, tungsten, lithium, and beryllium are in the right apex of the triangle.

The point for gold also has to be located in the vicinity of this apex at the base of the triangle. However, some difficulty arises with gold, since some of its reserves belong to that transitional group of epithermal deposits which develop at the end of the early and at the end of the late stages of evolution, i.e., exactly at the boundary between the stages during which the granitoid differentiates of basic magma and the independent granitoid magmas develop. The classification of this group of epithermal deposits as belonging to one or another stage of evolution cannot be effectively proved. Therefore they are shown on the diagram separately, as a fourth item, in the form of a bar (vector). Gold is represented on the diagram of distribution of reserves in the form of a small bar corresponding to 8% of the reserves, at the base of the triangle and located near its right apex. A considerably larger bar making up 61% of the molybdenum reserves is at the base of the triangle.

In the left apex of the triangle are the points representing platinum, chromium, titanium, and vanadium, which correspond entirely to the early stages of evolution. Near the left apex of the triangle is the point of

nickel (90%), with 10% deflection toward the upper apex of the triangle (the terminal stages). Barium is at half distance between the left and upper apexes, while cobalt is near the upper apex. The point for copper is represented by a bar of moderate length (31%) between the left and upper apexes. Lead and iron are near the left apex with some deflection toward the upper apex and the right apex. This happens because both metals develop during the late stages of evolution. Bismuth and fluorite are near the upper apex of the triangle with a slight deflection to the right. The distribution of reserves of metals among the separate stages of evolution (in percent) is shown in Table 3.

Analysis of the Diagram of Distribution of Metals. At first glance one may see the main features in the distribution of points and bars of separate metals. All of them gravitate toward the apexes of the triangle or to its lower base and left side. The right side of the triangle and its entire central part remain empty. This indicates that to some extent there are transitions from mineralization of early stages to mineralization of intermediate stages (gold, molybdenum) and from mineralization of the early stages to mineralization of terminal stages (nickel, cobalt, barium, copper, lead and iron). However, transitions from mineralization of the intermediate and late stages to mineralization of terminal stages are absent, if an insignificant deviation to the right of the points of lead and iron is disregarded.

It is quite natural to suppose that the transitions from mineralization of the early stages to mineralization of the terminal stages are caused by the presence in both cases of differentiates of basic magma, as the

TABLE 3

	Intermediate and Late Stages	Early Stages	Terminal Stages	Transitional Group
Tin, tungsten	100	–	–	–
Lithium, beryllium	100	–	–	–
Gold	76	16	–	8
Molybdenum	35	4	–	61
Platinum, chromium	–	100	–	–
Titanium, vanadium	–	100	–	–
Nickel	–	90	10	–
Cobalt	–	20	80	–
Barium	–	50	50	–
Copper	–	44	25	31
Lead	10	60	30	–
Iron	18	65	17	–
Bismuth	20	–	80	–
Fluorite	10	–	90	–

48

result of transition between mineralization of intermediate and early stages. Mainly, concentrations of
gold and molybdenum were produced, i.e., concentration of those metals which are part of the transitional
group of epithermal deposits. These deposits are connected predominantly with magmas of intermediate
acidity. To the same group of epithermal deposits
belong considerable reserves of copper. This makes it
possible to define a fourth, intermediate, complex of
magmas of mobile belts, with mineralization represented
mainly by copper, molybdenum, and gold. The intermediate position of this complex between the basic and
independent granitoid magmas from the point of view of
the time at which they develop and nature of mineralization, makes it possible to make a preliminary assumption that these magmas are hybrid.

Change of Magmas as the Result of the Geological
Evolution of Mobile Belts. At the present time it may
be considered proven by geophysical data that the earth's
crust in its upper zone consists of a sialic layer underlain by the basaltic layer. It is obvious that magma
producing the intrusive and extrusive rocks may penetrate the earth's solid crust, from the uppermost parts
of the abyssal magmatic reservoirs. This abyssal magmatic layer may be designated as the domain from which
the magmatic material was supplied. The penetration
of magma through a liquid magmatic layer of different
composition is naturally impossible.

Study of the magmatic complexes of the platforms
indicates that they are produced by magma supplied
from the reservoir in the basalt layer. In some
cases this is the nondifferentiated basaltic magma; in
other cases they are ultrabasic differentiates of basic
magma, corresponding to the deepest parts of magmatic
reservoirs; and in considerably less frequent cases,
they are granitoid differentiates of basic magma, corresponding to the uppermost parts of the abyssal reservoirs of basalt magma. In platform regions there is no
evidence of the activity of magmatic reservoirs which
would be located within the sialic layer. This indicates that under conditions existing in platform regions,
the sialic layer of the earth's crust is not affected by
the processes of melting.

The magmatic activity of mobile belts starts with
the predominant manifestation of basic and ultrabasic
magmas. In other words as the mobile belt starts
forming in the marginal part of an ancient platform, the
source of magma is initially within the same basaltic
layer. This layer supplies to the upper zones of the
earth's crust, at various times during the first major
orogenic phase, either the nondifferentiated basalt magma, or the granitoid differentiates of basaltic magma,
or its ultrabasic differentiates. Such magmatic processes take place within the earth's crust only during
the incipient stages in the evolution of the mobile belt.
After the first major phase of folding there is a strong
shift toward the granitoid differentiates of basic magma
which become predominant while the nondifferentiated
basalt magma remains strongly subordinate, and the
ultrabasic differentiates are entirely absent. Since the
lighter granitoid differentiates must occupy higher zones
within the magmatic reservoirs than the nondifferentiated basic magma, one may conclude that the zone
supplying magmas shifts to the higher levels. It is
quite possible that this is not only the result of a thermal flow directed upward, i.e., the progressive heating
of the abyssal parts of the earth's crust, but as well
the result of subsidence of the earth's crust itself, to

the deeper zones in the direction from which the heat
flow is coming both as a result of being covered by
the preceding accumulation of sediments as well as in
connection with the orogenic process.

The second major orogenic phase and the developing sedimentation contribute to the further evolution of
the same process. During the intermediate stages of
evolution of the mobile belt, the zone of magmatic supply passes into the sialic layer, as indicated by the
change of the granitoid differentiates of basic magma
to independent granitoid magmas accompanied by a typical group of metals. During the period of the third
major phase of orogeny, the zone of magmatic supply
occupies the highest level relative to the zones of the
earth's crust.

The third and last major phase of orogeny thereby
changes the trend in the process of evolution of the
mobile belt. The geosyncline is closed and is transformed into a folded belt. The tendency of the geosyncline toward downwarping is subsequently followed
by the tendency toward upwarping, while the tendency
to accumulate sediments is followed by the tendency
to erosion of the sediments accumulated earlier. As
the result of upwarping and erosion the abyssal parts
of the earth's crust are removed from the zone of melting, which is displaced to relatively deeper levels. On
the other hand, the transformation of the geosyncline
to a folded belt apparently sharply changes the thermal
regime of the earth's crust. Its progressive heating is
followed by progressive cooling. During some period of
time the domain of magmatic supply continues to remain
within the boundaries of the sialic layer, receding gradually to its deeper levels. Therefore during the late
stages of evolution of the mobile belt, the composition
of magmatic rocks and that of the endogenetic mineralization are inherited from the intermediate stages of
evolution developing in a quite different tectonic environment. At the end of the late stages of evolution the
magmatic rocks and the endogenetic mineralization
begin to show the effects of granitoid differentiation of
basic magma (concentration of iron, lead, and syenite
varieties of magmatic rocks), probably because the
domain of magmatic supply approaches the appropriate
depths in the earth's crust.

The terminal stages in evolution correspond to the
displacement of the region of magmatic supply again to
the upper parts of the basalt layer enriched with granitoid differentiates. Thus, the shift of the domain of magmatic supply of one layer of the earth's crust to another
is very regular and coincides with the predominant character of oscillating movements of the earth's crust
during the separate stages in the evolution of the mobile
belt.

The formation of a group of epithermal deposits at
the end of the early and at the end of the late stages
indicates that the source of magma at this time is in
the layer transitional from the sialic layer to the basaltic,
i.e., probably in the layer of mixed andesitic magma.

Causes for the Separation of Metals. The definite
and continuous division of metals among the complexes
of magmas corresponding to the different depths from
which magmatic supply was taking place creates a natural
question about the causes of such a separation. On the
other hand, it can be conjectured that this separation
is real, i.e., that the complexes of magmas corresponding to various depths have different initial contents of
separate metals, this being the main cause of the dif-

ference in their ore-bearing capacity. On the other hand, it can be supposed that under conditions of equal or similar contents of the same metals, different complexes of the magmas, due to some special features, do not possess the same capacities to produce ore deposits of economic grade.

Departing from the thesis that all features of magma, including its ore-bearing capacity, are the result of the entire preceding history of its geological evolution, let us consider what are the main tendencies in the evolution of those two layers of the earth's crust, the basaltic and the sialic, with the geological evolution of which the processes of endogenetic mineralization are connected.

The Main Tendencies in the Evolution of the Basaltic and Sialic Layers. The sialic layer of the earth's crust, located between the basaltic layer on the one side and the atmosphere and hydrosphere on the other side, develops in close relation with both. This relation is demonstrated mainly by the exchange of material (and, naturally, of energy) between them, resulting in the change of composition and of the total mass of each of these layers.

Through its interaction with the atmosphere and hydrosphere the sialic layer produces a complex of sedimentary formations of diverse compositions. These formations subside to considerable depths, produce a complex of metamorphic formations, and being drawn into the zone of melting may produce granitoid magmas. Thus the complex of sedimentary rocks always remains part of the sialic layer. However, during the process of transition of the complex magmatic rocks into the complex of sedimentary rocks, some components of the sialic layer are removed and become part of the atmosphere and hydrosphere. These are mainly salts which accumulate continuously in the world ocean, except the salts of potassium which are adsorbed by the clay minerals, and the main portion of which are not returned to the sialic layer. Thus, through the irreversible result of interaction between the sialic layer with the atmosphere and hydrosphere, there is a progressive re-

moval from the sialic layer into the sea of sodium and chlorine, and, to a lesser extent, potassium, magnesium, and sulfur.

The interaction of basaltic and sialic layers is quite complicated and it is different under different geological environments and during different stages in the evolution of the earth's crust. During those stages of evolution when the domain of the magmatic source is within the basalt layer, the latter supplies various magmatic products and ores in the form of extrusive and intrusive rocks of basic magmas and their differentiates accompanied by endogenetic mineralization to the sialic layer. As the result of this process the general amount of substance in the sialic layer grows at the expense of the basaltic layer, along with the appropriate change in its average composition. After it goes through the process of sedimentary differentiation, the substance of the basaltic layer loses its identity and becomes part of the total mass of the substance of the sialic layer.

The abyssal differentiation of basaltic magma resulting from gravitational processes during the stages of its cooling and crystallization may generate products of a granitoid composition, added through increment to the lower parts of the sialic layer. This is the second possible form of transition of the basalt layer substance to the substance of the sialic layer. On the other hand, the interaction between the basaltic and sialic layers during the stages of maximum heating and melting of the abyssal zones of the earth's crust may result in partial mixing of material and the formation of a transitional layer of andesitic magma.

The net irreversible result of the interaction between the basaltic and sialic layers of the earth's crust is the replenishment of the substance of the sialic layer at the expense of the basaltic layer with an accompanying change in the composition of the sialic layer. By making an extrapolation of this process into the remote geological past, it is possible to make the quite credible assumption that the entire sialic layer represents the product of differentiation of the basaltic layer, reworked to some extent by the processes of weathering and with the removal of some components into the sea.

Editor's Comments
on Paper 3

3   AUBOUIN
      Excerpts from *Geosynclines*

Aubouin's ideas on geosynclinal development are stripped to the bare bones in the figure reproduced in Paper 3. Stages are added to match the Bilibin theme, and dates are added from Harland (1969): initial stage = geosynclinal development; early stage = diapiric intrusion (early tectonic); intermediate stage = main orogeny, batholithic intrusion, and thrusting (syntectonic); late stage = vertical tectonism (late tectonic).

Some recent writers on plate tectonics would discard ideas on geosynclinal development. I have yet to see good reason to do so; one need only add a subduction zone. Geosynclinal development, as depicted by Aubouin, seems no more and no less than the mapping out of development in certain mobile belts. I have the impression that the broader aspects of the theme are commonly applicable, as Borrello has shown for the Baikalian and Paleozoic of Argentina, but that, as Aubouin himself showed, details vary from one belt to another, even in the same era.

That developing ideas on geosynclines and metallogeny were interwoven is clearly evident from Aubouin's review of the works of Stille, in which the thickness of sediments, facies, folding, magmatic evolution, and metamorphism are considered. Aubouin credited Hall (1859) as the acknowledged originator of the geosynclinal concept and reviews the contributions of J. D. Dana, Haug, Schuchert, Marshall Kay, and Peyve and Sinitzyn. Borrello (1972a, 1972b) was a leading disciple in Argentina.

Aubouin noted a suggestion by Kober (1921–1928) that "orogen" be defined as a chain of bilateral symmetry. I have previously discussed the relationship of bilateral symmetry to current plate-tectonic concepts at some length (Walker, 1972). I adhere to the model of Griggs (1939), who depicts the junction of the convecting cells as bilaterally symmetrical; in today's terms a subduction zone accompanies both cells where they meet. As a corollary, any two median rises are separated by a pair of subduction zones. Twin cordillera, like the Andean *cordillera occidentale* and northward extension through Central America to the Sierra Nevada and Coast Range on one limb, *cordillera orientale*, Antilles, and Rockies on the other limb, are separated by a *zwischengebirge* (a median or interior plateau on *continental* crust; an inland sea, such as the Caribbean and Scotia seas, on *oceanic* crust).

Under this concept, I suggest that the African rifts overlie the subduction zones between the median rises of the Atlantic and Indian oceans. As African crust plugged up the downflow zone, the upflowing material from the median rises could not circulate, and the median rises were forced away from the African coasts as inflowing material built up, causing the breakup of Gondwanaland. Gondwanide mantle cells expanded, ultimately by the width of the Atlantic and Indian oceans, overriding Pacific cells. Around the Pacific, therefore, two distinct phenomena must be considered: overriding related to expansion and subduction related directly to convecting cell margins.

The scheme requires about sixteen major plates rather than the six normally considered by plate tectonics. For example, the Americas are seen as four plates lying east of the cordillera; median rises are required between the north part of the Scotia Arc and the Venezuelan Andes (probably along the Amazon) and between the Greater Antilles and the Arctic (probably along Lancaster and Melville sounds). The mathematics of triple junctions merits attention with such a scheme.

Consideration has to be given to the interplay of north-south and east–west movements, as Laurasia and Gondwanaland closed on Tethys and were rent asunder in the Atlantic and Indian oceans. The subject probably holds more importance for oil and evaporite geologists concerned with sedimentation than it does for metalliferous geologists concerned with igneous activity.

3

Reprinted from J. Aubouin, *Geosynclines*, Elsevier Scientific Publishing Co.,
Amsterdam, The Netherlands, 1965, pp. 86-87

GEOSYNCLINES

J. Aubouin

Fig.23. Diagram showing the palaeogeographical evolution and tecto-orogenic development of a eugeosyncline–miogeosyncline couple as illustrated by the Alpine cycle (after AUBOUIN, 1961, fig.8).

This figure, based on the evolution of the Hellenides, is generalized; for a more detailed treatment, see AUBOUIN (1958).

1–9: Geosynclinal and late-geosynclinal periods

1, 2 = generative stage of the geosynclinal period: note that it commenced at an earlier date in the internal eugeosynclinal furrow than in the external miogeosynclinal furrow. In the Hellenides: *1* = Upper Triassic. *2* = Upper Lias. *3–5* = development stage of the geosynclinal period. *3* = emission of ophiolites—mainly in the Upper Jurassic—on the edge of the eugeanticlinal ridge; *4* = mid-Cretaceous orogenesis confined to the eugeanticlinal ridge, while flysch accumulated in the undisturbed eugeosynclinal furrow; *5* = Upper Cretaceous. In the Hellenides: *3* = Upper Jurassic; *4* = Barremian–Aptian. In addition to the eugeanticlinal ridge, the orogenesis now affected the more internal zones, i.e., the Rhodope Massif (intermediate hinterland = axial zone of divergence, of the Hellenides-Balkan bicouple) and the Vardar zone. *6–9* = orogenic stage of the geosynclinal period, and the late-geosynclinal period. The orogenic stage commenced progressively later towards the exterior; similarly, the development of late-geosynclinal molasse troughs was progressively more recent towards the exterior. In the Hellenides: *6* = Maastrichtian–Middle Eocene; *7* = Upper Eocene; *8* = Oligocene and Lower Miocene; *9* = Upper Miocene. The following should also be noted: *(a)* the outward migration of flysch sedimentation due to the twofold effect of infilling of furrows, which then ceased to function as "barrières en creux" (between *6* and *7*, the eugeosynclinal furrow was completely filled and the finest terrigenous material began to reach the more external miogeosynclinal zones, i.e., the miogeanticlinal ridge and miogeosynclinal furrow) and of progressive uplift, from interior to exterior, of the various isopic zones (*8* and *9*): the cordillera resulting from the buckling of the eugeosynclinal furrow supplied flysch sediment to the miogeosynclinal domain (and, in the Hellenides, to the molasse intra-deep). *(b)* Trachyandesitic volcanism, associated with granodioritic intrusions, in the molasse back-deeps *(9)* (in the Hellenides, this volcanism is best developed in the Vardar back-deep); numerous granodioritic intrusions also occur in the Rhodope intermediate hinterland (i.e., BRUNN's, 1960a, "noyau rhodopien").

10–11 = Post geosynclinal period

This period was characterized by positive and negative vertical movements, the latter mainly affecting the internal zones. In the Hellenides, there were two series of contemporaneous fault-subsidences—one in the internal zones ("Aegean fault-subsidences") giving rise to the Aegean Sea and to Plio-Quaternary lakes in northern Greece (e.g., the Plains of Thessaly), and the other in the external zones ("Ionian fault-subsidences") giving rise to the Ionian Sea. These movements were synchronous; the initial phase—affecting not only the Hellenides but the whole of the Mediterranean domain—dates from the end of the Miocene (Pontian) *(10)*, and repeated movements took place during the Plio-Quaternary *(11)*, especially at the time of the Pliocene–Quaternary transition. The resultant clastic sediments accumulated to considerable thicknesses in these fault-basins. In the Hellenides: the Pliocene of the Gulf of Corinth in the external zones; in the internal zones, Plio-Quaternary lacustrine basins appeared in which important deposits of lignite were formed. Note the terminal volcanic activity, essentially basaltic, which was no longer strictly confined to the internal zones; in the Hellenides, the maximum development occurred along the outer edge of the eugeanticlinal ridge; elsewhere, it extended very much further into the more external zones (cf. Fig.21).

[*Editor's Note:* In the original, material both precedes and follows the figure on p. 54.]   53

J. Aubouin

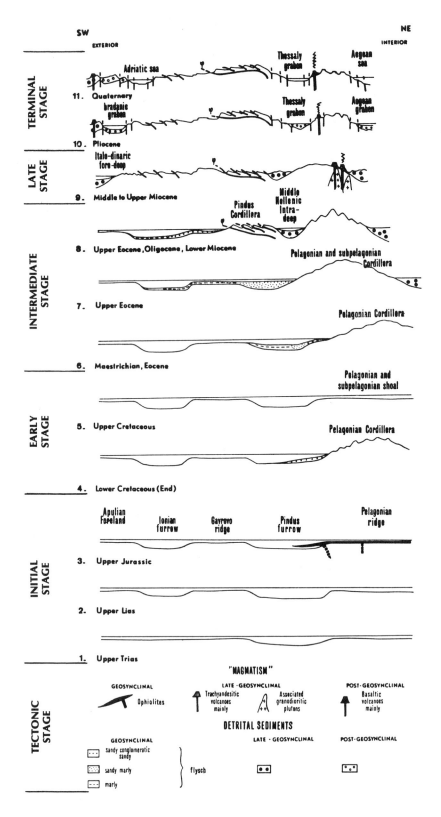

54

Editor's Comments
on Papers 4 Through 6

As one reads reviews of the early works, the contrast between the reviewers A. L. du Toit and B. C. King is most striking. Du Toit was so wholly immersed in his subject that even so great a man as he was obviously embroiled in controversy. King, writing an introduction to a new edition of Wegener thirty years later (Wegener, 1967), was clear of the woods. Thanks to Hess, continental drift was no longer controversial but almost universally accepted.

In 1937 du Toit described the history of the concept:

> Search for the germ of the vital concept of Continental Drift takes us back for three centuries at least, but, as it happens, the evolution of such ideas is more a matter of interest to the historian than the scientist. Regarding those who have expressed such opinions in the past, it will suffice to mention Francis Bacon (1620), Button (c. 1780), Young (c. 1810), R. Owen (1857), A. Snider (1859), H. Wettstein (1880), O. Fisher (1882), C. B. Warring (1887) and W. H. Pickering (1907). A crustal creep towards the equator forms the keynote of the hypothesis of D. Kreichgauer (1902), while shifting of the polar axis has been favored by quite a number of persons, though that view is generally discredited today. [Would that we were so sure today of the "hairpins and superintervals," the clustering and rapid movements of the paleomagnetic poles described by Irving and Park (1972).] It can be stated that the first definite and convincing presentation came from F. B. Taylor in 1908 (published in 1910), followed by H. B. Baker in 1911 and by A. Wegener in 1912. Taylor's able synthesis does not seem to have produced an undue stir, mainly because its heterodox character was then, as perhaps now, not fully realized. Wegener's opinions on the contrary constituted a more direct and flagrant challenge to recognized principles and his views have consequently been the more vigourously combated.

King, discussing Wegener in 1967, wrote;

> Although it is almost forty years since this edition was written it comes as a surprise to find how much of the argument is entirely acceptable today. The presentation of the evidence is masterly, but many of the ideas are unexpectedly modern, probably because so often they have been adopted by more recent writers, for example, the buckling of the orogenic belts to the north of Australia, the "lagging" island festoon between Tierra del Fuego and Grahamland, the separation of Africa from the Seychelles and Madagascar, the clockwise rotation of the lands around the Pacific, attested by strike-slip faulting, the dilation of the Red Sea coasts, the notion of sub-crustal currents, the down-buckling of oceanic crust to produce deep sea trenches with negative anomalies, and the notion of the widening of the Atlantic by welling up of substratum to form new ocean floor.

As the Dover and Methuen editions (1966 and 1967) of Wegener are still available, recourse here is only to his thoughts on the broad range of sciences affected by drift, given in the Author's Foreword to the 1928 edition (Paper 4).

Mild interest—and controversy—in the topic persisted until 1961. Meyerhoff (1968), Dietz, and Hess reviewed the situation in the *Journal of Geophysical Research* for October 15, 1968. Meyerhoff noted the voluminous literature since 1961, and it is clear that the reason for the renewed vitality and subsequent almost universal acceptance of the concept was Hess's paper on sea-floor spreading (Paper 5). Dietz wrote;

> It seems to me entirely correct to regard Wegener as the originator of continental drift as a respectable concept . . . . As regards sea-floor spreading, Hess deserves full credit for the concept . . . by reason of priority and for fully and elegantly laying down the basic premises. I have done little more than introduce the term sea-floor spreading [which is] creating new oceanic rind at the ocean swell and destroying it in trenches in the conveyor belt fashion.

The works of Vine, Wilson, Morgan, Le Pichon, Sykes, Atwater, and others were essential to the development of ideas on plate tectonics, but they were concerned with mechanisms and not with magmatism and such factors directly related to metallogeny. Many of the papers by these authors appear in the *Scientific American* and *Journal of Geophysical Research* compendium volumes. That of Isacks et al. (1968) is perhaps the most embracing.

4

THE ORIGIN OF CONTINENTS AND OCEANS

Alfred Wegener

Author's Foreword

SCIENTISTS still do not appear to understand sufficiently that all earth sciences must contribute evidence towards unveiling the state of our planet in earlier times, and that the truth of the matter can only be reached by combining all this evidence.

The well-known South African geologist du Toit wrote quite recently [78]: "As already stated, we must turn almost exclusively to the *geological* evidence to decide the probability of this hypothesis (continental drift), because arguments based on such matters as the distribution of fauna are not competent here; they can generally be explained equally well, even if less neatly, by the orthodox view that assumes the existence of extended land bridges, later sunk below sea level."

On the other hand, the palæontologist von Ihering [122] is short and to the point: "It is not my job to worry about geophysical processes." He holds to the "conviction that only the *history of life* on the earth enables one to grasp the geographical transformations of the past."

I myself in a weak moment once wrote of the drift theory [121]: "For all that, I believe that the final resolution of the problem can only come from *geophysics*, since only that branch of science provides sufficiently precise methods. Were geophysics to come to the conclusion that the drift theory is wrong, the theory would have to be abandoned by the systematic earth science as well, in spite of all corroboration, and another explanation for the facts would have to be sought."

It would be easy to add to the list of such opinions, each scientist deeming his own field to be the one most competent, or indeed the only one competent, to judge the issue.

In fact, however, the situation is obviously quite otherwise. At a specified time the earth can have had just one configuration. But the earth supplies no direct information about this. We are like a judge confronted by a defendant who declines to answer, and we must determine the truth from the circumstantial evidence. All the proofs we can muster have the deceptive character of this type of evidence. How would we assess a judge who based his decision on part of the available data only?

It is only by combining the information furnished by all the earth sciences that we can hope to determine "truth" here, that is to say, to find the picture that sets out all the known facts in the best arrangement and that therefore has the highest degree of probability. Further, we have to be prepared always for the possibility that each new discovery, no matter which science furnishes it, may modify the conclusions we draw.

This conviction gave me the stimulus to continue at times when my spirits failed me during the revision of this book. For it is beyond one man's power to follow up completely the details of the snowballing literature on drift theory in the various sciences. In spite of all my efforts, many gaps, even important ones, will be found in this book. That I was able to achieve the degree of comprehensiveness I did is due solely to the very large number of communications received from scientists in all the relevant fields, and I am most grateful for them.

The book is addressed equally to geodesists, geophysicists, geologists, palæontologists, zoogeographers, phytogeographers and palæoclimatologists. Its purpose is not only to provide research workers in these fields with an outline of the significance and usefulness of the drift theory as it applies to their own areas, but also mainly to orient them with regard to the applications and corroborations which the theory has found in areas other than their own.

Everything of interest concerning the history of this book, which is also the history of the drift theory, will be found in the first chapter.

The reader is referred to the Appendix for evidence of a shift of North America brought out by the new determinations of longitude in 1927; this result first appeared during the time the book was in proof.

Graz, November 1928

[*Editor's Note:* References can be found in the original volume.]

5

Reprinted from *Petrologic Studies: A Volume to Honor A. F. Buddington*,
Geological Society of America, Nov. 1962, p. 599

History of Ocean Basins

H. H. Hess

Princeton University, Princeton, N. J.

ABSTRACT

For purposes of discussion certain simplifying assumptions are made as to initial conditions on the Earth soon after its formation. It is postulated that it had little in the way of an atmosphere or oceans and that the constituents for these were derived by leakage from the interior of the Earth in the course of geologic time. Heating by short-lived radio nuclides caused partial melting and a single-cell convective overturn within the Earth which segregated an iron core, produced the primordial continents, and gave the Earth its bilateral asymmetry.

Mid-ocean ridges have high heat flow, and many of them have median rifts and show lower seismic velocities than do the common oceanic areas. They are interpreted as representing the rising limbs of mantle-convection cells. The topographic elevation is related to thermal expansion, and the lower seismic velocities both to higher than normal temperatures and microfracturing. Convective flow comes right through to the surface, and the oceanic crust is formed by hydration of mantle material starting at a level 5 km below the sea floor. The water to produce serpentine of the oceanic crust comes from the mantle at a rate consistent with a gradual evolution of ocean water over 4 aeons.

Ocean ridges are ephemeral features as are the convection cells that produce them. An ancient trans-Pacific ridge from the Marianas Islands to Chile started to disappear 100 million years ago. Its trace is now evident only in a belt of atolls and guyots which have subsided 1–2 km. No indications of older generations of oceanic ridges are found. This, coupled with the small thickness of sediments on the ocean floor and comparatively small number of volcanic seamounts, suggests an age for all the ocean floor of not more than several times 10^8 years.

The Mid-Atlantic Ridge is truly median because each side of the convecting cell is moving away from the crest at the same velocity, *ca.* 1 cm/yr. A more acceptable mechanism is derived for continental drift whereby continents ride passively on convecting mantle instead of having to plow through oceanic crust.

Finally, the depth of the M discontinuity under continents is related to the depth of the oceans. Early in the Earth's history, when it is assumed there was much less sea water, the continental plates must have been much thinner.

59

6

Reprinted from *Nature*, **190**(4779), 854–857 (1961)

CONTINENT AND OCEAN BASIN EVOLUTION BY SPREADING OF THE SEA FLOOR

By ROBERT S. DIETZ,

U.S. Navy Electronics Laboratory, San Diego 52, California

ANY concept of crustal evolution must be based on an Earth model involving assumptions not fully established regarding the nature of the Earth's outer shells and mantle processes. The concept proposed here, which can be termed the 'spreading sea-floor theory', is largely intuitive, having been derived through an attempt to interpret sea-floor bathymetry. Although no entirely new proposals need be postulated regarding crustal structure, the concept requires the acceptance of a specific crustal model, in some ways at variance with the present consensus of opinion. Since the model follows from the concept, no attempt is made to defend it. The assumed model is as follows:

(1) Large-scale thermal convection cells, fuelled by the decay of radioactive minerals, operate in the mantle. They must provide the primary diastrophic forces affecting the lithosphere.

(2) The sequence of crustal layers beneath the oceans is markedly different from that beneath the continents and is quite simple (Fig. 1). On an average 4·5 km. of water overlies 0·3 km. of unconsolidated sediments (layer 1). Underlying this is layer 2, consisting of about 2·0 km. of mixed volcanics and lithified sediments. Beneath this is the layer 3 (5 km. thick), commonly called the basalt layer and supposedly forming a world-encircling cap of effusive basic volcanics over the Earth's mantle from which it is separated by the Mohorovičić seismic discontinuity. Instead we must accept the growing opinion that the 'Moho' marks a change of phase rather than a chemical boundary, that is, layer 3 is chemically the same as the mantle rock but petrographically different with low-pressure phase minerals above the Moho and high-pressure minerals below. This change of phase may be either from eclogite to gabbro[1], or from peridotite to serpentine[2]; its exact nature is not vital to our concept, but we can tentatively accept the eclogite–gabbro transition as it has more adherents. Common usage requires that we reserve the term 'mantle' for the substance beneath the

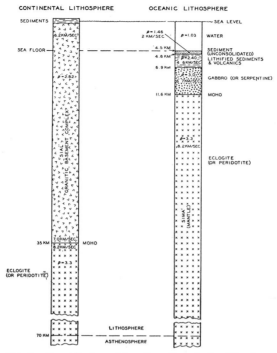

Fig. 1. Generalized crustal sections through the lithosphere beneath the continents and the ocean basins as presumed in this paper. Seismic velocities and densities are shown for the various layers

Moho, but in point of fact, the gabbro layer (as a change of phase) is also a part of the mantle—a sort of 'exo-mantle'. Except for a very thin veneer, then, the sea floor is the exposed mantle of the Earth in this larger sense.

(3) It is relevant to speak of the strength and rigidity of the Earth's outer shell. The term 'crust' has been effectively pre-empted from its classical meaning by seismological usage applying it to the layer above the Moho, that is, the sial in continental regions and the 'basaltic' layer under the oceans so that the continents have a thick crust and the ocean basins a thin crust. Used in this now accepted sense, any implications equating the crust with rigidity must be dismissed. For considerations of convective creep and tectonic yielding, we must refer to a lithosphere and an asthenosphere. Deviations from isostasy prove that approximately the outer 70 km. of the Earth (under the continents and ocean basins alike) is moderately strong and rigid even over time-spans

of 100,000 years or more; this outer rind is the lithosphere. Beneath lies the asthenosphere separated from the lithosphere by the level of no strain or isopiestic level; it is a domain of rock plasticity and flowage where any stresses are quickly removed. No seismic discontinuity marks the isopiestic level and very likely it is actually a zone of uniform composition showing a gradual transition in strength as pressure and temperature rise; and in spite of the lithosphere's rigidity, to speak of it as a crust or shell greatly exaggerates its strength. Because of its grand dimensions, for model similitude we must think of it as weak[3]. If convection currents are operating 'subcrustally', as is commonly written, they would be expected to shear below the lithosphere and not beneath the 'crust' as this term is now used.

(4) As gravity data have shown, the continents are low-density tabular masses of sial—a 'basement complex' of granitic rocks about 35 km. thick with a thin sedimentary veneer. Since they are buoyant and float high hydrostatically in the sima, they are analogous to icebergs in the ocean. This analogy has additional merit in that convection of the sima cannot enter the sial. But the analogy gives the wrong impression of relative strength of sial and sima; the continental lithosphere is no stronger than the oceanic lithosphere, so it is mechanically impossible for the sial to 'sail through the sima' as Wegnerian continental drift proposes. The temperature and pressure are too high at the base of the sial to permit a gabbroic layer above the Moho; instead, there may be an abrupt transition from granite to eclogite.

Spreading Sea Floor Theory

Owing to the small strength of the lithosphere and the gradual transition in rigidity between it and the asthenosphere, the lithosphere is not a boundary to convection circulation, and neither is the Moho beneath the oceans because this is not a density boundary but simply a change of phase. Thus the oceanic 'crust' (the gabbroic layer) is almost wholly coupled with the convective overturn of the mantle creeping at a rate of a few cm./yr. Since the sea floor is covered by only a thin veneer of sediments with some mixed-in effusives, it is essentially the outcropping mantle. So the sea floor marks the tops of the convection cells and slowly spreads from zones of divergence to those of convergence. These cells have dimensions of several thousands of kilometres; some cells are quite active now while others are dead or dormant. They have changed position with geological time causing new tectonic patterns.

The gross structures of the sea floor are direct expressions of this convection. The median rises[4,5] mark the up-welling sites or divergences; the trenches are associated with the convergences or down-welling sites; and the fracture zones[6] mark shears between regions of slow and fast creep. The high heat-flow under the rises[7] is indicative of the ascending convection currents as also are the groups of volcanic seamounts which dot the backs of these rises.

Much of the minor sea-floor topography may be even directly ascribable to spreading of the sea floor. Great expanses of rough topography skirt both sides of the Mid-Atlantic Rift; similarly there are extensive regions of abyssal hills in the Pacific. The roughness is suggestive of youth, so it has commonly been assumed to be simply volcanic topography because the larger seamounts are volcanic. But this interpretation is not at all convincing, and no one has given this view formality by publishing a definitive study. Actually, the topography resembles neither volcanic flows nor incipient volcanoes. Can it not be that these expanses of abyssal hills are a 'chaos topography' developed as strips of juvenile sea-floor (by a process which can be visualized only as mixed intrusion and extrusion) and then placed under rupturing stresses as the sea floor moves outward?

The median position of the rises cannot be a matter of chance, so it might be supposed that the continents in some manner control the convection pattern. But the reverse is considered true: conditions deep within the mantle control the convective pattern without regard for continent positions. By viscous drag, the continents initially are moved along with the sima until they attain a position of dynamic balance overlying a convergence. There the continents come to rest, but the sima continues to shear under and descend beneath them; so the continents generally cover the down-welling sites. If new up-wells do happen to rise under a continental mass, it tends to be rifted. Thus, the entire North and South Atlantic Ocean marks an ancient rift which separated North and South America from Europe and Africa. Another such rift has opened up the Mediterranean. The axis of the East Pacific Rise now seems to be invading the North American continent, underlying the Gulf of California and California[8]. Similarly, the Indian Ocean Rise may extend into the African Rift Valleys, tending to fragment that continent.

The sialic continents, floating on the sima, provide a density barrier to convection circulation—unlike the Moho, which involves merely a change of phase. The convection circulation thus shears beneath the continents so that the sial is only partially coupled through drag forces. Since the continents are normally resting over convergences, so that convective spreading is moving toward them from opposite sides, the continents are placed consequently under compression. They tend to buckle, which accounts for alpine folding, thrust faulting, and similar compressional effects so characteristic of the continents. In contrast, the ocean basins are simultaneously domains of tension. If the continental block is drifted along with the sima, the margin is tectonically stable (Atlantic type). But if the sima is slipping under the sialic block, marginal mountains tend to form (Pacific type) owing to drag forces.

Implications of the Concept

Ad hoc hypotheses are likely to be wrong. On the other hand, one which is consonant with our broader understanding of the history of the Earth may have merit. While the thought of a highly mobile sea floor may seem alarming at first, it does little violence to geological history.

Volumetric changes of the Earth. Geologists have traditionally recognized that compression of the continents (and they assumed of the ocean floors as well) was the principal tectonic problem. It was supposed that the Earth was cooling and shrinking. But recently, geologists have been impressed by tensional structures, especially on the ocean floor. To account for sea floor rifting, Heezen[10], for example, has advocated an expanding Earth, a doubling of the diameter. Carey's[11] tectonic analysis has resulted in the need for a twenty-fold increase in volume of the Earth. Spreading of the sea floor offers the less-radical answer that the Earth's volume has remained constant. By creep from median upwellings, the ocean basins are mostly under tension, while the continents, normally balanced against sima creepage from opposite sides, are under compression.

The geological record is replete with transgressions and regressions of the sea, but these have been shallow and not catastrophic; fluctuations in sea-level as severe as those of the Pleistocene are abnormal. The spreading concept does no violence to this order of things, unlike dilation or contraction of the Earth. The volumetric capacity of the oceans is fully conserved.

Continental Drift. The spreading concept envisages limited continental drifting, with the sial blocks initially being rafted to down-welling sites and then being stablized in a balanced field of opposing drag forces. The sea floor is held to be more mobile and to migrate freely even after the continents come to rest. The sial moves largely *en bloc*, but the sea floor spreads more differentially.

Former scepticism about continental drift is rapidly vanishing, especially due to the palæomagnetic findings and new tectonic analyses. A principal objection to Wegener's continental drift hypothesis was that it was physically impossible for a continent to 'sail like a ship' through the sima; and nowhere is there any sea floor deformation ascribable to an on-coming continent. Sea floor spreading obviates this difficulty: continents never move through the sima—they either move along with it or stand still while the sima shears beneath them. The buoyancy of the continents, rather than their being stronger than the sima, accounts for this. Drag associated with the shearing could account for alpine folding and related compressional tectonic structures on the continents.

Persistent freeboard of the continents. A satisfactory theory of crustal evolution must explain why the continents have stood high throughout geological time in spite of constant erosional de-levelling. Many geologists believe that new buoyancy is added to continents through the gravitative differentiation from the mantle. Spreading of the sea floor provides a mechanism whereby the continents are placed over the down-wells where new sial would tend to collect, even though the convection is entirely a mantle process and the role of the continents is passive. It also follows that the clastic detritus swept into the deep sea from the continents is not permanently lost. Rather, it is carried slowly towards, and then beneath, the continents, where it is granitized and added anew to the sialic blocks.

Youth of the ocean floor. It follows paradoxically from the spreading concept that, although the ocean basins are old, the sea floor is young—much younger than the rocks of the continents. Marine sediments, seamounts, and other structures slowly impinge against the sialic blocks and are destroyed by under-riding them. Pre-Cambrian and perhaps even most Palæozoic rocks should prove absent from the ocean floors; and Mohole drilling should not reveal the great missing sequence of the Lipalian interval (Pre-Cambrian to Cambrian) as hoped for by some. All this may seem surprising, but marine geological evidence supports the concept.

On his discovery of the guyots of the Pacific, Hess[12] supposed these were Pre-Cambrian features protected from erosion by the cover of the sea. But Hamilton[13] proved the guyots of the Mid-Pacific Mountains were Cretaceous, and these seem to be among the oldest of the seamount groups. In an analysis of the various seamount groups of the western Pacific, I was forced

65

to conclude that none of them was older than mid-Mesozoic. The young age of the seamounts has been puzzling; certainly they can neither erode away nor subside completely. Also, there seem to be too few volcanic seamounts, if the present population represents the entire number built over the past hundred million years or more. The puzzle dissolves if sea floor spreading has operated. Modern examples of impinging groups of seamounts may be the western end of the Caroline Islands, the Wake–Marcus Seamounts, and the Magellan Seamounts[14]. All may be moving into the western Pacific trenches. Seamount *GA*-1 south of Alaska may be moving into the Aleutian Trench[15].

The sedimentary layers under the sea also appear to be young. No fossiliferous rocks older than Cretaceous have yet been dredged from any ocean basin. Radioactive dating of a basalt from the Mid-Atlantic ridge gave a Tertiary age[16]. Kuenen[17] estimated that the ocean basins should contain on an average about 3·0 km. of sedimentary rocks assuming the basins are 200 million years old. But seismic reflexions indicate an average of only 0·3 km. of the unconsolidated sediments. Hamilton[18], however, believes that much of layer 2 may be lithified sediments. If *all* layer 2 is lithified sediments, Hamilton finds that the ocean basins may be Palæozoic or late Pre-Cambrian in age—but not Archæan. But very likely layer 2 includes much effusive material and sedimentary products of sea floor weathering. In summing up, the evidence from the sediments, although still fragmentary, suggests that the sea floors may be not older than Palæozoic or even Mesozoic.

Spreading and magnetic anomalies. Vacquier, V., *et al.* (in the press) recently have completed excellent sea-floor magnetic surveys off the west coast of North America. A striking north–south lineation shows up which seems to reveal a stress pattern (Mason, R. G., and Raff, A. D., in the press). Such interpretation would fit into spreading concept with the lineations being developed normal to the direction of convection creep. The lineation is interrupted by Menard's[6] three fracture-zones, and anomalies indicate shearing offsets of as much as 640 nautical miles in the case of the remarkable Mendocino Escarpment[19]. Great mobility of the sea floor is thus suggested. The offsets have no significant expression after they strike the continental block; so apparently they may slip under the continent without any strong coupling. Another aspect is that the anomalies smooth out and virtually disappear under the continental shelf; so the sea floor may dive under the sial and lose magnetism by being heated above the Curie point.

By considering an Earth crustal model only slightly at variance with that commonly accepted, a novel concept of the evolution of continents and ocean basins has been suggested which seems to fit the 'facts' of marine geology. If this concept were correct, it would be most useful to apply the term 'crust', which now has a confusion of meanings, only to any layer which overlies and caps the convective circulation of the mantle. The sialic continental blocks do this, so they form the true crust. The ocean floor seemingly does not, so the ocean basin is 'crustless'.

I wish to express my appreciation to E. L. Hamilton, F. P. Shepard, H. W. Menard, V. Vacquier, R. Von Herzen and A. D. Raff for critical discussions.

[1] Kennedy, G. C., *Amer. Sci.*, **47**, 4, 491 (1959).
[2] Hess, H. H., *Abst. Bull. Geol. Soc. Amer.*, **71**, Pt. 2, 12, 2097 (1960).
[3] Griggs, D. A., *Amer. J. Sci.*, **237**, 611 (1939).
[4] Ewing, M., and Heezen, B. C., *Amer. Geophys. Union Geophys. Mon. No.* 1, 75 (1956).
[5] Menard, H. W., *Bull. Geol. Soc. Amer.*, **69**, 9, 1179 (1958).
[6] Menard, H. W., *Bull. Geol. Soc. Amer.*, **66**, 1149 (1955).
[7] Von Herzen, R. P., *Nature*, **183**, 882 (1959).
[8] Menard, H. W., *Science*, **132**, 1737 (1960).
[9] Heezen, B. C., *Sci. Amer.*, Oct. 2, 14 (1960).
[10] Heezen, B. C., Preprints, First Intern. Ocean. Cong., 26 (1959).
[11] Carey, W. S., *The Tectonic Approach to Continental Drift: in Continental Drift—A Symposium*, 177 (Univ. Tasmania, 1958.)
[12] Hess, H. H., *Amer. J. Sci.*, **244**, 772 (1946).
[13] Hamilton, E. L., *Geol. Soc. Amer. Mem.*, **64**, 97 (1956).
[14] Dietz, R. S., *Bull. Geol. Soc. Amer.*, **65**, 1199 (1954).
[15] Menard, H. W., and Dietz, R. S., *Bull. Geol. Soc. Amer.*, **62**, 1263 (1951).
[16] Carr, D., and Kulp, J., *Bull. Geol. Soc. Amer.*, **64**, 2, 263 (1953).
[17] Kuenen, Ph., *Marine Geology* (John Wiley and Sons, New York, 1950).
[18] Hamilton, E. L., *Bull. Geol. Soc. Amer.*, **70**, 1399 (1959); *J. Sed. Petrol.*, **30**, 3, 370 (1960).
[19] Menard, H. W., and Dietz, R. S., *J. Geol.*, **60**, 3 (1952).

Editor's Comments
on Papers 7 Through 10

Dewey and Bird and Mitchell and Bell provided models, with Mitchell and Bell adding metallogeny to tectonic evolution. Zonenshain et al. and Scheibner gave applications, with Zonenshain et al. adding metallogeny in Paper 9 and Scheibner doing so in a paper due to appear shortly.

Paper 7, by Dewey and Bird, is probably the one most quoted in works on global tectonics and appears in full in the *Journal of Geophysical Research* compendium volume. As related to the present theme, their Figure 10 shows stages of development, related to subduction, comparable to those of Aubouin for geosynclines and to those of Bilibin for the several aspects of the geotectonic cycle: initial-stage geosyncline, early-stage diapiric differentiated intrusives, intermediate-stage batholithic intrusion and thrusting, and late-stage vertical tectonism. Flysch accompanied the developing orogenies and molasse followed their culmination.

Paper 8, by Mitchell and Bell, is the synthesis of a series of papers on metallogenic evolution according to the plate-tectonic model. Mitchell has early material with both Reading and Garson, Sillitoe has a series of papers, and these, with one by Sawkins tend to be analytical. Guild, Walker, and Snelgrove took different approaches to metallogeny as a global phenomenon.

I find Paper 8 to be difficult. I am not happy about Figures 1 and 2, in which Karig's ideas on crustal generation in marginal basins are incorporated. I prefer the Scheibner model of changes in the position of subduction zone. Perhaps when I see reason for an additional median rise, my view will change. The changing direction of subduction zones is also a difficulty. As an accommodation to the changing directions, perhaps we should revert to the ideas of Griggs, in which both flanks of the mobile belt were what we now call subduction zones, each associated with a convecting mantle cell. Nevertheless, Paper 8 represents a consensus of opinion, and the magmatic developments noted in the text are standard or representative.

Papers 9, by Zonenshain et al., and Paper 10, by Scheibner, both show repetitive development during the Alpine and Paleozoic eras, respectively. The particularly interesting concept utilized by Scheibner is that the Benioff bottoms against the mesosphere, buckles, breaks, and backs up. As these repetitions are localized along particular mobile belts, developments along mobile belts of a given mantle cell system are not in synchronization. It is interesting to compare the duration of cycles associated with specific Benioff zones. In the Hellenides, Aubouin showed just one cycle lasting from the Early Jurassic to the present, a span of 200 m.y. Haile (1969) depicted the development of the northwest Borneo geosyncline from Late Cretaceous (100 m.y.) to Pliocene (7 m.y.), a span of almost 100 m.y. Zonenshain and co-workers on West Pacific belts recognized three periods, Early Mesozoic (T–J1 or 225 to 178 m.y., a span of 47 m.y.), Late Mesozoic (J3–K1, or 157 to 106 m.y., a span of 51 m.y.), and Late Cretaceous including Early Paleogene times (or 100 to 45 m.y., a span of 55 m.y.). Scheibner recognized four rearrangements of Paleozoic plate margins in New South Wales, Early to Late Cambrian (or 570 to 515 m.y., a span of 55 m.y.), late Early Ordovician to the end of the Ordovician (or 500 to 430 m.y., a span of 70 m.y.), late Early Silurian to Middle Devonian (or 420 to 370 m.y., a span of 50 m.y.), and late Early Permian to early Late Permian (or 280 to 240 m.y., a span of 40 m.y.). All dates are from the Phanerozoic time scale (Harland et al., 1964).

I would like to apply the Scheibner concept of sequential development of cycles within an era to Superior Province, but at Rice Lake one cycle occupied the entire era (see p. 133). Scheibner made reference to the common twofold division of mobile belts, Mediterranean (continent–continent) and Pacific (continent–ocean). This division seems to me to be unduly simplified. I have previously illustrated various possibilities as subduction is initiated in oceanic and continental environments (Walker, 1971), and to these I would now add the effects of the expansion of the Gondwanide cells over the Pacific ones. Mitchell (1975) now considers a fourteen-fold division, which bears comparison.

MOUNTAIN BELTS AND THE NEW GLOBAL TECTONICS

J. F. Dewey and J. M. Bird

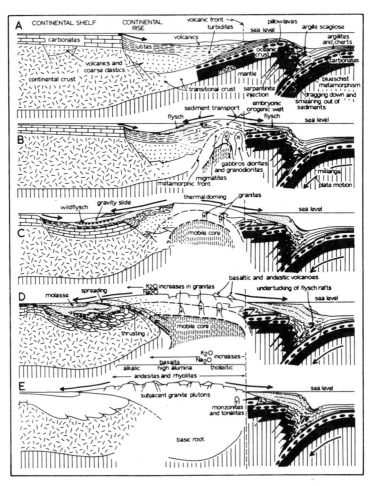

Fig. 10. Schematic sequence of sections illustrating a model for the evolution of a cordilleran-type mountain belt developed by the underthrusting of a continent by an oceanic plate.

8

ISLAND-ARC EVOLUTION AND RELATED MINERAL DEPOSITS[1]

ANDREW H. MITCHELL AND J. D. BELL

Department of Geology and Mineralogy, University of Oxford,
Parks Road, Oxford OX1 3PR, England

ABSTRACT

In ensimatic arcs, initial submarine eruptions of island-arc tholeiites are succeeded by subaerial and submarine volcanism which is either calcalkaline or island-arc tholeiitic. Besshi-type massive stratiform sulfides develop in deep water on the submarine flanks of islands. Pluton emplacement beneath waning volcanoes is accompanied by mercury, porphyry copper, and gold mineralization. Renewed calc-alkaline or island-arc tholeiitic volcanism commonly follows arc reversal or splitting; Kuroko-type massive sulfides form in shallow-water clastic dacitic rocks and gold is concentrated around monzonites and in meta-andesites. Reef limestone deposition, block faulting, and uplift may be followed by formation of bauxite on karstic limestones, and of stratiform manganese deposits near the limestone base. Upper mantle and ocean crust rocks, emplaced as ophiolites in mélanges on the arc side of the trench and as obducted slices during arc-arc collision, contain Cyprus-type stratiform massive sulfides, podiform chromite, and nickel sulfides; nickeliferous laterites may develop on the upper mantle rocks. Increase in island-arc crustal thickness and emplacement of granitic plutons is accompanied by tin-tungsten molybdenum-bismuth mineralization. Following arc-continent collision, massive sulfides, gold, tin, and ores associated with ophiolites are preserved.

INTRODUCTION

Island arcs, particularly those in the Western Pacific, have in the past few years attracted the interest of mining companies and state geological surveys as favorable prospects for metallic ore deposits. A number of publications have stressed the ore potential of these arcs, with varying emphasis on the origin of the ore bodies (e.g., Thompson and Fisher 1965; Liddy 1972; Stanton 1972).

Widespread acceptance of the plate tectonics hypothesis, together with increasing geological and geophysical data on island arcs, has led to new explanations of

volcanic, metamorphic, and tectonic processes and their relationship to arc evolution. The hypothesis requires that the inclined seismic plane, or Benioff zone, above which arcs are located, is the zone along which a descending rigid lithospheric slab consisting of ocean floor and upper mantle is consumed (Oliver and Isaacs 1967). This was implied by Hess (1962) and suggested by Coats (1962) for the Aleutian arc. A corollary of the hypothesis is that ancient island arcs occur within continents and interpretation of parts of orogenic belts within continents in terms of island arc and ocean floor successions is now quite commonly encountered. Suggestions that some ore deposits occurring within continents developed initially in island arcs (e.g., Stanton 1960, 1972) are thus now becoming widely accepted.

[1] Manuscript received January 4, 1973; revised March 28, 1973.

[JOURNAL OF GEOLOGY, 1973, Vol. 81, p. 381–405]

The aims of this paper are to consider relationships between the stages of arc evolution and the formation of types of metallic mineral deposit for which an island arc forms a particularly favorable environment. Examples of most of these deposits are found in modern island arcs lying within the oceans, but a few are known only in ancient arcs now located within continents.

MODERN ISLAND ARCS AND NATURE OF UNDERLYING CRUST

Modern island arcs are separated from continents by marine basins underlain by oceanic crust. Active island arcs (Mitchell and Reading 1971) include a belt of active volcanoes, are underlain by a seismic zone, and are bordered by a submarine trench. Inactive arcs lack these features but contain older volcanic or volcaniclastic rocks (e.g., Greater Antilles, Lau Islands).

Some modern arcs contain rocks which indicate that they lay initially adjacent to, or perhaps on, a continental margin, and moved oceanward with development of oceanic crust on the continental side of the arc. Examples of these ensialic arcs are Japan, where pebbles of mid-Pre-Cambrian rock occur in a Permian conglomerate (Sugisaki et al. 1971), and New Zealand (Fleming 1969) and New Caledonia, where continent-derived sediments occur. In most other arcs the nature of the oldest exposed rocks suggests that the arcs originated on oceanic crust, and are ensimatic.

Cenozoic igneous rocks in Japan and New Zealand resemble those in many arcs which lack evidence of continental crust. This suggests that the development of initially ensialic arcs containing fragments of continental crust is similar to that of ensimatic arcs. In this paper we consider the evolution of an ensimatic arc, but in discussing ore bodies emplaced late in its development we use examples from Japan and from other arcs which could have formed initially on, or adjacent to, continental crust.

STAGE 1: PRE-ARC GENERATION OF OCEANIC CRUST AND UPPER MANTLE

GEOLOGICAL EVENTS

The plate tectonics hypothesis requires that oceanic crust form above ascending upper mantle generated at ocean ridge spreading centers and in marginal basins (figs. 1A and 2A). Generation of marginal basin crust is subject to different interpretations (e.g., Karig 1971a; Matsuda and Uyeda 1971), but may resemble that of normal oceanic crust (Sclater et al. 1972).

Igneous rocks are emplaced either continuously or intermittently within a narrow axial zone along an ocean ridge crest. Tholeiitic magma, derived from the partial melting of mantle peridotite, forms layered gabbroic intrusions passing upward into dolerite sheet complexes. These intrude and are overlain by pillow lavas, hyaloclastites, and local thin pelagic sediments. "Burial metamorphism" of pillow basalts to greenschist and zeolite facies and of dolerites and gabbros to amphibolites (Miyashiro et al. 1971) probably takes place soon after emplacement in water-rich and water-deficient environments, respectively (Cann 1970); spilitization of basalts could result from post-cooling hydrothermal alteration (Cann 1969).

During lateral movement of this rock pile down the ridge flanks, it is locally intruded and overlain by alkali basalts and thinly mantled by cherts, pelagic mudstones, and limestones. Local volcanic and carbonate turbidites are derived from oceanic volcanic islands, either single seamounts or chains, which often consist of pedestals of tholeiitic or transitional basalt topped by alkali basalt and related fractionates.

FORMATION OF MINERAL DEPOSITS

The formation of mineral deposits within ophiolites, now interpreted as tectonically emplaced oceanic crust or upper mantle (table 1), is unrelated to arc formation, but exposure of the deposits may occur only subsequent to arc development. Deposits of this type include some of the basaltophilic metals considered typical of an early stage of geosynclinal evolution (Smirnov 1968).

Cyprus-type massive sulfides.—The origins of some massive sulfide deposits associated with submarine or "eugeosynclinal" volcanism (e.g., Anderson 1969) have recently been related to oceanic ridge environments (Pereira and Dixon 1971). Well-known examples of ore bodies in this setting are those of the Troodos ophiolite complex in Cyprus (Sillitoe 1972*a*; Dunham 1972); we term massive sulfides located within similar ophiolite associations "Cyprus-type deposits."

Cyprus-type ore bodies lie either within the tholeiitic pillow lava, metabasalt, or spilite succession—representing layer 2 of the oceanic crust—or between the pillow lavas and overlying pelagic sediments and turbidites of layer 1. The deposits in Cyprus contain massive pyrite together with chalcopyrite, sphalerite, and marcasite, and minor galena, pyrrhotite, gold, and silver (Hutchinson 1965). Strong (1972) suggested that central Newfoundland ores in oceanic tholeiites—such as Whalesback, Little Bay, and Tilt Cove—are characterized by simple mineralogy of pyrite-pyrrhotite-chalocpyrite. Other examples are those of Ergani Maden and Kure in Turkey and Island Mountain in California (Hutchinson 1973).

Evidence that Cyprus-type deposits are emplaced syngenetically during ocean ridge or marginal basin volcanism is provided by the Red Sea metal-rich brines and muds. Unless these originate on the adjacent continental margins (Davidson 1966), their presence in the Red Sea and the occurrence of concentrations of metals in oceanic sediments (Anon 1970; Cronan et al. 1972) suggest that some sulfides originate either at spreading centers or on the sea floor. Alternatively, possible early epigenetic, or subsurface, formation of Cyprus-type deposits might take place in aquagene tuffs, pillow breccias, and volcaniclastic sediments, which could form traps susceptible to magmatic hydrothermal or metamorphic hydrothermal mineralization (Smitheringale 1972). A possible factor accounting for the formation of Cyprus-type deposits is high trace-metal discharge associated with periods of exceptional global volcanism resulting from active plume convection (Vogt 1972).

Massive sulfides in Hawaii-type volcanoes.—The probable formation of Cyprus-type ore bodies at ocean ridges suggests that stratiform massive sulfides might also develop near submarine vents in intraoceanic tholeiitic shield volcanoes. Possible environments include the Hawaiian ridge, where minor quantities of iron, copper, and nickel sulfides occur within phenocrysts in subaerially erupted tholeiites (Desborough et al. 1968). Ore minerals of Hawaii-type deposits would probably be indistinguishable from those of Cyprus-type; Smitheringale (1972) suggested that the volcanic rocks and copper sulfide deposits of the early Ordovician Lush's Bight Group in Newfoundland probably formed either at an intraoceanic volcano or at an oceanic ridge.

Podiform chromite.—In island arcs, chromite or chrome spinel occurs mainly as podiform deposits within deformed Alpine-type dunites or harzburgite bodies (Thayer 1964), some of which are overlain by gabbros, basaltic lavas, and cherts. Economic chromite deposits in arcs are known only in Cuba and the Philippines.

In Cuba, refractory chromite deposits occur in the northeast of Oriente Province, where sacklike layered bodies of massive chromite within dunite pods are surrounded by peridotite and locally cut by gabbro dikes (Park and MacDiarmid 1964). In the Philippines, refractory podiform chromite ores of metallurgical grade occur in the Zambales ultramafic complex on Luzon as lenticular layered bodies within dunite, and are surrounded by saxonite and intruded by dolerite dike swarms (Bryner 1969).

Both the Cuban and Philippine deposits

TABLE 1

Mineral Deposit Settings in Island Arcs

Mineral Deposit	Formation			Tectonic Emplacement		Exposure			
	Active Magmatic Arc	Ocean Marginal Basin Rise	Ocean Marginal Basin Floor	Within Ophiolites in Mélange	Within Ophiolites in Obducted Slices	Magmatic Arc Active	Magmatic Arc Inactive	Within Ophiolites in Mélange	Within Ophiolites in Obducted Slices
Endogenous deposits:									
Cyprus-type massive sulfides	...	1–7	?1–7	2–7	6	...	...	2–7	7
Podiform chromites	...	1–7	?1–7	2–7	6	...	...	2–7	7
Nickel sulfides	...	?1–7	?1–7	?2–7	?6	...	...	?2–7	?7
Hawaii-type massive sulfides	...	...	1–7	2–7	6	...	...	2–7	7
Island-arc tholeiite-type massive sulfides	2	...	...	...	...	...	5–7	...	...
Mercury	4(5, 7)	...	...	...	...	4(5, 7)	5–7	...	...
Besshi-type massive sulfides	3(4, 5, 7)	...	...	...	...	...	5–7	...	...
Porphyry copper	4(5, 7)	...	...	...	...	...	5–7	...	...
Gold around granodiorites	4(5, 7)	...	...	...	...	...	5–7	...	...
Pyrometasomatic deposits	4(5, 7)	...	...	...	...	...	5–7	...	...
Kuroko-type massive sulfides	5, ?3, (7)	...	...	...	...	...	6, 7	...	...
Gold in andesites and around monzonites	5, ?4, (7)	...	...	...	...	...	6, 7	...	...
Tin-tungsten-wolfram-molybdenum	7	...	...	...	...	...	After 7	...	...

Mineral Deposit	Formation and Exposure
Exogenous deposits:	
Bauxite on karstic limestones	Commonly overlying inactive magmatic arc 5–7
Stratiform manganese	Commonly overlying inactive magmatic arc 5–7
Niockeliferous laterites	Overlying ultrabasic rocks, mostly obducted 7

NOTE.—1, 2, 3, etc. indicate stage of arc evolution with mineralization; (5, 7), etc. indicate stage of evolution with possible mineralization; ?3, etc. indicate stage of evolution with doubtful mineralization.

are considered to be of early magmatic origin (Guild 1947; Bryner 1969). Their presence in rocks interpreted as upper mantle slices suggests formation at either an oceanic or marginal basin spreading center beneath contemporaneously erupted abyssal tholeiites. Recent work involving the discovery of the incongruent melting of chromian diopside (Dickey et al. 1971) indicates that chromium might be released from the silicate phases of lherzolitic mantle rock by incongruent partial fusion. Ore genesis by crystal fractionation of basic magma leading to the formation of ultra-basic cumulates is favored by Thayer (1969); this could occur either beneath an ocean ridge, or beneath an oceanic volcano as shown in figure 2B.

Nickel sulfides.—Economic deposits of nickel sulfides in island arcs are known only in the Acoje Mine in the Philippines, within the ultrabasic complex on Luzon. The ore occurs with platinum sulfides as irregular blebs in sepentinized dunite, and contains pyrrhotite, troilite, pentlandite, and variolarite (Bryner 1969). Like the chromite, this ore presumably developed near the contact of upper mantle rocks with oceanic crust. Its origin probably differs from that of nickel sulfides in Archaean and early Proterozoic shield areas which are characteristically associa-

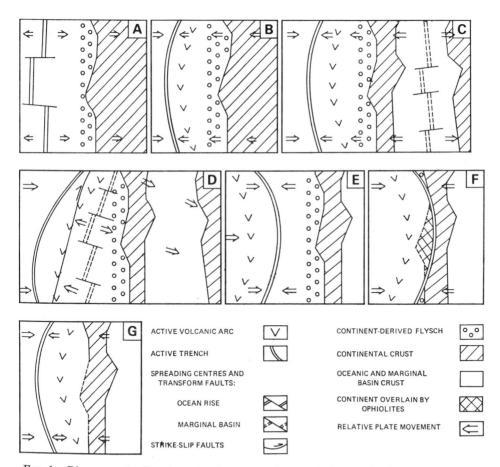

Fig. 1.—Diagrammatic plan views showing stages of arc evolution: *A–G* refers to corresponding cross section in fig. 2*A–G.*

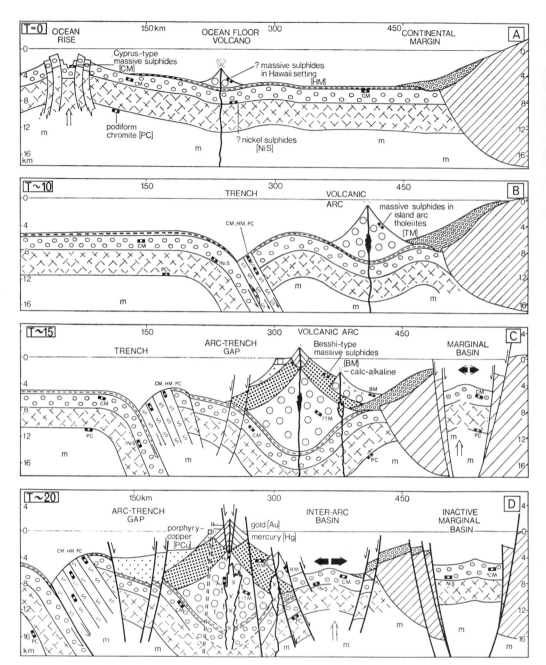

Fig. 2.—Diagrammatic cross sections through an evolving island arc; $T \sim 15$, etc. = time in m.y. since start of arc evolution: A, Stage 1.—Pre-arc emplacement of oceanic crust and upper mantle. Spreading ocean ridge migrates away from passive continental margin: formation of Cyprus-type and Hawaii-type massive sulfides, podiform chromite, and possibly nickel sulfides (ocean ridge after Osmaston 1971). B, Stage 2.—Submarine volcanism and initial arc development. Ensimatic arc develops on ocean floor near continental margin; possible massive sulfides formed in island-arc tholeiite lavas. C, Stage 3.—Subaerial and submarine volcanism. Volcanic arc builds up to sea level; Besshi-type massive sulfides formed on flanks of volcanic arc; rifting near continental margin and development of marginal basin with associated mineralization. D, Stage 4.—Plutonic activity, waning volcanism, faulting and arc rifting. Rise of granodioritic plutons in volcanic arc; caldera development with formation of porphyry copper, gold, and mercury deposits; sedimentation in arc-trench gap; development of interarc basin. (Note that porphyry copper [PCu] should be shown occurring below mercury [Hg] deposit.) E, Stage 5.—Arc reversal and development of new volcanic arc. Reversal of Benioff zone and loss of marginal basin crust; formation of Kuroko-type massive sulfides in dacites, and

76

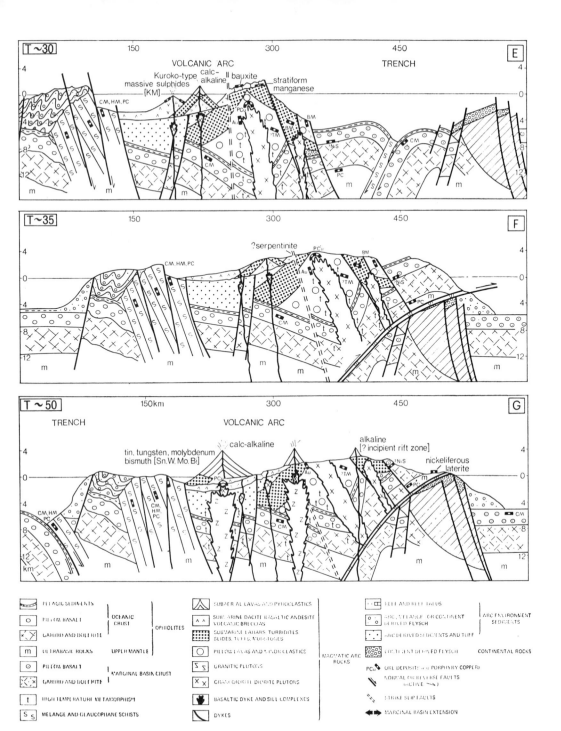

gold mineralization; elevation of trench-fill deposits; bauxite and manganese deposits form on uplifted volcanic arc. *F, Stage 6.—Collision of arcs.* Arc-arc collision follows loss of marginal basin crust; emplacement of Cyprus-type massive sulfides, podiform chromite, and nickel sulfides in obducted ophiolites. *G, Stage 7.—Development of new volcanic arc and emplacement of granites.* Erosion of obducted ocean floor to expose upper mantle rocks; uplift and erosion of mélange and old volcanic arc to expose paired metamorphic belts; development of new volcanic arc and trench; shoshonitic or alkaline volcanism in incipient rift zone prior to marginal basin development; formation of some types of ore body also formed in older volcanic arc; rise of granite plutons with associated tin-tungsten-molybdenum-bismuth mineralization. (Note that this mineralization [Sn, Wo, Mo, Bi] should be shown occurring at apices of granitic plutons.)

ted with highly magnesian silicate magma (Hudson 1972).

STAGE 2: SUBMARINE VOLCANISM DURING INITIAL ARC DEVELOPMENT

GEOLOGICAL EVENTS

In some arcs, the oldest exposed volcanic rocks are basaltic (Baker 1968) with the composition of island-arc tholeiites (Jakeš and White 1972). If island-arc tholeiitic volcanism can continue intermittently for several tens of millions of years (Gill 1970), only the oldest rocks of this composition will be erupted on the ocean floor. Submarine island-arc tholeiites and ocean floor basalts cannot be easily distinguished, but there is evidence of significant differences in the proportions of Ti, Zr, and Y (Pearce and Cann 1971) and in the K/Ba and Sr^{87}/Sr^{86} ratios (Hart et al. 1972).

Stratigraphic contacts between island-arc successions and oceanic crust are rarely exposed; consequently, it is uncertain whether the oldest exposed volcanic unit in an arc represents the first episode of island-arc volcanism, or is underlain by older arc rocks. A probable example of initial arc volcanism is the Water Island Formation, the oldest stratigraphic unit in the Lesser Antilles, with a "chemically primitive" or island-arc tholeiite composition (Donnelly et al. 1971). The Water Island Formation consists of spilitic and quartz keratophyre flows and minor volcaniclastic rocks, erupted in deep water (Donnelly 1964).

Eruption of basalts and basaltic andesites on the ocean floor results in a thick succession of pillow lavas (fig. 2B). Breccias, formed by gravitational collapse of some pillows, move as mass flows down the volcano flanks and accumulate as talus cones (Jones 1969). Changes in the erupted basaltic rocks as the volcano nears sea level probably resemble those described from basalts in Hawaii (Moore 1965; Moore and Fiske 1969) and from Icelandic intra-glacial olivine basalts (Jones 1966). Vesicle size increases upward, and vitric palagonitic tuff and breccia, with pillow breccia and peperites, become abundant in rocks erupted within a few hundred meters of the surface. Within the volcanic pile, anastomosing dikes and sills form an intrusive complex which may develop into a high-level reservoir.

Figures 1B and 2B show a volcanic arc bordered by a submarine trench developing above a Benioff zone near a continental margin.

FORMATION OF MINERAL DEPOSITS

Although stratiform sulfide deposits are not known from volcanic successions interpreted as early island-arc rocks, their presence in ocean crust tholeiites suggests that they could also occur in deep-water island-arc tholeiites. Similarities in composition of the basalts erupted in the two settings suggests that the ore minerals from each setting also would be similar.

STAGE 3: SUBAERIAL AND SUBMARINE VOLCANISM

GEOLOGICAL EVENTS

As a submarine volcano attains sea level, subaerial lava flows and tuffs are erupted, overlying predominantly clastic, shallow marine rocks (fig. 2C). Subaerially erupted successions are common in island arcs, both as stratovolcanoes in present active volcanic chains, and as block-faulted successions in active and inactive arcs (Mitchell and Reading 1971). However, conformable stratigraphic contacts between these and the oldest exposed submarine volcanic succession are rare. The examples below indicate that rocks erupted during the first subaerial episode in an arc are mostly either andesites or basalts.

In the Virgin Islands (Donnelly et al. 1971) the submarine Water Island Formation is overlain unconformably by the Louisenhoj Formation—a thick series of subaerially erupted porphyritic augite-andesite breccias similar to the pre-Robles succession in Puerto Rico. The Louisenhoj and pre-Robles Formations have chemically primitive characteristics, resembling those of the underlying Water Island Formation.

The New Hebrides islands of Malekula and Espiritu Santo contain very thick early Miocene successions of volcaniclastic basaltic andesites (Robinson 1969; Mitchell 1971). On Malekula, rocks of deep-water facies are abundant and include carbonized tree trunks and reef limestone detritus, suggesting accumulation on the submarine flanks of subaerially active volcanoes. Rapid erosion led to mass downslope transport of rock debris—as subaerial and submarine lahars, slide, and turbidites—into deep water (Mitchell 1970). Consequently, relatively small subaerial volcanoes become surrounded by a much larger volume of submarine volcaniclastic rocks.

The South Sandwich islands in the Scotia arc consist mostly of late Cenozoic subaerially erupted lavas forming basaltic shield volcanoes and are at an early stage of arc volcanism (Baker 1968). Submarine slopes probably resemble those of intraoceanic basaltic volcanoes where flow-foot breccias, hyaloclastites, and tuff from phreatic eruptions move downslope into deep water (Moore and Fiske, 1969; Jones 1969).

Subaerial and submarine volcanism and sedimentation in the volcanic arc are accompanied by sedimentation in the arc-trench gap, commonly a topographic trough, located between the trench inner margin and the active volcanic front. Possible examples of these successions occur in the Cretaceous Median Zone of southwest Japan and in the Mesozoic Hokonui facies of New Zealand (Dickinson 1971), although distinction between these and interisland volcanic arc deposits is difficult.

The distribution of the recent subaerially erupted rocks suggests that the magmas originate along or above a Benioff zone. Migration and probable change in slope of a Benioff zone during volcanism may have occurred in the Sunda arc; there is evidence that the trench has been forced oceanward as scraped-off continent-derived turbidites (Hamilton 1972) were tectonically emplaced on the arc side of the trench (e.g., Oxburgh and Turcotte 1971), forming the Mentawai-Nicobar-Andaman Islands and the Indoburman ranges. The resulting tectonic mélange of ocean floor sediments, high-pressure metamorphic rocks, and upper mantle material forms a belt parallel to the trench and is commonly partly overlain by arc-derived sediments.

Accumulation of a thick pile of volcanic rocks, accompanied by subsidence due partly to isostatic adjustments (Moore 1971), results in deep burial and regional high-temperature metamorphism of the volcanic arc rocks and underlying oceanic crust. Possible examples of metamorphosed oceanic crust are the amphibolites of the Bermeja Complex, Puerto Rico (Donnelly et al. 1971), amphibolite rafts in serpentinite on Pentecost Island, New Hebrides (Mallick 1970), and "basement" greenschists and amphibolites on Yap Island in the Palau arc (Shiraki 1971). Metamorphosed submarine island-arc tholeiitic rocks form part of the Wainimala Group of greenschist facies in Fiji (Gill 1970).

In some arcs, the presence of rocks typical of continents suggests that the arc has migrated oceanward away from the continental margin. Examples of rifted continental fragments within active arcs occur in New Zealand (Landis and Bishop 1972) and probably in Japan. In Figures 1C and 2C, rifting near the continental margin is followed by development of a marginal basin (Karig 1972) and oceanward migration of the arc and continental fragment.

FORMATION OF MINERAL DEPOSITS

Mineral deposits formed at this and subsequent stages of arc evolution include some copper-zinc deposits considered typical of early stages of geosynclinal development (Smirnov 1968).

Besshi-type massive sulfides.—Many massive sulfide deposits occur in association with andesitic or basaltic volcanic rocks together with significant thicknesses of carbonaceous mudstones, clastic limestones, quartz-rich sediments, or clastic volcanic rocks showing evidence of deep-water accumulation. These nonophiolitic components indicate deposition adjacent to a land mass, volcanic islands, or shallow-water volcanoes, rather than on an oceanic ridge.

Metamorphism and structural complexity of ores and host rocks are common features of these deposits. Both host-rock lithology and ore mineralogy are very variable and the deposits could probably be divided into subgroups. We adopt Kato's (1937) term "Besshi-type deposits," applied to bedded cupriferous iron sulfide deposits at Shikoku on Honshu Island, Japan.

Deposits at Besshi occur in the high-temperature Sanbagawa metamorphic belt of late Mesozoic age, mostly within a succession of isoclinally folded alternations of basic, pelitic, and quartzose schists (Kanehira and Tatsumi 1970). The sedimentary environment has been interpreted as a continental slope and shelf (Ernst 1972), although mafic tuffs and subsequent metamorphism suggest a volcanic arc environment. The stratiform sulfide ores are of three types: (1) compact pyrite, chalcopyrite, sphalerite, and gangue; (2) banded sulfides and silicates; and (3) copper-rich ore, containing chalcopyrite, minor pyrite, and gangue. Pyrrhotite is abundant toward the base of the deposits.

Similar massive sulfide deposits occur in the high-temperature Hida and Abukuma metamorphic belts of Japan (Kanehira and Tatsumi 1970). These have been interpreted as either volcanic sedimentary or early epigenetic deposits; evidence of slumping and mass flow of host sediments and ore (Jenks 1971) suggests a deep-water environment.

Besshi-type deposits in other arcs probably include the Hixbar and Bagacay deposits in the Philippines (Bryner 1969). In intracontinental greenstone belts, ore bodies in the Bathurst-Newcastle area of New Brunswick and at Captains Flat in New South Wales—considered by Stanton (1960) to be strata-bound in island arc rocks—are probably of Besshi-type. Other ore bodies in similar settings occur at Stekenjokk (Zachrisson 1971) and Menstrask (Grip 1951) in Sweden, and possibly Mt. Isa in Queensland (Hutchinson 1973). In the Iberian pyrite belt, the presence of resedimented mudstones, slates, and conglomerates interbedded with massive sulfides (Schermerhorn 1970) suggests that these are Besshi-type deposits, rather than Kuroko-type (Sillitoe 1972a) discussed below.

Besshi-type deposits include varying assemblages of ore minerals, some of which may resemble those of Kuroko type. Strong (1972) suggested that massive sulfides associated with intermediate to basic calc-alkaline volcanic rocks—for example, Betts Cove in Central Newfoundland—comprise polymetallic ores (Cu, Pb, Zn, Ag, and Au); Hutchinson (1973) and Sillitoe (1972a) considered that deposits in calc-alkaline rocks include more Pb, Zn, Ag, and Ba than those in ophiolite complexes.

Support for a syngenetic volcanic exhalative origin for deposits of Besshi-type is provided by sulfur isotope ratios in the Bathurst-Newcastle deposits (Sangster 1968, Lusk 1969). Lusk (1972) showed that the ratios of these and some similar deposits elsewhere can be explained by mixing of sulfur in rising hot igneous fluids with marine or connate sulfate below the sediment-water interface.

Deposits associated with ophiolites.— Since similarities exist between the crust and upper mantle of the oceans and of marginal basins, it is possible that mineral deposits comparable with those considered in Stage 1 may also form during generation of marginal basin crust and upper mantle in Stages 3 and 4. However, it is not yet possible to distinguish between either rocks or mineral deposits formed in these two settings.

STAGE 4: PLUTONIC ACTIVITY, WANING VOLCANISM, FAULTING, AND ARC RIFTING

GEOLOGICAL EVENTS

Islands in many arcs contain stocks of gabbroic to granodioritic composition intruding volcanic successions (fig. 2D). Examples are the Utuada Pluton intruding Middle Cretaceous rocks in Puerto Rico (Donnelly 1964), Upper Tertiary granodiorites intruding Miocene volcanic rocks in the Aleutians (Coats 1962), Pliocene dioritic plutons intruding early Miocene andesitic rocks in Bougainville (Blake and Miezitis 1966; Macnamara 1968), pre-Middle Miocene diorites intruding lower Miocene volcaniclastic rocks on Malekula, New Hebrides (Mitchell 1966), and diorite and granodioritic plutons intruding Miocene and Pliocene rocks in Guadalcanal and New Georgia, Solomon Islands (Stanton and Bell 1969).

Stratigraphic relations rarely indicate whether the plutons were emplaced during or following eruption of the overlying volcanic rocks. Although the presence of calderas suggests that large magma chambers lie beneath volcano summits, magnetic anomalies in the New Hebrides indicate that perched magma chambers are only a few kilometers in diameter, and could consist of dike complexes (Malahoff 1970). Possibly island-arc volcanoes, like those in anorogenic areas (Thompson 1972), are underlain by small reservoirs as in Hawaii (Wright and Fiske 1971), which develop into large magma chambers only after the main volcanic activity has ceased. Magma ascent by stoping, as in the coastal batholith of Peru (Cobbing 1972), and its passive emplacement high in the volcanic pile, are probably accompanied by regional metamorphism of the subsiding volcanic and sedimentary prism around the deeper levels of the rising plutons.

In most arcs, intense volcanic activity along any one belt during the Mesozoic or Cenozoic lasted no longer than about 10 m.y. However, after an interval volcanism was commonly either renewed along the same belt, or commenced along a different belt. These variations in volcanic activity could result from stress changes accompanying arc rotation relative to the subducting plate during marginal basin opening—for example, clockwise rotation of the New Hebrides arc, during the late Cenozoic. Major strike-slip faults trending approximately parallel to the arc, as in the Philippines or Sumatra, could also result from stress changes related either to arc rotation or to changes in ocean ridge spreading direction relative to the arc. Strike-slip faults approximately parallel to complex arcs are also known in Honshu, Japan (Miyashiro 1972), Taiwan (Biq 1971), and Sulawesi (Sarasin 1901).

Strike-slip faults oblique to the arc—for example, in Guadalcanal in the Solomon Islands (Coleman 1970) and in the west of the New Hebrides (Malahoff and Woollard 1969)—are possibly continuations of oceanic transform faults along which ultrabasic and basic rocks can be emplaced (Thompson and Melson 1972). Fault slices and pods of serpentinite, commonly showing a linear distribution, such as those on Guadalcanal and the "filons" in New Caledonia (Lillie and Brothers 1969), may have been emplaced along faults of this type.

Vertical displacements along high-angle faults in volcanic arcs, common both during and following volcanism, may result either from block faulting accompanying isostatic adjustment or from vertical movements along strike-slip faults. Control of eruptive center locations by high-angle faults is indicated by the linear distribution of parasitic cones and major craters (Kear 1957; Warden 1967, 1970), by the location of major centers of eruption and calderas where two or more linear features intersect as in Hawaii (Woollard and Malahoff 1966), and by the distribution of volcanoes along en echelon fractures oblique to the arc trend as in Tonga (Bryan et al. 1972.

Figures 1D and 2D show rifting and subsequent interarc basin development

(Karig 1972) between the volcanic arc and the continental fragment. This process may explain the intraoceanic location of rafts of continental rocks or continent-derived sediments lacking a volcanic arc—for example, the Mesozoic succession on New Caledonia (Lillie and Brothers 1969).

FORMATION AND EXPOSURE OF MINERAL DEPOSITS

Mineral deposits formed at this stage and during the preceding Stages 2 and 3 can be exposed following uplift and erosion either during or subsequent to the fault movements described above.

Porphyry copper.—Porphyry copper, or copper-molybdenum and copper-gold, deposits are emplaced in island arcs and on Andean-type continental margins in the belt of andesitic to dacitic igneous activity above a Benioff zone (Pereira and Dixon 1971; Mitchell and Garson 1972; Taylor 1972; Sawkins 1972; Sillitoe 1972b). They occur at the summit, or around the margins, of stocks or small plutons intruded beneath contemporaneously erupted volcanic rocks.

In island arcs, the intruded host rock forms part of a thick volcanic and clastic succession which mostly accumulated below sea level, although it was probably overlain by subaerial volcanoes during mineralization. Occurrence of Miocene and younger ore bodies at a high elevation—as at Mamut in Sabah and Ok Tedi in New Guinea—suggests that a thick column of intruded host rock favors mineralization, perhaps by permitting differentiation in the rising pluton. Later uplift and erosion expose the ore body in the submarine host rocks.

Whether the volcanic host rock succession is underlain by rocks erupted in an earlier stage of arc volcanism, by oceanic crust, or by continental fragments is uncertain. Porphyry copper deposits may, therefore, develop not during the first episode of subaerial volcanism in an arc, but only during a later episode (Stage 5). Alternatively, emplacement of the deposits may be independent of the stage of

arc evolution, and occur beneath any belt of andesitic or dacitic subaerial volcanoes. It has yet to be demonstrated that porphyry coppers are associated only with calc-alkaline volcanic rocks (e.g., Sillitoe 1972b), and not with island-arc tholeiitic volcanic rocks of intermediate composition.

Controls on location of porphyry deposits within a volcanic belt are poorly understood. The occurrence of many active volcanoes along faults or at fault intersections suggests tectonic control of related porphyry copper mineralization. Intense shattering of the ore host rocks has been attributed to mineralization stoping (Locke 1926), pulsating magma movements (Perry 1961), fault and joint development related to strike-slip faults (Bryner 1968), and pressure release beneath a caldera (Taylor 1972)—for example, by explosive emission of nuée ardentes. The common development of both calderas and strike-slip faults toward the close of arc volcanism suggests that related porphyry copper deposits are emplaced late in the development of a magmatic arc.

Mercury deposits.—Deposits of cinnabar and minor amounts of quicksilver occur in the Philippines, Japan, and New Zealand, but are not known in the less complex arcs. Most deposits are in Cenozoic volcanic rocks and are located near either active or old volcanic centers.

Restriction of the known ore bodies to the more complex arcs is probably of genetic significance, but does not indicate whether the metal is magmatic or non-magmatic in origin. If the mercury is magma-derived, the presence of thick crust typical of complex arcs may be necessary to allow differentiation of a rising high-level pluton and concentration of volatiles. If the metal is sediment-derived, a thick stratigraphic prism including fine-grained sediments may be necessary to form a source from which mercury is expelled by magmatic heat (Moiseyev 1971). White et al. (1971) suggested that mercury, separated from less volatile metals in vapor-dominated reservoirs, could be deposited above boiling brine zones in which porphyry copper deposits develop. A mercury deposit in this position beneath a subaerial volcano is shown in Figure 2D.

Gold associated with granodioritic plutons. —Problems in relating the origins of some types of gold deposit to geological environments above a Benioff zone have been discussed by Sawkins (1972). Two of the most important settings for gold mineralization in island arcs are gold-quartz veins associated with granodioritic plutons, and gold in andesitic volcanics; the latter is discussed under Stage 5 below.

In the Solomon Islands, and probably in the western belt of the New Hebrides, small quantities of placer gold have been derived from lodes associated with the margins of dioritic or granodioritic plutons. Limited evidence suggests that these magmatic-hydrothermal deposits are emplaced at a deeper structural level than porphyry copper ore bodies and are related to nonporphyritic intrusions. Deposits in this type of environment are widespread in western North America, but the related plutons were probably emplaced mostly in an Andean-type igneous belt on a continental margin rather than in island arcs.

Pyrometasomatic deposits.—Pyrometasomatic deposits are common around plutons in the more complex arcs, such as Japan and the Philippines. However, deposits of this type are neither restricted to, nor particularly characteristic of, island arcs, and are, therefore, not considered here.

STAGE 5: ARC REVERSAL AND DEVELOPMENT OF NEW VOLCANIC ARC

GEOLOGICAL EVENTS

In many arcs, successive volcanic belts have not necessarily developed in the same place. Change in volcanic arc position can result from arc rifting and interarc basin development (Karig 1971a, 1971b; 1972). Thus, the relative positions of the active Tonga arc and the Lau-Colville inactive or remnant arc have been explained by late Cenozoic splitting and eastward migration of the active volcanic belt (Karig 1972; Sclater et al. 1972). Similarly, the Mariana arc has probably migrated eastward away from the west Mariana remnant arc (Karig 1971a).

Change in position of a volcanic arc related to changes in Benioff zone inclination have been suggested in the Peruvian Andes (J. Cobbing, personal communication, 1972), but have yet to be convincingly demonstrated in an island arc. In Japan, where the distribution of Quaternary volcanoes coincides approximately with the Miocene "Green Tuff" volcanic belt, the two volcanic episodes possibly resulted from intermittent descent of lithsphere along a Benioff zone.

Changes in position of the volcanic arc, or arc reversal, related to changes in direction of dip of the Benioff zone (McKenzie 1969) shown in figures 1E and 2E may have occurred in the New Hebrides (Mitchell and Warden 1971), the New Ireland-Bougainville part of the Solomon Islands arc (Mitchell and Garson 1972), and Taiwan (Murphy 1972). Arc reversals probably result from attempted subduction of continental, island-arc, or oceanic island crust. Changes in Benioff zone dip may be related to emplacement of mélanges and consequent oceanward migration of the trench, to changes in rate of lithosphere descent, or to lithosphere drift relative to underlying deep mantle (Hyndman 1972).

Development of a new volcanic arc is probably preceded by associated trench formation. The old inactive trench becomes filled with sediments and rises isostatically as belts of thick folded and faulted flysch-type rocks, perhaps bordered on the arc side by mélanges and glaucophane schists. The Mentawai-Nicobar-Andaman Islands and the Indoburman Ranges could be interpreted as trench fill rather than mélange deposits.

The belt of flysch, glaucophane schists, and mélanges commonly lies parallel to uplifted and eroded rocks of an extinct volcanic arc, showing high-temperature metamorphism (see fig. 2F). These form the paired metamorphic belts (Miyashiro 1961) sometimes separated by a tectonic line along which major strike-slip movements may have occurred (Miyashiro 1972).

Block faulting of inactive arc segments can result in raised atolls, such as Rennell Island south of the main Solomons chain, and extensive raised carbonate platforms, as in Jamaica. Volcanic arc rocks lying between a younger active trench and volcanic arc can be elevated due to upward flexure of the arc plate above the downgoing plate (Fitch and Scholz 1971), resulting in tilted terraces, as in the western belt of the New Hebrides (Mitchell and Warden 1971), southwest Japan, and Eua in Tonga.

The younger volcanic arcs may develop on oceanic crust, on submarine sediments derived from an older arc (e.g., the active Central Chain in the New Hebrides), or on a subaerial or submerged erosion surface of older arc deposits and plutons (e.g., the Japanese "Green Tuff" succession). These younger volcanic successions may resemble either the subaqueous or subaerially erupted successions of Stages 2 or 3 described above.

The composition of these later volcanic rocks varies widely both within an arc and between different arcs. In many late Cenozoic volcanic arcs, the potash content increases with increasing depth to the Benioff zone (Kuno 1966; Hatherton and Dickinson 1969). In some arcs (e.g., Honshu), tholeiites nearest the trench pass laterally into calc-alkaline rocks and finally into shoshonites (Jakeš and Gill 1970). Variations from calc-alkaline to tholeiitic volcanism along the length of the arc are known in the Central Islands of the New Hebrides (Mitchell and Warden 1971) and also in the Lesser Antilles (Donnelly et al. 1971), where calc-alkaline rocks are associated with under-saturated basaltic lavas (Sigurdsson et al. 1973). Ignimbrites are common in the late Cenozoic calc-alkaline belt of some complex arcs

because the islands are larger and the lava more acidic than in arcs at earlier stages of development.

Figures 1E and 2E show arc reversal with "flipping" of a Benioff zone, and development of a related volcanic arc.

FORMATION AND EXPOSURE OF MINERAL DEPOSITS

The endogenous Kuroko-type massive sulfide and gold deposits described below may be exposed during the subsequent Stages 6 and 7 of arc development. Deposits formed in the older magmatic arc during Stages 2, 3, and 4 may be exposed during Stages 6 and 7, and also during tilting and faulting described above.

Kuroko-type massive sulfides.—Among the best-known massive sulfide ore bodies in a modern island arc are those at Kuroko in northeast Japan. They are all associated with predominantly clastic dacitic or more rarely andesitic volcanic rocks interpreted as shallow near-shore marine deposits. The host rock therefore differs considerably from that of both Cyprus-type and Besshi-type deposits. Massive sulfides associated with clastic andesitic or more acidic rocks emplaced mostly in shallow water are here termed "Kuroko-type" deposits.

In the Kosaka deposits (Horikoshi 1969; Horikoshi and Sato 1970), the stratiform ore bodies are vertically layered with an upper zinc-rich layer of black (= Kuroko) ore and a lower zinc-poor yellow layer. The main minerals are pyrite, chalcopyrite, sphalerite, galena, and minerals of the tetrahedrite group. The ores have been interpreted as volcanic exhalative deposits formed by hydrothermal activity during the last stages of volcanism. This activity followed phreatic explosions which accompanied emplacement of dacite domes and flows and formed lenticular units of lithic dacitic fragments. Graded bedding in the upper levels of the ore bodies indicates a syngenetic sedimentary origin for at least part of the deposit.

Examples of Kuroko-type ore bodies are probably fairly common in orogenic belts now lying within continents. In the Archaean Keewatin lithofacies of Canada, Hutchinson et al. (1971) have stressed the association of massive pyrite base metal ores with felsic extrusives and pyroclastics. Possible examples are the Horne Mine (Sinclair 1971) and the Delbridge deposit in the Noranda area (Jenks 1971). It is just possible that certain deposits associated with earlier, more mafic, extrusives should be included within the Besshi-type, for example, deposits near Asmara in Ethiopia (Anon. 1971), and the pyrite-chalcopyrite-sphalerite massive sulfides in Precambrian rocks at Jerome, Arizona (Anderson and Nash 1972).

The Kosaka deposits lie in the "Green Tuff" belt of intense early Miocene volcanic activity in which andesitic and dacitic rocks predominate (Sugimura et al. 1963). This succession overlies a basement of Mesozoic age or older, and is bordered on the east by an older paired metamorphic belt. It therefore postdates the first period of island-arc volcanism in Honshu, although Kuroko-type deposits might also be expected within calc-alkaline rocks of the first episode of shallow marine volcanism of Stage 3; a possible example is the deposit at Undua in Fiji (H. Colley, personal communication, 1973).

Gold associated with andesites and mon-zonites.—Gold deposits, mostly associated with quartz veins, are common in thick successions of andesitic lavas and meta-sedimentary rocks in complex arcs. For example, in the Hauraki Peninsula in New Zealand, gold occurs in early Tertiary propylitized andesitic and dacitic flows overlain by Pliocene lavas (Lindgren 1933).

Host rocks in deposits of this type are commonly metamorphosed and folded but not necessarily cut by intrusions. The metamorphism indicates relatively deep burial and suggests that the succession was overlain by younger rocks, possibly of volcanic arc facies. Migration and concentration of gold together with quartz probably accompanied metamorphism and deformation of the andesites (e.g., Helgeson and Garrels 1968). We therefore place the origin of gold occurring within andesites or meta-andesites in this stage of arc evolution.

Gold deposits at Vatukoula in Fiji occur in brecciated andesitic rocks of Pliocene age, closely associated with a caldera boundary fault. Infilling of the caldera with sedimentary rocks and andesites was followed by intrusion of trachyandesite and monzonite plugs. Tellurides and auriferous sulfide mineralization followed plug emplacement (Denholm 1967). The more basic Vatukoula rocks are island-arc alkali basalt or shoshonites (Dickinson et al. 1968; Gill 1970), and locally lie unconformably on older rocks of calc-alkaline or island-arc tholeiite composition which form much of the island. Hence they were erupted subsequent to the first major volcanic episode in Fiji.

Deposits broadly similar in structural setting and mineralogy to those at Vatukoula occur at Antamok and Acupan in the Philippines (Bryner 1969; Callow and Worley 1965).

Bauxite.—Economic deposits of bauxite on elevated limestones in island arcs are known in Jamaica and on Rennell Island (De Weisse 1970) south of the Solomon Chain; minor deposits occur in the Dominican Republic, Haiti, and the Lau Islands.

In Central Jamaica, large gibbsitic bauxite deposits occur in solution pockets, sinkholes, and troughs on a karstic surface. The limestone, of Oligicene and early Miocene age, overlies upper Eocene carbonates and accumulated after the final major volcanic episode in the island. The bauxite developed in post-Miocene time, following mid-Miocene faulting and uplift, in well-drained areas at elevations of 700–1,000 m. The hypothesis that the ore is a residual deposit resulting from weathering of several hundred feet of limestone (e.g., Hose 1963) is supported by trace element data (Sinclair 1967). Some authors (Zans 1954; Burns 1961) favored derivation from weathered volcanic rocks above the limestone, a hypothesis now coming back into favor.

Stratiform manganese.—Manganese deposits associated with raised limestones are largely restricted to recently elevated arcs in or near the tropics. These deposits are distinct from raised ocean floor manganese nodule deposits which are not currently economic. Economic deposits associated with raised limestones occur only in the New Hebrides and Cuba, although other types of occurrence are known, for example on Hanesavo island in the Solomon Islands (Grover et al. 1962).

In the New Hebrides, manganese occurs at or near the contact of raised late Cenozoic reef limestone with clastic volcanic rocks— as, for example, in Erromango, the Torres Islands, and Malekula. The Erromango deposits are probably syngenetic, but those on the other islands are considered to be largely or entirely epigenetic. The recently exploited Forari deposits on Efate Island, which mostly occur at the contact of laterites or limestones with underlying Pliocene volcaniclastic rocks. were possibly precipitated from solutions leached from volcanic rocks (Warden1970).

In Cuba, economic deposits of manganese are largely restricted to the southwest of Oriente Province. The ores occur near the top of an Upper Cretaceous to Middle Eocene thick marine volcaniclastic and sedimentary unit, and are concentrated within a few tens of meters of the contact between pyroclastic rocks and an overlying limestone member. Psilomelane, pyrolusite, and wad are the chief ore minerals. Simons and Straczek (1958) considered the deposits to be syngenetic and related to hot submarine springs.

STAGE 6: COLLISION OF ARCS

Reversal of arcs and loss of marginal basin crust along a younger Benioff zone can result in approach of the active and remnant arc and their eventual collision.

GEOLOGICAL EVENTS

Collision of arcs in the late Oligocene or early Miocene probably took place in New Caledonia (Dewey and Bird 1971; Karig 1972), where blueschist metamorphism accompanied southwestward thrusting of peridotites (Avias 1967) over older rocks including Permo-Triassic greywackes (Lillie and Brothers 1969). Derivation of these greywackes from the Australian continent presumably preceded late Mesozoic rifting and northeastward drift of the New Caledonia ridge. Attempted subduction of the ridge, perhaps beneath the Loyalty Islands, resulted in thrusting or "obduction" (Coleman 1971) of the Loyalty Island plate margin onto the New Caledonia plate.

In the Philippines, Tertiary arc collision is suggested by the discontinuous arcuate belt of layered ultrabasic rocks and diabase-gabbro dike swarms exposed on Palawan and western Luzon. These rocks were possibly emplaced during attempted subduction of the Palawan-western Luzon arcuate ridge along a Benioff zone which dipped east beneath an island arc, since fragmented by opening of the Sulu sea and by tectonic movements in western Luzon.

In the Solomon Islands, the ultrabasic rocks, basic lavas, and folded pelagic sediments of the northeastern Pacific Province (Coleman 1970) may be remnants of a late Mesozoic-early Cenozoic marginal basin thrust or obducted southwestward onto the Central Province during the Oligocene.

In the Oriente Province of Cuba, an ultrabasic complex interlayered with gabbroic rocks lies at the eastern end of a belt of northward-thrust ultrabasic slices of probable late Jurassic age (Meyerhoff and Hatten 1968). To the south, Cretaceous quartz diorite and granitic plutons (Khudoley 1967) were probably intruded beneath a volcanic arc. Upper mantle and ocean crust forming the layered complex were either emplaced in a mélange or obducted northward onto the Bahama Bank carbonate platform during southward subduction of a marginal basin. Emplacement of the layered complex may thus have involved collision of a Cretaceous island arc with the Bahama Bank continental margin (M. Itturalde-Vinent, personal communication, 1973).

Figures 1*F* and 2*F* show collision of the reversed arc with the previously rifted continental fragment. Obduction of ophiolites over continent-derived greywackes resembles that inferred for New Caledonia.

EMPLACEMENT AND EXPOSURE OF MINERAL DEPOSITS

Mineral deposits formed on the ocean floor, in oceanic crust, or in upper mantle are tectonically emplaced within ophiolites in island arcs; they occur in obducted slices, in mélanges, and possibly in fault-elevated blocks. Mineral deposits are not known from serpentinized ultrabasic rocks, perhaps mantle-derived, which are locally emplaced along strike-slip faults. Although only obducted ophiolites are emplaced during collisions, we consider here the emplacement of mineral deposits in any of these settings.

Cyprus-type massive sulfides.—Deposits of Cyprus-type could occur in ocean floor basalts emplaced either as obducted slices or within mélanges (Sillitoe 1972*a*). At present no deposits are known from mélanges; those in Newfoundland, for example, probably occur in obducted slices later deformed by continent-continent collision.

Podiform chromite.—The podiform chromites in upper mantle rocks in Cuba and the Philippines were probably emplaced together with their host rocks as obducted slices during collision of the island arcs with, respectively, a continental margin and another arc. Exposure of the ore followed erosion or tectonic removal of overlying ocean crustal rocks. Podiform chromite within upper mantle rocks could also occur in mélanges, or in fault slices or diapirs emplaced along major faults, but no deposits are known in these settings.

Nickel sulfides.—Nickel sulfides in the Philippines were presumably emplaced together with the chromite with which they are associated. Like podiform chromites, they are likely to be exposed only at deep structural levels, probably in obducted ophiolites.

Formation of nickeliferous laterites.— In southeast Sulawesi, nickel-bearing laterites overlie partly serpentinized harzburgite and lherzolite possibly emplaced as mélange during early Tertiary westward descent of lithosphere (Hamilton 1972). The ore is best developed as nickeliferous serpentine in the lower level of the laterite profile (PT International Nickel Indonesia 1972), which presumably formed during late Cenozoic weathering. Economic deposits of nickeliferous laterites on rocks possibly emplaced in mélanges also occur on Obi and Gube Islands in the Moluccas (Anon. 1972).

In Burma, nickeliferous laterite deposits approaching economic grade overlie ultrabasic rocks in the early Tertiary flysch belt forming the Chin Hills (Gnau Cin Pau, personal communication, 1972). These and other elongate ultrabasic bodies exposed along the eastern margin of the Indoburman Ranges are probably related to eastward subduction of lithosphere prior to late Cenozoic sedimentation in the Central Valley of Burma.

The nickeliferous laterites of the Dominican Republic also probably belong to this environment (F. J. Sawkins, personal communication, 1973).

STAGE 7: DEVELOPMENT OF COMPLEX ARCS AND CHANGES IN MAGMA COMPOSITION WITH TIME

GEOLOGICAL EVENTS

Repeated arc riftings, marginal basin spreading, and arc reversals lead to development of successive magmatic arcs, flysch belts, and paired metamorphic belts (fig. 1*G*). Arc collisions and strike-slip fault movements result in juxtaposition of tectonic blocks of different age and lithology. An original simple arc thus becomes increasingly complex with magmatic and tectonic addition of younger arcs. For example, the late Cenozoic simple Scotia arc might progress through an arc reversal to the stage of the New Hebrides arc; collision with remnant arcs, obduction of ophiolites, and major strike-slip movements would result in an arc with the complexity of the Philippines. Alternatively, with intermittent loss of lithosphere and no reversal, an arc of Andaman-Nicobar type could eventually reach the complexity of Honshu Island in Japan.

As stated earlier, the development through time of a complex arc involves changes in magma composition. Jakeš and White (1972) considered that initially tholeiitic, predominantly basaltic, lavas are overlain by both tholeiites and minor calc-alkaline andesitic and dacitic rocks, and that finally tholeiitic, calc-alkaline, and shoshonitic rocks are erupted. Although such changes in composition with time have been described from Puerto Rico (Donnelly et al. 1971) and Viti Levu in Fiji (Gill 1970), they are evidently not universally found. In the Japanese "Green Tuff" belts, thick Miocene successions of largely dacitic and andesitic rocks (Sugimura et al. 1963) are overlain by Quaternary, predominantly tholeiitic, basalts. In the southern Kitakami Massif of eastern Honshu, late Palaeozoic and Cretaceous high-alumina basalts are overlain by Quaternary low-alkali tholeiites (Sugisaki and Tanaka 1971). In the new Britain-Schouten Islands, andesitic clastic rocks of probable early Tertiary age are cut by porphyries (Thompson and Fisher 1965) and overlain in the north by predominantly tholeiitic Quaternary lavas lacking contemporaneous calc-alkaline rocks (Jakeš and White 1969; Lowder and Carmichael 1970). Moreover, evidence that alkaline or shoshonitic rocks in arc environments are related to extensional tectonics typical of marginal basins (Martin and Piwinskii 1972) suggests that they are unrelated to this stage of arc evolution.

There are several possible controls on changes in composition of island-arc magma with time. Donnelly et al. (1971)

suggested that island-arc tholeiites result from partial melting of primitive upper mantle prior to development of a Benioff zone; they considered that later calc-alkaline rocks originate from partial melting of tholeiites metamorphosed to amphibolite or eclogite along a Benioff zone, and that as the arc develops descent of increasing volumes of arc-derived sediment would contaminate the tholeiitic layer. Armstrong (1971) suggested that descent of continent-derived ocean floor sediments and mixture with partially melting tholeiite, could explain the lead isotope ratios and high proportion of Pb, Ba, Th, K, Rb, and Cs in some calc-alkaline arc magmas.

The composition of arc magma may be related to rate of lithosphere descent (Sugisaki 1972) which could control the inclination of the Benioff zone (Luyendyk 1970). Benioff zones beneath active arcs erupting tholeiitic magma—such as the Marianas, Izu Islands, and Scotia arcs—mostly dip at more than 40°, while those beneath some arcs erupting predominantly calc-alkaline magma—such as the western Honshu arc and the Aleutians—are less steeply inclined. Marginal basin crust is not developing above the shallow dipping zones which underlie the Peruvian Andes and Central America; in the Marianas arc system, the generation of marginal or interarc basin crust may be dependent on the dip of the Benioff zone which possibly varies cyclically (Bracey and Ogden 1972).

There is a possibility that calc-alkaline magmas may be generated above a Benioff zone. Rise of volatiles from the descending plate into the zone of isotherm inversion above the descending cold slab of oceanic lithosphere could reduce melting points below the ambient temperature (McBirney 1969), facilitating partial melting of wet peridotite to produce liquids of calc-alkaline composition (Yoder 1969). Consequent magma compositions would be largely independent of the nature of the descending lithospheric plate, and would change with time due either to depletion in the low-temperature melting fraction or to addition of descending ocean crust material to the upper mantle (Arculus and Curran 1972). Magma composition may also be determined partly by the increasing thickness of crust as the arc develops (Hamilton 1972), which could control the degree of differentiation of, and partial melting around, rising magma.

The composition of plutons is related partly to the stage of arc development. Quartz diorite and granodiorite occur in relatively simple arcs with crust of moderate thickness (such as the Solomon Islands, Puerto Rico, and Fiji), but large bodies of alkali granite or adamellite have been described only from the complex Japanese arcs where they form batholiths, some of late Cretaceous age (e.g., Murakami 1970). The granites could either be highly differentiated products of partially melted mantle or they could have resulted from partial melting or anatexis of the lower part of the island-arc crust. A time difference of 40 m.y. between emplacement of diorites and granites has been demonstrated in a calc-alkaline association of Palaeozoic age at Yeoval in New South Wales (Gulson and Bofinger 1972).

FORMATION OF MINERAL DEPOSITS

Successive volcanic arcs developing in the same or adjacent localities may each be accompanied by formation of similar ore bodies—for example, Kuroko and perhaps Besshi-type massive sulfides, porphyry coppers, and mercury deposits.

As the thickness of the island-arc crust increases, emplacement of adamellitic and granitic plutons may be accompanied by mineralization associated with alkali and, particularly, soda granites, commonly believed to be largely restricted to continental crust. This mineralization includes deposits of the granitophile elements, sometimes considered to have originated in the crust together with palingenetic granitic plutons (Smirnov 1968) during the later stages of geosynclinal evolution (McCartney and Potter 1962).

Tin - tungsten - molybdenum - bismuth.—
These are of economic importance only in Japan, occurring mainly around batholiths in southwest Honshu (Shunso 1971). However, tin ores are also known in the Philippines (Bryner 1969). Fluorine, invariably present with tin and tungsten deposits (e.g., Rub 1972), may originate in the upper mantle or lower crust, or it may be derived with other volatiles from downgoing oceanic crust at depths of 200–400 km (Mitchell and Garson 1972).

ANCIENT ARCS ON AND WITHIN CONTINENTS

Continued addition to an island arc of igneous rocks and of tectonically emplaced ophiolites results in development of crust with a thickness approaching that of continents. However, before an arc complex can grow to continental thickness it usually becomes attached to, and forms part of, an older continental mass. Tectonic, and less common sedimentary, accretion of arcs and related ore bodies to a continental margin have been discussed elsewhere (e.g., Dewey and Bird 1971; Mitchell and Garson 1972).

Island arc-continent accretion may eventually be followed by collision with another continent, resulting in further orogeny and deformation of the arc succession. Thrust movements exceeding 100 km during collision have been explained by crustal "flaking" whereby continental crust on the descending plate splits into an upper overriding and lower underriding slab (Oxburgh 1972). Late rifting of the continent over a spreading center may take place in a zone different from that of the collision, resulting in continental fragments each containing arc complexes lying between older shields.

Many orogenic belts of Palaeozoic and Proterozoic age within continents contain metamorphosed greenstone belts. Parts of these resemble in lithology and chemistry the successions in Cenozoic island arcs, with submarine volcanic rocks of calc-alkaline and island-arc tholeiite composition, and calc-alkaline plutons. Examples are the late Precambrian Harbour Main, Conception, and Holywood rock groups of the Avalon Peninsula Newfoundland (Hughes and Bruckner 1971).

Archaean shield areas also include low-grade metamorphic greenstone belts with volcanic rocks broadly resembling those of island arcs in composition and, in some cases, in lithology. Examples include the igneous rocks of the Slave Province in the Yellowknife area of Canada (Folinsbee et al. 1968), part of the Kalgoorlie System in Western Australia (Glikson 1970), and the Onverwacht Group in South Africa (Viljoen and Viljoen 1969). Despite minor chemical differences between some of these successions and modern island arcs (Glikson 1971; Jakeš and White 1971; Hart et al. 1970), many recent workers consider that tectonic accretion of island arcs to continental margins has continued for at least 3.5×10^9 years (e.g., Engel and Kelm 1972).

Although the formation of some types of massive sulfide deposits may be characteristic of certain periods of the earth's development (Hutchinson 1973), we consider that similarities between ancient and modern arc rocks together with similarities in associated mineral deposits suggest that ore-forming processes in arcs have changed little during the last 2×10^9 years.

CONCLUSIONS

Major mineral deposits in island arcs can be divided into three main groups according to their environment of formation: (1) deposits formed in magmatic arcs, (2) deposits formed in oceanic crust or upper mantle and tectonically emplaced within ophiolites, and (3) exogenous deposits formed in or on raised limestones and on ultrabasic rocks.

Magmatic arc deposits consist of stratiform massive sulfides and deposits related to plutons. Stratiform massive sulfides of Kuroko-type form in a shallow nearshore marine environment together with acidic volcaniclastic rocks; those of Besshi-type form in a deep-water environment together with sediments and minor amounts of intermediate to basic volcanic rocks; possibly massive sulfides also form in submarine island-arc tholeiites. Magmatic hydrothermal porphyry copper, gold, and some mercury deposits form around the upper margins of dioritic and granodioritic plutons emplaced beneath volcanoes. Gold mineralization also occurs around monzonitic intrusions and in andesites.

Deposits formed in or on oceanic crust comprise stratiform massive sulfides of Cyprus-type, and similar deposits may form on Hawaii-type oceanic volcanoes. Near the mantle-crust boundary, podiform chromites and possibly nickel sulfides develop. These are emplaced tectonically with the host rocks in mélanges or in obducted slices.

Exogenous deposits include bauxites developed on karstic raised limestones, stratiform manganese deposits formed near limestone-tuff contacts, and nickeliferous laterites developed on tectonically emplaced upper mantle rocks.

Ore deposits likely to be preserved in island arcs now within continents are magmatic arc deposits of deeply eroded porphyry copper and of gold and in some cases tin and tungsten deposits, and island-arc tholeiite-type, Besshi-type, and Kuroko-type massive sulfides; preserved deposits in ophiolites within arcs include chrome deposits, Cyprus-type massive sulfides, and possibly nickel sulfides.

ACKNOWLEDGMENTS.—We thank Dr. H. G. Reading and Mr. E. Eadie of Oxford University, Dr. M. S. Garson of the Institute of Geological Sciences, London, and Dr. Frederick J. Sawkins for critically and constructively reviewing the manuscript. The first author is grateful to the Rio Tinto Zinc Co., Ltd. for permission to publish the paper.

REFERENCES CITED

ANDERSON, C. A., 1969, Massive sulfide deposits and volcanism: Econ. Geology, v. 64, p. 129–146.

———, and NASH, J. T., 1972, Geology of massive sulfide deposits at Jerome, Arizona—a reinterpretation: Econ. Geology, v. 67, p. 845–863.

ANON., 1970, Deep-sea drilling project: Leg 11: Geotimes, v. 15, p. 14–16.

——— 1971, Ethiopia Geol. Survey, 1970, Ann. Rept., p. 31–38.

——— 1972, Financial Times, October 10.

ARCULUS, R. J., and CURRAN, E. B., 1972, The genesis of the calc-alkaline rock suite: Earth and Planetary Sci. Letters, v. 15, p. 255–262.

ARMSTRONG, L. A., 1971, Isotopic and chemical constraints on models of magma genesis in volcanic arcs: Earth and Planetary Sci. Letters, v. 12, p. 137–142.

AVIAS, J., 1967, Overthrust structure of the main ultrabasic New Caledonian massives: Tectonophysics, v. 4, p. 531–541.

BAKER, P. E., 1968, Comparative volcanology and petrology of the Atlantic island arcs: Bull. volcanol., v. 32, p. 189–206.

BIQ, CHINGCHANG, 1971, Some aspects of post-Ordovician block tectonics in Taiwan, in Recent crustal movements: Royal Soc. New Zealand Bull., v. 9, p. 19–24.

BLAKE, D. H., and MIEZITIS, Y., 1966, Geology of Bougainville and Buka Islands, Territory of Papua and New Guinea: Australia Dept. Nat. Devel., Bur. Mineral Resources Geol. Geophys., Records No. 62.

BRACEY, D. R., and OGDEN, T. A., 1972, Southern Mariana arc: geophysical observations and hypothesis of evolution: Geol. Soc. America Bull., v. 83, p. 1509–1522.

BRYAN, W. B.; STICE, G. D.; and EWART, A., 1972, Geology, petrography and geochemistry of the volcanic islands of Tonga: Jour. Geophys. Research, v. 77, p. 1566–1585.

BRYNER, L., 1968, Notes on the geology of the porphyry copper deposits of the Philippines: Mineral Eng. Mag., v. 19, p. 12–23.

——— 1969, Ore deposits of the Philippines—an introduction to their geology: Econ. Geology, v. 64, p. 644–666.

BURNS, D. J., 1961, Some chemical aspects of bauxite genesis in Jamaica: Econ. Geology, v. 56, p. 1297–1303.

CALLOW, K. J., and WORLEY, B. W., 1965, The occurrence of telluride minerals at the Acupan Gold Mine, Mountain Province, Philippines: Econ. Geology, v. 60, p. 251–268.

CANN, J. R., 1969, Spilites from the Carlsberg Ridge Indian Ocean: Jour. Petrology, v. 10, p. 1–19.

——— 1970, New model for the structure of the ocean crust: Nature, v. 226, p. 928–930.

COATS, R. R., 1962, Magma type and crustal structure in the Aleutian arc: Australian Geophys. Union Geophys. Mon. 6, p. 92–109.

COBBING, J., 1972, Tectonic elements of Peru and the evolution of the Andes: Internat. Geol. Cong., 24th, Montreal 1972, Rept., p. 306–315.

COLEMAN, P. J., 1970, Geology of the Solomon and New Hebrides Islands, as part of the Melanesian re-entrant, Southwest Pacific: Pacific Sci., v. 24, p. 289–314.

COLEMAN, R. G., 1971, Plate tectonic emplacement of upper mantle periodotites along continental edges: Jour. Geophys. Research, v. 76, p. 1212–1222.

CRONAN, D. S.; VAN ANDEL, T. H.; HEATH, G. H.; DINKELMAN, M. G.; BENNETT, R. H.; BULENY, D.; CHARLESTON, S.; KANEPS, A.; RODOLFO, K. S.; and YEATS, R. S., 1972, Iron-rich basal sediments from the eastern equatorial Pacific: Leg 16, Deep-Sea Drilling Project: Science, v. 175, p. 61–63.

DAVIDSON, C. F., 1966, Some genetic relationships between ore deposits and evaporites: Inst. Mining and Metallurgy Trans., sec. B, v. 75, p. B216–225.

DENHOLM, L. S., 1967, Geological exploration for gold in the Tavua Basin, Viti Levu, Fiji: New Zealand Jour. Geology and Geophysics, v. 10, p. 1185–1186.

DESBOROUGH, G. A.; ANDERSON, A. T.; and WRIGHT, T. C., 1968, Mineralogy of sulfides from certain Hawaiian basalts: Econ. Geology, v. 63, p. 636–644.

DE WEISSE, G., 1970, Bauxite sur un atoll du Pacifique: Mineralium Deposita, v. 5, p. 181–183.

DEWEY, J. F., and BIRD, J. M., 1971, Origin and emplacement of the ophiolite suite: Appalachian ophiolites in Newfoundland: Jour Geophys. Research, v. 76, p. 3179–3206.

DICKEY, J. S., JR.; YODER, H. S.; and SCHAIRER, J. F., 1971, Chromium in silicate oxide systems: Carnegie Inst. Washington Year Book 70, p. 118–122.

DICKINSON, W. R., 1971, Clastic sedimentary sequences deposited in shelf, slope and trough settings between magmatic arcs and associated trenches: Pacific Geology, v. 3, p. 15–30.

———; RICHARD, M. J.; COULSON, F. I.; SMITH, J. G.; and LAWRENCE, R. L., 1968, Late Cenozoic shoshonitic lavas in northwestern Viti Levu, Fiji: Nature, v. 219, p. 148.

DONNELLY, T. W., 1964, Tectonic evolution of eastern Greater Antillean island arc: Am. Assoc. Petroleum Geologists Bull., v. 48, p. 680–696.

———; ROGERS, J. J.; PUSHKAR, P.; and ARMSTRONG, R. L., 1971, Chemical evolution of the igneous rocks of the eastern West Indies: an investigation of thorium, uranium, and potassium distributions, and lead and strontium isotopic ratios: Geol Soc. America Mem. 30, p. 181–224.

DUNHAM, K. C., 1972, Basic and applied geochemistry in search of ore: Inst. Mining and Metallurgy, Trans., sec. B, v. 81, p. 13–18.

ENGEL, A. E. J., and KELM, D. L., 1972, Pre-Permian global tectonics: a tectonic test: Geol. Soc. America Bull., v. 83, p. 2225–2340.

ERNST, W. G., 1972, Possible Permian oceanic crust and plate junction in Central Shikoku, Japan: Tectonophysics, v. 15, p. 233–239.

FITCH, T. J., and SCHOLZ, C. H., 1971, Mechanism of underthrusting in southwest Japan: a model of convergent plate interactions: Jour. Geophys. Research, v. 76, p. 7276–7292.

FLEMING, C. A., 1969, The Mesozoic of New Zealand: chapters in the history of the Circum-Pacific mobile belt: Geol. Soc. London Quart. Jour., v. 125, p. 125–170.

FOLINSBEE, R. E.; BAADSGAARD, H.; CUMMING, G. L.; and GREEN, D. C., 1968, A very ancient island arc, in KNOPOFF, L.; DRAKE, C. L.; and HART, P. J., eds., The crust and upper mantle of the Pacific area: Am. Geophys. Union Geophys. Mon. 12, p. 441–448.

GILL, J. B., 1970, Geochemistry of Viti Levu, Fiji, and its evolution as an island arc: Contr. Mineralogy and Petrology, v. 27, p. 179–203.

GLIKSON, A. Y., 1970, Geosynclinal evolution and geochemical affinities of early Pre-Cambrian systems: Tectonophysics, v. 9, p. 397–433.

——— 1971, Primitive Archaean element distribution patterns: chemical evidence and geotectonic significance: Earth and Planetary Sci. Letters, v. 12, p. 309–320.

GRIP, E., 1951, Geology of the sulfide deposits at Menstrask: Sveriges Geol. Undersokning, ser. C., no. 515: Stockholm, Norstedt and Soner, 52 p.

GROVER, J. C.; THOMPSON, R. B.; COLEMAN, P. J.; STANTON, R. L.; and BELL, J. D. et al., 1962, The British Solomon Islands geological record, v. 2, 1959–62: London, H.M.S.O., 208 p.

GUILD, P. W., 1947, Petrology and structure of the Moa district, Oriente Province, Cuba: Am. Geophys. Union Trans., v. 28, p. 218–246.

GULSON, B. L., and BOFINGER, V. M., 1972, Time differences within a calc-alkaline association: Contr. Mineralogy and Petrology, v. 36, p. 19–26.

HAMILTON, W., 1972, Tectonics of the Indonesian region: U.S. Geol. Survey Project Rept., Indonesian Inv. (IR) IND-20, 13 p.

HART, S. R.; BROOKS, C.; KROGH, T. E.; DAVIS, G. L.; and NAVA, D., 1970, Ancient and modern volcanic rocks: a trace element model: Earth and Planetary Sci. Letters, v. 10, p. 17–28.

———; GLASSLEY, W. E.; and KARIG, D. E., 1972, Basalts and sea floor spreading behind the Mariana island arc: Earth and Planetary Sci. Letters, v. 15, p. 12–18.

HATHERTON, T., and DICKINSON, W. R., 1969, The relationship between andesitic volcanism and seismicity in Indonesia, the Lesser Antilles and other island arcs: Jour. Geophys.Research, v. 74, p. 5301–5310.

HELGESON, H. C., and GARRELS, R. M., 1968, Hydrothermal transport and deposition of gold: Econ. Geology, v. 63, p. 622–635.

HESS, H. H., 1962, History of ocean basins, in Petrologic studies (Buddington volume): New York, Geol. Soc. America, p. 599–620.

HORIKOSHI, EI., 1969, Volcanic activity related to the formation of the Kuroko-type deposits in the Kosaka District, Japan: Mineralium Deposita, v. 4, p. 321–345.

———, and SATO, TAKEO, 1970, Volcanic activity and ore deposition in the Kosaka mine, in TATSUMI, T., ed., Volcanism and ore genesis: Tokyo, Tokyo Univ. Press, p. 181–195.

HOSE, H. R., 1963, Jamaica-type bauxites developed on limestones: Econ. Geology, v. 58, p. 62–69.

HUDSON, D. R., 1972, Evaluation of genetic models for Australian sulfide nickel deposits: Australasian Inst. Mining and Metallurgy Newcastle Conf., p. 59–68.

HUGHES, C. J., and BRUCKNER, W. D., 1971, Late Pre-Cambrian rocks of the eastern Avalon Peninsula Newfoundland—a volcanic island complex: Canadian Jour. Earth Sci., v. 8, p. 899–915.

HUTCHINSON, R. W., 1965, Genesis of Canadian massive sulfides reconsidered by comparison to Cyprus deposits: Canadian Inst. Mining and Metallurgy Trans., v. 68, p. 266–300.

———, 1973, Volcanogenic sulfide deposits and their metallogenic significance: Econ. Geology (in press).

———; RIDLER, R. H.; and SUFFEL, G. G., 1971, Metallogenic relations in the Abitibi Belt, Canada: a model for Archeaan Metallogeny: Canadian Mining and Metall. Bull., v. 64, p. 49–57.

HYNDMAN, R. D., 1972, Plate motions relative to the deep mantle and the development of subduction zones: Nature, v. 238, p. 263–265.

JAKEŠ, P., and GILL, J., 1970, Rare earth elements and the island arc tholeiitic series: Earth and Planetary Sci. Letters, v. 9, p. 17–28.

————, and WHITE, A. J. R., 1969, Structure of the Melanesian arcs and correlation with distribution of magma types: Tectonophysics, v. 8, p. 233–236.

————, —————— 1971, Composition of island arcs and continental growth: Earth and Planetary Sci. Letters, v. 12, p. 224–230.

————, —————— 1972, Major and trace element abundances in volcanic rocks of orogenic areas: Geol. Soc. America Bull., v. 83, p. 29–40.

JENKS, W. F., 1971, Tectonic transport of massive sulphide deposits in submarine volcanic and sedimentary host rocks: Econ. Geology, v. 66, p. 1215–1224.

JONES, W. G., 1966, Intraglacial volcanoes of southwest Iceland and their significance in the interpretation of the form of the marine basaltic volcanoes: Nature, v. 212, p. 586–588.

———— 1969, Pillow lavas as depth indicators: Am. Jour. Sci., v. 267, p. 181–195.

KANEHIRA, K., and TATSUMI, T., 1970, Bedded cupriferous iron sulphide deposits in Japan, a review, in TATSUMI, T., ed., Volcanism and ore genesis: Tokyo, Tokyo Univ. Press, p. 51–76.

KARIG, D. E., 1971a, Structural history of the Mariana island arc system: Geol. Soc. America Bull., v. 82, p. 323–344.

———— 1971b, Origin and development of marginal basins in the western Pacific: Jour. Geophys. Research, v. 76, p. 2542–2561.

———— 1972, Remnant arcs: Geol. Soc. America Bull., v. 83, p. 1057–1068.

KATO, T., 1937, Geology of ore deposits [in Japanese] (new ed.): Tokyo, Fuzambo.

KEAR, D., 1957, Erosional stages of volcanic cones as indicators of age: New Zealand Jour. Sci. and Technology, sec. B, v. 38, p. 671–682.

KHUDOLEY, K. M., 1967, Principal features of Cuban geology: Am. Assoc. Petroleum Geologists Bull., v. 51, p. 668–677.

KUNO, H., 1966, Lateral variation of basalt magma types across continental margins and island arcs: Bull. volcanol., v. 29, p. 195–222.

LANDIS, C. A., and BISHOP, D. G., 1972, Plate tectonics and regional stratigraphic-metamorphic relations in the southern part of the New Zealand geosyncline: Geol. Soc. America Bull., v. 83, p. 2267–2284.

LIDDY, J. C., 1972, Mineral deposits of the southwestern Pacific: Mining Mag. (London), March 1973, p. 197–203.

LILLIE, A. R., and BROTHERS, R. N., 1969, The Geology of New Caledonia: New Zealand Jour. Geography and Geophysics, v. 13, p. 145–183.

LINDGREN, W., 1933, Mineral deposits: New York, McGraw-Hill, 930 p.

LOCKE, A., 1926, The formation of certain ore bodies by mineralization stoping: Econ. Geology, v. 21, p. 431–453.

LOWDER, G. G., and CARMICHAEL, I. S., 1970, The volcanoes and caldera of Talasea, New

Britain: geology and petrology: Geol. Soc. America Bull, v. 81, p. 17–38.

LUSK, J., 1969, Base metal zoning in the Heath Steele B-1 orebody, New Brunswick, Canada: Econ. Geology, v. 64, p. 509–518.

———— 1972, Examination of volcanic exhalative and biogenic origins for sulphur in the stratiform massive sulfide deposits of New Brunswick: Econ. Geology, v. 67, p. 169–183.

LUYENDYK, B. P., 1970, Dips of downgoing lithospheric plates beneath island arcs: Geol. Soc. America Bull., v. 81, p. 3411–3416.

McBIRNEY, A. R., 1969, Compositional variations in Cenozoic calc-alkaline suites of Central America, in McBIRNEY, A. R., ed., Proceedings of the Andesite Conference: Upper Mantle Proj., Sci. Rept. 16, Oregon State Bull. 65, p. 185–189.

McCARTNEY, W. D., and POTTER, R. F., 1962, Mineralisation as related to structural deformation, igneous activity and sedimentation in folded geosynclines: Canadian Mining Jour., v. 83, p. 83–87.

McKENZIE, D., 1969, Speculations on the causes and consequences of plate motions: Royal Astron. Soc., Geophys. Jour., v. 18, p. 1–32.

MACNAMARA, P. M., 1968, Rock types and mineralisation at Panguna porphyry copper prospect, Upper Kaverong Valley, Boungainville Island: Australasian Inst. Mining and Metallurgy Proc., v. 228, p. 71–79.

MALAHOFF, A., 1970, Gravity and magnetic studies of the New Hebrides Island arc: New Hebrides Condominium Geol. Survey Rept., 67 p.

————, and WOOLLARD, G. P., 1969, The New Hebrides Islands' gravity network, pt. 1, Final Rept.: Honolulu, Hawaii Inst. Geophysics, Univ. Hawaii, 26 p.

MALLICK, D. I. J., 1970, South Pentecost, in MALLICK, D. I. J., ed., Annual Report of the Geological Survey for the year 1968: New Hebrides Anglo-French Condominium, p. 22–27.

MARTIN, R. F., and PIWINSKII, A. J., 1972, Magmatism and tectonic setting: Jour. Geophys. Research, v. 77, p. 4966–4975.

MATSUDA, T., and UYEDA, S., 1971, On the Pacific-type orogeny and its model-extension of the paired belts concept and possible origin of marginal seas: Tectonophysics, v. 11, p. 5–27.

MEYERHOFF, A. A., and HATTEN, C. W., 1968, Diapiric structures in Central Cuba: Am. Assoc. Petroleum Geologists Mem. 8, p. 315–357.

MITCHELL, A. H. G., 1966, Geology of South Malekula: New Hebrides Condominium Geol. Survey Rept., 42 p.

———— 1970, Facies of an early Miocene volcanic arc, Malekula Island, New Hebrides: Sedimentology, v. 14, p. 201–243.

———— 1971, Geology of Northern Malekula:

New Hebrides Condominium Geol. Survey Regional Rept., 56 p.

———, and GARSON, M. S., 1972, Relationship of porphyry copper and circum-Pacific tin deposits to palaeo-Benioff zones: Inst. Mining and Metallurgy Trans., sec. B, v. 81, p. B10–B25.

———, and READING, H. G., 1971, Evolution of island arcs: Jour. Geology, v. 79, p. 253–284.

———, and WARDEN, A. J., 1971, Geological evolution of the New Hebrides Island arc: Geol. Soc. London Jour., v. 127, p. 501–529.

MIYASHIRO, A., 1961, Evolution of metamorphic belts: Jour. Petrology, v. 2, p. 277–331.

——— 1972, Metamorphism and related magmatism in plate tectonics: Am. Jour. Sci., v. 272, p. 629–656.

———; SHIDO, F.; and EWING, M., 1971, Metamorphism in the Mid-Atlantic Ridge near 24° and 30°N: Royal Soc. (London) Philos. Trans., v. A268, p. 589–603.

MOISEYEV, A. N., 1971, A non-magmatic source for mercury ore deposits: Econ. Geology, v. 66, p. 591–601.

MOORE, J. G., 1965, Petrology of deep-sea basalt near Hawaii: Am. Jour. Sci., v. 263, p. 40–52.

——— 1971, Relationship between subsidence and volcanic load, Hawaii: Bull. volcanol., v. 4, p. 562–576.

———, and FISKE, R. S., 1969, Volcanic substructure inferred from dredge samples and ocean-bottom photographs, Hawaii: Geol. Soc. America Bull., v. 80, p. 1191–1202.

MURAKAMI, N., 1970, An example of the mechanism of emplacement of the Chugoku Batholith—the Kuga Granites, southwest Japan: Pacific Geology, v. 3, p. 45–56.

MURPHY, R. W., 1972, The Manila Trench—West Taiwan foldbelt: a flipped subduction zone (Abs.): Regional Conf. Geology Southeast Asia, 1st, Geol. Soc. Malaysia.

OLIVER, J., and ISAACS, B., 1967, Deep earthquake zones, anomalous structures in the upper mantle, and the lithosphere: Jour. Geophys. Research, v. 72, p. 4259–4275.

OSMASTON, M. F., 1971, Genesis of ocean ridge median valleys and continental rift valleys: Tectonophysics, v. 11, p. 387–405.

OXBURGH, E. R., 1972, Flake tectonics and continental collision: Nature, v. 239, p. 202–204.

———, and TURCOTTE, D. L., 1971, Origin of paired metamorphic belts and crustal relation in island arc regions: Jour. Geophys. Research, v. 76, p. 1315–1327.

PARK, C. F., and MACDIARMID, R. A., 1964, Ore deposits: San Francisco, Freeman, 475 p.

PEARCE, J. A., and CANN, J. R., 1971, Ophiolite origin investigated by discriminant analysis using Ti, Zr and Y: Earth and Planetary Sci. Letters, v. 12, p. 339–349.

PEREIRA, J., and DIXON, C. J., 1971, Mineralisation and plate tectonics: Mineralium Deposita, v. 6, p. 404–405.

PERRY, V. D., 1961, The significance of mineralised breccia pipes: Mining Eng., v. 13, p. 367–376.

P. T. INTERNATIONAL NICKEL INDONESIA, 1972, Laterite deposits in the southeast arm of Sulawesi (Abs): Regional Conf. Geology Southeast Asia, 1st, Geol. Soc. Malaysia, p. 32.

ROBINSON, G. P., 1969, The geology of North Santo: New Hebrides Geol. Survey Rept., 77 p.

RUB, M. G., 1972, The role of the gaseous phase during the formation of ore-bearing magmatic complexes: Chemical Geology, v. 10, p. 89–98.

SANGSTER, D. F., 1968, Relative sulphur isotope abundances of ancient seas and strata-bound sulphide deposits: Geol. Assoc. Canada Proc., v. 19, p. 79–91.

SARASIN, P., 1901, Entwurf einer geografischen und geologischen Beschreibung der insel Celebes: Wiesbaden.

SAWKINS, F. J., 1972, Sulfide ore deposits in relation to plate tectonics: Jour. Geology, v. 80, p. 377–397.

SCHERMERHORN, L. J. G., 1970, The deposition of volcanics and pyritite in the Iberian pyrite belt: Mineralium Deposita, v. 5, p. 273–279.

SCLATER, J. G.; HAWKINS, J. W.; MAMMERICKX, J.; and CHASE, C. G., 1972, Crustal extension between the Tonga and Lau ridges: petrological and geophysical evidence: Geol. Soc. America Bull., v. 83, p. 505–518.

SHIRAKI, K., 1971, Metamorphic basement rocks of Yap Islands, Western Pacific: possible oceanic crust beneath an island arc: Earth and Planetary Sci. Letters, v. 13, p. 167–174.

SHUNSO, ISHIHARA, 1971, Major molybdenum deposits and related granitic rocks in Japan: Geol. Survey Japan, Rept. 239, 183 p.

SIGURDSSON, H.; BROWN, G. M.; TOMBLIN, J. F.; HOLLAND, J. G.; and ARCULUS, R. J., 1973, Strongly undersaturated magmas in the Lesser Antilles island arc: Earth and Planetary Sci. Letters, v. 18, p. 285–295.

SILLITOE, R. H., 1972a, Formation of certain massive sulphide deposits at sites of sea-floor spreading: Inst. Mining and Metallurgy Trans., sec. B, v. 81, p. B141–148.

———, 1972b, A plate tectonic model for the origin of porphyry copper deposits: Econ. Geology, v. 67, p. 184–197.

SIMONS, F. S., and STRACZEK, J. A., 1958, Geology of the manganese deposits of Cuba: U.S. Geol. Survey Bull. 1057, 289 p.

SINCLAIR, I. G. L., 1967, Bauxite genesis in Jamaica: new evidence from trace element distribution: Econ. Geology, v. 62, p. 482–486.

SINCLAIR, W. D., 1971, A volcanic origin for the No. 5 zone of the Horne Mine, Noranda, Quebec: Econ. Geology, v. 66, p. 1225–1231.

SMIRNOV, V. I., 1968, The sources of ore-forming fluids: Econ. Geology, v. 63, p. 380–389.

SMITHERINGALE, W. G., 1972, Low-potash Lush's Bight tholeiites: ancient oceanic crust in Newfoundland?: Canadian Jour. Earth Sci., v. 9, p. 574–588.

STANTON, R. L., 1960, General features of the conformable "pyritic" ore bodies: Canadian Inst. Mining and Metallurgy Trans., v. 63, p. 22–36.

———, 1972, Ore petrology: New York, McGraw-Hill, 713 p.

———, and BELL, J. D., 1969, Volcanic and associated rocks of the New Georgia Group, British Solomon Islands Protectorate: Overseas Geology and Mineral Resources, v. 10, p. 113–145.

STRONG, D. F., 1972, The importance of volcanic setting for base metal exploration in Central Newfoundland (Abs.): Canadian Mining and Metallurgy Bull., v. 65, p. 45.

SUGIMURA, A.; MATSUDA, T.; CHINZBI, K.; and NAKAMURA, K., 1963, Quantitative distribution of late Cenozoic volcanic materials in Japan: Bull. volcanol., v. 26, p. 125–140.

SUGISAKI, RYUICHI, 1972. Tectonic aspects of the Andesite Line: Nature, v. 240, p. 109–111.

———; MIZUTANI, S.; ADACH, M.; HATTORI, H.; and TANAKA, T., 1971, Rifting in the Japanese late Palaeozoic geosyncline: Nature, v. 233, p. 30–31.

———, and TANAKA, T., 1971, Magma types of volcanic rocks and crustal history in the Japanese pre-Cenozoic geosynclines: Tectonophysics, v. 12, p. 393–413.

TAYLOR, D., 1972, The liberation of minor elements from rocks during plutonic igneous cycles and their subsequent concentration to form workable ores, with particular reference to copper and tin: Geol. Soc. Malaysia, 5th Presidential Address, unpub. rept.

THAYER, T. P., 1964, Principal features and origin of podiform chromite deposits and some observations on the Guleman-Soriday district, Turkey: Econ. Geology, v. 59, p. 1497–1524.

——— 1959, Alpine-type sensu strictu (ophiolitic) peridotites: refractory residues from partial melting or igneous sediments?—a contribution to the discussion of the paper: "The origin of ultramafic and ultrabasic rocks," by P. J. Wyllie: Tectonophysics, v. 7, p. 511–516.

THOMPSON, G., and MELSON, W. G., 1972, The petrology of oceanic crust across fracture zones in the Atlantic Ocean: evidence of a new kind of sea-floor spreading: Jour. Geology, v. 80, p. 526–538.

THOMPSON, J. E., 1972, Evidence for a chemical discontinuity near the basalt-"andesite" transition in many anorogenic volcanic suites: Nature, v. 236, p. 106–110.

———, and FISHER, N. H., 1955, Mineral deposits of New Guinea and Papua and their tectonic setting: Commonwealth Mining and Metallurgy Cong., 8th Proc., A.N.Z. Preprint 129, p. 59.

VILJOEN, M. J., and VILJOEN, R. P., 1969, The geochemical evolution of granitic rocks of the Baberton region, in Upper Mantle Project: Geol. Soc. South Africa Spec. Pub. 2, 189 p.

VOGT, P. R., 1972, Evidence for global synchronism in mantle plume convection and possible significance for geology: Nature, v. 240, p. 338–342.

WARDEN, A. J., 1967; Geology of the Central Islands: New Hebrides Condominium Geol. Survey Rept. No. 5, 108 p.

———, 1970, Evolution of Aoba Caldera Volcano, New Hebrides: Bull. volcanol., v. 34, p. 107–140.

WHITE, D. E.; MUFFLER, L. J. P.; and TRUESDELL, A. H., 1971, Vapor-dominated hydrothermal systems compared with hot water systems: Econ. Geology, v. 66, p. 75–97.

WOOLLARD, G. P., and MALAHOFF, A., 1966, Magnetic measurements over the Hawaii Ridge and their volcanological implications: Bull. volcanol., v. 29, p. 725–760.

WRIGHT, T. L., and FISKE, R. S., 1971, Origin of the differentiated and hybrid lavas of Kilauea Volcano, Hawaii: Jour. Petrology, v. 12, p. 1–65.

YODER, H. S., 1969, Calc-alkaline andesites: experimental data bearing on the origin of their assumed origin, in McBIRNEY, A. R., ed., Proceedings of the Andesite Conference: Upper Mantle Proj., Sci. Rept. 16, Oregon State Bull. 65, p. 77–89.

ZACHRISSON, E., 1971, The structural setting of the Stekenjokk ore bodies, Central Swedish Calendonides: Econ. Geology, v. 66, p. 641–652.

ZANS, V. A., 1954, Bauxite resources of Jamaica and their development: Colonial Geology and Mineral Resources, v. 3, p. 307–333.

9

Reprinted from *Earth Planetary Sci. Letters*, **22**, 96–109 (1974)

MESOZOIC STRUCTURAL-MAGMATIC PATTERN AND METALLOGENY OF THE WESTERN PART OF THE PACIFIC BELT

L.P. ZONENSHAIN, M.I. KUZMIN, V.I. KOVALENKO and A.J. SALTYKOVSKY

Science Research Laboratory of Geology of Foreign Countries, Moscow (USSR)
Institute of Geochemistry, Siberian Branch of Academy of Science, Irkutsk (USSR)
Institute of Physics of the Earth, Moscow (USSR)

Received January 1974

The distribution of magmatism and related endogenous metallogeny within short intervals of geological time displays strong lateral zonal pattern governed by the positions of contemporaneous eugeosynclines, i.e. previous oceanic basins. This pattern includes: (1) the eugeosyncline with ultramafics and mafics, and with Cu, Au, Cr, Pt; (2) the amagmatic back troughs filled by clastic sediments; (3) a zone of granite-granodiorite batholiths with Au, Mo; (4) a zone of diorite-monzonite with Pb–Zn; (5) a zone of standard and Li–F granites with Sn, W, Mo; and (6) a zone of alkaline plutons.

The zones in (3)–(5) correspond to calc-alkaline volcanism, and the zone in (6) to alkaline volcanism. The zonal pattern is related to the activity along fossil Benioff zones. Great transversal faults displaced structural-magmatic and metallogenic zonality far inside continents. They are interpreted as transform faults. The existance of a zonal pattern is discussed in terms of plate tectonics.

1. Objectives and methods of the investigation

The theory of plate tectonics suggests a close relation between greatly different but coeval events occurring in different parts of the globe. It has been established that extensional and compressional features are distributed in space and time according to the relative motion of lithospheric plates. The distribution of magmatic and, consequently, metallogenic products may also be strictly determined by this theory.

The approach, offered to investigate the relationship between tectonics on one hand, and magmatism and endogenic metallogeny on the other hand, is based on an analysis of both paleotectonic conditions and spatial distribution of magmatic series within the comparatively short periods of time, where each period corresponds to an "outburst" of magmatic activity, and the boundaries between them being, as a rule, marked by structural reconstructions. In these periods, which we shall call epochs, magmatic formations of extremely different types develop almost simultaneous-ly from products of so-called initial magmatism (ophiolites) in eugeosynclinal zones to various "synorogenic" and "postorogenic" granitoids and alkaline rocks in the adjacent areas. All these magmatic complexes prove to be regularly related.

The analysis offered continues Smirnov's conception [1] which recognized two zones in the Pacific Belt: an "Inner" zone with dominantly a sulphide mineralization and an "Outer" zone chiefly characteristic of rare-metal mineralization. In our studies the Smirnov's "Inner" and "Outer" zones are also distinguished for each period. The available data [2–11, 24] make it possible to outline three epochs which are significant for the whole of Eastern Asia; these are: (1) Early Mesozoic ($T - J_1$); (2) Late Mesozoic ($J_3 - K_1$); and (3) Late Cretaceous including Early Paleogene times. In the northeast of the U.S.S.R. the additional epoch coincident with the time of formation of the Okhotsk volcanic series is distinguished, i.e., uppermost Lower Cretaceous and lowermost Upper Cretaceous times.

For all the epochs mentioned paleotectonic schemes

have been compiled involving the distribution of magmatic rocks and leading endogeneous ore component. A special scheme (see Fig. 3) has been compiled for the additional epoch in the northeast of the U.S.S.R. Data for each epoch are listed in Tables 1–3.

The eugeosynclinal zones are shown in all schemes. The importance of their role in the development of orogenic belts is beyond any doubt. The eugeosynclines correspond to Smirnov's "Inner" zones for each of the epochs. They represent active axes of orogenic belts from which tectonic and magmatic activities spread to the adjacent areas of the continental blocks [25, 26]. They originated from previous oceanic basins and are apparently of composite structure. Inside them elements may be present corresponding to present-day trenches, volcanic island arcs and marginal basins. Unfortunately, a detailed palinspastic restoration of the eugeosynclinal zones, although very important in metallogeny, is at present impossible, due to inadequate study. However, this problem deserves thorough investigation.

In the areas adjacent to the eugeosynclinal zones from the continental side, mainly the distribution fields of various magmatic rocks are shown; of the structural elements only the outlines of back marine troughs (back-deeps) are indicated. These back troughs (of Mongolian–Okhotsk or West Sakhalin type) are usually filled with terrigenous flysch-like sequences. It is remarkable that they are usually, but not necessarily, devoid of coeval magmatic products. Tectonic uplifts seem to be the framework of back troughs. This is evident from the molasse accumulation of intramountain basins (the basins are not shown in the schemes). These uplifts formed the main areas of magmatic activity. All intrusive formations of given age are shown. They are subdivided into several geochemical types, each type is associated with a certain complex of ore deposits. The principle of recognition and the characterization of geochemical types of granitoids have been discussed in detail elsewhere [11, 27]. The following geochemical types of granitoids are distinguished in the schemes: (a) granite-granodiorite, (b) diorite-monzonite, (c) regular and lithium-fluoric granites, and (d) agpaitic granites and alkaline rocks.

The *granite-granodiorite* type of rock consist of mainly batholith-like bodies (for instance, the massifs of the Uda series of the Stanovoi range) of both

granodiorites and granites; the earlier phases are represented by gabbro-diorites. This rock type is characterized by a normal alkaline content and a low (not higher than clarke) content of lithofilic rare elements. Gold and molybdenum mineralizations are generally related to the granitic rocks of this type. The *diorite-monzonite* type of rock is usually represented by massifs stretched along fractures. Among them, diorites, gabbro-diorites, monzonites, and in places, adamellites, tonalites and granodiorites are most abundant. All these rocks are characterized by a higher alkaline content (earlier appearance of potash feldspar together with amphibole and even pyroxene) and a higher magnesium content in both rocks and femic minerals [28]. The diorite-monzonites from the Sikhote-Alin tin-bearing areas are representative of this type. Diorite-monzonite bodies are associated with polymetallic, sometimes tin-polymetallic and gold mineralization. The *granites of standard type*, together with the *lithium-fluoric granites*, are widespread, for instance in the Transbaikal tin–tungsten belt. Their main features are the high silicic acidity of rocks, the abundance in both two micas and tourmaline varieties and a higher content of rare alkaline and fluorine along with slightly lower Ba and Sr concentrations. These granites are associated with tin–tungsten and other rare-metal mineralization. The *alkaline* rock type is chiefly characteristic of the presence of alkaline minerals. The $K_2O + Na_2O$ content is as high as 10%. These rocks often bear rare-earth–niobium–zircon mineralization. When granitoids of different geochemical type but belonging to the same epoch, occur together, this is an indication that their emplacement took place in the order in which they are listed, the granite-granodiorite age relationship being the only uncertain fact. When the inverse relationship is the case (i.e. granodiorite intruding alkaline rocks), it may be stated that the younger intrusive rocks belong to the subsequent epoch.

The various terrestrial volcanic rocks have been distinguished into two groups: those with a calc-alkaline composition and those with an alkaline composition. The first group includes rocks from basalts to liparites of a different basicity and a relatively low alkaline content. The second group includes volcanics which, according to their composition, vary widely from trachybasalts to trachyliparites, but have a relatively high alkaline content. The alkaline content is the most sensitive factor when laterally tracing the change of the composition of volcanics.

The main endogenic ore deposits of the Pacific belt, pointed out previously by Smirnov [1], are shown in the schemes. They are tin, tungsten, molybdenum, lead–zinc, sometimes mercury and antimony; and copper, chromium and nickel in some areas of the "Inner" zone.

Tables 1–3 give the available data for the following regions: (1) northeast U.S.S.R., (2) the Mongolian Okhotskian region including North China, (3) South China, Korea and Japan, and (4) South East Asia. For all these regions are shown: (a) the eugeosynclinal zones, (b) the back troughs, (c) intrusive massifs of the following geochemical types: granite-granodiorite, diorite-monzonite, regular granits and alkaline rocks, and (4) the volcanic series of calc-alkaline and alkaline composition.

Before we analyse the structural-magmatic pattern of each epoch, it should be noted that all schemes assume that the Japan Sea, like the other marginal seas of the Western Pacific, originated due to Late Tertiary extension [29]. Therefore, the Japanese islands merged with the Asian continent, the northeastern part of Honshu Island being displaced along the Fossa–Magna fault by approximately 200 km westward with respect to southwest Honshu. This reconstruction matches the Fossa–Magna fault with the Yanshanian fault.

2. Description of the schemes

2.1. Early Mesozoic epoch $(T-J_{1-2})$ (Fig. 1, Table 1)

In the Early Mesozoic epoch four separate sectors are recognized in East Asia: (1) northeast U.S.S.R., (2) the Mongolian–Okhotskian region, (3) South China, and (4) the Malaya – Kalimantan region. Northeast U.S.S.R. falls out of the general scheme, possibly due to lack of study and the fact that the eugeosynclines of that time did not yet belong to the Pacific ring. In the remaining three sectors the eugeosynclinal zones are clearly displayed along the eastern inner flanks and the surrounding outer magmatic zones. The magmatic zonality, which is recognized according to the prevailing leading geochemical types, is presented as follows (from the east to the west): (1) the zone of granite-granodiorite batholiths with gold metallization; (2) the

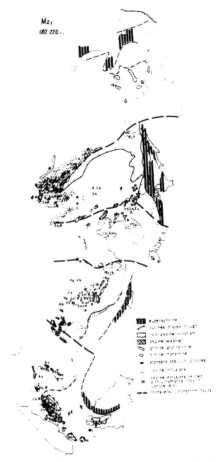

Fig. 1. Early Mesozoic structural-magmatic pattern and main ore deposits of the western part of the Pacific belt.

zone of granites of standard and Li–F-type with tin–tungsten metallization; and (3) the zone of alkaline intrusive rocks. The first two zones are characteristic of calc-alkaline volcanism; the third zone is more or less coincident with the areas of alkaline volcanics. The zonal pattern as a whole follows the strike of eugeosynclinal zones. However, along the large

TABLE 1
Main tectonic units and magmatic types of the Lower Mesozoic $(T-J_1)$ of the Western Pacific

	Northeast U.S.S.R.	Mongolian−Okhotskian region	South China, Korea and Japan	Southeast Asia
Eugeosynclinal zones	Anujian zone [12], Alazeijan plateau [13]	eastern Sikhote-Alin, west of Hokkaido (Lower Khidaka group); intersected by great transcurrent faults (Mongol−Okhotsk and Fossa-Magna)	supposed underlying Alishan trough of West Taiwan	central part of Kalimantan
Back troughs	?	Mongolian−Okhotsk trough [15]	marine terrigenous sediments of the Hong Kong region [5]	marine terrigenous sediments of the edge of Malaya Peninsula
Granite-granodiorite	not recognized	batholiths of Kyrinean granite-granodiorites in Zabajkalje and northeast Mongolia, Khessanian of North Korea [16]	Nanlin granite-granodiorite batholiths [5]	granite-granodiorite batholiths of West Kalimantan and East Malaya [18]
Diorite monzonite	distributed on large areas [14]	Amakan, Tegeduj, and early complexes, Kuden complexes in Zabajkalje	not recognized	not recognized
Standard and Li−F granites		late Kudun and Ilenguj granites in Zabajkalje [17] and granites of Central Mongolia [11]	granites of "Granite Domes" region [16]	granites of South Malaya and tin islands
Calc-alkaline vulcanism	andesites of Koni peninsula	volcanism of the Orkhon area	lower parts of the Kathasian volcanic belt	?
Alkaline intrusive rocks	?	Malokunalej and Kunalej alkaline rocks in Zabajkalje [17] Pihenganian in Korea [16]		
Alkaline volcanism		alkaline volcanism of the Selenginian belt		

transversal faults, the zonality is sharply displaced (to 1,000 km) inside the continent and at the margins it has a horseshoe-like form embracing the faults. The transversal faults also form the boundaries between the sectors. One of the main features of these faults is that they cross-cut the eugeosynclinal zones as is clearly seen in the example of the Sikhote-Alin. Such transversal faults include the Mongolian−Okhotsk, the Yanshanian (together with the Fossa−Magna fault), the Tzin-Lin and, possibly, the Black River faults.

The metallogeny is in close agreement with the distribution of intrusive rocks of appropriate types and do not depend directly upon the composition of the country rocks. For example, the tin-bearing deposits of Transbaikal region and Mongolia occur in terrigenous and crystalline rocks, while the Chinese deposits are concentrated within carbonate series. Only the type of ore deposits changes, i.e. a skarn deposit, appears instead of the quartz-cassiterite one.

2.2. Late Mesozoic epoch $(J_{2-3} - K_1)$ (Fig. 2 Table 2)

The sectors of Early Mesozoic times remain the same

Mz₂
110-150 m.

Fig. 2. Late Mesozoic structural-magmatic pattern and main ore deposits of the western part of the Pacific belt. For legend see Fig. 1.

in the Late Mesozoic. The eugeosynclinal zone (the Pengian—Anadyr zone) typical for the Pacific belt appear in the northeast of the U.S.S.R., and the magmatic zonality in the adjacent areas are subordinated to this eugeosyncline. The development of the eugeosyncline in southeast Honshu took place at that time. In this region magmatism spread over the territory between the Yanshanian and Tzin-Lin faults. The structural-magmatic zonality, as a whole, remains the same as in Early Mesozoic times, except for some complication in the Mongolian—Okhotskian sector where an additional diorite-monzonite intrusive subzone has developed along the batholith framework. The Late Mesozoic zonal pattern is displaced eastward compared to the Early Mesozoic zones. This resulted in the overlapping of inner Early Mesozoic zones by outer Late Mesozoic ones. The relationship is reversed only in Korea where Late Mesozoic granite-granodiorites intrude the earlier alkaline rocks. This may be due to the fact that just in Late Mesozoic times the adjacent eugeosyncline of southeast Honshu came into existence.

The nepheline syenites associated with ultramafics along the Main Sikhote-Alin fault stand apart [24]. They are similar to the alkaline rocks of old platforms and may be the result of processes which are beyond the scope of this paper.

2.3. Late Cretaceous–Early Paleogene epoch (Fig. 4, Table 3)

In Late Cretaceous times the structural-magmatic zonality preserves its main features but is displaced eastward. This time no appreciable influence of the transversal faults can be noticed. All the zones are directly traceable longitudinally and the zonality itself becomes linear, losing its complicated shape when long tongues extend deep into the continent (with the only possible exception in northern Vietnam and the Yunnan province). At the same time, the magmatic area narrows sharply and, as a result, in places granitoids of standard type coincide with diorite-monzonite granitoids.

3. Structural-magmatic and metallogenic zonal pattern

All the periods considered display the same structural-magmatic zonal pattern which is subordinate, ac-

TABLE 2

Main tectonic units and magmatic types of the Upper Mesozoic (J_2-K_1) of the Western Pacific

	Northeast U.S.S.R.	Mongolian–Okhotskian region	South China, Korea and Japan
Eugeosynclinal zones	Pengian-Anadyr [8] and South Anujian zones [9]	East Sakhalin, Central Hokkaido	Southeast Honshu and East Taiwan
Back troughs	Olei and Kolymian zones [9]	Sikhote-Alin and Udian trough	median zone of Honshu
Granite-granodiorite	Kolymian batholiths and early granitoids in Chuckchee peninsula [9]	Udian batholiths of the Stanovic range [21]	Dsoezu and Abukuma granites of North Honshu [22], Tanchkhonian granites of Korea [16]
Diorite-monzonite	not recognized	Acatuj, Amudjikano-Sretensk and Shakhtama complexes in Zabajkalje [17]	not recognized
Standard and Li–F granites	peripheral Kolymian granites similar to the Okhandjin massif [9]	Kukulbei and Gudjir granites in Zabajkalje [17] rare-metal granites in northeast Mongolia [11]	late Tanchkhon granites in Korea [16] coast granites of southeast China [6].
Calc-alkaline volcanism	Kolymian volcanic belt	volcanism of the Khingan, northeast Mongolia and eastern Zabajkalje	Kathasian volcanic belt in southeast China, Kimmou group of Honshu
Alkaline intrusive rocks	Negolakh massif [9]	alkaline rocks of the Aldan area, northeast Mongolia and upper Amur area [11, 21].	alkaline rocks of Laonin, Shansi and Peking areas [5].
Alkaline volcanism	liparites of Sarychev range [20]	alcaline volcanism of Djida and Onon regions	?

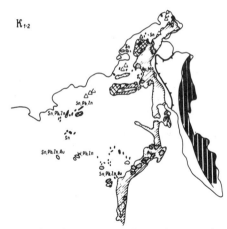

K_{1-2}

Fig. 3. Middle Cretaceous structural-magmatic pattern and main ore deposits of the northeast of the U.S.S.R. For legend see Fig. 1.

cording to its setting, to the contemporaneous eugeosynclinal zones and transversal faults. This pattern involves the following (Fig. 5): (1) the eugeosynclinal zone with spilite-basalt volcanic rocks, ultramafics and gabbro-plagiogranites; (2) the back troughs with terrigenous and frequently turbidite sedimentation and limited calc-alkaline volcanism.

The zones of intensive terrestrial calc-alkaline volcanism (including volcanic belts) and granitoid magmatism which is subdivided into: (3) the inner zone of granite-granodiorite batholiths, and (4) the outer zone of small bodies of rare-metal granites (of Li–F and standard types) and rocks of diorite-monzonite series; (5) the zone of alkaline effusive and intrusive magmatism.

The zones are generally elongated parallel to the eugeosyncline zone replacing each other in the order in which they are listed. The linear arrangement is distorted near the largest transversal faults which bring the zonality deep inside the continent. This is clearly displayed in the case of the Mongolian–Okhotsk fault. Early and Late Mesozoic magmatism is displaced along

TABLE 3

Main tectonic units and magmatic types of the Upper Cretaceous–Lower Paleogene of the Western Pacific

	Northeast U.S.S.R.	Mongolian–Okhotskian region	South China, Korea and Japan	Southeast Asia
Eugeosynclinal zones	East Kamchatka	East Hokkaido	Shimanto zone, Philippines island	Arakan
Back troughs	Penjian trough	West Sakhalinian trough	Alishan trough (West Taiwan)	
Granite-grano-diorite	Eropol granite-grano-diorites [9]	Amirian and coast granite-granodiorites [21]	Kitikami granites	batholiths of Burma
Diorite-monzonite-	early Omsukchan rocks [9]	early Maochan rocks [21]	not recognized	
Standard and Li–F granites	late Omsukchan granites	late Maochan granites	Hiroshima granites (Japan), granites of South Junnnan and North Vietnam	Malayan tin-bearing granitic belt
Calc-alkaline volcanism	Okhotskian–Chukkchian volcanic belt (Evenian series)	Sikhote-Alinian volcanic belt	calc-alkaline volcanics of Honshu	
Alkaline intrusive rocks	Omolonian complex	some small plutons of Djugdjur range	Amnokanian complex in Korea [23]	
Alkaline volcanism	alkaline liparites assossiated with the Omolonian complex, trachybasalts of Alazean plateau [41]			

this fault by as much as 1,500 – 2,000 km inward Asia and the zonality loses its linearity; it becomes concentric and spreads over the wide area. In the band, which is affected by the Mongolian–Okhotsk fault, a type of "core" composed of granite-grano-diorite batholiths and an area of dispersed magmatic rocks surrounding the core from outside are distinguished. The latter displays more or less clearly: (a) the zone of diorite-monzonite intrusive rocks adjacent to the batholiths; and (b) the outer zone of granites of Li–F and standard types. The outer periphery is formed by a horseshoe-like strip of alkaline intrusive rocks.

The metallogenic zonality of each epoch conforms with the distribution of magmatic rocks. The following zones are recognized (from inner to outer areas).

(1) The zone of copper sulphide, chrome–nickel

and platinum and often gold and mineralization; this zone is typical for the eugeosyncline and is associated with spilite-basalt volcanism as well as with gabbro-ultramafic and gabbro-plagiogranite complexes.

(2) The gold–molybdenum zone coincident with both intrusive rocks of granite-granodiorite type and volcanic belts represented by calc-alkaline volcanics.

(3) The polymetallic zone related with diorite-monzonite intrusive rocks. It is displayed in the areas of "dispersed" magmatism along the transversal faults. In other places this zone coincides frequently with the adjacent zones and this results in complex gold-bearing polymetallic ore subzones.

(4) The tin–tungsten zone associated with granites of standard and Li–F types, the latter also contains another rare-metal (tantalum–niobium, lithium, etc.) mineralization.

(5) The rare-earth niobium–zircon zone related with alkaline intrusive rocks.

K_2-P_1
$60-90$

Fig. 4. Late Cretaceous–Early Paleogene structural-magmatic pattern and main ore deposits of the western part of the Pacific belt. For legend see Fig. 1.

Toward the outer zone there are telethermal deposits of the mercury–antimony group which often also contain gold, arsenic and polymetals. It should be emphasized that the boundaries between the zones are recognized statistically and therefore they should not be understood as clearly marked lines. Both intrusive rocks of mixed geochemical composition and mixed complex ore deposits occur on the zone boundaries.

As is mentioned above, the zonality is restored within the limits of the separate epochs only. The younger zonality overlaps the older one. In East Asia the zonality is shifted eastwards with time, thus the younger outer zones superimpose, as a rule, the earlier inner zones. However, inverse relationships are possible, as for example in Korea, where Triassic alkaline rocks are cut by Upper Jurassic granite-granodiorites. These relationships are associated with the appearance of new eugeosynclinal zones and, consequently, with the inclusion of additional areas into magmatic areas.

All mentioned above shows that Mesozoic magmatism and metallogeny of Eastern Asia should not be considered as a whole as we have here the result of the repeated superimposing magmatism and metallogeny of several epochs, and not a single one. Metallogenic zones drawn a manner that involve the informations of different age are mostly artificial. At the same time, the analysis of magmatism and metallogeny within the separate epochs enables the recognition of the remarkably strict order in which magmatism and metallogeny change with increasing distance from the contemporaneous eugeosynclinal zones. This approach of metallogenic analysis evidently represents the basis of mineral prospecting.

The distribution of structural magmatic and metallogenic zones over the region is only governed by the eugeosynclines and transversal faults and is, in fact, independent of the composition and structure of country rocks. The zones may overlap the old platforms (the Siberian and the Chinese Platforms), the Paleozoic orogenic belts (the Central Asiatic belt), the middle massifs (the Omolonsk, the Indonisian massifs), the earlier eugeosynclines (Sikhote-Alin) the areas composed of thick clastic sequences (Verkhoyansk Mountains), the carbonate cover of the platform (southern China), etc. The type

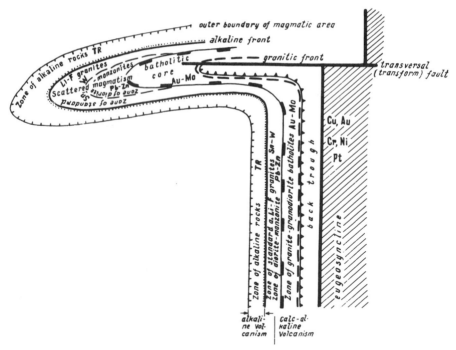

Fig. 5. General distribution of structural-magmatic and metallogenic zonality in a frame of eugeosynclines and along transversal faults.

of both magmatic products and ore deposits is always only defined by the position of a given zone with respect to the contemporaneous eugeosyncline. This suggests that the flow of juvenile (mantle) material plays an important role in the formation of magmatic rocks and associated ore elements. At the same time, the environment defines, to some extent, the compositional features of intrusive rocks as well as the formational type of ore deposits. Quartz-vein and greisen types tin- and tungsten-bearing deposits occur usually in terrigenous sequences, whereas skarn deposits are in the carbonate sequences. In a similar manner, the structural features govern in places the granitoid emplacement and determine the location of ore knots.

The regular structural-magmatic zonal distribution under consideration implies a causal relationship be-

tween the terrestrial magmatism and the phenomena occurring within the eugeosynclinal zones. This enables to properly relate the geosynclinal process with the phenomenon of so-called tectono-magmatic activation (or revivification). The tectono-magmatic activation represents a part of the entire geosynclinal process which is fully displayed in the eugeosyncline zones. Recently it became clear that the similar structural-magmatic zonality is not unique, but was peculiar only to the Mesozoics of East Asia. Similar structural-magmatic pattern subordinated to the previous eugeosynclines (oceanic basins) is restored, for instance, for the Middle Paleozoic (Devonian) of the Central Asiatic belt [15, 26]. The replacement of calc alkaline volcanics by alkaline ones with the distance from the eugeosyncline was also shown by Mossakovsky [30] for the Late Paleozoic volcanic belt of Eurasia. Consequently, we may speak of the general regularity inherent to all geosyncline systems and the structural-magmatic zonality itself appears universal to some extent. This zonality has, of course, some individual features due to the peculiarity of certain epochs and regions and, these features need to be recognized and examined. However, the change of granite-granodiorite batholiths by granite and alkaline intrusions as well as the change of calc-alkaline volcanism by alkaline volcanism remains invariable with increasing distance from the eugeosyncline.

4. Discussion and possible interpretations

When analyzing the pattern of structural-magmatic zones the following points should be emphasized: (1) the interrelation between different events which occur within the eugeosynclinal zones and their framework; (2) the appearance of terrestrial magmatism under the conditions of strongly dissected topography; (3) the nature of transversal faults; and (4) the causes of zonal distribution of magmatism. Possible interpretations naturally cannot go beyond the scope of the most favourable hypotheses.

The interrelation of events occurring simultaneously in the eugeosynclinal zones and the adjacent areas of the continental blocks is likely to be best explained with the existence of a single deep-seated active zone belonging to the type of present-day Benioff zones or to that of the tectonosphere after Sheinmann [31] which is dipping under the continent. Conjugation and subordination of the calc-alkaline volcanic belts with eugeosynclinal ophiolite zones displayed clearly in Mesozoic times of East Asia are known to be evidence for the existence of fossil Benioff zones. Therefore, the presence of such fossil Benioff zones appears to be most evident in this conception. The Benioff zone out crops in the trench, i.e. in the eugeosyncline, and all other stuctural-magmatic zones are arranged above it (Fig. 6). The shifting of the Benioff zone in any direction causes the displacement of the whole zonality in the same direction.

Considering the fact that the Benioff zones play an important role in recent plate tectonics, it is desirable to explain the observed picture from the standpoint of interaction between the moving lithosphere plates, as was done to interpret the metallog-

enetic zonality of the eastern Pacific [32—33]. According to the theory of plate tectonics, the oceanic lithosphere slab sinks along the Benioff zone and undergoes friction, heating and partial melting at depth; the resulting melts elevate to give the whole range of calc-alkaline rocks. This is accompanied by sialic masses accumulating in rear parts, and melting of these masses to yield granitic masses. However, this rather simple and harmonious explanation encounters some difficulties and requires still some corrections.

An extremely large amount of magmatic products originate above the Benioff zone. Their abundance in acidic and intermediate varieties, enriched in potassium and other lithophilic elements including ore components, is not compatible with their origin from partial melting of the oceanic plate. This is pointed out by many geologists. Consequently, along the Benioff zone there should occur an additional upflow of material not associated with subduction. This upflow is supposed to be rich in lithophilic components which might have provided for the formation of rocks above the Benioff zone. The necessity of such an upflow, which moves in the direction opposite to the subduction, is a new and important addition to the mechanism of plate tectonics. This also follows from some other data. The recent Benioff zones are a composite part of the island-arc systems, the marginal oceanic basins being one of the most important members of these systems. These basins are characteristic of a high heat flow which does not agree with the subduction concept. Moreover, as has recently been established [33], the marginal oceanic basins have extensional, not compressional, features as they should have had according to the orthodox variant of plate tectonics. They were originated by sea-floor spreading which is analogous (but evidently not identical) to the spreading in mid-oceanic ridges. It has been shown that eugeosynclines contain remnants of oceanic crust of the geological past [34]; with great reason they may be compared to both the earlier island-arc systems and the marginal oceanic basins. The eugeosynclines also originated due to extension, and old spreading processes are reconstructed within them [26]. Consequently, the analogy with recent marginal basins may be considered right. This implies that high heat flow was generated during the development of the eugeosynclines.

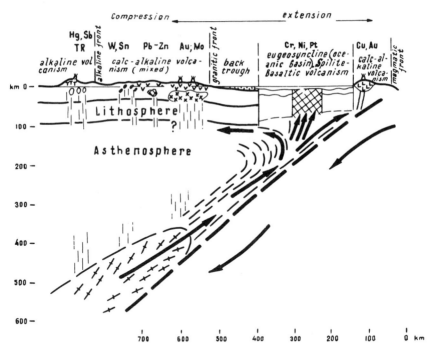

Fig. 6. Structural-magmatic and metallogenic pattern related to the activity of the Benioff zone (after Zonenshain [26] with some corrections). Elongated crests indicate light lithophilic substance, which rises up along Benioff zone. Arrows show the movement of energy and substance. Dashed lines correspond to supposed thermal and, possible, hydrothermal flows. Oblique crests indicate granite-granodiorite batholiths, the thin cross-hatching indicates diorite-monzonite, black corresponds to granites of standard and Li–F types. Oceanic lithosphere, newly created by sea-floor spreading, is thick cross-hatched.

The extension in eugeosynclines and the oceanic basin formation through sea-floor spreading had to be compensated, according to the principles of plate tectonics, by compression in other places. Traces of such compressions are expressed by intensive deformations which occurred during the entire Mesozoic. The unconformities within the Mesozoic sequences seems to reflect certain periods of highly intensive spreading in the eugeosynclines. This suggests that due to the eugeosyncline spreading, the adjacent continental plate shifted slightly aside and underwent a strong warping. This may explain the formation of back troughs as downwarp and surrounding uplifts as upwarp features. The general fracturing

facilitated the supply of magma-forming fluids. The accumulation of sialic mass might promote melting of eutectic granites. The deformation dies with distance giving way to quiet tectonic conditions that favour the intrusion of alkaline rocks.

Plate tectonics may be helpful in the attempts to explain the nature of the transversal faults. These largest faults appear to intersect the entire lithosphere. At the same time, they bring the zonality inside the continent without breaking its continuity. Only the eugeosynclines – the zones of spreading and new-creation of the crust – are offset by the transversal faults. This suggests a certain analogy with the mid-oceanic ridges where transversal faults of a transform nature are clearly developed [35].

TABLE 4

Analysis of Late Cretaceous volcanics of the Sikhoto-Alin belt[*]

Location	K_{55}	K_{60}	h (km)	d (km)
Eastern part of the Sikhote-Alin belt	1.43	–	160	400
Central part	1.69	–	180	450
Western part	2	2.50	220	525
	–	2.67	230	525

[*] K_{55} = the K_2O content (%) at 55% of SiO_2; K_{60} = the same for 60% of SiO_2; h = the depth to the Benioff zone (after Hatherton and Dickinson [37]). d = the distance to the Benioff-zone exposure (to previous trench).

The transversal faults of East Asia are interpreted as transform faults extending inside the continent. They are not, of course, associated with mid-oceanic ridges but with the previous systems of island arcs and marginal oceanic basins. It is not incidental that these faults separate the sectors, each sector developing independency and being possibly associated with the activity of the Benioff zone involved. Separate segments of lithosphere plates were also sliding along these faults. These faults were evidently the best channels to transfer the material and energy of the earth's interior to the surface.

The main problem is to explain the magmatic zonality itself. Its main feature is a gradual increase of the alkaline content from normal in granite-granodiorites through higher in diorite-monzonites and to the highest in alkaline rocks. The alkaline content also increases in volcanic rocks.

The increase of alkaline content coincides with that of the potassium content in magmatic rocks as the distance from the continental margin increases. Vistelius et al. [36] demonstrated this feature by example of Late Mesozoic granites of northeastern Asia (the potassium content in contemporaneous granites increases from 2.7 to 4.5%). Hatherton and Dickinson found that recent island-arc lavas displayed a sympathetic relationship between the potassium content of different rocks (according to their silica content), and the depth to the Benioff zone or the distance to the Benioff-zone exposure [37]: the

greater the distance, the higher is the potassium content. The same was shown by our calculations for Late Cretaceous volcanics of the Sikhoto-Alin belt, chemically analysed by Popkova et al. [38] (Table 4.)

The data mentioned above show that the K-activity in the magma-forming fluids increases with distance from the eugeosynclinal zones (continental margins).

There may be at least two explanations for this regular variation in the composition of magmatic rocks. It may be thought that fluids of different composition were simultaneously separated from the main upflow at different depths, the fluids branched at a greater depth being of higher alkaline (potassium) potential and highly saturated with lithophilic element. However, it may also be assumed that the upflow travelling upward along the Benioff zone reached the lower boundary of the lithosphere and spread under it, at a depth of about 100 km. A portion of the flow cropped out near the Benioff-zone exposure and contributed to the formation of a eugeosynclinal oceanic basin, the other portion flowed subhorizontally under the continental plate and was responsible for the entire continental magmatism. This horizontal branch may be considered, to some extent, to be analogous with transmagmatic solutions from Korzhinsky [39]. It is evident that the composition of these solutions has to vary, as they move inwards continents, due to the precipitation of a number of components (which go upwards) and changes in the temperature regime. The alkaline content as a whole has to increase with distance from the eugeosyncline. Undoubtedly, there are also other mechanisms worth discussing.

5. Conclusion

(1) The structural-magmatic and metallogenic zonality of the East Asian margin restored for short intervals of the geological past is closely associated with the contemporaneous eugeosynclinal zones. The zonality remains uniform for all the epochs and appears as follows: (a) the eugeosynclinal zone with spilite-basaltic volcanism, ultramatics and gabbro-granodiorite intrusive rocks, and with gold-bearing copper sulphide, chrome-nickel, and platium metallization; (b) the back trough with terrigenous sedimentation and limited calc-alkaline volcanism.

Zones of intensive terrestrial volcanism and granitic magmatism which are subdivided into: (c) the inner zone characterized by the development of granite-granodiorite batholiths with gold–molybdenum mineralization; and (d) the outer zone of rare-metal granites of standard and Li--F type with associated tin–tungsten and other rare-metal deposits; rocks of diorite-monzonite composition associated mainly with polymetallic mineralization are also concentrated here; (e) the zone of alkaline effusive and intrusive magmatic rocks with characteristic rare-metal manifestations.

(2) Continental magmatism (tectono-magmatic reactivation) and the development of eugeosynclinal zones are closely interrelated representing different forms of the single geosynclinal process caused probably by the existence of the Benioff zones dipping under the continent.

(3) Metallogenic analysis should be, consequently, carried out within the limits of the short intervals corresponding to the active periods of the given eugeosynclinal zones.

The structural-magmatic and metallogenic zonality recognized for different Mesozoic epochs of East Asia is peculiar to not only Mesozoic times. Therefore, the regularities considered are evidently of general importance.

Acknowledgements

Materials concerning magmatism and metallogeny of South East Asia were presented to the authors by Drs. J.G. Gatinskii, I.V. Vinogradov and A.V. Mishina. They have also taken part in the compilation of the paleotectonic schemes. Dr. L.M. Natapov helped us very much in compiling the schemes for the northeast of the U.S.S.R. Drafts were drawn by V.A. Novikova. We would like to thank all the persons mentioned above.

References

1 S.S. Smirnov, About the Pacific ore-bearing belt, Izvest. Akad. Nauk U.S.S.R. Ser. Geol. 2 (1946) 3–17.
2 S.B. Brandt, V.N. Smirnov, V.I. Kovalenko, and M.I. Kyzmin, Geologic and isotopic interpretation of isotopic age values (Abstracts), 24 Sess. Comiss. Isotopic Age, Moscow (1969) 42–44.
3 M.S. Markov, Meso-Cenozoic History and Crustal Structure of the Ohotsk Region (Nauka, Moscow, 1967) 222.
4 Geological Development of the Japan Islands (Mir, Moscow, 1968) 1–352.
5 The Main Features of the Tectonics of China (Gosgeoltehizdat, Moscow, 1962) 1–524.
6 Geology of Tin Occurrences of the Foreign Countries (Nedra, Moscow, 1969) 1–327.
7 N.V. Babkin and A.A. Sidorov, Gold–silver occurrences in the northeast of U.S.S.R., Razved. Okehrana nedr, M 10 (1972) 15–27.
8 G.E. Nekrasov, The place of the hyperbasites, basic effusives and radiolarites in the history of the development of the peninsula Tajganos and Penginsk Mountains, Geotectonics M5 (1971) 37–44.
9 Geology of the U.S.S.R., Vol. XXX (1970) Part 1: 1–547, Part 2: 1–434.
10 M.S. Nagibina, Tectonic and magmatism of the Mongol–Ohotskian belt, Akad. Nauk U.S.S.R., Moscow (1963) 1–462.
11 V.I. Kovalenko, M.I. Kuzmin, L.P. Zonenshain, M.S. Nagibina, A.S. Pavlenko, N.V. Vladykin, C. Ceden, C. Goondsambu and A.V. Goreglyad, Rare-Metal Granitoids of Mongolia, (Nauka, Moscow, 1971) 1–230.
12 S.M. Tilman, Tectonics of the Mesozoides of the northern part of the Pacific belt, Novosibirsk (1970) 1–40.
13 N.A. Shilo and V.M. Merslyakov, Eugeosynclinal zones of the central part of Mesozoides in the northeast of the U.S.S.R. Dokl. Akad. Nauk U.S.S.R., 204, 5 (1972) 1202–1204.
14 L.V. Firsov, Occurrences of Triassic magmatism in Verkhojan-Chukotka region, Izvest. Akad. Nauk U.S.S.R., Ser. Geol. 8 (1962) 38–45.
15 L.P. Zonenshain, The Geosynclinal Theory and its Application to the Central Asiatic Orogenic Belt, (Nedra, Moscow, 1972) 1–243.
16 Geology of Korea (Nedra, Moscow, 1964) 1–261.
17 The Intrusive Complexes of the Transbaikalic (Gosgeoltehizdat, Moscow, 1964), 1–214.
18 C.S. Hutchison, Invalidity of the Billiton granite, Indonesia, for defining the Jurassic – Upper Triassic boundary in the Thai–Malayan Orogen, Geol. Mynb. 47, 1 (1968) 28–37.
19 K.B. Seslavinsky, The structure and development of the South-Anuyi suture trough (Western Chukotka), Geotectonics 5 (1970) 56–68.
20 K.N. Rudich, The Volcano-Plutonic Formations of the Chersky Mountains, (Nauka, Moscow, 1966) 1–124.
21 Geology of the U.S.S.R., Vol. XIX (Nedra, Moscow, 1966) Part 1: 1–560.
22 Geology and Mineral Resources of Japan (1961).
23 V.K. Putincev and S.E. Sinitsky, A brief description of the geology of the northeastern part of KNDR, in: The Materials of the Regional Geology and Metallogeny of the Korea Peninsula and Mongolia, Leningrad (1963) 33–57.
24 Geology of U.S.S.R., Vol. XXXII (Nedra, Moscow, 1969) 1–695.
25 L.P. Zonenshain, Geosynclinal process and the new global tectonics, Geotectonics 6 (1971) 3–26.

26 L.P. Zonenshain, The evolution of Central Asiatic geosynclines through sea-floor spreading, Tectonophysics 19
(1973) 213–232.

27 M.I. Kuzmin, The geochemical types of the Mesozoic
granitoids of the western part of the Mongolian–
Ohotsky belt, Proc. Geochem. Congr., (Nauka, Moscow,
1971) 11.

28 P.V. Koval, M.I. Kuzmin, V.S. Antipin, M.N. Zaharov,
E.B. Znamensky, G.S. Gormasheva and S.A. Yurchenko,
The composition of the biotites of the granitoids of
the Eastern Trans-baikalian, Geochemistry 8 (1972)
15–26.

29 P.N. Kropotkin and K.A. Shahvarsfova, The geological
feature of the Pacific belt, (Nauka, Moscow, 1965)
1–365.

30 A.A. Mossakovsky, About the Upper Paleozoic volcanic
belt of Europe and Asia, Geotectonics 4 (1970) 9–25.

31 Ju.M. Sheinmann, Problems of Deep-Seated Geology,
(Nedra, Moscow, 1968) 1–231.

32 R.H. Sillitoe, Relation of metal provinces in Western
America and the subduction of oceanic lithosphere,
Bull. Geol. Am. 83 (1972) 813–818.

33 D.E. Karig, Origin and development of marginal basins
in the Western Pacific, J. Geophys. Res. 76 (1971)
2542–2561.

34 A.V. Peive, Oceanic crust of the geologic past, Geotectonics 4 (1969), 5–23.

35 J.T. Wilson, A new class of faults and their bearing
on continental drift, Nature 207, Nr. 4995 (1965)
343–347.

36 A.B. Vistelius and A.I. Aralina, The general law of
the distributions of potassium in the post-Jurassic
granitoids of the northeast of Asia and adjacent part
of the Pacific, Dokl. Akad. Nauk U.S.S.R., 184,
2 (1969) 441–443.

37 T. Hatherton and W.R. Dickinson, The relationship
between andesitic volcanism and seismicity in
Indonesia, the Lesser Antilles and other island arcs,
J. Geophys. Res. 74 (1969) 5301–5310.

38 T.I. Popkova and E.F. Kaidolova, Book of the chemical
analyses of the rocks of the southern part of the Far
East Khabarovsk, DVJU (1961) 1–57.

39 D.S. Korzhinsky, The flows of the transmagmatic solutions and the processes of the granitization, in: Magmatism, Formations of the Crystalline Rocks and
Depths of the Earth, 1, (Nauka, Moscow, 1972) 144–
152.

40 V.I. Smirnow, The Notes of the Metallogeny, (Gosgeoltehizdat, Moscow, 1963) 1–163.

41 B.L. Flerov, L.H. Indolev, J.V. Jakovlev and B.J. Bigus,
Geology and Genesis of the Deposits of Jakutia. (Nauka,
1971) 1–315.

10

Reprinted from *Jour. Geol. Soc. Australia*, **20**, Pt. 4, 405–426 (Dec. 1973)

A PLATE TECTONIC MODEL OF THE PALAEOZOIC TECTONIC HISTORY OF NEW SOUTH WALES

by ERWIN SCHEIBNER

(With 16 Figures)

(*Received May 1972; revised MS received April 1973*)

ABSTRACT

An updated* tectonic model for the Palaeozoic tectonic history of New South Wales, based on actualistic models of plate tectonics, has resulted from tectonic analyses and syntheses during the compilation of the Tectonic Map of New South Wales.

Most emphasis is given to marginal seas, which characterize Pacific marginal mobile zones. Marginal seas form in the regime of lithospheric tension under the influence of retrograde motion of the Benioff zone, and during the formation of a new Benioff zone, or during formation of spheno- and rhombochasms. Additional mechanisms for obduction zone formation are suggested: the development of an obduction zone at the originally coupled continental margin is important. Three basic types of volcanic chains are distinguished: volcanic arcs, volcanic arches, and volcanic rifts, in the formation of which both primary subduction zones at major plate margins, and secondary subduction zones which develop in marginal seas may be concerned.

During the Palaeozoic the eastern part of the Australian continent evolved in a complicated process of continental crust accretion. By interaction of the Australian plate (Proterozoic continent) with the neighbouring oceanic, Palaeo-Pacific plate, systems of marginal seas, volcanic chains, island arcs, microcontinents, flysch wedges or trench complexes, and primary and secondary Benioff zones were formed, and also deformed, episodically. The most active role during the structural deformation was played by the continental plate moving towards the adjacent primary Benioff zone.

INTRODUCTION

During the recent study of the tectonics of New South Wales in connexion with the compilation of the State Tectonic Map the following topics have been considered:

1. Tectonic mapping and the problems of presentation of tectonic units based on plate tectonics (Scheibner, 1972a).
2. The plate tectonics theory, the original form of which should be modified (Scheibner, 1972a, b); only a few relevant points are mentioned below.
3. Study of present structures; only a brief review is given below.
4. Interpretation of the tectonic history of New South Wales from the plate tectonic standpoint†. A brief updated version of a plate tectonic model (Scheibner, 1972c) of the Palaeozoic tectonic development of New South Wales is presented below. This

is based on new data and reinterpretation of some older geological data.

REVIEW OF THE STRUCTURE AND GEOLOGY OF NEW SOUTH WALES

Structural development in general is governs the differentiation of the mantle; this determined by a basic irreversible trend which trend is the accretion of continental crust and lithosphere. The ultimate product is a continental craton. Cratons are formed by the process of cratonization, which is episodic and not cyclic. During cratonization, new continental crust is accreted by igneous activity, sedimentation, metamorphism, and structural deformation, and four types of tectonic province can be distinguished: *oceanic, pre-cratonic, transitional,* and *epi-cratonic* (Scheibner, 1972a, and in prep.). Rock bodies observable on the surface of the earth can be classified as being

---

* The first version of this model has been published in the Records of the 24th International Geological Congress (Scheibner, 1972). New data have been considered in this paper and also in the palinspastic maps. A detailed discussion of the tectonic history will be published in the Explanatory Notes to the Tectonic Map of New South Wales. The citation of publications has been kept to a minimum in the present paper to restrict its size.

† Independently, several workers have recently presented plate tectonic models for eastern Australia.

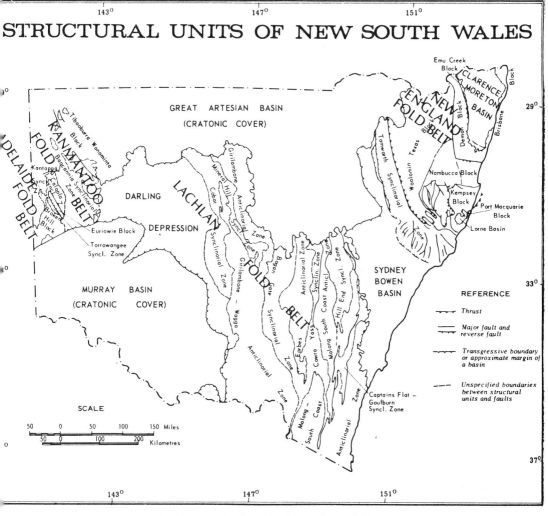

Fig. 1. Structural units of New South Wales.

formed in a certain tectonic province. Rocks of pre-cratonic provinces were formed in marginal mobile zones. Rocks of transitional provinces were formed in the environment of terminal cratonization, and are characterized by platformal facies, post-kinematic intrusives, rift-valley or volcanic rift volcanism during initiation, and generally weak deformation (examples are the Alpine Molasse, and in New South Wales the Upper Devonian Lambie Group, cf. Doutch, 1972). Rocks of epi-cratonic provinces represent platforms, or covers of cratons (the outcropping part of the craton is a shield).

The rock bodies which belong to oceanic, pre-cratonic, and transitional tectonic prov-

inces accumulate at plate margins, and through structural deformation they build *fold belts* (a preferred English term synonymous with orogenic belts). Rock bodies of, epi-cratonic provinces occurring within continental plates are also sometimes deformed into fold belts, especially during the deformation of adjacent marginal mobile zones (Scheibner, 1972a).

Structurally the area of New South Wales is not very complicated (Fig. 1) and is composed of four fold belts: the Adelaide Fold Belt, the Kanmantoo Fold Belt, the Lachlan Fold Belt, and the New England Fold Belt. The last three form the composite Palaeozoic Tasman Fold Belt System. Between the

Lachlan and New England Fold Belts is the Sydney-Bowen Basin, and within the New England Fold Belt the Lorne and Clarence-Moreton Basins. The fold belts are covered extensively by cratonic cover in the Great Australian (Artesian) Basin and the Murray Basin.

New South Wales is composed of rock bodies belonging to four pre-cratonic, three transitional, and two epi-cratonic provinces (Figs. 2 and 3).

The Gawler-Willyama Pre-Cratonic Province of early Proterozoic age forms anticlinorial basement blocks or internal massifs (Willyama Complex) in the Adelaide Fold Belt, and anticlinorial blocks (Wonominta Complex) in the Kanmantoo Fold Belt. The Adelaide Epi-Cratonic Province of Late Proterozoic to Early Cambrian age, which developed above the Gawler-Willyama Pre-Cratonic Province, was deformed into the Adelaide Fold Belt during the Delamerian Orogeny (Thomson, 1970). Only a very small part of this fold belt occurs in New South Wales. Rock bodies of the Kanmantoo Pre-Cratonic Province of Cambrian age were deformed during the Delamerian Orogeny into the Kanmantoo Fold Belt. This fold belt as it exists at present is composite, and contains elements of the Gnalta Transitional Province of Late Cambrian and Ordovician age, and also elements of the Lambian Transitional Province of Late Devonian to early Carboniferous age deformed in the Kanimblan Orogeny. The Lachlan Pre-Cratonic Province started in Early Ordovician time, although some older elements (the Girilambone Beds) were involved. It ended in the Carboniferous; however, the complexes west of the present Sydney Basin were cratonized during the Tabberabberan Orogeny to such an extent that the Lambian Transitional Province developed above them. During the Kanimblan Orogeny the Lachlan Fold Belt started to form, and as a new craton it was attached to the already existing Australian Craton. Above the Lachlan Fold Belt in its eastern sector the Newcastle Transitional Province began as an intra-deep. In lowermost Permian the New England Pre-Cratonic Province started. Elements of the Lachlan Fold Belt were separated and remobilized, and a new marginal mobile zone formed. After intensive and rapid development, cratonization occurred during the Hunter Orogeny,

after which the Newcastle Transitional Province spread over the deformed mobile zone. Cratonization was completed and the New England Fold Belt formed during the Bowen Orogeny.

The inland epi-cratonic (platformal) basins were initiated during the Permian (Permian basins under the Tertiary Murray Epicontinental Basin, and in the area of the Great Australian (Artesian) Basin), and they represent the start of the Tasman Epi-Cratonic Province, which is practically still in existence.

Tectonic provinces overlap in space and time, as is illustrated on the palinspastic table (Fig. 3).

SOME BASIC ARGUMENTS USED IN THE RECONSTRUCTION OF THE TECTONIC HISTORY OF NEW SOUTH WALES

At the outset of the study the plate tectonics theory (McKenzie, 1969) was applied; but it became apparent that the theory had to be modified (Scheibner, 1972a, b).

The plate tectonics theory is widely known, and points relevant to the Palaeozoic tectonic history of New South Wales have recently been discussed (Scheibner, 1972a, b); therefore, only some basic arguments will be mentioned. Some of these arguments are hypothetical and speculative, and they should be considered as models used to try to explain past and present features.

1. Plates are rigid and they deform at their margins; however, rigidity has limitations. If the motion of plates exceeds the limit which could be accommodated at their margins, either the major plates break down into smaller ones, or they deform internally. During the evolution of plates it is possible to define both relatively large plates (megaplates) and small plates (subplates) (McKenzie, 1970b).

2. The internal deformation of continental plates is expressed in the formation, and sometimes also in the deformation, of intracratonic mobile zones (aulacogenes), rift valleys, and basins (syneclises), which commonly have extended or thinned crust and upwelled upper mantle. With the internal deformation (extension) of continental plates, extrusion of plateau basalts and cratonic igneous activity are commonly coupled.

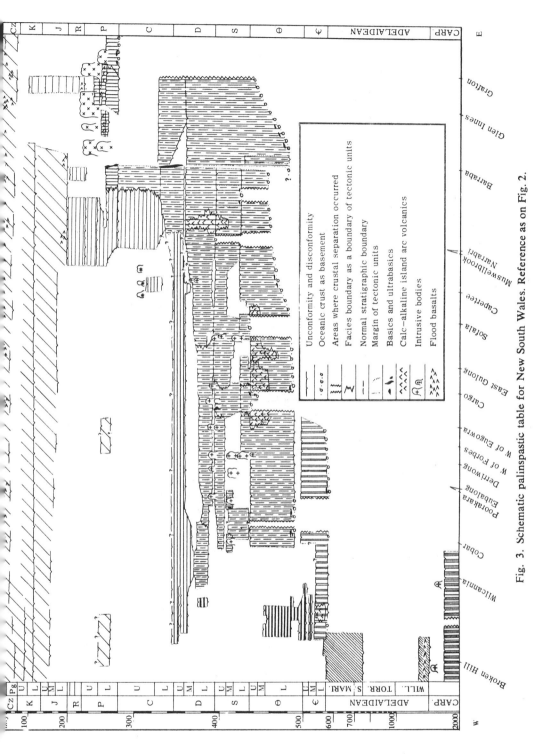

Fig. 3. Schematic palinspastic table for New South Wales. Reference as on Fig. 2.

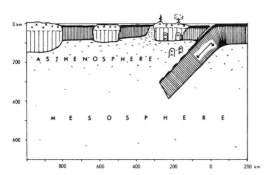

Fig. 4. Model of formation and growth of marginal seas under the tensional influence of retrograde motion of the Benioff zone (Elsasser, 1971). Reference as on Fig. 7.

3. The earth's surface can be subdivided into mobile areas with high rates of seismicity, volcanism, plutonism, and structural deformation, and cratonic areas partly or entirely lacking mobility. The mobility is caused by uneven distribution of crustal/mantle properties (i.e. areas with contrasting lithosphere — oceanic, continental, or transitional type — occur close together causing gravitational imbalance) and by interaction of plates.

4. Accretion of continental-type crust takes place most effectively at plate margins in marginal mobile zones. Two types of marginal mobile zones develop, depending on the type of interacting plates. Mediterranean marginal mobile zones form during the interaction of continental plates; pacific marginal mobile zones form during interaction of continental and oceanic plates.

5. Crustal or lithospheric separation can occur practically anywhere, but preferentially along inherited zones of weakness, if tensile stresses are exerted. Depending on the character of tensile stresses, continental crust is extended or thinned, or in extreme cases new basic crust is formed between the separated plates. Such crustal separation is the cause of mobilization and remobilization.

6. Depending on the speed and amount of lithospheric separation, different types of new crust and lithosphere are generated from the upwelled asthenosphere. In rift valleys and intracratonic mobile zones (aulacogenes) basic magmas are derived at depth (within or under the continental lithosphere); therefore, they are alkaline and peralkaline (Green, 1971). At centres of sea-floor spreading

abyssal tholeiites are derived at shallow depth (Kay et al., 1970).

7. The oceanic lithosphere of major oceans is formed by axial sea-floor spreading at the margins of diverging megaplates (true mid-oceanic ridge). The oceanic lithosphere of small ocean basins is formed at the margins of diverging subplates by axial or non-axial (i.e., diffused) sea-floor spreading (cf. Dewey & Bird, 1971).

8. Marginal seas characterize pacific mobile zones. Marginal seas (the active form is named inter-arc basin) form in a regime of lithospheric tension (Karig, 1970, 1971; Packham & Falvey, 1971) under the influence of retrograde motion of the Benioff zone (Fig. 4), during formation of a new Benioff zone (Fig. 5), and during formation of sphenochasms and rhombochasms (Carey, 1955, 1958; Rodolpho, 1969).

9. From the analyses of the Palaeozoic tectonic history of New South Wales it follows that when a Benioff zone terminates, possibly by bottoming of the mesosphere (Le Pichon, 1968) (Fig. 5a), the new trench is formed a few hundred kilometres oceanward. It can be speculated that a major synclinal bend subsequently develops in the oceanic plate oceanward of the old Benioff zone (Fig. 5b). This synclinal bend will have the form of a deep oceanic basin adjacent to the leading plate margin. Because the strongest distension develops in the centre of such a synclinal form, a new Benioff zone will develop there by shearing (Fig. 5c). This model can explain why the flysch wedge is partly accumulated above an oceanic crust, and why some volcanic island chains (arcs) form on oceanic crust.

10. During plate convergence, processes of subduction (sinking of oceanic plates), overriding or underthrusting (subduction-underthrusting or 'Verschluckung' of continental plates), and obduction may develop.

11. Convergence of plates causes not only compression, but also tension in the lithosphere. The major source of tension in the lithosphere is Benioff zones, along which the heavy oceanic lithosphere sinks (Elsasser, 1967, 1970, 1971; Laubscher, 1969; Ringwood, 1969, 1972).

12. Local tension and local compression can develop contemporaneously in marginal mobile zones. For example, on the leading edge of a subplate (microcontinent) rotating towards

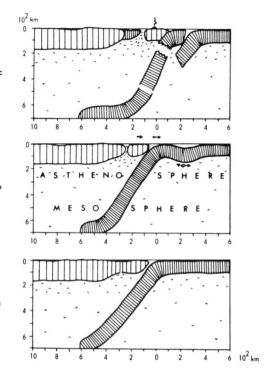

Fig. 5. Model of formation of a new Benioff zone and its causal influence on the formation of marginal seas. a. Termination of a Benioff zone by bottoming of the meso-sphere (Le Pichon, 1968). b. Formation of a synclinal bend in the oceanic plate; arrows show tension in centre of bend. c. A new Benioff zone forms, and under tension (dilatation) a marginal sea (inter-arc basin) develops. The volcanic arc develops on the margin of the microcontinent or on the oceanic crust. Closely spaced lines denote oceanic lithosphere, more widely spaced lines show continental lithosphere.

The Coolac Serpentinite Belt in New South Wales can be taken as an example (Fig. 6). Along an easterly dipping thrust plane a secondary Silurian subduction zone, which formed in a marginal sea, was upthrust and turned over. Upside-down facing of upper mantle and oceanic crust (layers 2 and 3) can be observed east of Coolac.

Another mechanism of obduction zone formation is by upthrust of oceanic crust and upper mantle at an originally coupled continental margin. This means that during the process of decoupling not only subduction (Dewey, 1969), but also obduction can develop, depending on the relative movement of involved plates. Active movement of the continental plate will cause obduction. Subduction (sinking of oceanic lithosphere) develops probably as a function of aging of oceanic lithosphere leading to gravitational imbalance (Ringwood, 1969). The oceanic lithosphere can be obducted or upthrust by the adjacent continental plate as a result of differential rate of movement of involved plates.

One particular type of obduction can develop if the oceanic basement of the flysch wedge is coupled to the continental plate, and a Benioff zone develops in the oceanic plate (cf. the model of new Benioff zone formation, Fig. 5). As the Benioff zone retreats ocean-ward (Elsasser, 1971), the adjacent leading plate follows sympathetically. If the leading plate moves faster than the sinking plate, then the leading plate will encounter the closest

the Benioff zone, sediments of the flysch wedge are deformed in a regime of compressional stresses, while on the trailing edge of the subplate, in the inter-arc basin, tensile stresses exist.

13. An obduction zone (Coleman, 1971) can be defined as an oceanic plate thrust over or underthrust by a continental plate (New Caledonia), or another oceanic plate (Macquarie Ridge). However, obduction zones defined as oceanic crust and upper mantle material lying on continental crust can form by other mechanisms than those mentioned in the literature. One of these involves upthrust of a terminated Benioff zone, and this is typical of the terminal stages of orogeny.

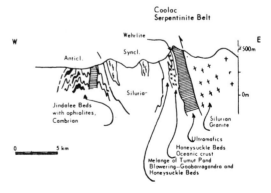

Fig. 6. Schematic geological section across the Coolac Serpentinite Belt, New South Wales, which represents an upthrust oceanic lithosphere.

oceanic crust (i.e. the basement of the flysch wedge), and at the interface of continental and oceanic lithosphere or geosuture a thrust or an obduction zone will develop. The Peel Thrust in New South Wales can be taken as an example (Scheibner & Glen, 1972).

14. Orogenic volcanism and plutonism are connected with plate subduction (Kuno, 1966; Dickinson, 1968, 1971; Hamilton, 1969; and others). Volcanic chains are situated 150 to 400 km from the axis of the trench, depending on the dip of the Benioff zone (Hamilton, 1969; and others). The potassium content of volcanic chain lavas increases away from the trench; in other words, it is possible to establish the polarity of volcanic chains (Kuno, 1966; Dickinson, 1968, 1971).

15. Depending on the tectonic realm or on the character of the basement, orogenic volcanism and plutonism vary. Three basic types of *volcanic chains* or volcanic tectonic units occur in marginal mobile zones of the pacific type: *volcanic arcs* (volcanic island arcs, volcanic rises, and ridges), which accrete on oceanic crustal basement and seldom on the margins of microcontinents; *volcanic arches*, which are closely connected with orogenic plutonism and accrete on continental crustal basement of microcontinents, earlier volcanic arcs, or continental margins; and *volcanic rifts*, which develop on large continental islands (microcontinents) or on continental margins in a setting of lithospheric tension similar to that of inter-arc basins (Karig, 1970). Volcanic arcs are characterized by island-arc volcanism (Jakes & White, 1972), volcanic arches by 'andean-type' calc-alkaline volcanism (Jakes & White, 1972) and by orogenic plutonism (Hamilton, 1969), and volcanic rifts by calc-alkaline volcanism and widespread ignimbrites.

16. The model of orogenic granite derivation as proposed by Hamilton (1969, 1970) accounts for the majority of orogenic granites in New South Wales. However, because many orogenic granites and coupled intermediate and acid volcanics (volcanic arches) were formed too far from the primary Benioff zone at the megaplate margin (over 400 km from the trench), it is necessary to suggest that these granites and volcanic arches were derived by subduction at secondary subduction zones which developed in marginal seas. Orogenic granites in New South Wales only

exceptionally have formed in troughs; most are situated in anticlinorial zones (Figs. 1 and 2).

17. Deposits of marginal seas are characterized by high-temperature and low-pressure metamorphism (Packham & Falvey, 1971), pelagic and flysch facies, and the presence of basic crust (ophiolites). Marginal sea complexes are bounded and surrounded by older structural elements.

18. Flysch wedges or trench complexes are characterized by the presence of diachronous oceanic sediments lying on, interbedded with, and sometimes intruded by ophiolites (basic submarine volcanics, basic intrusives, and ultrabasics), and by Mn oxides, rhodonite, stratiform sulphide deposits of exhalative syngenetic type, etc. Higher in the section volcanic or terrestrially derived flysch facies occur. Slump structures, repeated or continuous deformation, isoclinal and recumbent folds, refolded folds, and burial metamorphism are typical. Rare slices of limestone and other shallow-water sediments, possibly representing sedimentary capping of guyots, occur in the flysch wedge. Typical mélanges and products of 'blue-schist' metamorphism can be observed when the deeper parts of the trench complex are uplifted.

PALAEOZOIC TECTONIC HISTORY OF NEW SOUTH WALES

The Australian Proterozoic craton, covered to a large extent by epicontinental sediments, probably formed part of a larger Australian-Antarctic plate. The geological data from the Proterozoic complexes in eastern Australia lend support to a conclusion that the continental margin was coupled at that time, i.e. the oceanic crust of the hypothetical Pacific Ocean was coupled to the continental one.

The nature of Lower Cambrian rock complexes, the newly formed deep troughs with ophiolites, and volcanic arc volcanicity, indicate that in Early Cambrian time the Palaeo-Pacific oceanic lithosphere was decoupled from the continental lithosphere. The decoupling led to the formation of a trench, which created intensive tensile stresses, and therefore break-up has occurred at the newly established margin of the Australian-Antarctic plate. Small plates (microcontinents) were separated from the Australian-Antarctic plate. These small plates were the Wonominta Complex (Fig.

116

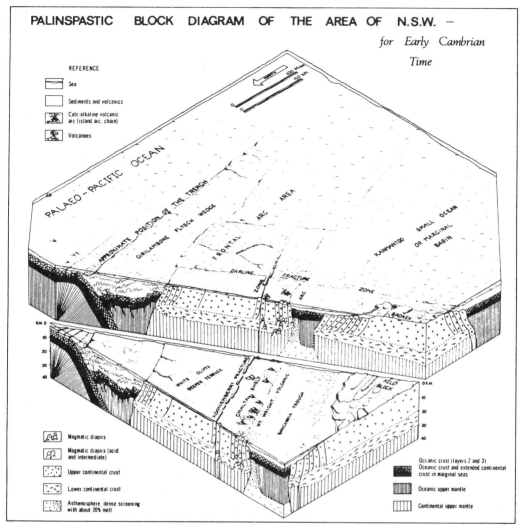

Fig. 7. Palinspastic block diagram of the area of New South Wales for the Early Cambrian; from
Scheibner (1972d).

7*) in New South Wales, the Proterozoic elements of the Ravenswood-Lolworth Block in Queensland (Heidecker, 1972), the Proterozoic complexes cropping out in the Central Highlands in Tasmania, and the possible Proterozoic complexes on the west coast of the South Island of New Zealand. Between the microcontinents and the major plate, marginal seas originated having oceanic and/or extended continental crust: these are, in South Australia, New South Wales, and western Victoria, the Kanmantoo Trough and the Bancannia Trough, its northern continuation; the Dundas Trough was the marginal sea in Tasmania; and the Cape River Beds and Charters Towers Metamorphics formed in a marginal sea in Queensland (Heidecker, 1972). These marginal seas were inter-arc basins in Early and Middle Cambrian time, when on the western side of the microcontinents volcanic island arcs originated: a volcanic arc in New Zealand, the Mount Read

---

* During the production of the palinspastic maps (Figs 7 to 16) the original minimal pre-deformational
  width of tectonic units was deduced from structural information, maximal distance of volcanic chains
  from the axis of the trench, sense of strike-slip displacements, and other considerations (Scheibner,
  in prep.).

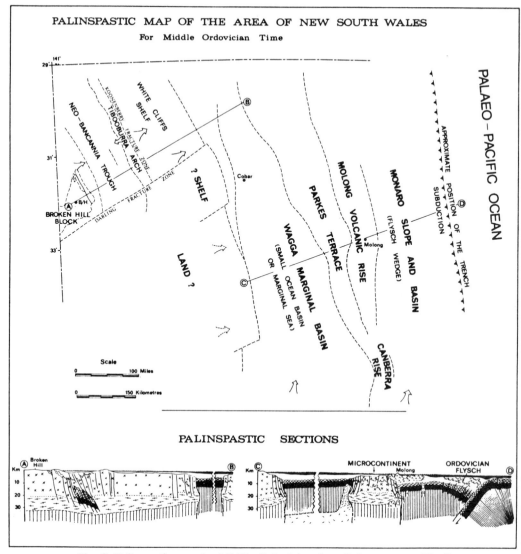

Fig. 8. Palinspastic map for Middle Ordovician time. Reference as on Fig. 7.

Volcanic Arch in Tasmania, the Mount Wright Volcanic Arc in New South Wales (Fig. 7), and the Argentine Metamorphics in Queensland (Heidecker, 1972). Above the Wonominta Block (microcontinent) in New South Wales lay the Gnalta Shelf, and east of it, separated by the Koonenberry Fracture Zone, the White Cliffs Deeper Terrace, where flysch-like sediments (the Copper Mine Range Beds) accumulated (Fig. 7). The Proterozoic basement continued for some distance to the east, and at the leading edge of the microcontinent the flysch wedge, i.e., a trench complex (the Girilambone Beds), was collected. These beds lie unconformably under the Ordovician sediments, have interbedded basic volcanics, and are extremely deformed (polyclinal folding, several episodes of deformation) and metamorphosed. Some basics and ultrabasics show intrusive and tectonic relationships, resembling recent situations in oceanic areas of slow spreading.

The Cambrian tectonic units at the Australian plate margin formed the Kanmantoo (-Tyennan) Pre-Cratonic Province, which in Victoria is represented by the Heathcote Zone. Several more fracture zones than those shown in Figure 7 probably existed in this tectonic

118

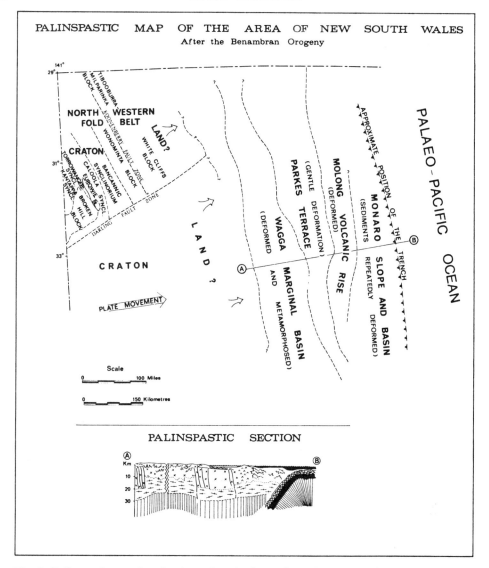

Fig. 9. Palinspastic map for the time after the Benambran Orogeny. Reference as on Fig. 7.

province. The presence of the mentioned tectonic units helps to position the old core of New Zealand for Cambrian time either south of Tasmania, or elsewhere in the Kanmantoo Pre-Cratonic Province.

The structural information from New South Wales and other states of eastern Australia indicates that most of the structural deformation has been caused by the eastward rotation of the Australian plate. The main plate collided episodically with the micro-continents and accreted volcanic arcs at its margin (Scheibner, 1972c, and in prep.).

During the Middle to Late Cambrian and Early Ordovician the Delamerian-Tyennan Orogeny occurred. The Kanmantoo Pre-Cratonic Province (a marginal mobile zone) was deformed and cratonized, as well as the Proterozoic to Early Cambrian Adelaide Aulacogene — an intracratonic mobile zone. The Adelaide Aulacogene developed within the Australian-Antarctic cratonic plate and represents part of the cratonic cover (the Adelaide Epi-Cratonic Province). Kanmantoo granites intruded not only the rocks of the Kanmantoo Pre-Cratonic Province, but also the neighbouring Adelaide Epi-Cratonic

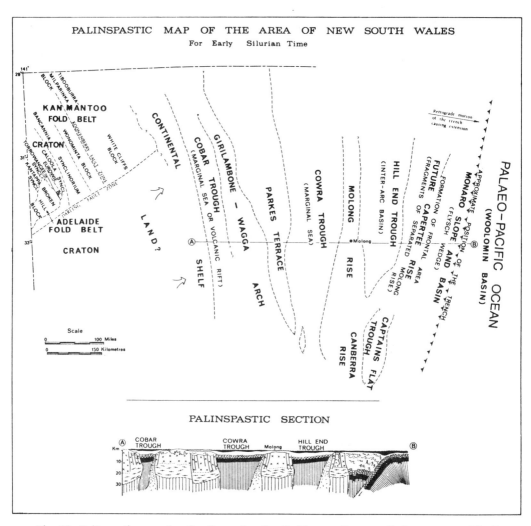

Fig. 10. Palinspastic map for the time after the Quidongan Orogeny. Reference as on Fig. 7.

Province and its basement. These granites were connected with a primary subduction zone, and possibly also derived by subduction along a secondary subduction zone and by fusion of the oceanic crust of the Kanmantoo Trough. The Kanmantoo Fold Belt (or the Kanmantoo-Tyennan Fold Belt) and also the neighbouring Adelaide Fold Belt were formed during the same Delamerian-Tyennan Orogeny. In Antarctica up to now only one fold belt, the Ross Orogen (Craddock, 1970), has been distinguished in continuation of these two very different Australian fold belts. It is hoped that more detailed investigations will correct this anomaly.

The Gnalta Transitional Province developed in New South Wales above the Kanmantoo

Fold Belt (Figs. 8 and 3). On the basis of outcropping rock bodies three tectonic units can be distinguished: the Neo-Bancannia Trough, Tibooburra Arch (land), and White Cliffs Shelf.

Judging from the development of a new marginal sea, a volcanic island arc, and a new flysch wedge or trench complex during the Early Ordovician, a new Benioff zone formed. This caused the tensile stresses necessary for creation of a marginal sea. Along a zone of weakness, probably the original margin of the Proterozoic continental crust (or close to it), the Girilambone flysch wedge was separated from the newly accreted plate margin and formed a microcontinent (section in Fig. 8). The Wagga Marginal Basin (a

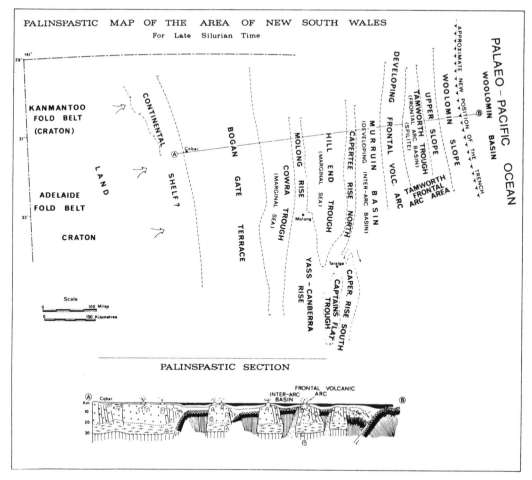

Fig. 11. Palinspastic map for the Late Silurian. Reference as on Fig. 7.

marginal sea) was formed between the microcontinent and the major plate (Fig. 8). I suggested earlier that the Mount Dijou Volcanics, a suite of basaltic pillow-lavas, represent oceanic crust of the Wagga Marginal Basin. The recently established very high alkali content (A.J.R. White, pers. commun.) suggests that the Mount Dijou Volcanics represent either products of rift-valley volcanism during the initial rifting in the Wagga Marginal Basin, or oceanic island volcanism.

On the leading edge of the microcontinent (separated Girilambone Beds) the Molong Volcanic Rise (an island arc) developed in the late Early Ordovician, with typical calc-alkaline island arc volcanism. The volcanism prograded eastward with time over the adjacent flysch wedge. Farther to the east the new flysch wedge accumulated in the Monaro

Slope and Basin. These sediments were scraped together into a trench complex. Basic volcanics are known at present only in the Wagonga Beds. During the Late Ordovician the western part of the Molong Volcanic Rise which accreted at the margin of the micro-continent has split under tension and the Initial Cowra Trough (a marginal sea) originated; however, data for this are available only in the area east of Parkes and north of Grenfell. The western segment of the Molong Volcanic Rise was left attached to the Parkes Terrace. The Benambran Orogeny at the end of the Ordovician was caused by the relative eastward rotation of the Australian plate (in the sense of present orientation). The sedimentary filling of the Wagga Marginal Basin was deformed and metamorphosed and some syngenetic anatectic granite emplaced (Brooks & Leggo, 1972). The Molong Volcanic Rise and adjacent areas to the east were also deformed, but with decreasing intensity. No evidence for the Benambran unconformity is known just west of the present Sydney Basin. Less intensive deformation occurred in the Gnalta Transitional Province (Fig. 9).

Strong extension after the Benambran Orogeny caused the Cobar Trough (a marginal sea or a volcanic rift) to form in the west, and further extension occurred in the Cowra Trough. During the later part of the Early Silurian, the Quidong or Quidongan Orogeny (Crook *et al.*, 1973) occurred. During this orogeny further high-level and synkinematic granites were emplaced (Crook *et al.*, 1973), and the Capertee Rise formed. Acid volcanism started (volcanic arches), and orogenic pluton-ism lasted throughout the Silurian and Early-Mid Devonian in the Lachlan Pre-Cratonic Province (Fig. 10).

During the late Early and early Late Silurian possibly a new Benioff zone formed; but it may have been the previous one, moved farther to the east. The uncertainty is caused by the lack of evidence to the east, where stratigraphical data are available only since the Late Silurian (cf. Figs. 11 and 2). Strong extension has widened the Cowra Trough, and a new marginal sea, the Hill End Trough, was formed after Campbells Group time, i.e. late Early Silurian. The *en echelon* equivalent of the Hill End Trough to the south was the Captains Flat Trough (a narrow marginal sea or volcanic rift). The Hill End Trough originated by a new splitting of the Molong Volcanic Rise. The eastern split-away segment formed by the Sofala and Rockley Volcanics became part of the Capertee Rise. Important from the metallogenic point of view are the peripheral zones of marginal seas, where in volcanic rifts and grabens stratiform Kuroko-type sulphide deposits have formed. Examples are the stratiform deposits on the western margin of the Hill End Synclinorial Zone (Fig. 1). The Cobar Trough and the Captains Flat Trough were transitional be-tween a marginal sea and volcanic rift setting, and in them Kuroko-type deposits formed also. The history of progressive formation of marginal seas from Late Ordovician to early Late Silurian presented here is a modification of my earlier ideas (Scheibner, 1972c), when it was suggested that they formed during one extensional event.

During the Late Silurian a new inter-arc basin can be identified east of Capertee Rise, the Murruin Basin (a marginal sea). Judging from the presence of extensive volcanic detritus in the Tamworth Synclinorial Zone a frontal volcanic arc, now concealed, developed at the end of the Silurian east of the Murruin Basin (Fig. 11). From the character of pre-Devonian sediments and presence of spilite volcanism, it can be suggested that a trough existed in the Tamworth Frontal Arc Area. This can be called the Tamworth Trough; it is different from the Tamworth Trough as described in Australian geological literature. Farther east on the Woolomin Slope and in the trench, the Woolomin Beds were collected. The Woolomin Beds as usually described represent the flysch wedge accumulated at the plate margin and scraped-off oceanic sedi-ments. The basal sections of the Woolomin Beds, which represent the Woolomin Beds *sensu stricto*, are typically oceanic. They are composed of cherts and basic submarine volcanics. Mn oxide, rhodonite, and stratiform sulphide deposits are common. The pelagic facies may be diachronous: they were scraped off progressively in the trench from the oceanic crust, which was progressively younger and carried younger sediments. The lenses (tectonic slices) of Silurian limestone and the slice of the Ordovician Trelawney Beds occur-ring in the Woolomin Beds may possibly represent sedimentary cappings of guyots and oceanic islands. These sediments are closely associated with basic volcanics, which could be parts of volcanic islands, the basement of which sank into the trench and the top parts scraped off.

During the late Late Silurian and Early Devonian (Fig. 12) the Australian plate rotated oceanward. The oceanic crust of the marginal seas was subducted and fused to give rise to orogenic plutonism, but in the terminal episodes of orogeny by upthrust of oceanic lithosphere, obduction zones have developed, such as the Coolac Serpentinite Belt (Fig. 6) and the Tumut Pond Serpentinite Belt. Shelf sedimentation reflecting the progress of cratonization spread over the areas of deformed troughs, while the Hill End Trough and the Murruin Basin further continued the flysch sedimentation of terrestrial and volcanic provenance. The frontal arc shed quantities of volcanic debris into the Tamworth Frontal Arc Area (Fig. 12).

122

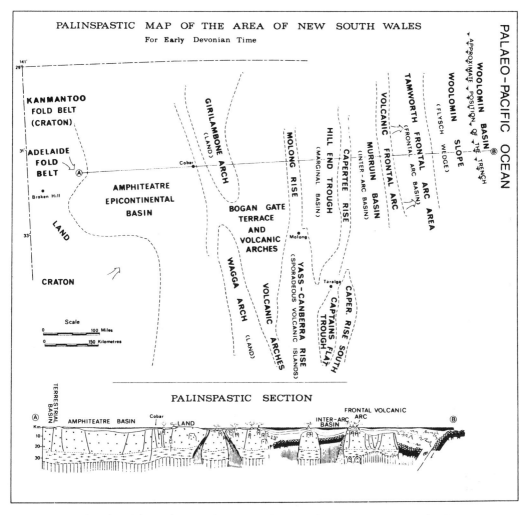

Fig. 12. Palinspastic map for the Early Devonian. Reference as on Fig. 7.

The Woolomin flysch wedge further accreted at the leading plate margin. West of the frontal volcanic arc, extensive volcanic arch volcanism and associated plutonism has developed. These centres of igneous activity were so far from the main or primary subduction zone that it is necessary to suggest that they were derived from secondary subduction zones and by fusion of the oceanic crust of marginal seas. Partial melting of newly accreted lower crust could also have contributed to the igneous activity (cf. sections in Figs. 11 and 12).

During the Middle Devonian Tabberabberan Orogeny extensive areas of previous marginal seas and rises, which formed the Lachlan Pre-Cratonic Province, were cratonized. The cratonization proceeded by structural deformation, metamorphism, and igneous activity, and was more or less continuous. During certain intervals the intensity of these processes must have increased or effects of small gradual changes accumulated into marked revolutionary events. The intensity of cratonization was less pronounced in the Tamworth Frontal Arc Area and in the Woolomin flysch wedge, which were far away from the advancing continental plate, and continued in their pre-cratonic development. Two tectonic units can be recognized: the Mandowa Unstable Shelf and the Texas Unstable Shelf and (?) Slope — a frontal area (Fig. 13). Above the

123

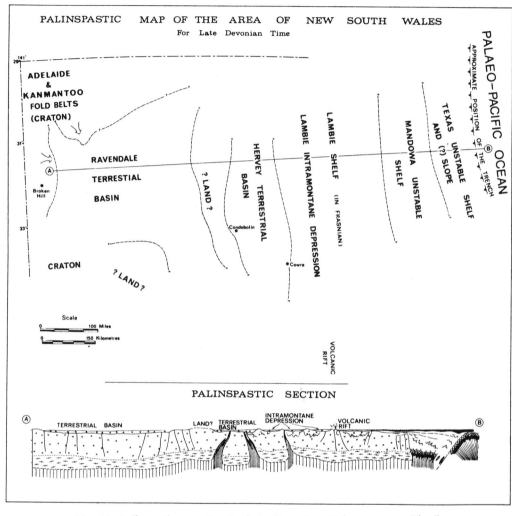

Fig. 13. Palinspastic map for the Late Devonian. Reference as on Fig. 7.

strongly cratonized western part of the Lachlan Pre-Cratonic Province the Lambian Transitional Province developed, in the eastern part of which the continental Lambian Shelf developed during Frasnian time by incursion of the sea from the east. After brief marine sedimentation the sea receded. Terrestrial sedimentation continued in terrestrial basins and intramontane depression (Fig. 13).

During the Kanimblan Orogeny the Lachlan Fold Belt was formed, and attached as a new craton to the already existing Australian craton. The Kanmantoo Fold Belt with the overlapping Lambian Transitional Province was also gently deformed. Rapid movement of the Australian plate caused the oceanic

crust and upper mantle under the Woolomin flysch wedge to be encountered and upthrust. The Peel Thrust, an obduction zone, developed by this mechanism (Figs. 13 and 14) (Scheibner & Glen, 1972). This thrust may have been initiated during the Tabberabberan Orogeny.

The discordant Bathurst granites were emplaced during the Namurian in the eastern marginal part of the Lachlan Fold Belt. Subsequently, subduction probably terminated or slowed down for some time along the Woolomin subduction zone, because volcanic activity ceased or substantially decreased in New South Wales. A landmass, the New England Arch, formed by the cratonized

Woolomin-Texas complexes, could be recognized. To the west and south of the New England Arch the Kullatine Shelf* was formed.

Farther west of the terrestrial-paralic Ayr Basin marks the beginning of the Newcastle Transitional Province, which had at that time the tectonic setting of an intra-deep. Judging from the initiation of extension in the future Sydney-Bowen Basin, a new episode of plate interaction began at the Australian plate margin after the Kanimblan Orogeny. An island arc possibly started to develop in Queensland (the lower part of the Lizzie Creek Volcanics). Large quantities of ignimbrite, alkali basalt, and, later, minor andesite (Fig. 14) in New South Wales indicate the presence of a volcanic rift. The Queensland portion of the Newcastle Transitional Province was separated from the New South Wales portion by an active fracture zone, possibly part of the Darling River Fracture Zone. This northeast-trending fracture zone enabled different intensity of crustal separation to exist in the two portions of the Newcastle Transitional Province.

Fracture zones enabling strongly contrasting areas to form can become boundaries between tectonic provinces. In this case the New South Wales portion has the characteristics of a transitional tectonic province, while the Queensland portion is more similar to a pre-cratonic province. This problem needs further study.

A renewed episode of dilatational stresses is expressed in the definitive formation of the Sydney-Bowen Basin. Also, farther west, especially in the future Murray Basin and the Great Australian (Artesian) Basin tensional stresses caused formation of epicontinental basins by thinning of the continental crust. Crustal separation at the margin of the Australian plate had cut across the older structures†, as it did during the later Mesozoic separation. This was the reason that the Nambucca Marginal Basin was in contact with two units: the Drake Shelf and the Yessabah Shelf (Fig. 15). These units were introduced because of differences in their basement and later deformation. The Drake Shelf was developed around the Texas Arch

---

* In the Kullatine Formation west of Kempsey (in the Parrabel Anticline), products of subaqueous mass-movements (turbidites, slumping) suggest a southerly palaeoslope (Lindsay, 1966). Supply of clastics was from the present northeast. About 7 percent of the clastics in the tilloid conglomerates is made up of intrusives, mostly pink granite and several varieties of orthoclase porphyry (Lindsay, 1966). Yet this granite was derived from an area where no granites of pre-Permian age are known to occur (Fig. 14).
Several alternative hypotheses can be considered:
1. There was a source of pre-Permian granite in the Late Carboniferous northeast of the Kempsey Block (Figs. 14 and 3); this area was separated during the opening of the Tasman Sea and at present forms part of the Lord Howe Rise.
2. The clastics were derived from the New England Arch (Fig. 14) and represent reworked clastics, or the Woolomin-Texas complex contains some older intrusives not yet discovered.
3. The Kempsey Block was originally so oriented that clastics were supplied from the foreland (Lachlan Fold Belt) from the south and southwest. Later, as a result of intensive block movements during the Hunter Orogeny, the Kempsey Block was rotated anticlockwise. If this happened, it should be possible to determine the rotation with the help of palaeomagnetic measurements (Idnurm & Scheibner, in prep.). However, certain difficulties are imposed by the closeness of the palaeopole.

† Recently, probable Precambrian detrital zircons have been found in Cretaceous arkosic sandstone from New Caledonia. It was suggested that the Lord Howe Rise is the likely source and that it represents a submerged continental link between New Zealand and Australia (Aronson & Tilton, 1971). On other grounds it was suggested that the Lord Howe Rise is a Proterozoic continent which collided with the Australian plate in Permian time (Solomon & Griffiths, 1972). However, geophysical evidence indicates that the crustal structure of the Lord Howe Rise is the same as that of the Australian continental crust (Shore et al., 1971); yet there are no Proterozoic elements present at its immediate eastern margin. The crust of Lord Howe Rise, some 28 km thick, (Shore et al., 1971) may represent a separated part of the cratonized Woolomin flysch wedge and also the younger Nambucca flysch wedge. The present continental margin is clearly discordant as it cuts older structures (cf. Fig. 2).
To explain the source of old zircons several hypotheses may be considered:
1. The Lord Howe Rise represents a microcontinent with Proterozoic rocks present (Aronson & Tilton, 1971; Solomon & Griffiths, 1972).
2. Proterozoic rocks are absent, and old zircons are recycled deposits.
3. The old zircons were supplied from Proterozoic complexes in Queensland, New Zealand, or Tasmania in Cretaceous time, when New Caledonia was much closer to the main plate than it is now.

125

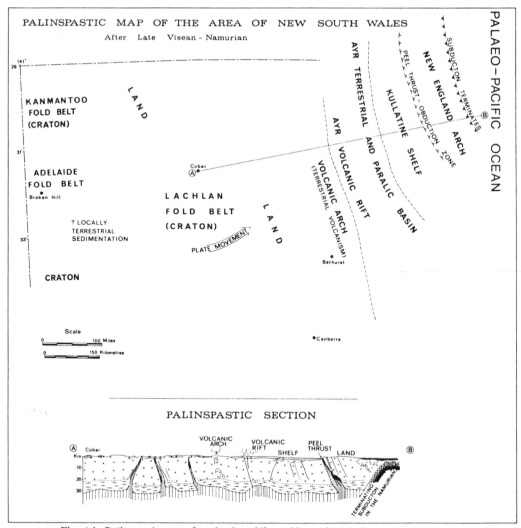

Fig. 14. Palinspastic map for the late Visean-Namurian. Reference as on Fig. 7.

above the basement formed by the Woolomin-Texas Beds, while the Yessabah Shelf developed above the basement built by the Tamworth Group. The Nambucca Beds accumulated on oceanic crust, probably in a marginal sea, and at a secondary subduction zone the Nambucca Beds were repeatedly deformed and metamorphosed in a regime of burial metamorphism. The basement of the Nambucca Beds — oceanic layers 2 and 3 — crops out south of the Crossmaglen Fault, south of Dorrigo.

The existence of pre-Permian sediments in New Caledonia favours an interpretation of the Nambucca flysch wedge as marginal sea deposits. A Benioff zone dipping towards the continent can be assumed from the presence of island arc volcanism in Queensland, and volcanic arch volcanism and later orogenic plutonism in New England, as well as from the character of the Nambucca flysch wedge. The Nambucca flysch wedge was plastered against the margin of the continental plate as it rotated eastward and relatively anti-clockwise.

During the Hunter Orogeny, mainly in the late Early and early Late Permian, the Carboniferous Peel Thrust — an obduction zone — — was displaced dextrally. The strike-slip movement amounted to about 150 km in some sectors; this amount was calculated from strike-slip displacements of the Peel Thrust on the assumption that it was formerly continuous (Scheibner, in prep.). The study of

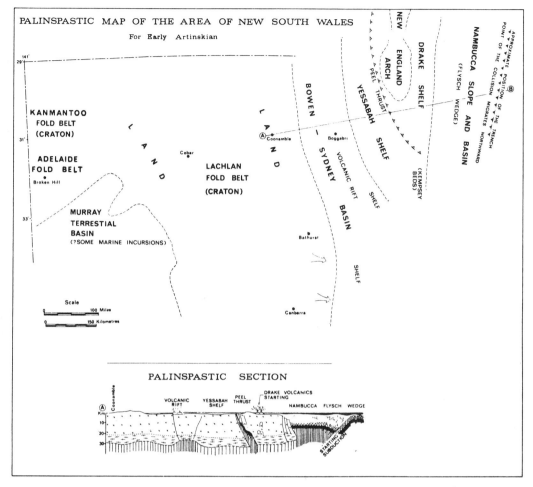

Fig. 15. Palinspastic map for the early Artinskian. Reference as on Fig. 7.

structures of the New England Fold Belt suggests large-scale horizontal displacement of crustal blocks. This can account for all structural peculiarities of this region, like the torsional stresses, radial block tectonics, and shearing. Extensive horizontal displacement has been taken up also on the Hunter-Mooki Thrust System and in the area close to it. Thrusting and folding occurred partly during sedimentation (Stuntz, 1972). Along one of the largest strike-slip faults, the Demon Fault, the Demon Block was displaced dextrally. The Nambucca Beds were caught between this and a southern (Kempsey Block) rigid block, which probably caused their final intensive deformation. South of the Demon Block a minor obduction zone was formed along the Crossmaglen Fault (Fig. 16). New thrust

movement (contemporaneously with the strike-slip movement) along the Peel Thrust theoretically amounted in some sections up to 80 km. The overriding, or underthrusting, of the western block (the present Tamworth Synclinorial Zone) and the continent-ward dipping Benioff zone account for the enormous amount of granites forming the New England Batholith and for the Drake Volcanic Arch. Most of the granites are post-kinematic and were emplaced into the Woolomin-Texas complexes; only a few were syn-kinematic and occur in the Permian complexes. (For a review of plutonism in New England Fold Belt see Leitch, 1969). The post-kinematic granites are also later than the strike-slip movement, because, with few exceptions, they intrude the strike-slip faults (Leitch, 1969).

127

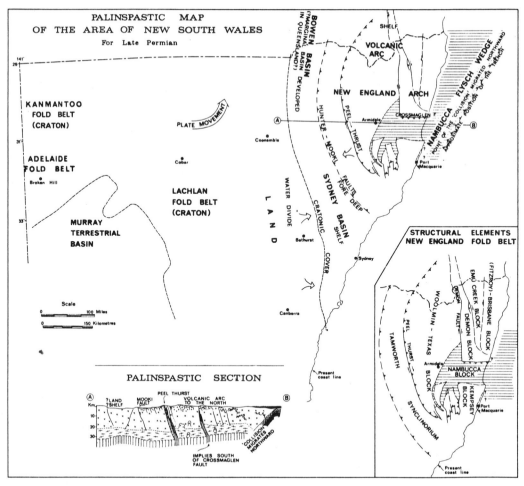

Fig. 16. Palinspastic map for the early Late Permian. Reference as on Fig. 7.

Some strike-slip movement on the Demon Fault also occurred later, as indicated by displacement of some granites.

From the distribution of granites and volcanics, we may deduce the existence of an unstable migrating plate junction in late Late Permian and Early Triassic, as the granites become younger from south to north. The structural deformation followed a similar migratory pattern from south to north in time. Differences in the intensity of orogenic deformation exist south and north of the Darling River Fracture Zone.

The Newcastle Transitional Province spread over the deformed New England Pre-Cratonic Province, and during the Bowen Orogeny the New England Fold Belt assumed its final shape.

CONCLUSIONS

The Palaeozoic tectonic history of New South Wales shows that the eastern part of the continent evolved in a complicated process of crust accretion. By interaction of the Australian plate (Proterozoic continent) with the neighbouring oceanic plate (Paleo-Pacific plate), systems of marginal seas, volcanic chains, island arcs, microcontinents, and primary and secondary Benioff zones formed episodically. Because these tectonic units were situated at major plate margins, and are characterized by strong mobility, they formed mobile zones which can be named marginal mobile zones of pacific type (Scheibner, 1972a). These marginal mobile zones were episodically formed and deformed, and the symmetry of structures indicates that the most

128

active role was played by the continental plate moving towards the adjacent Benioff zone. The structural deformation began close to the continental plate and proceeded towards the plate margin. The continent collided with the microcontinents and accreted volcanic chains at its margin. The megastructures (for example the Upper Devonian synclinoria) are often asymmetric, with the sides closer to the continent more deformed. The eastward asymmetry of episodic continental accretion is reflected in the volcanism, plutonism, and structural deformation, which becomes younger from west to east (Figs. 1 to 3).

It is worth noting that 'island arc type' and 'andean type collisions' (Mitchell & Reading, 1969; Dewey & Bird, 1970) developed contemporaneously from time to time at the margin of the Australian plate during the Palaeozoic. The island arcs were mostly associated with primary Benioff zones which formed at the margins of the major plate, while the 'andean' volcanic chains were associated with secondary Benioff zones which developed in marginal seas.

The development of marginal mobile zones is more dynamic than has been suggested in the past, and the Palaeozoic tectonic history of New South Wales supports the views about uniqueness of fold and mountain belts (Trümpy, 1960; Coney, 1970; Dickinson, 1971). During Palaeozoic time in New South Wales the plate margin was rearranged four times. These rearrangements are partly in accordance with earlier recognized episodes of orogenic activity (Packham, 1969). Four flysch wedges or trench complexes can be recognized. Episodic tension (dilatation) at the plate margin caused the formation of marginal seas, often several progressively. The marginal seas became the sites of intensive sedimentation, and during the orogeny sites of high temperature/low pressure metamorphism (Packham & Falvey, 1971). The oceanic crust of marginal seas was subducted or fused or both. The complicated array of marginal seas and intervening volcanic chains is unique, and resembles the present existing complex situations in the Southwest Pacific. Unique also is the fragmentation of the long marginal mobile zones by fracture zones, enabling different development in adjacent segments of the zones. This problem will be discussed elsewhere (Scheibner, in prep.).

The model of Palaeozoic tectonic history of New South Wales presented is based on actualistic models of plate tectonics. These models have been modified, and new ones suggested, to explain the observed data. The presented model (hypothesis) is therefore specific; it is dynamic and genetic. It shows the dynamism in development. It enables a genetic insight into the Palaeozoic tectonic history of the eastern margin of the Australian continent as it can be deciphered from the record in rock complexes. New observations, i.e. new data, will probably require modifications to the model. However, it seems to have practical value in the understanding of metallogenic processes. It is believed that, besides integration of new data, it could best be verified in metallogenic studies. Metallogenic accumulations in New South Wales seem to follow the pattern of tectonic units and their history as described here.

ACKNOWLEDGMENTS

The co-operation of all who supplied unpublished data, discussed tectonic problems, or demonstrated the geology of certain areas is gratefully acknowledged.

I wish to thank Dr James Gilluly for valuable discussions and comments on the manuscript.

I would like to thank all who participated in discussion after the presentation of this paper at the Specialist Groups Meeting of the Geological Society of Australia in Canberra in February 1972, and Miss Helena Basden, Messrs J. N. Cramsie, W. S. Chesnut, D. H. Probert, G. Rose, and Dr N. L. Markham for critically reading the manuscript.

Permission to publish this paper was given by the Director of the Geological Survey of New South Wales, Department of Mines.

REFERENCES

ARONSON, J. L., & TILTON, G. C., 1971: Probable Precambrian detrital zircons in New Caledonia and southwest Pacific continental structure. *Bull. geol. Soc. Am., 82,* pp. 3449-3456.

BROOKS, C., & LEGGO, M. D., 1972: The local chronology and regional implications of a Rb-Sr investigation of granitic rocks from Corryong District, Southeastern Australia. *J. geol. Soc. Aust., 19,* 1-19.

CAREY, S. W., 1955: The orocline concept in geotectonics, Part I. *Pap. Proc. Roy. Soc. Tasm., 89,* pp. 255-288.

CAREY, S. W., 1958: The tectonic approach to continental drift: in CAREY, S. W. (Ed.), *Continental Drift: A Symposium,* pp. 177-355 Univ. Tasmania, Hobart.

COLEMAN, R. G., 1971: Plate tectonics emplacement of upper mantle peridotites along continental edges. *J. geophys. Res., 76,* pp. 1212-1222.

CONEY, P. J., 1970: The geotectonic cycle and the new global tectonics. *Bull. geol. Soc. Am., 81,* pp. 739-748.

CRADDOCK, C., Ed., 1970: Antarctic Map Folio Series, 12: geologic maps of Antarctica. *Am. geogr. Soc., N.Y.*

CROOK, K. A. W., BEIN, J., HUGHES, R. J., & SCOTT, P. A., 1973: Ordovician and Silurian history of the southeastern part of the Lachlan Geosyncline. *J. geol. Soc. Aust., 20,* pp. 113-144.

DEWEY, J. F., 1969: Continental margins: a model for the conversion of Atlantic type to Andean type. *Earth planet. Sci. Lett., 6,* pp. 189-197.

DEWEY, J. F., & BIRD, J. M., 1970: Mountain belts and the new global tectonics. *J. geophys. Res., 75,* pp. 2625-2647.

DEWEY, J. F., & BIRD, J. M., 1971: Origin and emplacement of the ophiolite suite: Appalachian ophiolites in Newfoundland. *J. geophys. Res., 76,* pp. 3179-3206.

DEWEY, J. F., & HORSFIELD, B., 1970: Plate tectonics, orogeny and continental growth. *Nature, 225,* pp. 521-525.

DICKINSON, W. R., 1968: Circum-Pacific andesite types. *J. geophys. Res., 73,* pp. 2261-2269.

DICKINSON, W. R., 1971: Plate tectonic models for orogeny at continental margins. *Nature, 232,* pp. 41-42.

DOUTCH, H. F., 1972: The concept of transitional tectonics in Australia. *Joint Spec. Groups Mtg, Canberra, Abstr., geol. Soc. Aust.,* pp. E10-12.

ELSASSER, W. M., 1967: Convection and stress propagation in the upper mantle. *Princeton Univ. tech. Rep. 5,* 15 June 1967.

ELSASSER, W. M., 1970: The so-called fold mountains. *J. geophys. Res., 75,* pp. 1615-1618.

ELSASSER, W. M., 1971: Sea-floor spreading as thermal convection. *J. geophys. Res., 76,* pp. 1101-1112.

GREEN, D. H., 1971: Magmatic consequences of plate tectonics. *12th Pacif. Sci. Cong., Canberra, Abstr.,* p. 446.

HAMILTON, W., 1969: The volcanic central Andes — a modern model for the Cretaceous batholiths and tectonics of western North America: in McBIRNEY, A. R., Ed., Proc. of the Andesite Conference. *S. Oregon Dep. Geol. Bull. 65,* pp. 175-84.

HAMILTON, 1970: Tectonic map of Indonesia — a progress report. *U.S. geol. Surv. open file Rep.* (unpubl.), 29 p.

HEIDECKER, E., 1972: Evolution of the Ravenswood-Lolworth Block: influence upon Devonian tectonism in northeastern Queensland. *Joint Spec. Groups Mtg, Canberra, Abstr., geol. Soc. Aust.,* pp. F7-11.

JAKES, P., & WHITE, A. J. R., 1972: Major and trace element abundances in volcanic rocks of orogenic areas. *Bull. geol. Soc. Am., 83,* pp. 29-40.

KARIG, D. E., 1970: Ridges and basins of the Tonga-Kermadec island arc system. *J. geophys. Res., 75,* pp. 239-254.

KARIG, D. E., 1971: Origin and development of marginal basins in the western Pacific. *J. geophys. Res., 76,* pp. 2542-2561.

KAY, R., HUBBARD, N. J., & GAST, P. W., 1970: Chemical characteristics and origin of oceanic ridge volcanic rocks. *J. geophys. Res., 75,* pp. 1585-1613.

KUNO, H., 1966: Lateral variations of basalt magma type across continental margins and island arcs. *Bull. volcanol., 29,* pp. 195-222.

LAUBSCHER, H., 1969: Mountain building. *Tectonophysics, 7,* pp. 551-563.

LEITCH, E. C., 1969: Igneous activity and diastrophism in the Permian of New South Wales. *Geol. Soc. Aust. spec. Publ., 2,* pp. 21-37.

LE PICHON, X., 1968: Sea-floor spreading and continental drift. *J. geophys. Res., 73,* pp. 3661-3697.

LINDSAY, J. F., 1966: Carboniferous subaqueous mass-movement in the Manning-Macleay Basin, Kempsey, New South Wales. *J. sediment. Petrol., 36,* pp. 719-732.

McKENZIE, D. P., 1969: Speculation on the consequence and cause of plate motion. *Geophys. J. R. astr. Soc., 18,* pp. 1-32.

McKENZIE, D. P., 1970a: Plate tectonics and continental drift. *Endeavour, 29,* pp. 39-44.

McKENZIE, D. P., 1970b: Plate tectonics of the Mediterranean region. *Nature, 226,* pp. 239-243.

MITCHELL, A. H., & READING, H. G., 1969: Continental margins, geosynclines and ocean floor spreading. *J. Geol., 77,* pp. 629-646.

PACKHAM, G. H., Ed., 1969: Geology of New South Wales. *J. geol. Soc. Aust., 16(1)*, pp. 1-654.

PACKHAM, G. H., & FALVEY, D. A., 1971: An hypothesis for the formation of marginal seas in the Western Pacific. *Tectonophysics, 11*, pp. 79-109.

RINGWOOD, A. E., 1969: Composition and evolution of the upper mantle: *in* HART, P. J., Ed., The Earth's Crust and Upper Mantle. *Am. geophys. Un., geophys. Monogr. 13*, pp. 1-17.

RINGWOOD, A. E., 1972: Phase transformation and mantle dynamics. *Earth planet. Sci. Lett., 14*, pp. 233-241.

RODOLFO, K. S., 1969: Bathymetry and marine geology of the Andaman Basin, and tectonic implications for Southeast Asia. *Bull. geol. Soc. Am., 80*, pp. 1203-1230.

SCHEIBNER, E., 1972a: Tectonic concepts and tectonic mapping. *Rec. geol. Surv. N.S.W., 14*, pp. 37-87.

SCHEIBNER, E., 1972b: To reject or accept and modify the theory of plate tectonics? *Tecton. Struct. Newsletter, geol. Soc. Aust., 1*, pp. 2-13.

SCHEIBNER, E., 1972c: Actualistic models in tectonic mapping. *24th int. geol. Cong., Sec. 3*, pp. 405-422.

SCHEIBNER, E., 1972d: The Kanmantoo Pre-Cratonic Province in New South Wales. *Quart. Notes geol. Surv. N.S.W., 7*, pp. 1-10.

SCHEIBNER, E., & GLEN, R. A., 1972: The Peel Thrust and its tectonic history (New South Wales). *Quart. Notes geol. Surv. N.S.W., 8*, pp. 2-14.

SHORE, G. G., Jr, KIRK, H. K., & MENARD, H. W., 1971: Crustal structure of the Melanesian area. *J. geophys. Res., 76*, pp. 2562-2568.

SOLOMON, M., & GRIFFITHS, J. R., 1972: Tectonic evolution of the Tasman Orogenic Zone. *Joint Spec. Groups Mtg, Canberra, Abstr., geol. Soc. Aust.*, pp. F6-7.

STUNTZ, J., 1972: The subsurface distribution of the "Upper Coal Measures", Sydney Basin, New South Wales. *Aust. Inst. Min. Metall., Conf., Newcastle*, pp. 1-9.

THOMSON, B. P., 1970: A review of the Precambrian and Lower Proterozoic tectonics of South Australia. *Trans. R. Soc. S. Aust., 94*, pp. 193-221.

TRUMPY, R., 1960: Palaeotectonic evolution of the Central and Western Alps. *Bull. geol. Soc. Am., 71*, pp. 843-908.

Erwin Scheibner
Geological Survey of New South Wales,
Department of Mines,
A.D.C. Building, 189-193 Kent Street,
Sydney, N.S.W. 2000

Editor's Comments
on Papers 11 Through 15

The theme of this group of papers is the geotectonic cycles and cyclical and linear change. Gastil's histograms (Paper 11) of mineral dates accumulated on a time scale provide a lynch pin in the concept that eras, represented here by dated thermal events, are episodic and worldwide. Equally, one notes intervals between the eras. One false guide is the 1400-m.y. peak that represents the Elsonian, that is, early-stage Grenvillian plutons: oversampling in North America resulted in worldwide representation on the histogram.

Before going on to show examples of cyclicity, it is wise to consider what happens when continental scum is removed from the mantle, a linear, nonreversible phenomenon. On a smaller scale, Wager and Deer (Paper 12), from their study of the Skaergaard layered basic intrusive, showed the type of interplays of cumulate and residuum to be anticipated. As an example of the result, initial-stage ultramafic to felsic vulcanism may be expected to differ in trace-element content from one cycle to another. This may be a reason why copper–zinc (of Superior Province) and nickel (of the Yilgarn) are dominant features only in the 2800- to 2350-m.y.-old Superior era.

Pereira and Dixon (Paper 13) attempted to quantify serial, nonre-versible changes, and their work has yet to be surpassed. It may be time-ly now, ten years after the paper was written, to follow the same themes. Some of the changes that will be seen relate to an improved understand-ing of the tectonic cycles and the rocks and mineral deposits ascribable to them. In particular, their Table I can be replaced by that given in the introduction to this volume, and many important deposits are now known to be older than 2700 b.y. As examples, the Isua iron deposit in Greenland is at least 3750 b.y. old. The Bomvu Ridge iron and Barber-ton gold deposits belong to the Swazilandian cycle. In the Pilbara, Mt. Goldsworthy iron belongs to the Pilbaran cycle, and the "Group 5, Arch-aean" probably started at about 2800 b.y., thus including the copper-zinc ores of the Canadian Shield and nickel of the Yilgarn. It would be interesting to see Figure 1 plotted in accord with the concept of ten eras, for valency changes within the range of biochemical process is an expanded topic now that we are learning more of ancient life (see Paper 28).

Trends associated with related rock types is a topic in which the re-lationship is interpretative and subject to change. Ten years ago ideas on the volcanogenic origin of copper–zinc and nickel–copper in Super-ior Province (and later in the Yilgarn) were relatively new, and descrip-tions in the literature included interpretations as to structural emplace-ment and intrusive relationships. The theme of Pereira and Dixon is the matter of interest, but the data merit reconsideration with changing ideas.

Janet Watson (1973a) has continued the work of Pereira and Dixon, the latter her colleague at the Royal School of Mines. It becomes evident that in any quantitative assessment by time it is necessary to weigh the tonnage by a factor related to the remnant area of mobile belts from each era. A weighted value for iron, for example, results in a very early peak, rather than the generally considered Early Proterozoic, because very little mobile belt remains from the earliest Godthaabian and Swa-zilandian eras, when the Isua and Bomvu Ridge iron deposits were laid down in Greenland and Swaziland.

In Paper 14, Weber, with experience of the terms of flysch and mo-lasse from his native Alps, was well able to compare the evolution of an ancient mobile belt with an Alpine-type geosynclinal development dur-ing one orogenic cycle. One may apply Bilibin's stages to Weber's Table I: the initial stage is pre- and early orogenic activity"; the early stage is the "Edmunds Lake Formation" and "Rice Lake greenstone belt plu-tonic rocks"; the intermediate stage is the "Manigotagan gneissic belt" and "San Antonio Formation"; and the late stage is "late orogenic plu-tonism, thermal metamorphism," to which in a personal communication Weber adds cataclasis and mylonites. These he dates at 2345 $\pm$ 160 m.y., and the oldest dated rock of the cycle is the biotite quartz monzonite of the intermediate stage, 2735 $\pm$ 55 m.y. With due allowance for the initial and early stages, therefore, the cycle extended from about 2800 m.y. to 2345 $\pm$ 160 m.y. Weber's stratigraphic column is a replica of that in almost any regional work in northern Manitoba, northern Ontario, and northwestern Quebec. The breakthrough was his direct comparison with the Alpine cycle. It led directly to comparison with yet another cycle by Walker (1973) and to ongoing use of Weber's ideas in Manitoba (Bailes, 1971). McGregor (1973) deciphered the sequence of older rocks in Greenland, dated by the Oxford school. His synthesis is amenable to interpretation following the theme of tectonic cycles.

In Africa, Cahen and Snelling (1966; Paper 15) recognized the cyclicity, and Cahen takes his place as a long-time leader, and co-worker with Arthur Holmes (Holmes and Cahen, 1955), in relating age groups and orogenic cycles. To me the highlight is the paragraph,

> It would be premature to draw further conclusions at present [regarding the Kibaran "era"—their term] but what is already known leaves little doubt that in Equatorial Africa, North American (Grenville) and Southern Scandinavia (Dalslandian–Gothian) there are three portions of a major "world-wide" orogenic cycle, in the same way that the Circum-Pacific Orogeny and the Alpine and Himalayan Orogenies all belong to one major cycle.

Further evidence is given to support the inclusion of the Satpura of India in the cycle. A synthesis such as this is readily integrated into a world-wide review of cycles based on Gastil's graphic depiction.

In a subsequent paper, Cahen (1970) concluded that "the major tectono-stratigraphic subdivisions of Precambrian rocks in Africa often comprise sediments belonging to two successive orogenic cycles—namely the late stage(s) of a particular cycle, and the early and median stages of the succeeding cycles." He specifically grouped molasse of the Kibaran cycle and early stages in the deposition of the Katangan cycle as having together been considered Katangan. A similar problem arises from the definition of the terms Aphebian, Helikian, and Hadrynian; these periods commence immediately after the main orogeny and so related molasse and late-tectonic phenomena fall in the next classification. Helikian, for example, according to definition would include late orogenic material in Saskatchewan with the initial-, early-, and intermediate-stage materials of the Grenville Province; but late-stage material in the Grenville Province is Hadrynian, which essentially refers to mobile belt material of the Avalon Peninsula in Newfoundland.

Writing on the Grenvillian, Wynne-Edwards (1972) noted that Rb/Sr values of 1200 m.y. marked the approximate culmination of metamorphism and granitic intrusion, whereas he believed that K/Ar values 250 m.y. younger were related to cooling through a critical isotherm about

200°C. In the main body of their text, Cahen and Snelling (p. 33), writing about the Baikalian cycle, noted concordant U/Pb ages of 600–650 m.y. from posttectonic pegmatites setting an upper limit, although K/Ar apparent ages on micas and hornblendes are younger.

Burke and Dewey (1972) carried the theme into West Africa. Clifford (1970) had a minor modification, considering cratonization the earmark. Nevertheless, he reviewed the entire African continent in structural terms readily amenable to the cyclical theme of geological evolution, and also showed (1966) that metal abundances vary with time.

11

THE DISTRIBUTION OF MINERAL DATES IN TIME AND SPACE

R. G. Gastil

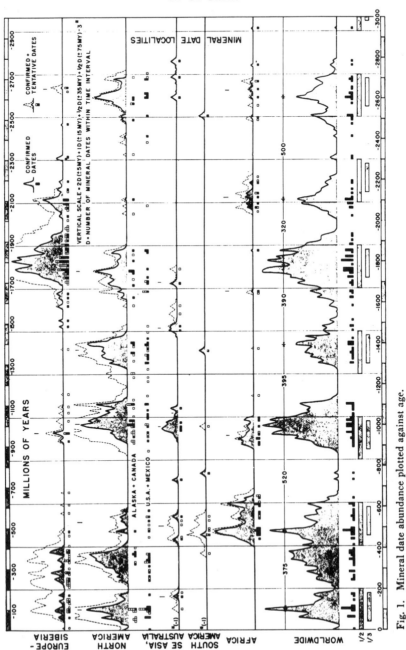

Fig. 1.   Mineral date abundance plotted against age.

[*Editor's Note:*   In the original, material both precedes and follows this figure.]

12

Copyright © 1939 by C. A. Reitzels Forlag

Reprinted from *Meddelelser om Grønland,* **105**(4), 64–67 (1939)

GEOLOGICAL INVESTIGATIONS IN EAST GREENLAND: PART III. THE PETROLOGY OF THE SKAERGAARD INTRUSION, KANGERDLUGSSUAQ, EAST GREENLAND

L. R. Wager and W. A. Deer

VI. PETROLOGY OF THE LAYERED SERIES

(a) General Features and Nomenclature.

The rocks of the layered series vary in two independent ways. The variation most conspicuous in the field is in the relative abundance of the light and heavy minerals. This gives rise to the rhythmic layering and the melanocratic facies near the inner contact. The less conspicuous, but more significant variation, which gives rise to what has been called the cryptic layering, is due to gradual changes in the composition of certain minerals which are solid solutions; these are high temperature varieties in the lowest exposed horizons and they change to lower and lower temperature varieties as the layered series is ascended. At certain horizons of the layered series a new mineral suddenly begins to be present, and at others, a mineral which has been present, suddenly ceases to occur. In the evolution of the rock series such abrupt changes in the minerals present were apparently related to the continuous changes in composition of the minerals, which are solid solutions, and the two sets of changes are grouped together and called the major variation. The term minor variation will be used for that which gives rise to the rhythmic layering and the increase in melanocratic constituents near the inner contact. The minor variation is an effect superimposed on the major and therefore the latter will be dealt with first. This involves treatment of average rocks which were carefully selected in the field as free from abnormal concentration of either the light or heavy constituents.

The composition and other features of the layered rocks may be conveniently plotted against their height in the intrusion as calculated from their position and the dips given on the map. The height is measured conveniently from the lowest layered rock at present exposed, 4087. Ten rocks representative of the whole sequence have been particularly fully investigated and chemically analysed. Results for these and for others lying near or at the limit of certain phases, are summarised diagrammatically in figure 18.

The average rock of the lower horizons is a hypersthene-bearing, olivine-gabbro of a fairly normal type. In ascending the layered series

137

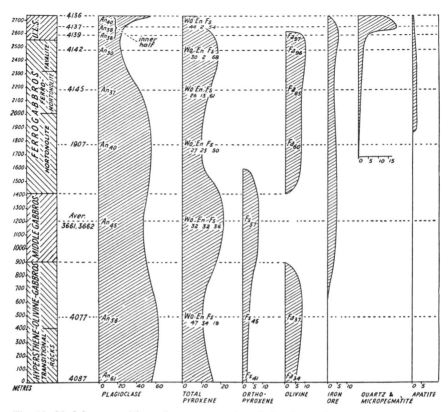

Fig. 18. Modal composition of average rocks of the layered series and the composition of certain of the minerals.

the plagioclase shows a slow decrease in anorthite content and the pyroxenes and olivine a slow increase in the amount of the iron component. At about 900 metres in the layered series the rocks suddenly become olivine-free and remain so for 500 metres, above which olivine again becomes an important constituent. Over this range the plagioclase and pyroxene continue to become richer in their more fusible component. The 500 metres which are described as olivine-free, are not strictly so, as a little olivine is developed in reaction rims between the pyroxene and iron ore. This, however, we regard as a secondary feature, developed after the accumulation of the primary phases and at a slightly lower temperature. At 1400 metres olivine reappears as a primary phase; a little higher hypersthene disappears, and a little higher still quartz comes in for the first time. The rocks here are of thoroughly abnormal composition, being unusually rich in ferrous iron which is mainly present in the pyroxene and olivine. Above this horizon, changes in the composition of the various minerals become rapid and by the level of the purple band, 2500 metres, the plagioclase contains 30 per cent. anorthite, the pyroxene only 2 per cent. $MgSiO_3$, and the olivine only 2 per cent. Mg_2SiO_4. The highest rocks of the layered series on Basistoppen are unlaminated, and seem to have been produced by rather different physical processes from those responsible for the differentiation of the rest of the layered series.

This summary of the characteristics of the layered series is sufficient to show that it includes some hitherto unrecorded rock types, and the difficulty of finding satisfactory names at once arises. The rocks of the lowest 900 metres may legitimately be called hypersthene-olivine-gabbro. By the time the olivine disappears the rocks are beginning to be abnormal,—the pyroxene is already rich in iron, and the absence olivine is not due to increasing acidity but to some other cause connected with crystal fractionation. It is not easy to find an adequate name for this part of the series, and for the present the non-committal term "middle gabbro" will be used. Above this, the rocks fall definitely outside the range of any normal gabbros. They keep the low silica percentage of olivine gabbro but the plagioclase is andesine and the ferromagnesian minerals are abnormally rich in iron. The analyses, when plotted on a triangular diagram (Fig. 19), showing normative amounts of the important constituents arranged in a way discussed below (p. 231), lie in a region which is well outside the normal range of the gabbros, and indeed, outside the range of any known igneous rocks. Since these rocks are related to gabbros and are characterised by richness in iron, we have ventured to use for them the new name ferrogabbro, a term more precisely defined below (pp. 98—91).

The further subdivision of the ferrogabbros may reasonably follow two directions. The classification may be based on significant special phases present in the rock, giving for example, quartz-free ferrogabbro or quartz-ferrogabbro. Such names will sometimes be used and the limit of these two particular types has been put down on the map. Since quartz first appears in very small amounts, this boundary could not be drawn without recourse to microscopic examination. On the other hand, the nomenclature might take into consideration the composition of the ferromagnesian minerals as determined by analysis or by optical data. It has proved convenient to use the composition of the olivine for subdivisions of this kind and such names as hortonolite-ferrogabbro, ferrohortonolite-ferrogabbro[1]) and fayalite-ferrogabbro will be used.

In this nomenclature, the relative abundance of the different minerals is not considered, the subdivisions being based on variation in the minerals present or their composition, that is, on the major variation. Variation in relative abundance of the minerals, the rhythmic variation, is not of such fundamental genetic significance and should

---

[1]) The new name ferrohortonolite here adopted for a certain range of iron-rich olivines is defined in the succeeding section.

only play a secondary role in nomenclature, a general principle which
is recognised in those classifications of igneous rocks which make the
composition of the plagioclase the prime factor of subdivision. From the
point of view of general appearance, mineralogical composition and bulk

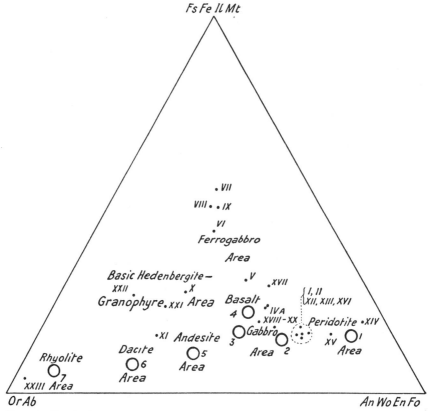

Fig. 19. The field of the ferrogabbros, basic hedenbergite-granophyres, and other
Skaergaard rocks compared with Daly's averages of calc-alkaline rocks.

chemical composition, the extreme rock types developed in some of the
layers are very different from the average rocks adjacent to them, and
they might legitimately be called peridotite, anorthosite and andesinite
etc. For the most part, however, such names will not be used and instead
the rocks will be described as melanocratic or leucocratic varieties of the
average rock.

Plate 8.

Fig. 1. Gravity differentiated layers separated by layers of average rock, 300 metres
west of the Main House. Towards the top centre incipient trough banding
can be seen.

Fig. 2. Differentiated layers separated by layers of average rock, 400 yards west-
north-west of Main House.

Fig. 1.

Fig. 2.

13

Reprinted from *Trans. Inst. Mining Mettallurgy*, **B74**, 505–527 (June 1965)

Evolutionary Trends in Ore Deposition*

J. PEREIRA,† A.R.C.S., ASSOCIATE MEMBER, and
C. J. DIXON,‡ B.Sc., A.R.S.M.

SYNOPSIS

A study is made of the relationship between different types of orebodies and their significance in the geological time scale and changes in environment. Part 1 describes the problems involved in obtaining a representative world-wide sample of orebodies and the factors taken into account in the analysis which followed. Part 2 gives the principal results obtained from data collected from a total sample of some 200 mineral deposits.

It is concluded that with the passage of time there is an increase in the diversity of morphological types; that particular combinations of associated rocks or environments are significantly more characteristic of some geological periods than others; and that major deposits of particular metals tend to have a maximum abundance in definable geological periods.

Some specific suggestions are made to explain these changing relationships, the principal deduction being that the factors which lead to and control ore deposition reflect irreversible sequences and that an increase in the number and complexity of factors involved demonstrates an evolving world and hence evolutionary trends in the sequences of orebodies formed.

FASHIONS IN ECONOMIC GEOLOGY have led successive geologists to emphasize different aspects of ore deposits and, consequently, to devise classifications of deposits based on the different factors controlling metal concentration. At one extreme, classes of ore deposits result in which the only common factor may be that they have ores with the same metal. Among the systems of classification may be included physico-chemical systems, structural systems, environmental systems and genetic systems; all have a logical basis and lead to groupings which have useful aspects from the exploration point of view but, as Routhier[3] has pointed out, every one of these classifications has limitations.

Any geological phenomenon is the total expression of so many interdependent parameters that in the present state of knowledge it is only possible to move towards what appear to be the natural geological groupings of mineral deposits, without any assurance that the perfect classification exists. The subject is thus a highly empirical one, too many of the factors involved being still imprecisely definable.

*Paper received by the Institution of Mining and Metallurgy on 7th January, 1965, and published on 3rd June, 1965; for discussion at a General Meeting to be arranged.
†R.T.Z. Services, Ltd., London.
‡Department of Geology, Imperial College, London.
³etc. See list of references at the end of the paper.

This present paper is to some degree the by-product of thoughts on these aspects of the geology of ore deposits. That certain types of deposits are characteristic of particular geological epochs is well known; and the intention in analysing some 200 selected examples was to examine whether those special characteristics are due to progressive changes in environment that can be considered an evolutionary sequence. In so doing it is necessary to avoid, as far as possible, descriptive terms, classifications or criteria which impose any preconceived or subjective views on ore genesis. The purpose of the investigation was to achieve a form of 'world sampling' that would show changes in the processes of ore deposition with time— without, in the first instance, attempting to discuss the causes or implications of such changes.

It is important to emphasize that the use of the term 'evolution' here does not imply that there were periods in geological time when the evidence suggested that the usually accepted laws of chemistry and physics were inapplicable; but rather that there has been a succession of changes in the overall environment in which orebodies were formed which, in its turn, has produced types not repeated in later geological epochs. There appears to have been a series of changes in the controls affecting ore deposition and thus new factors arose which may have played no, or very little, part in previous epochs.

The evolution of living organisms—of a biosphere—was of critical importance to the formation of deposits of coal, phosphate, chalk, etc.; the range of the present paper has, however, been limited to exclude the majority of deposits where biological factors played a primary role.

Numerous authors have commented on probable changes in the composition of the atmosphere in the past; in particular, Holland,[4] and Berkner and Marshall[5] have shown how an evolutionary model for these changes is consistent both with theoretical physico-chemical requirements and with geological observations. Such changes must have affected ore deposition either directly—in the case of elements chemically sensitive to oxygen or carbonate formation—or indirectly—through biochemical processes.

On a basis which is 'economic', and hence unavoidably arbitrary, data have been restricted to the main metalliferous deposits; it is suggested, however, that at every point of time in which the deposits were formed there has been a series of newly introduced factors affecting all forms of deposition and ore formation, even though mineral concentrations in which biological processes played a major part show more emphatic and characteristic changes. It is further suggested that, in the widest sense, there is almost no feature of ore geology which lacks an evolutionary background, for it seems reasonable to suppose that the crust has always undergone progressive geochemical modifications and that these in themselves brought about the more obvious changes in the environment at the surface and, ultimately, in the biosphere. On this basis, significant differences between deposits formed in different geological epochs must be expected, which apply not only to orebodies formed at or near the surface, but also in depth.

Many authors have drawn attention to the fact that deposits of particular metals tend to characterize particular geological epochs, and numerous examples are given in a recent paper by Meng Hsien-Min.[6] Moreover, many papers deal with specific instances and the particular controls which may have had a major, formative, influence on their localization. It is not proposed to review the literature on this subject; the following brief list does, however, indicate some of the factors believed by the present authors to have played an important role.

(1) Trends which cause minerals of lower specific gravity to form nearer the surface of the crust where the pressures are less than in the deeper regions: such trends are related to the evolution and growth of continents and possible thickening of the sial.

(2) Geochemical changes largely involving an increase in molecular complexity at or near the surface and leading ultimately to the formation of the biosphere: early differentiation produced an oxygen-rich atmosphere with corresponding and complementary changes in meteoric water and in the oceans.

(3) The appearance and evolution of living organisms, probably on a microbiological scale in the first instance, and of significant importance in the concentration and formation of orebodies produced by sedimentary processes: marine environments with the deposition of carbonaceous or graphitic shales would mark an early phase, and the first development of soils must also have greatly affected all forms of alluvial and residual mineral concentrations. Recent work by Temple[18] has drawn attention to the possible significance of bacterial action. In particular, he has demonstrated that objections based upon toxicity of metals are not valid.

(4) An increase in the abundance and range of living organisms, and hence of biochemical factors in general.

(5) Persistent trends in crustal evolution: geological literature abounds in references to such processes as 'oceanification', described by Belousov,[16] the 'Chellogenesis' of Sutton,[8] 'continental drift' and 'the expanding earth' postulated by Hilgenberg and Egyed,[13,14] and, most recently, by Creer who has suggested a form of evolution that involves both expansion and crustal drift.[19] Approaching the problems of crustal tectonics from a very different viewpoint, Brock[12] also propounds an evolutionary interpretation for orogenic processes. A recent paper by Leonov[20] makes the important point that while geosynclinal–orogenic phases such as the Caledonian, Hercynian and Alpine may be largely cyclic, there is evidence of 'super-phases' which are links in an irreversible evolution rather than repeating cycles.

All of these assumptions may be questioned, but it is beyond the scope of the present paper to dwell upon them or to embark on an elaborate analysis of all the problems of 'ore genesis'. Instead, an attempt has been made to collate a large amount of information on ore deposits which it is believed represents a world-wide and chronologically complete sample; comment on some of the contrasts displayed then follows. As the selection of these data presented special problems, it is necessary to explain briefly the basis on which it was made and the assumptions that had to be drawn.

It must be stressed that the present paper can be no more than a preliminary essay, the gaps in the information ideally required being still too large to allow any great degree of precision or assurance. It should also be mentioned that many of the underlying ideas given here were discussed in advance with Haddon F. King,[7] who also has an interest in this particular topic, but who bears no responsibility for the conclusions presented.

Part 1—Selection of Data

Owing to the lack of information on orebodies in China, the U.S.S.R. and the Polar regions, the sample selected is not a truly world-wide one. It is assumed, however, that the area which has been covered is sufficiently large to offset the danger of giving an unbalanced picture.

There are other problems which cannot be handled in any ideal way at present and a brief note on these is necessary to explain the limitations to the scheme on which the data were selected and assembled.

Mine Density and Selection

As soon as one begins to plot the position of all the recorded mines on any particular continent, two quite different forms of related problems arise. There are obvious mineral provinces or mineral zones where the density of mines is a natural expression of the extent of numerous but related orebodies as, for example, in the belt of Palaeozoic rocks which crosses the southern end of the Iberian peninsula with its host of pyritic deposits. In the same way, a dense belt of mines may mark the outcrops and sub-outcrops of major stratigraphic deposits, such as the Witwatersrand conglomerates of South Africa with their immense number of gold mines. One mine may work many deposits or, conversely, many mines may represent a single mineralized unit or series of units.

Large areas of the world also begin to show up which are noticeably deficient in mines and it is clear in a number of instances that that deficiency is most probably related to such factors as accessibility, climate, population distribution and such historical accidents as the extent to which different parts of the world have become industrialized and developed. For example, the world's major deserts are generally areas with few mines, as are such inaccessible regions as the Amazon basin.

It is difficult to find satisfactory ways of balancing these contrasts in the density of mine distribution—which are suspected to be due to non-geological causes—other than by a system of rating designed to give a relatively even pattern of samples. To achieve this, a world map was divided into 347 squares, each measuring 5° of latitude by 300 nautical miles. In some parts of the world a single square covered major contrasts in the type of deposit and in many cases it was therefore thought necessary to include two examples. Iron ore deposits are an exception to the general rule since there is hardly any country in the world without potential reserves of iron ore, many of which are exploited for political or strategic reasons rather than economic ones. In general, iron ore deposits were not chosen to the exclusion of other types of mineral deposit and they frequently represent the second deposit in a particular sample square. Furthermore, in order to make the analysis relevant to 'economic geology', only those deposits capable of sustaining an annual production in excess of 100 000 tons of crude ore were included, thus giving a total sample of about 200 deposits.

An attempt was made to check the extent to which the sampling method eliminated the effects arising from different parts of the world having been explored to a greater or lesser degree (see section on *Age Relationships*).

Size of Deposit

Ideally, total reserves might seem to be the best measure of size, but few mining companies disclose their complete reserve figures and, in any case, more often than not such figures constitute the known reserves at a given moment in time and are probably far smaller than the actual tonnage that will eventually be mined from a particular deposit. A further difficulty commonly arises in cases such as Broken Hill, New South Wales, where many separate mines are working a single deposit. The solution favoured has been to work on the assumption that the larger the deposit, the larger the scale on which it is worked. Although only partly true, such an assumption seems to provide a more even and generalized result than can be obtained by attempting to make an almost unending series of very inexact estimates of total reserves. Even with this framework, however, some exceptions to the scale set out on the data sheets have been necessary in order to include some mines which are planned but not yet in production (see I of Appendix for size classification).

Grade of Deposit

Some squares offered a large and often difficult choice of possible representatives and the general rule has been to select the largest mining operation on which adequate data were available. The present-day trend is towards the exploitation of large, low-grade deposits, rather than small, rich orebodies which may be expensive to mine and short-lived. Such a trend helps to narrow the gap between the 'geological' grade and that which has to be maintained to provide an economic mining operation. Ideally, an analysis such as this should be based on grades representing the geochemical concentration of the metal, but it is impossible on the information available to know what this figure should be, and hence an arbitrary factor is introduced here; it is felt, however, that while it affects the degree of accuracy, it does not seriously distort the general picture. While it is important for the purposes of this survey to know whether different types of deposit are relatively higher or lower in grade, it is unnecessary to put an exact figure on the difference.

Age Relationships

Where possible, both absolute age determinations and stratigraphic ages have been collected, and the views of authors whose works have supplied most of the data set out on the individual sheets compiled for each mine have been accepted. It must be admitted that in some instances the estimate of the age of a deposit depends on personal opinion as to its probable genesis, and exact dating is therefore seldom possible. For this reason only five age brackets have been used, thus allowing for the limitations of available information while still providing an adequate time scale to show up the type of evolutionary changes it is hoped to demonstrate.

Owing to the paucity of absolute age determinations in certain parts of the world, the age groupings used for plotting the curves (Figs. 1–19) are based on stratigraphic or orogenic ages modified as appropriate by absolute determinations. In formulating the five age groups the work of Gastil[9] and others on the time distribution of absolute dates has largely been followed. The sub-division of the Precambrian owes something to many authors, including Stockwell,[10] Voitkevich,[11] and Sutton,[8] the final grouping being shown in Table I.

TABLE I

Group 1	Upper Phanerozoic	.	.	$0 \cdot 0$ to $0 \cdot 3 \times 10^9$ years
Group 2	Lower Phanerozoic	.	.	$0 \cdot 3$ to $0 \cdot 6 \times 10^9$ years
Group 3	Upper Proterozoic	.	.	$0 \cdot 6$ to $1 \cdot 1 \times 10^9$ years
Group 4	Lower Proterozoic	.	.	$1 \cdot 1$ to $1 \cdot 9 \times 10^9$ years
Group 5	Archaean	.	.	$1 \cdot 9$ to $2 \cdot 7 \times 10^9$ years

A further period, older than $2 \cdot 7 \times 10^9$ years and called by various authors Katarchaean, obviously exists, but it was omitted from the analysis since there are no important deposits that can be ascribed with certainty to it.

The distribution of rocks which fall within the divisions shown in Table I was a further factor to be taken into account, since in judging the results as a whole it is necessary to know whether they are, or are not, biased by the age distribution of the rocks at present cropping out at the surface. An ideal presentation would give a sample in which every time division was represented by an equivalent amount of outcrop. Table II shows the proportions of rocks in each age group actually occurring within the parts of the world sampled.

TABLE II

Age group				Percentage in parts of the world sampled*
1	.	.	.	57
2	.	.	.	15
3	.	.	.	7
4	.	.	.	12
5	.	.	.	9

*i.e. Non-communist world less Polar regions.

Similar proportions of the five age groups were worked out for countries grouped according to their state of development (on the basis of annual income per head of the population) in order to determine the extent to

147

which the age distribution of the sample might be biased. It was found that the bias was not generally significant, although the sample may have a slight positive bias in the Lower Phanerozoic owing to the relative abundance of this age group in the industrial countries.

Morphology of Deposits and Associated Rocks

The problems presented here arise from the need to use terms descriptive of different types of orebody without implying any commitment about the question of their genesis. For example, to the present authors the majority of stratiform deposits are either sedimentary or sedimentary–volcanic, but for the purposes of the present survey it was clearly necessary to avoid making a series of personal judgements open to alternative interpretations.

If the figures presented in the final section are accepted as providing evidence of evolutionary changes owing to evolving environments, then they indeed have a bearing on the subject of classification and ore genesis; but at the outset this aspect has to be excluded.

Minerals and Metals Listed

The selection is largely governed by the need to place limits to the range of data to be collected, and the simplest way of achieving this was to restrict the choice to the major 'economic' metals. Naturally, use of the term 'economic' automatically introduces its own forms of arbitrary limitation, but such a limitation applies to the science of economic geology as a whole, and its effect on the larger general conclusions is beyond the scope of these introductory remarks.

It also seemed desirable to cater for a range of minor elements, but since reliable quantitative data concerning such elements are, in many cases, not available, their consideration has been restricted merely to an indication of the significant presence or absence of the element in question.

Part 2—Analysis and Interpretation

DIVERSITY IN MORPHOLOGY OF ORE DEPOSITS

Figs. 1 and 2 are related to a theoretical assumption that, in a hypothetical situation where conditions were constant throughout time, the distribution of the deposits over the nine morphological types into which the various ore deposits have been classified (see II of Appendix) would show no change. Fig. 1 refers to the whole sample and is the curve given by plotting the mean square deviation from the theoretically assumed level in each of the five age groups. It shows an increasing diversity of the morphology of ore deposits with the increase in geological time apart from a marked break between the Archaean and the Lower Proterozoic. It is considered that this break is largely due to the number of iron ore deposits in the total sample and that the interruption was caused by the change to an oxygen atmosphere, together with related factors that accompanied the beginning of organic life. Iron is sensitive to valency changes within the range of biochemical processes, as noted by, among others, Garrels[17] and Kuznetsov.[1]

The same is not true of the base and precious metals. Fig. 2 demonstrates a progressive increase in the morphological diversity of deposits of these metals with geological time, with no interruption at the change from Archaean to Lower Proterozoic.

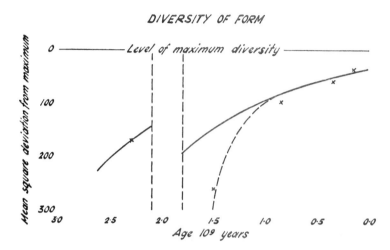

Fig. 1.—*Diversity of form of whole sample.*

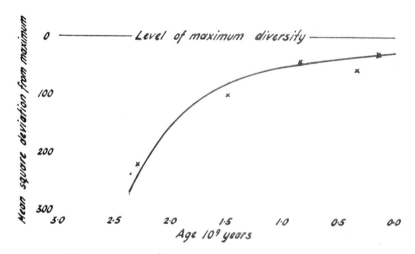

Fig. 2.—*Diversity of form of base- and precious-metal deposits.*

TRENDS ASSOCIATED WITH RELATED ROCK TYPES

Fig. 3 illustrates the rated number of deposits associated with volcanic rocks and the change which occurs between successive age groups. The types of volcanic association taken into account and the mathematical basis of Figs. 3, 4 and 5 are shown in I and IV of the Appendix.

From the Archaean up to the present time there is a steady increase in the number of deposits which have associated volcanicity. The increase in volcanic associations may be attributed to two different factors: an increase in the amount of extrusives derived from products of granitization which contain iron; and an increase in the number of deposits associated with both geosynclinal volcanicity and with volcanicity accompanying rifting and block faulting.

Fig. 4 is a composite of three curves derived, respectively, from the stratiform deposits, the stratiform and lenticular banded deposits, and the stratiform, lenticular banded and lenticular massive deposits combined. These three series are plotted as the rated number of deposits related to argillites, shales, greywackes and other types of predominantly geosynclinal sedimentation. The curves show a marked peak in the Lower Phanerozoic— probably the first geological epoch in which geosynclines became a distinctive tectonic feature. Thereafter, the number of deposits associated with altogether different tectonic environments began to mask the effect of geosynclinal deposits; in other words the increase in diversity of type begins to be a greater factor than the number of deposits associated with a particular tectonic form.

Fig. 5, using the same three groups of morphological types as Fig. 4, shows the change in the rated number of deposits associated with limestones and dolomites. It demonstrates a general increase throughout the time scale which is considered to be due to the increasing importance of sediments in which biochemical factors played a dominant role.

The result of a count made on the number of veins, stockworks and related deposits, recorded as related to intrusive plutonic rocks, is given in Table III. The figures giving the proportion of the sample are set out in this form because the results, while not revealing a constant trend, nevertheless appear to be significant and to show that although a fairly high proportion of the vein deposits in the youngest age group are closely associated with plutonic intrusives, that is not true of the Archaean. This fact may indicate that while some of the classic hydrothermal opinions on ore genesis may be valid for many of the more recent vein-type deposits, they become less tenable when applied to older periods of geological time.

TABLE III.—*Vein, fissure, stockwork and related deposits*

Age group					Percentage of deposits associated with plutonic rocks
1	.	.	.	.	60
2	.	.	.	.	31
3	.	.	.	.	75
4	.	.	.	.	67
5	.	.	.	.	19
Rated average of whole	.		.		44

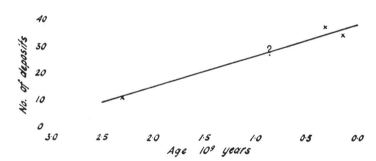

ROCK ASSOCIATION

Fig. 3.—*Rated number of deposits associated with volcanic rocks.*

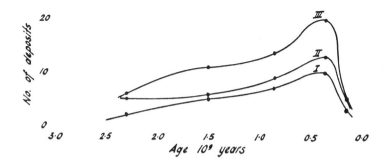

Fig. 4.—*Rated number of deposits associated with shales. (In Figs. 4 and 5 curve I refers to stratiform deposits; II refers to stratiform and lenticular banded deposits; and III refers to stratiform, lenticular banded and lenticular massive deposits.)*

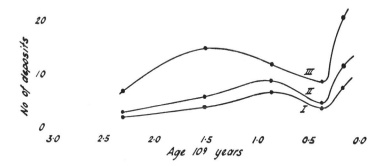

Fig. 5.—*Rated number of deposits associated with carbonate rocks. (See also caption to Fig. 4.)*

TRENDS IN THE GRADE OF ORE DEPOSITS

Base and Precious Metals

Throughout this section it must be noted that the scale against which grade is measured is relative and not quantitative. This is partly owing to the lack of sufficiently precise data on grade and partly to the fact that the samples include every deposit in which the particular metal under consideration occurs, not just the deposits in which it is the principal metal being mined. The grade factor is expressed as a proportion of unity; the higher the figure, the higher the grade. In the case of Cu, Pb and Zn (shown in Figs. 6 and 7), there is an increase in grade with the passage of geological time. It is of particular note that in the case of Cu this relationship is an exponential one which is characteristic of a simple evolutionary trend, and that since primary copper minerals are sulphides, biochemical factors probably played an important part in their deposition.

In contrast to the graphs for Cu, Pb and Zn, that for Au (Fig. 8) shows a marked exponential decrease, which is presumed to be related to the fact that Au neither occurs as a sulphide nor is its precipitation related to biochemical reactions, and the overall effect of geological processes causes dispersal rather than concentration. (This point, of course, does not apply to alluvial deposits, which have not been included in the sample.)

Fig. 9, which shows the same data in the case of Ag, is less easy to interpret. Presumably the decrease in relative grade which shows up in the early Precambrian age groups is associated with the dispersion of Au, whereas the increase which occurs after the Upper Proterozoic is largely related to Ag found in association with Pb—a possibility also referred to in connection with a separate series of curves explained later (p. 518).

Iron Ore Deposits

The grade factors taken from all the sampled iron ore deposits proved to be too erratic to suggest particular dominating controls or trends. However, by eliminating the deposits with close igneous and volcanic associations and by subjecting the stratiform iron ore deposits to the same analysis as the deposits in the preceding section, the graphs shown in Figs. 10–13 were obtained.

Fig. 10 suggests that there is a fairly constant grade of iron throughout the time scale, but it is thought likely that this must also to some extent reflect the economics of mining, since the range in the grades of different iron ore deposits which are mined is much less significant than that of the other metals considered in this analysis. The curve indicating the grade of Mn in iron ores drops off sharply between the Archaean and the Proterozoic, and increases to a lesser extent in subsequent age groups. In a detailed paper Lepp and Goldich[2] comment on this feature, attributing it to an oxygen-deficient atmosphere in the Archaean tending to immobilize the Mn, the process becoming reversed in the Lower Proterozoic with the development of an oxygen-rich atmosphere; thus Fe and Mn tended to separate into distinct, though adjacent, deposits.

Fig. 11 seems to show a fixation of manganese in early iron ore deposits as the atmosphere became more and more oxygenated, reaching a minimum

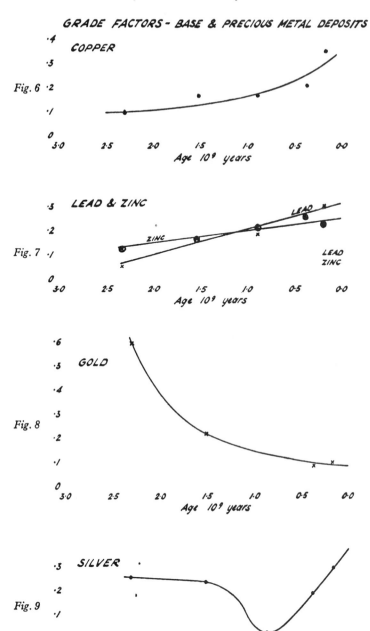

Fig. 6

Fig. 7

Fig. 8

Fig. 9

GRADE FACTORS - IRON DEPOSITS

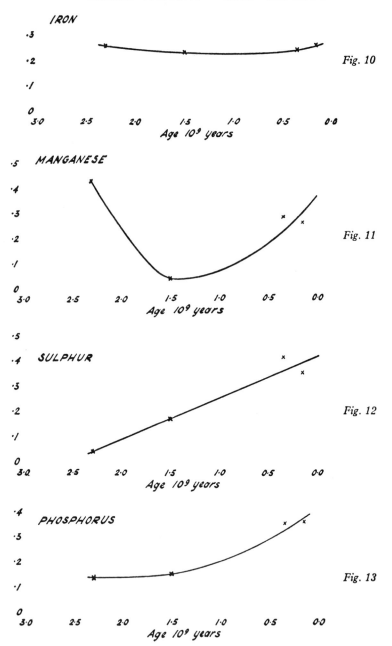

IRON

Fig. 10

MANGANESE

Fig. 11

SULPHUR

Fig. 12

PHOSPHORUS

Fig. 13

in the Lower Proterozoic followed by a partial recovery in later times due, perhaps, to the evolution of anaerobic environments.

Figs. 12 and 13 demonstrate the grades of sulphur and phosphorus, both of which increase with time—a feature to be expected since both elements are involved in organic processes.

RELATIVE ABUNDANCE OF MAJOR DEPOSITS

Figs. 14–19 show the relative abundance of major deposits of particular metals within the different age groups. The mathematical basis on which these figures were rated is explained in V of the Appendix. In each case the factors are expressed as a proportion of one.

The graph for Cu (Fig. 14) is thought to be distorted owing to the method used to sample the world's deposits. Many of the major copper producers, such as the Zambian Copperbelt and the Porphyry Copper deposits, occur within relatively small geographical areas so that on the present writers' scale of sampling they are represented by too small a number of examples. This gives undue weight to the important but more evenly distributed deposits in age groups 2 and 4, and hence in the case of Cu a more detailed examination is required.

This geographical factor does not apply to the Pb and Zn deposits shown in Fig. 15. As might be expected, the two curves are very similar. Both show a marked increase in the Lower Phanerozoic, much of which is due to deposits occurring in geosynclinal environments and also in environments of the Mississippi type.

Between age groups 2 and 3 an important phase in biological evolution took place which led to the widespread appearance of most of the marine phyllae and, in association with this, new sedimentary environments. Similarly, there was an equally important phase of evolution between age groups 1 and 2 spanning the period when the land was colonized by plant life; this, in its turn, must have affected the processes of weathering and created still more diverse sedimentary environments. Although outside the scope of the present paper, the formation of coalfields was one obvious result.

Au and Ag (shown in Fig. 16) both have a peak in the Lower Proterozoic when large numbers of important deposits accompanied the formation of banded ironstone formations and the auriferous conglomerates. After this there is a decline, less rapid with Ag than with Au, because silver tends to be associated with Pb–Zn deposits in later geological periods (see Fig. 15).

Fig. 17 shows the relative abundance of Ni and Co during geological time. The high Ni in age group 4 is largely due to the important sulphide deposits in Canada. Although the Co curve has a similar peak, owing to Co which is associated with copper deposits, it does not decline at the same rate. The absence of Ni in the younger gabbros has often been noted, but no satisfactory explanation has been produced to explain this absence.

The curve for Sn (Fig. 18) shows an exponential increase, with a maximum in age group 1, which is considered to be due to the fact that,

RELATIVE ABUNDANCE FACTORS

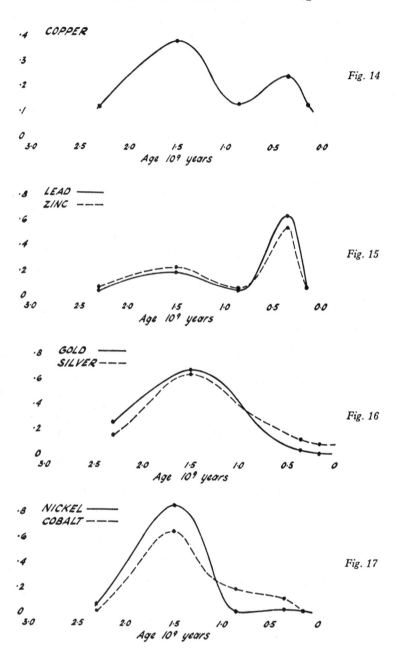

Fig. 14

Fig. 15

Fig. 16

Fig. 17

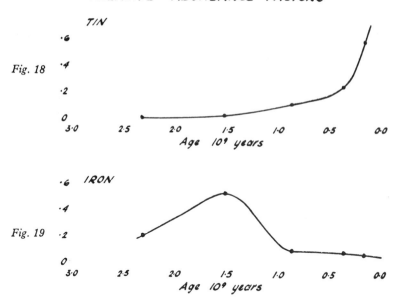

although not associated with any biochemical cycles, it does appear to be progressively concentrated by granitization processes.

Fig. 19, which shows the curve for Fe, has a maximum in the Lower Proterozoic corresponding with the change in atmospheric composition previously noted (see p. 515). In economic terms it also corresponds with the large number of important bedded iron ore deposits which are mined all over the world from formations of a similar age.

CONCLUSIONS

It will be clear from many of the preceding remarks that one of the more important conclusions is that there are limitations to the doctrine of 'Actualism' and its corollary 'Uniformitarianism'. A recent paper by Belousov[16] makes the same point, but it must be added that the present writers' objections to these doctrines are limited and do not invoke chance or accidental events in the earth's history.

Some of the relationships demonstrated here show a progressive change which must reflect an evolutionary sequence, e.g. the copper and gold 'grade factor curves' (Figs. 6 and 8). Others, although outwardly discontinuous, can be interpreted in terms of the interaction between several evolutionary trends, e.g. the manganese 'grade curve' (Fig. 11) and the lead–zinc 'abundance factor curves' (Fig. 15).

Little or no evidence has been found to indicate that random external events affected the development of ore deposition; the concept is rejected,

Plate I

J. PEREIRA AND C. J. DIXON: *Evolutionary Trends in Ore Deposition*

Fig. 20.—Map showing location of deposits included in the sample.

Scale 1 : 48 000 000.

Outline map reproduced by permission of Oxford University Press.

Communist block not sampled

158

as is the idea that geological processes have not changed during the course of geological time.

Figs. 1 and 2, which show the trend towards increasing diversity in deposits of younger age, reflect the progressive way in which the sum of newly introduced factors affecting ore deposition produced an evolving and increasingly complex series. Very similar conclusions were reached recently by Gorzhevsky and Kozerenko[21] and Salop.[22]

These factors are widely different in their nature, ranging from bio-chemical trends related to the evolution of the biosphere to factors such as the progressive changes in the nature of the crust due to granitization, volcanicity and other physico-chemical processes. Individual metals are affected in different ways according to the processes which most closely control their geochemical behaviour—hence the dissimilarity between the age curves for the various elements. These curves are believed to demonstrate effectively that evolving processes underlie the formation of ore deposition. Although the data at present available are limited, it is to be hoped that in the future this theme can be greatly expanded and made much more precise. It is intended to continue this work by trying to obtain more and better information over a wider range of mineral deposits.

The pioneer work of Lafitte and Rouveyrol[15] is a very valuable step which will greatly assist and stimulate studies of this nature, and their plea for greater international co-operation in this field is one which the present writers strongly endorse.

REFERENCES

1. KUZNETSOV, S. I., et al. *Introduction to geological microbiology* (New York, London: McGraw-Hill, 1963), 252 p. (English translation.)

2. LEPP, H., and GOLDICH, S. S. Origin of Precambrian iron formations. *Econ. Geol.*, **59**, no. 6, 1964, 1025–60.

3. ROUTHIER, P. *Les gisements métallifères: géologie et principes de recherche* (Paris: Masson et Cie, 1963), 2 vols.

4. HOLLAND, H. D. On the chemical evolution of the terrestrial and cytherean atmospheres. In *The origin and evolution of atmospheres and oceans*, edited by P. J. Brancazio and A. G. W. Cameron (New York, London: Wiley, 1964), chapter 5, 86–101.

5. BERKNER, L. V., and MARSHALL, L. C. The history of growth of oxygen in the earth's atmosphere. Chapter 6, 102–26, of reference 4.

6. MENG HSIEN-MIN. The problems of genesis and classification of ore deposits. *Scient. sin.*, **11**, June 1962, 837–58.

7. KING, Haddon F. Personal discussion.

8. SUTTON, J. Long-term cycles in the evolution of continents. *Nature, Lond.*, **198**, May 25, 1963, 731–5.

9. GASTIL, R. G. The distribution of mineral dates in time and space. *Am. J. Sci.*, **258**, 1960, 1–35.

10. STOCKWELL, C. H. Third report on structural provinces, orogenies, and time-classification of rocks of the Canadian Precambrian Shield. *Pap. geol. Surv. Can.* 63–17, 1963, 125-31.

11. VOITKEVICH, G. H. Unit of the geochronology of the Precambrian. *Priroda, Mosk.*, no. 5 1958, 77–9. (Russian text.)

12. BROCK, B. B. On orogenic evolution, with special reference to southern Africa. *Trans. geol. Soc. S. Afr.*, **62**, 1959, 325–72.

13. HILGENBERG, O. C. *Vom wachsenden Erdball* (Berlin: Verf.-Verl., 1933), 56 p.

14. EGYED, L. A new dynamic conception of the internal constitution of the earth. *Geol. Rdsch.*, **46**, 1957, 101-21.

15. LAFITTE, P., and ROUVEYROL, P. *Carte minière du globe sur fond tectonique* (Paris: B.R.G.M., 1964).

16. BELOUSOV, V. V. Ways of development of geological science. *Sov. Geol.*, no. 1 1963, 11–28. (Russian text.) English transl. in *Int. Geol. Rev.*, **6**, Oct. 1964, 1750–60.

17. GARRELS, R. M. *Mineral equilibria* (New York: Harper and Brothers, 1960), 254 p.

18. TEMPLE, K. L. Syngenesis of sulphide ores: an evaluation of biochemical aspects. *Econ. Geol.*, **59**, no. 8, Dec. 1964, 1473–91.

19. CREER, K. M. An expanding earth? *Nature, Lond.*, **205**, Feb. 6 1965, 539–44.

20. LEONOV, G. P. The problem of cycles in geological history. *Vest. Mosk. Univ.*, no. 4 1962, 3–12. (Russian text.) English transl. in *Int. Geol. Rev.*, **6**, Dec. 1964, 2093–9.

21. GORZHEVSKY, D. I., and KOZERENKO, V. M. On the irreversible character of geological and metallogenic development of the earth's crust. Paper presented to *22nd Int. geol. Congr., Delhi, 1964.*

22. SALOP, L. I. Pre-Cambrian geochronology and some peculiarities of the early stage of evolution of the earth. Paper presented to *22nd Int. geol. Congr., Delhi, 1964.*

APPENDIX

I.—*List of Recorded Features for Each Deposit in the Sample*

Information was recorded on a *pro forma* containing the variables listed below, and then punched on to a standard I.B.M. 80-column data card. Information was obtained from the deck of cards with the aid of a card sorter and desk calculator or an electronic computer.

Country

Name of deposit

Mine reference number
Mining field 1
Mine   2
Absolute age
  Upper Phanerozoic   0–0·3 by 1
  Lower Phanerozoic 0·03–0·06 by 2
  Upper Proterozoic   0·6–1·1 by 3
  Lower Proterozoic   1·1–1·9 by 4
  Upper Archaean   1·9–2·7 by 5
Stratigraphic or Orogenic age
  Permian to Recent (Alpine, etc.)   1
  Cambrian   to   Carboniferous
    (Hercynian and Caledonian)   2
  U. Proterozoic (Grenville, etc.)   3
  L. Proterozoic (Hudsonian, etc.)   4
  U. Archaean (Kenoran, etc.) .   5
Size in annual production of mine(s)
  Very large mine, over
    10 000 000 t per annum   .   1
  Large mine, 10–1 000 000 t per
    annum   2
  Medium mine, 1–0·1 000 000 t
    per annum   .   .   .   3
Morphology
  Stratiform
  Lenticular banded
  Lenticular massive
  Diatremes and cupolas
  Pipes
  Veins
  Fissures, mantos and stockworks
Associated rocks—Volcanic
  Andesites and dacites
  Rhyolites and trachites
  Porphyroids and keratophyres
  Spilites and greenstones
  Serpentines
  Gabbros
  Norites and anorthosites
  Alkaline rocks and carbonatites
  Basalts
Associated rocks—Plutonic
  Granite
  Granodiorite and diorite
  Syenite
Associated rocks—Sediments
  Geosynclinal 2 Shales, etc. 1
  Red beds
  Carbonaceous
  Evaporites

Associated rocks—Sediments (cont.)
  Quartzites and conglomerates
  Limestones, dolomites and marbles
  Lagoonal
Associated rocks—Metamorphic
  Supposed meta-volcanics
    (leptites, etc.)
  Amphibolites and pyroxenites
  Skarns
  Other schists
  Gneisses and migmatites
  Granulites and charnockites
Grade—Total content of ore minerals
  Solid ore 100–10% .   .   .   1
  Rich disseminated 10–1% .   .   2
  Lean disseminated under 1% .   3

Copper	over 10%	.	1
	10–5%	.	2
	5–1%	.	3
	under 1%	.	4
Lead	over 10%	.	1
	10–1%	.	2
	under 1%	.	3
Zinc	over 10%	.	1
	10–1%	.	2
	under 1%	.	3
Gold	over 1 oz (35 g)	.	1
	1 oz–2 dwt	.	2
	under 2 dwt	.	3
Silver	over 1 oz (35 g)	.	1
	under 1 oz	.	2
Nickel	over 1%	.	1
	under 1%	.	2
Cobalt	over 1%	.	1
	under 1%	.	2
Tin	over 25 lb (1·25%)		1
	under 25 lb	.	2
Iron as oxide	over 50%	.	1
	50–30%	.	2
	under 30%	.	3
Sulphur	over 30%	.	1
	under 30%	.	2

161

Other Elements (presence or absence)

Antimony	Phosphorus
Arsenic	Platinum metals
Barium	Selenium
Bismuth	Strontium
Cadmium	Boron
Fluorine	Uranium
Germanium	Vanadium
Manganese	Tungsten
Mercury	Tellurium
Molybdenum	Chromium
	Titanium

II.—*Calculation of the Morphological Diversity Factors*

For the specific group the numbers of deposits in each age group showing each of the nine morphological features were found by card sorting. The results were recorded as below.

Morphological type

Age group	10	11	12	13	14	15	16·1	16·2	16·3	Row totals
1										
2										
3										
4										
5										

The figures were re-calculated so that each of the row totals was 90. At the ideal level of maximum diversity each figure should therefore be 10. The diversity factor given is the mean of the squares of deviations of each figure from 10 along each row.

III.—*Calculation of the Rock Association Figures*

The required set of deposits was selected by sorting on the appropriate card column and those associated with a certain rock type were similarly selected. The cards were then sorted into the five age groups and counted. The numbers of deposits were rated by the total number of deposits in each age group in order to eliminate the effect of the total age variation of the deposit type.

IV.—*Calculation of Grade Factors*

Deposits containing a certain metal were sorted into age and grade groups with a card sorter. In each age group in turn the grade figures were multiplied by a factor for the magnitude of each grade group to give an index of average grade. The resulting figures were rated by the total number of deposits in each age group and reduced so that the total of the factors over the five age groups was one.

In cases where the points plotted seemed to conform to some simple relationship a curve was fitted by the method of least squares, and the function tested by

162

applying the χ^2 test. If this proved significant the curve was drawn; if not, the points were connected by eye to produce a smooth curve. Both linear and exponential curves were plotted in appropriately significant cases.

Example

Age group	Number of deposits			Total deposits in group	Index of assay	Index total	Grade factor
	> 10% × 30	10–1% × 10	< 1% × 1				
1	5	14	8	40	298	7·450	0·23
2	7	11	12	39	332	8·513	0·25
3	3	1	0	14	100	7·143	0·22
4	1	3	2	11	62	5·636	0·17
5	2	2	8	21	88	4·190	0·13
					Sum =	32·932	1·00

V.—*Calculation of Relative Abundance Factors*

Each deposit containing a selected metal was sorted according to age, grade and size. For each age group a total was derived according to the following formula:

Total metal index for age group $n = \Sigma G(g) \times S(s)$

where g = grade group from card
  s = size group from card
  G = grade magnitude constant for the selected metal
  S = size magnitude constant

These total index figures were rated by the area of the age group on the parts of the world sample (Table II) and reduced so that the total over the five age groups was one. The calculation was repeated for the nine major metals. This simple but lengthy calculation was carried out on an I.B.M. 1401/7094 computer system.

Some curves were fitted mathematically to the data points, as with the grade factors (see IV of Appendix), but in most cases points were connected by eye.

Example

Size groups on the cards are 1, > 10 000 000 tons per annum
  2, 10–1 000 000 tons per annum
  3, 1–0·1 000 000 tons per annum

If the size group s of a deposit was 2, the size constant $S = 10$; if the size group of a deposit was 1, the size constant $S = 100$, etc. Similarly, if the metal grade groups were

  1, > 10%
  2, 10–1%
  3, < 1%

the grade magnitude constant would be 30, 10 or 1 depending on whether the grade group was 1, 2 or 3 respectively. For instance, in the case of a deposit producing 1 500 000 tons per annum (size group 2) at 0·9% (grade group 3),

$$G(g) \times S(s) = 1 \times 10 = 10$$

This figure would be added to the total for the particular metal.

VI.—List of Mineral Deposits in the Sample

(Numbers refer to locations given in Fig. 20 (Plate I))

North America
1. Yellowknife
2. Pine Point
3. Lynn Lake
4. Thompson
5. Labrador Iron
6. Elke River
7. Jedway
8. Texada
10. Flin Flon
12. Red Lake
13. Mattagami Lake
14. Chibougamau
15. Allard Lake
16. Quebec–Cartier
17. Kennedy Lake
18. Sullivan
19. Craigmont–Merritt
20. Butte
21. Cœur d'Alene
22. Hartville
23. Gas Hills
24. Homestake
25. Thunder Bay
26. Steep Rock
27. Mesabi Range
28. Michipicoten
29. Porcupine
30. Sudbury
31. Murdockville
32. Mount Pleasant
33. Buchans
34. Wabana
35. Bingham
36. Tintic
37. Upper Mississippi Pb–Zn
38. Marmorton
39. Suffield, Sherbrook
40. St. Lawrence Company, N.Y.
41. Cornwall, Pennsylvania
42. Stirling Hill
43. Ely
44. Cedar City
45. Climax
46. Leadville
47. Questa
48. Tri-State Field
49. South-east Mississippi Field
50. Iron Mountain (Pilot Knob)
51. East Tennessee Field
52. Ducktown
53. Eagle Mountain
54. Globe
55. Chino
56. Bisbee
57. La Cananea

58. Frisco
59. Naica, Saucillo
60. Texas Iron Ores Field
61. Birmingham, Alabama
62. Boleo
63. Plateros
64. Cerro de Mercado
65. Providencia
66. Agaltica
67. Rosita
68. Mochito

South America
1. Silencio
2. El Callao
3. Cerro Bolivar
4. Cerro de Pasco
5. Corocoro District
6. Ouroro (Itos)
7. Caraiba
8. Itabira (Caué Peak)
9. Morro Velho
10. Casa de Pedra
11. Niquelandia
12. Jacupiranga
13. Zapla
14. Chuquicamata
15. El Salvador
16. El Teniente
17. Aguilar Jujuy
18. Sierra Grande

Western Europe
1. Kiruna
2. Laisvall
3. Boliden
4. Vihanti
5. Otenmaki
6. Outokumpu
7. Fosdalen
8. Løkken
9. Stråssa
10. Folun
11. Knaben
12. Tellness
13. Silvermines
14. Avoca
15. Northampton Field
16. Rammelsberg
17. Grund Hartz
18. Moresnet
19. Lorraine
20. Soumont et de May-Orne
21. Anjou, Bretagne
22. Raibl
23. Majdanpek

Western Europe (cont.)
24. Ljnbija
25. Cassandra
26. Aosta
27. Elba
28. L'Argentière
29. Bilbao
30. Panasqueira
31. Tuella
32. San Domingos
33. Rio Tinto
34. Ponente Masua
35. Marrovourrie
36. Engani
37. Divrik
38. Murgul
39. Timna

Southern Asia and Australasia
1. Angouran
2. Bafq
3. Bafq, Kerman
4. Zawar
5. Khetri
6. Bailadila Field
7. Orissa
8. Mosaboni
9. Kola
10. Salem, Trichenopoly
11. Bawdwin
12. Pelepah Kanan
13. Sungei Lembing
14. Hitachi
15. Kucchan
16. Kamioka
17. Besshi
18. Lepanto
19. Larap
20. Tennant Creek
21. Constance Range
22. Mt. Goldsworthy
23. Mt. Isa
24. Mary Kathleen
25. Mt. Morgan
26. Koolyianobing
27. Ravensthorpe
28. Sons of Gwallia

29. Kalgoorlie (Lake View and Star)
30. Central Norseman Field
31. Ediacara
32. Iron Knob
33. Broken Hill
34. Radium Hill
35. Ardlethan
36. Cobar
37. Captain's Flat
38. Ballarat Field
39. Mt. Lyell
40. Roseberry

Africa
1. Ait Amar
2. Bou Bekir
3. Buenza
4. Djerissa
5. Assouan
7. Whadi Fatima
8. Mahad Dhabad
9. Bom Gheib
10. Gara Djebilet
11. Akjoujit
12. Marampa
13. Nimba
14. Ashanti
15. Abakaliki
16. Kilembe
17. Shinkolobwe
18. Bancroft
19. Roan Antelope
20. Broken Hill
21. Sange
22. Tsumeb
23. Sinoia
24. Cam and Motor
25. Empress
26. Thabazimbe
27. Rooiberg
28. Palabora
29. Bomvu Ridge
30. Central Rand Goldfield
31. O.F.S. Field
32. Postmansburg Field
33. Namaqualand Field

14

Reprinted from *Geol. Assoc. Canada Spec. Paper 9*, 97–103 (1971)

THE EVOLUTION OF THE RICE LAKE – GEM LAKE GREENSTONE BELT, SOUTHEASTERN MANITOBA

Werner Weber
Mines Branch, Manitoba Department of Mines, Resources and Environmental Management, Winnipeg, Manitoba

ABSTRACT

The evolution of the Rice Lake-Gem Lake greenstone belt is compared with an Alpine-type geosynclinal development during one orogenic cycle.

The geosyncline started as a eugeosyncline with a small miogeosynclinal zone. The volcanogenic, lower section of the Rice Lake Group represents the eugeosynclinal phase. It consists of two (ophiolitic) volcanic piles and derived sedimentary rocks. The older pile (Bidou Lake Subgroup) is bimodal, composed predominantly of basaltic flows and dacitic pyroclastics. The younger pile (Gem Lake Subgroup) forms a complete basalt-andesite-dacite-rhyolite cycle. Evidence of a volcanic vent and a felsite plug, comagmatic to the rhyolite, are found in the Gem Lake Subgroup.

The ophiolitic rocks are overlain by the sedimentary Edmunds Lake Formation which, in the southern part of the greenstone belt, undergoes a facies change from eugeosynclinal to flyschoid. This uppermost section of the Rice Lake Group is characterized by the increased deposition of turbidity greywacke. The main flysch trough lies between the centre axis of the greenstone belt and the tectonically most active part of the area, the Manigotagan gneissic belt to the south.

Molasse deposits are represented by the San Antonio Formation. The syn- and late-orogenic plutonism is accompanied by complex deformations and regional metamorphism in the south, and largely thermal metamorphism in the north. The orogenic events lasted approximately 400 m.y. and ceased with the formation of mylonite zones which reflect isostatic readjustments.

INTRODUCTION

The area studied comprises the Rice Lake greenstone belt from Wanipigow Lake to the Manitoba-Ontario border (Weber, Figure 1, in pocket at rear). It forms part of an Archean greenstone belt, exposed from the east shore of Lake Winnipeg into the Bee Lake area, just east of the Ontario-Manitoba border. This belt is probably related to the Uchi Lake belt (Ontario), because both belts lie north of the major Lake St. Joseph fault zone (Parkinson, 1962).

The Rice Lake region was investigated from 1966 to 1969 during "Project Pioneer", a joint project of the Manitoba Mines Branch and the University of Manitoba (Davies, 1966). Studies were undertaken in several fields of the earth sciences, (geology, geophysics and geochemistry); the results are contained in Manitoba Mines Branch Publication 71-1.

Perspective was given to the project by conducting the geological investigations on both detailed and regional scales. The writer conducted his main studies south and west of Bissett, and in the Long Lake-Gem Lake-Flintstone Lake area (Weber, 1971a, b; McRitchie and Weber, 1971a, b). Although the present paper deals mainly with observations in these areas and the writer's interpretations, it could not have been written without the results of other geologists on this project.

THE GEOLOGICAL MODEL

The development of the Rice Lake-Gem Lake greenstone belt is summarized in Table I, which lists the succession of various rock types. The evolution of the belt shows some similarities with the Alpine orogeny, e.g. the development and sequence of geosynclinal stages (eugeosyncline-miogeosyncline pair, flysch, molasse) and related igneous activity (pre- and early-orogenic ophiolitic extrusions, and syn- and late-orogenic granitic intrusions).

Other characteristics in the evolution of the Rice Lake greenstone belt are different from Alpine, e.g. the types of metamorphism and deformation. The metamorphism in the Rice Lake region was of lower pressure and had a higher thermal gradient than that of the Alpine orogeny, which indicates a thinner crust with higher heat flow in the Archean (McRitchie, 1971c). The deformation was therefore more plastic and resulted in isoclinal folding rather than thrust faulting.

Alpine terminology is generally not applied to Archean geology, because of differences between Alpine and Archean geology, and because of various misuses of terminology which lead to misconceptions. The questionable validity of applying Alpine terminology to the Archean has recently been discussed in Anhaeusser, et al., (1969). Goodwin (1970) only recently adopted some Alpine terms, i.e., flyschoid or flysch-like. Nevertheless, the Alpine geosynclinal model has been applied to the Rice Lake belt in order to illustrate certain changes in and inter-relations between igneous and metamorphic-deformational processes, and sedimentary patterns, e.g. the development of the flysch facies towards the tectonically more active part of the area. The Alpine model has recently been applied also to Archean sequences of Australia (Glikson, 1970), and observations and interpretations of the Rice Lake area are similar. The Alpine event is also an ideal model because it is well documented and known to a degree that will never be possible in most of the Precambrian shield areas because of inferior exposure conditions, particularly in the third dimension, and the lack of biostratigraphical control.

EUGEOSYNCLINAL STAGE

The eugeosynclinal deposits of the Rice Lake greenstone belt are represented by volcanic (ophiolitic) and derived sedimentary rocks of the Rice Lake Group. This group is subdivided (from bottom to top) into Bidou Lake Subgroup, Gem Lake Subgroup, and Edmunds Lake Formation. The Bidou Lake Subgroup (Weber, 1971a) consists of a basalt-dacite sequence and associated sedimentary rocks. It represents the oldest exposed extrusive event. Rocks of this subgroup have a large areal extent (Figure 1, from north of Gem Lake to Wanipigow Lake) and comprise most of the greenstones (Church and Wilson, 1971; Weber, 1971b). The total thickness of the Bidou Lake Subgroup is approximately 20,000 to 25,000 feet.

The Gem Lake Subgroup (Weber, 1971a) represents a younger period of volcanic activity (Table I). It consists of a complete basalt-andesite-dacite-rhyolite cycle and minor associated sedimentary rocks. Rocks of the Gem Lake Subgroup are exposed in the Gem Lake area (Weber, 1971a) and similar rocks occur south of Rice Lake (Weber, 1971b). The Gem Lake Subgroup is of limited extent and is only half as thick as the Bidou Lake Subgroup.

The Edmunds Lake Formation is an extensive sedimentary sequence which overlies both volcanogenic subgroups. Its basal section consists of argillite and oxide facies iron formation

TABLE I: Table of Formations, Rice Lake area, southeastern Manitoba

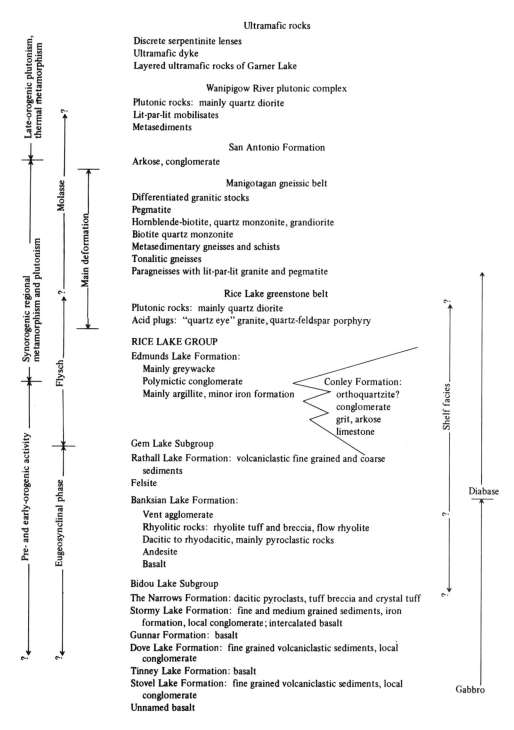

Ultramafic rocks
Discrete serpentinite lenses
Ultramafic dyke
Layered ultramafic rocks of Garner Lake

Wanipigow River plutonic complex
Plutonic rocks: mainly quartz diorite
Lit-par-lit mobilisates
Metasediments

San Antonio Formation
Arkose, conglomerate

Manigotagan gneissic belt
Differentiated granitic stocks
Pegmatite
Hornblende-biotite, quartz monzonite, grandiorite
Biotite quartz monzonite
Metasedimentary gneisses and schists
Tonalitic gneisses
Paragneisses with lit-par-lit granite and pegmatite

Rice Lake greenstone belt
Plutonic rocks: mainly quartz diorite
Acid plugs: "quartz eye" granite, quartz-feldspar porphyry

RICE LAKE GROUP
Edmunds Lake Formation:
    Mainly greywacke
    Polymictic conglomerate
    Mainly argillite, minor iron formation

Conley Formation:
    orthoquartzite?
    conglomerate
    grit, arkose
    limestone

Gem Lake Subgroup
Rathall Lake Formation: volcaniclastic fine grained and coarse sediments
Felsite

Banksian Lake Formation:
    Vent agglomerate
    Rhyolitic rocks: rhyolite tuff and breccia, flow rhyolite
    Dacitic to rhyodacitic, mainly pyroclastic rocks
    Andesite
    Basalt

Bidou Lake Subgroup
The Narrows Formation: dacitic pyroclasts, tuff breccia and crystal tuff
Stormy Lake Formation: fine and medium grained sediments, iron formation, local conglomerate; intercalated basalt
Gunnar Formation: basalt
Dove Lake Formation: fine grained volcaniclastic sediments, local conglomerate
Tinney Lake Formation: basalt
Stovel Lake Formation: fine grained volcaniclastic sediments, local conglomerate
Unnamed basalt

Diabase

Gabbro

Shelf facies

Late-orogenic plutonism, thermal metamorphism

Synorogenic regional metamorphism and plutonism

Pre- and early-orogenic activity

Molasse

Flysch

Eugeosynclinal phase

Main deformation

167

and is similar to the older volcaniclastic rocks. This section is also part of the eugeosynclinal sedimentation. The rocks are very similar to the Alpine, eugeosynclinal "Buenderschiefer" (Truempy, 1960).

The surface geology indicates that the eugeosynclinal trough was filled predominantly by ophiolitic material and that sediments were subordinate. This ratio is the reverse of that found in the Alpine geosyncline.

Bidou Lake Subgroup

The basalts of the Bidou Lake Subgroup outcrop south of Wallace Lake (Figures 1, 2, in map pocket at rear) in the core of a dome structure plunging northwest and southeast (see also Stockwell, 1945).

There are three formations of basalt (Table I), each overlain by a formation of derived sedimentary rocks (Figure 2; Campbell, 1971). Each basalt formation is thought to represent a period of extrusion. The basalts are commonly pillowed, indicating subaqueous extrusion. Taking the size and abundance of vesicles in the pillows as an indication of water depth (Moore, 1965), depths have been estimated to be in excess of 800 m for most of the rocks (Weber, 1971a). The basalts contain sills of fine grained gabbro that are similar to the basalts and are interpreted to be comagmatic, shallow intrusions. The three basalt units have a maximum exposed thickness of approximately 4,000 feet (Figure 2, E-E') One of the characteristic features of the interlayered sediments is their lack of detrital quartz, which reflects the basic composition of the source area. The sediments are mainly fine grained, thinly bedded and often graded. They are partly pyroclastic and partly epiclastic. Arenitic and conglomeratic beds occur locally. The conglomeratic horizons contain dacitic pebbles. This indicates, that in addition to the basalt, felsic differentiates were also extruded during this early volcanism, however the source rocks of these felsic fragments are not exposed. The maximum thickness of each sedimentary formation is approximately 1,000 feet (Figure 2, E-E').

The rate of sedimentation was generally slow during deposition of the Bidou Lake Subgroup. In places, however, soft rock deformation and graded bedding are evidence of a more rapid sedimentation by turbidity currents.

The change from basaltic to more acidic volcanism was rather abrupt. The Narrows Formation, which overlies the basalts and derived sedimentary rocks, comprises dacitic pyroclastic rocks (approximately 10,000 feet thick). The initial phase of this eruption was also basaltic, but of much less magnitude than any of the older basalt formations. Andesites have not been found. Although rhyolitic lapilli occur in tuff breccia of The Narrows Formation, the bulk of the fragments (approximately 95%) are dacitic in composition (Plate I, figure 1). It was concluded (Weber, 1971a) that most of The Narrows Formation represents submarine pyroclastic flows, similar to those described by Fiske (1963).

Gem Lake Subgroup

The Gem Lake Subgroup (Table I) comprises the volcanic rocks of the Banksian Lake Formation, a comagmatic intrusive felsite, and the volcaniclastic rocks of the Rathall Lake Formation.

The composition of the volcanic rocks varies from basalt, to andesite, dacite and rhyolite. Basalt and andesite occur as volcanic flows, whereas dacite and rhyolite are represented mainly by pyroclastic rocks (Plate I, figure 2). The chemical compositions of the volcanic sequence have been established by chemical analyses (Weber, 1971a). The basic and intermediate volcanics resemble corresponding rocks of the

older bimodal cycle. Andesite occurs only locally, north of Gem Lake (figure 1). The rhyolitic rocks occur as breccia, flows (Plate I, Figure 3) and brecciated flows. They are soda rhyolites, as are most of the Archean rhyolites. The amount of rhyolitic material is very small compared to the intermediate and basic volcanic rocks. They are of restricted extent and occur only at Gem Lake near a postulated volcanic vent (Figures 1, 2). This vent is marked by the concentric arrangement of rhyolitic units about a central vent agglomerate and a felsite, a shallow seated intrusion. On the basis of chemical and textural similarities this felsite appears to be comagmatic with the rhyolite.

In addition to the volcanic vent at Gem Lake, there is evidence for other possible vents of intermediate to acidic volcanic composition in the following areas:

1) Long Lake area: the largest breccia fragments (up to 3 m long) have been observed there, particularly in the basal part of The Narrows Formation. The volcanic vent may have been situated north of Long Lake and replaced subsequently by the quartz diorite batholith.

2) Bissett area: two areas, 3 miles southeast of Bissett (Gold Lake area) and 3 miles southwest of Bissett (Gilbert shaft), are characterized by the association of quartz feldspar porphyry (acid plug on Figures 1,2) and dacitic to rhyolitic, massive and fragmental volcanic rocks. The similarity of rock types to those in the Gem Lake area and their spatial relationship suggests that these two areas may also be volcanic vents. The quartz feldspar porphyry would represent a comagmatic plug, comparable with the felsite in the Gem Lake area.

3) Wanipigow Lake area: the quartz feldspar porphyry (acid plug in Figure 1) may be a shallow intrusion in a volcanic vent area, similar to those in the Bissett area.

At least some of the volcanic piles have probably reached above sea level (island arcs, geanticlines?) as indicated by the coastal facies of the late-volcanic Rathall Lake Formation. It consists of volcaniclastic conglomerate, arkose and minor siltstone up to 1,000 feet thick. The only exposures occur between Gem and Flintstone Lakes, but pebbles in the Edmunds Lake Formation indicate that similar sedimentary rocks originally extended over wider areas and have been reworked subsequently during the Flysch phase.

Edmunds Lake Formation, basal part

The volcanic and volcaniclastic rocks of the Gem Lake and Bidou Lake Subgroups are overlain by sedimentary rocks of the Edmunds Lake Formation (Table I, figure 2). Its lower section (mainly argillite, minor iron formation, Table I) displays a eugeosynclinal "Buendnerschiefer" facies (Truempy 1960), consisting mainly of argillite (mudstone, siltstone shale), and minor fine- to medium-grained greywacke beds. Near the base of the Edmunds Lake Formation one or more layers of oxide facies iron formation are commonly present. It is partly developed as banded magnetite-chert layers, as on Wallace Lake, but generally it is exposed as layers of black magnetite-rich argillite.

The presence of iron formation indicates oxidizing conditions due to good circulation of aerated water and not necessarily shallow water. On the other hand the local occurrence of pyritic shale near the oxide facies iron formation is indicative of reducing conditions, suggesting that certain parts of the geosynclinal basin were temporarily isolated from the main trough.

MIOGEOSYNCLINAL ZONE

Only the area north of Wallace Lake on the northern

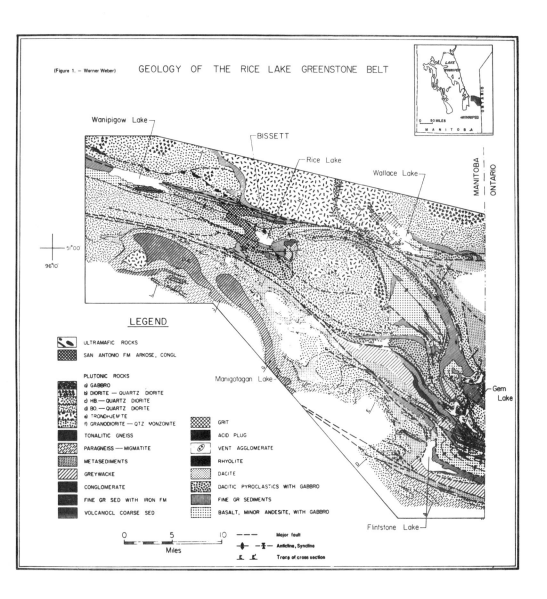

(Figure 1. – Werner Weber) GEOLOGY OF THE RICE LAKE GREENSTONE BELT

Wanipigow Lake

BISSETT

Rice Lake

Wallace Lake

MANITOBA

ONTARIO

51°00'
96°10'

LEGEND

| | ULTRAMAFIC ROCKS |
| | SAN ANTONIO FM ARKOSE, CONGL. |

PLUTONIC ROCKS
a) GABBRO
b) DIORITE — QUARTZ DIORITE
c) HB.— QUARTZ DIORITE
d) BO.— QUARTZ DIORITE
e) TRONDHJEMITE
f) GRANODIORITE — QTZ MONZONITE

	TONALITIC GNEISS
	PARAGNEISS — MIGMATITE
	METASEDIMENTS
	GREYWACKE
	CONGLOMERATE
	FINE GR SED WITH IRON FM
	VOLCANOCL COARSE SED

	GRIT
	ACID PLUG
	VENT AGGLOMERATE
	RHYOLITE
	DACITE
	DACITIC PYROCLASTICS WITH GABBRO
	FINE GR SEDIMENTS
	BASALT, MINOR ANDESITE, WITH GABBRO

Manigotagan Lake

Gem Lake

Flintstone Lake

0   5   10
Miles

– – – Major fault
Anticline, Syncline
E E' Trace of cross section

169

PLATE I

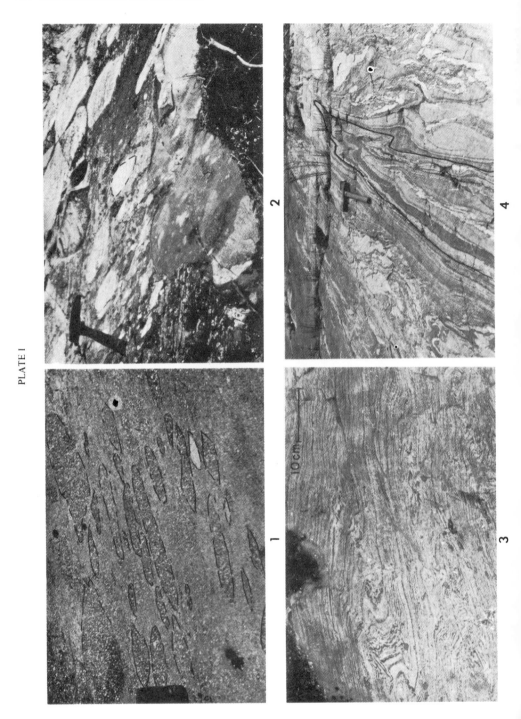

margin of the greenstone belt shows a miogeosynclinal facies (shelf facies, Conley Formation, Table I). It is composed of grit and sandstone with minor tonalite-boulder conglomerate, siltstone, euxinic shale and limestone. The detritus is predominantly of granitic origin and was derived from the northern craton. This miogeosynclinal facies grades along strike and upward into a eugeosynclinal facies (McRitchie, 1971b).

Rocks of miogeosynclinal facies probably occurred along the entire northern boundary of the greenstone belt but are no longer exposed due to the later intrusion of the plutonic Wanipigow River Suite and the subsequent faulting in this area (Figure 1; McRitchie, 1971b).

FLYSCH

The typical Alpine flysch (Truempy, 1960) consists of a thick sequence of rhythmically interlayered, immature sandstone and shale, indicating rapid and continuous marine sedimentation, largely by turbidity currents. The submarine relief is greater than during the eugeosynclinal stage as a result of the start and continuation of tectonic movements. The flysch is therefore synorogenic. It marks the filling of the geosyncline.

Most of the Edmunds Lake Formation is similar to the Alpine flysch (except the eugeosynclinal basal part). In contrast to the eugeosynclinal sedimentary rocks, the flyschoid sequence is characterized by a predominance of greywacke over argillite, a lithic (instead of feldspathic) greywacke, increased quartz content and generally thicker-bedded sequences. Turbidity action is indicated by graded greywacke sequences, load casts and slump breccias.

The flyschoid facies is thickest and most characteristic in the southern part of the greenstone belt, grading into metasedimentary gneisses and schists of the Manigotagan gneissic belt (Figure 1). The Edmunds Lake Formation along the eastern and northern boundary of the greenstone belt shows predominantly eugeosynclinal facies.

The contact between flyschoid and eugeosynclinal facies is commonly gradational. In the area between Long Lake and Gem Lake, however, a polymictic conglomerate occurs at the facies change, resulting in a sharp contact. Most fragments of this conglomerate are intraformationally derived, perhaps from escarpments within the basin; some, however, are exotic, e.g., the orthoquartzite pebbles and well rounded tonalitic fragments. The exotic clasts are interpreted as reworked shallow water deposits of unknown origin. The exotic character and the mixing with intraformational fragments suggests a sudden increase in relief in the flanks of the geosyncline, which is interpreted to indicate the start of major tectonic movements. The tonalitic fragments have been found in the Gem Lake, Wanipigow Lake and Wallace Lake areas. In the latter area they occur in the shelf facies, which may be similar to the source rocks of the exotic fragments (Table I).

The thickness of the Edmunds Lake Formation can only be roughly estimated, due to the isoclinal folding (Figure 2), the incertainty about the position of the southern and northern boundaries of the depositional trough due to subsequent metamorphism and plutonism. A conservative estimate of the maximum thickness of the Edmunds Lake Formation along the southern part of the belt would be 20,000 feet. Of this, a maximum of 5,000 feet would be eugeosynclinal, and the rest, or 15,000 feet, flyschoid facies rocks.

MOLASSE

The San Antonio Formation has many characteristics of typical molasse (Truempy, 1960): (a) it fills the marginal depression of the main orogenic belt; (b) it is a continental deposit; (c) it is a syn- to late-tectonic sedimentary deposit, because it postdates the early orogenic quartz diorite batholith (Davies, 1963; Weber, 1969) and the isoclinal folding (D_1, McRitchie and Weber, 1971b), but is deformed by a later fold phase (D_2). The San Antonio basin could also be termed a exogeosyncline.

The San Antonio Formation is dominantly pebbly arkose, with minor interlayered conglomerate or pebble bands (Weber, 1969, 1971a). Similar sediments were probably, once, more widespread, but are only preserved in the deeper part of a graben structure. The facies of the San Antonio Formation exposed in the Bissett area is possibly not typical, as it is composed mainly of granitic detritus with only minor erosion products of the volcanic and/or sedimentary rocks originally overlying the granites. The granitic detritus was derived from the quartz diorite to the south and from granitic rocks which were situated probably to the north. The southern basal section of the San Antonio Formation is composed of coarse landslide material and boulder conglomerate (Weber, 1969; McRitchie and Weber, 1971a) suggesting considerable relief to the south near the shore of a shallow-water basin (shallow water is indicated by limestone matrix in the conglomerate). The landslide may have been caused by uplift (start of isostatic readjustment?), or an advancing mountain front. In addition to the facies, sedimentary structures (cross bedding, channel filling and lack of graded bedding), characterize the San Antonio Formation as a continental delta deposit.

PRE- AND EARLY-OROGENIC IGNEOUS ACTIVITY

The pre- and early-orogenic igneous activity is largely extrusive. The intrusive phase is partly represented by gabbros. On the basis of the close spatial relationship with the basalts and similarities in chemical composition, they are considered to be consanguineous (Weber, 1971a). Gabbros continued to form during the felsic volcanism. They probably were derived from the same magma chamber that produced the earlier basalts and gabbros. One of the later gabbro sills is anorthositic in the core (Weber, 1971a) thus indicating a possible genetic link with the ultramafic rocks of the area.

The intrusion of gabbros terminated with the end of volcanism, and, subsequently, only a few diabase dykes were emplaced.

The felsite plug in the Gem Lake area, and possibly also the quartz feldspar porphyry plugs in the Bissett region, are the comagmatic intrusive equivalents of the felsic volcanism. These plugs in turn are related to the quartz diorite plutons within the greenstone belt by similarities in structural setting. This, and the chemical affinities of the felsic volcanic and plutonic rocks (Weber, 1971a), are the basis for the hypothesis that the

---

PLATE I

Figure 1: Dacite tuff breccia of the Bidou Lake Subgroup, The Narrows Formation. Note the similar composition of most of the fragments and the matrix.

Figure 2: Tuff breccia of the Gem Lake Subgroup. Rhyolitic fragments in dacitic groundmass.

Figure 3: Flow-banded and -folded rhyolite of the Gem Lake Subgroup.

Figure 4: D_1 isoclinal fold, refolded by D_2-Z fold in layered paragneiss of the Manigotagan gneissic belt.

quartz dioritic plugs and plutons are comagmatic to the volcanism. There is evidence that the quartz diorite is younger than the volcanogenic sequence of the Rice Lake Group and was emplaced under a weak stress field, but prior to the main metamorphism and deformation of the Manigotagan gneissic belt. The San Antonio Formation is transgressive onto one of the quartz diorite plutons. The quartz diorite is therefore early-orogenic.

SYNOROGENIC REGIONAL METAMORPHISM AND PLUTONISM; DEFORMATION

The first major orogenic events are indicated by the polymictic conglomerate of the Edmunds Lake Formation. The main deformation took place in the southern part of the greenstone belt and in the Manigotagan gneissic belt, and involved probably the entire area between the Rice Lake and Bird River greenstone belts. The deformation was associated with a regional metamorphism, with the higher amphibolite facies as the highest grade (McRitchie and Weber, 1971b), and the emplacement of granitic rocks (McRitchie, 1971a). Several phases of deformation and metamorphism have been differentiated (McRitchie and Weber, op. cit.), but collectively, they represent one orogenic event, the Kenoran orogeny. The deformation started with plastic deformation (D_1, isoclinal folding, Figure 2; D_2, similar folding, Plate I, figure 4) during a period with increased thermal gradient. became more brittle with decreasing temperatures, and ended with fracturing. The main fractures are developed along the margins of the greenstone belt and along batholiths within the belt, and reflect isostatic readjustments related to the imbalance produced by the rising of granitic melts during the orogeny.

The flyschoid facies of the Edmunds Lake Formation became progressively metamorphosed towards the south (metasediments, Figure 1). Adjoining to the south are paragneisses (paragneiss-migmatite, Figure 1) of unknown origin. They are either higher metamorphosed rocks of the Edmunds Lake Formation, or part of an older, the eugeosynclinal Edmunds Lake Formation underlying, basement which was remobilized during the Kenoran orogeny. Although there is no direct evidence, such as basement could be the source of much detritus and would explain the increased sedimentation of more quartz-rich greywacke in the flyschoid trough to the north.

Granitic rocks were emplaced in the form of diapiric plutons, *lit-par-lit* mobilisates and metasomatic processes (McRitchie, 1971a) partly in the paragneisses and partly in the metasediments. Most of the plutons are granodiorites and quartz monzonites, probably representing late-kinematic melts. The tonalitic gneisses, which occupy large areas in the Manigotagan gneissic belt and farther south (Figure 1; McRitchie, 1971a), were derived from sedimentary or older plutonic rocks.

Based on Rb-Sr isochron ages (Turek and Peterman, 1968) the igneous-metamorphic portion of this orogenic cycle lasted approximately 400 m.y. During the first 200 m.y. (2730 m.y. to 2530 m.y., Turek and Peterman, op. cit.), the syn- and late-orogenic plutons crystallized and the main deformation and metamorphism took place. The thermal gradients decreased afterward and the last event, the mylonitization, occurred 2350 m.y. ago (Turek and Peterman, op. cit.).

LATE-OROGENIC PLUTONISM

The Wanipigow River plutonic complex, composed of several phases of dominantly quartz dioritic composition (Figure 1; Marr, 1971), represents the youngest granitic activity in the area (Turek and Peterman, 1968). This activity is probably also younger than the San Antonio Formation

(Turek and Peterman, op. cit.; Davies, 1963). The intrusion o the plutonic complex has produced a 1 mile to 4 miles-wide zone of contact metamorphism in the adjacent rocks of the greenstone belt. The similarity of these batholiths to the quartz diorites within the greenstone belt raises the questions whether these intrusions may be genetically related, an whether the Wanipigow River plutonic complex may have been emplaced earlier (early-orogenic) and partly remobilize later (Figure 1). Recent lead isotope studies by Ozard an Russell (1971) support this hypothesis.

The ultramafic rocks of the Rice Lake region (Scoates 1971) appear to be, at least in part, late-orogenic. Th ultramafic rocks at Garner Lake (Figures 1, 2) form a layere body, consisting of serpentinized peridotite an clinopyroxenite. A contact aureole was developed in the roc zone of the ultramafic intrusion on the north shore of Garne Lake, but not below the intrusion on the south shore.

A late ultramafic dyke, emplaced along a shear zone in th Wadhope area (Figures 1, 2, E-E'), is possibly related to th Garner Lake body, but no relationship is evident for othe ultramafic and ultrabasic rocks in the area. These are th gabbro sills with anorthositic phases (Weber, 1971a) and th discrete serpentinite lenses which occur all along the nort margin of the greenstone belt (Figures 1, 2, L-L'). They a possibly related to the early ophiolitic volcanism; th serpentinite lenses represent, perhaps, dislocated parts o feeder channels that led into the mantle. This would impl that an early, deep fracture existed along the northern margi of the greenstone belt, and that this fracture was reactivate later to form one of the main fault zones in the area (Wallac Lake-Wanipigow Lake fault).

CONCLUSIONS

The Rice Lake greenstone belt started as an east-we: striking, most likely intracratonic, subsiding trough. Th margin of the northern craton was near the (now exposed contact of the granitic rocks to the north, the position of th margin of the southern craton is uncertain. The troug possibly started as a graben structure such as described b Illies (1969), by faulting of the sialic crust and upsurging differentiated mantle material. The crust became thinne finally broke and drifted apart.

The history of the trough is comparable with th Alpine-type geosynclinal evolution. In the Rice Lake area, th evolution is related to one orogenic cycle, the tectonic style an metamorphism of which, however, are not typically Alpin The subsiding trough is comparable with an eugeosynclin flanked by a small miogeosynclinal zone, exposed at th margin of the northern craton. The eugeosyncline characterized by the extrusion of predominantly basaltic an dacitic ophiolites, and associated, relatively quie sedimentation. There is no evidence that much detritus fro the craton reached this part of the trough; it was deposited the miogeosynclinal zone.

In the southern part of the greenstone belt th eugeosynclinal stage is superseded by a more flyschoid phas characterized by increased deposition of sediments. In contra to the earlier epiclastics these are less mature, more lithi richer in quartz, and are entirely turbidite deposits. Parts the clastic material are recognizable as of volcanic origin, b sources of sedimentary and plutonic rocks are also indicate It is, however, not possible to determine the source of most the epiclastic material.

The flysch facies indicates an increase in relief which probably related to orogenic tectonism, such as folding an uplift, etc. With the initial movements, the volcanism cease and was superseded by plutonic intrusions.

The tectonic activity and metamorphism reached its peak

the Manigotagan gneissic belt—to the south of the greenstone belt—with the intrusion of synkinematic, granitic melts.

Molasse facies developed in a subsidiary basin near the northern craton. The detritus is mainly granitic and was derived, at least partly, from greenstone belt-internal plutons.

Late-orogenic quartz diorites (remobilized?) and ultramafic rocks represent the latest igneous activity; both occur along the northern margin of the greenstone belt, the ultramafic rocks mostly at the contact between greenstones and the late-orogenic quartz diorites. Some of the ultramafic rocks are re-emplaced within even later faults and may be part of the ophiolitic volcanism, others, however, are not and their late appearance in the orogenic history of the greenstone belt is not typically Alpine.

ACKNOWLEDGEMENTS

I am grateful to W. D. McRitchie, R.F.J. Scoates and F.H.A. Campbell, Manitoba Mines Branch, and H. Zwanzig, Queen's University, for fruitful discussions and for permission to include some of their not yet published work in this paper.

Published with permission of the Minister of Mines, Resources and Environmental Management, Province of Manitoba.

SELECTED REFERENCES

Anhaeusser, C. R., Mason, R., Viljoen, M. J., and Viljoen, R. P. (1969) — A reappraisal of some aspects of Precambrian shield geology; Bull. Geol. Soc. Amer., vol. 80, pp. 2175-2200.

Campbell, F.H.A. (1971) — Stratigraphy and sedimentation of part of the Rice Lake group; Manitoba Mines Br. Publ. 71-1, pp. 135-188.

Church, N. B. and Wilson, H.D.B. (1968) — Volcanology of the Wanipigow Lake-Beresford Lake area; Manitoba Mines Br. Publ. 71-1, pp. 127-134.

Davies, J. F. (1963) — Geology and gold deposits of the Rice Lake-Wanipigow River area; Ph.D. thesis, Univ. Toronto.

. . . . . (1966) — Project Pioneer—A new approach to the study of Precambrian geology in Manitoba; Can. Mining J., vol. 87, pp. 86-104.

Fiske, R. S. (1963) — Subaqueous pyroclastic flows in the Ohanapecosh Formation, Washington; Bull. Geol. Soc. Amer., vol. 74, pp. 391-406.

Goodwin, A. M. (1968) — Growth and early crustal history of the Canadian shield; XXIII Internatl. Geol. Congress, vol. 1, pp. 69-89.

. . . . . (1970) — Preliminary reconnaissance of the Flin Flon volcanic belt; Manitoba and Saskatchewan; Geol. Surv., Canada, Paper 69-1A.

Glikson, A. Y. (1970) — Geosynclinal evolution and geochemical affinities of early Precambrian systems; Tectonophysics, vol. 9, pp. 397-433.

Illies, J. H. (1969) — An intercontinental belt of the world rift system; Tectonophysics, vol. 8, No. 1, pp. 5-29.

Marr, J. M. (1971) — Petrology of the Wanipigow River suite, Manitoba; Manitoba Mines Br. Publ. 71-1, pp. 203-214.

McRitchie, W. D. (1969) — Project Pioneer (7 and 8); Summ. geological fieldwork, Manitoba Mines Br., Geol. Paper 4/69, pp. 107-114.

. . . . . . (1971a) — Petrology and environment of the acidic plutonic rocks of the Wanipigow-Winnipeg Rivers region; Manitoba Mines Br. Publ. 71-1, pp. 7-62

. . . . . . (1971b) — Geology of the Wallace Lake-Siderock Lake area: a reappraisal; Manitoba Mines Br. Publ. 71-1, pp. 107-126.

. . . . . . (1971c) — Metamorphism in Precambrian rocks of Manitoba; Geol. Assoc. Can., Sp. Paper 9 (this volume).

McRitchie, W. D. and Weber, W. (1971a) — Geology of the Manigotagan River-Moose River area; Manitoba Mines Br. Publ. 71-1, Maps 69-1, 2, 3 and 4.

. . . . . . (1971b) — Metamorphism and deformation in the Manigotagan gneissic belt, southeastern Manitoba; Manitoba Mines Br. Publ. 71-1, pp. 235-284.

Moore, J. A. (1965) — Properties of Hawaiian submarine basalts; Geol. Soc. Amer., Sp. Paper 82, p. 267.

Ozard, J. M. and Russell, R. D. (1971) — Lead isotope studies of rock samples from the Superior geological province; Can. J. Earth Sci., vol. 8, No. 4, pp. 444-454.

Parkinson, R. N. (1962) — Operation overthrust; the tectonics of the Canadian Shield; Roy. Soc. Can., Sp. Publ. No. 4.

Scoates, R.F.J. (1971) — Ultramafic rocks of the Rice Lake greenstone belt; Manitoba Mines Br. Publ. 71-1, pp. 189-202.

Stockwell, C.H. (1938) — Rice Lake-Gold Lake area, southern Manitoba; Geol. Surv., Canada, Mem. 210.

. . . . . . (1945) — Can. Dept. Mines Resources, Maps 809A (Beresford Lake), 810A (Rice Lake), 811A (Gem Lake).

Stockwell, C. H. and Lord, C. S. (1939) — Halfway Lake-Beresford Lake, Manitoba; Geol. Surv., Canada, Mem. 219.

Truempy, R. (1960) — Paleotectonic evolution of the central and western Alps; Bull. Geol. Soc. Amer., vol. 71, pp. 843-908.

Turek, A. and Peterman, Z. E. (1968) — Preliminary Rb-Sr geochronology of the Rice Lake-Beresford Lake area, southeastern Manitoba; Can. J. Earth Sci., vol. 5, No. 6, pp. 1373-1380.

Weber, W. (1969) — Project Pioneer (5 and 6); Summ. geological fieldwork, Manitoba Mines Br., Geol. Paper 4/69, pp. 102-106.

. . . . . . (1971a) — Geology of the Long Lake-Gem Lake area, southeastern Manitoba; Manitoba Mines Br. Publ. 71-1, pp. 63-106.

. . . . . . (1971b) — Geology of the Wanipigow River-Manigotagan River region, southeastern Manitoba; Manitoba Mines Br. Publ. 71-1, Map 71-1/4.

Zwanzig, H. V. (1971) — Structural geology at Long Lake, Manitoba; Manitoba Mines Br. Publ. 71-1, pp. 285-298

15

Reprinted from L. Cahen and N. J. Snelling, *The Geochronology of Equatorial Africa*, North-Holland Publishing Co., Amsterdam, The Netherlands, 1966, pp. 150-161

THE GEOCHRONOLOGY OF EQUATORIAL AFRICA

L. Cahen and N. J. Snelling

Chapter 17

SOME CONCLUSIONS

17.1. *Geologic cycles and correlations*

In 1955 and 1957, Holmes and Cahen recognized a number of age groups and considered that most of them represented independent orogenic cycles. At that time it was thought that the difference in age between post-tectonic plutonic episodes and syntectonic episodes was slight; consequently it appeared impossible for pegmatites of 900 m.y. and of 1100 m.y. to belong to the same cycle. The data assembled in the foregoing pages show conclusively that post-tectonic plutonic episodes of a given cycle may be considerably younger than syntectonic episodes and that they may spread over a long span of time. The present, much more abundant data, coupled with continued field work, allow a more correct grouping of ages, with a corresponding reduction of the number of cycles to five (see fig. 17.1).

In Equatorial Africa, only two cycles are more or less adequately defined from a geochronological point of view; the others are still rather sketchy.

A) Katangan Cycle

In 1957, Holmes and Cahen distinguished an "Early Cambrian" Cycle of which pegmatites were circa 485 m.y. old, and a Katangan Cycle of which late tectonic events were circa 620 m.y. old. Holmes [1960] and Cahen [1961] remarked upon the fact that in most of Africa, ages of circa 485 m.y., of circa 620 m.y., and intermediate ages, co-existed in restricted areas, and presumably dated different episodes of a single cycle. Furthermore, recent work has shown that there may be a long interval of time between syntectonic and post-tectonic events of the same cycle. The data at present available for Equatorial Africa show that in several regions the younger ages (465 m.y. to 525 m.y.) are those of post-tectonic events, whereas ages between circa 600 and 750 m.y. are of syntectonic events, all in the same belt. The two 'cycles' distinguished by Holmes and Cahen should thus, in Equatorial Africa at least, be merged into one. It has been shown that, in Katanga, subsidence and sedimen-

tation must have started a little after 1250 m.y. (± 70 m.y.). To this cycle belong the Katangan proper, the Lindian, the West Congolian and the Mozambiquian as well as other units for which no age data are as yet available. This important cycle will be discussed further in section 17.2.

B) Kibaran-Burundian-Karagwe-Ankolean Cycle

In 1957, Holmes and Cahen had distinguished two cycles in the range of this cycle, the "875-900 m.y." cycle and the Karagwe-Ankolean Cycle of 1040 m.y. Formerly, a Gordonia Cycle of 1025 m.y. and a Karagwe-Ankolean Cycle of 1200-1400 m.y. had been distinguished (Holmes and Cahen, 1955). It is now known that all these ages and apparent ages fall into the range characterizing the Kibaran-Burundian-Karagwe-Ankolean Belt with ages ranging from 1290 m.y. to 850 m.y. This Belt will be commented upon in section 17.3.

C) 2150 m.y. to 1650 m.y.

In Equatorial Africa, this range covers post-tectonic events of different belts from 2050 m.y. to 1650 m.y., and syntectonic events 2100 m.y. old and older. If all the belts (Buganda-Toro-Kibalian, Ubendian-Rusizian, Luizian, Mayumbian) with ages in this range are truly contemporaneous, this is a major and widespread cycle. However, such a conclusion is premature at this stage. Previously Holmes and Cahen [1957] had referred to the Limpopo Cycle at circa 2000 m.y. and had tentatively grouped comparable ages representing the closing plutonic phases of various other cycles with this event. Only indistinct indications had been obtained of events during the period of 1000 m.y. separating the Karagwe-Ankolean and Limpopo Cycles.

D) 2700 m.y. to 2300 m.y.

At least four, more or less composite, units of Equatorial Africa might belong to the Shamvaian Cycle (Holmes and Cahen, 1955, 1957). In the type area of Rhodesia, the closing stages of this cycle have been dated by pegmatites at approximately 2650 m.y. However, these four

174

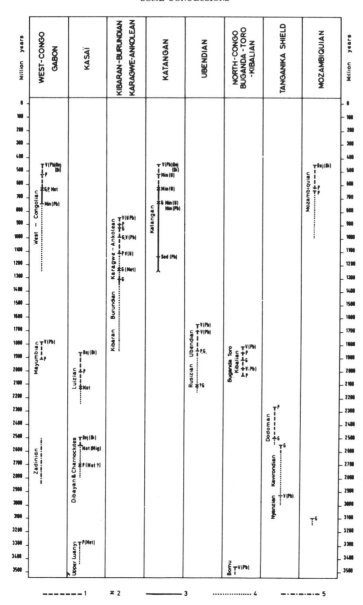

Fig. 17.1. Orogenic cycles in Equatorial Africa. 1: post tectonic time range; 2: tectonic phase; 3: known time range of deposition; 4: inferred time range of deposition; 5: imprecise post-tectonic time range. (Bi): biotite; G: granite; Met: metamorphism; Mig: migmatite; Min: mineralisation; P: pegmatite; (Pb): lead; Rej: rejuvenation; Sed: syngenetic; V: vein; (U): uranium.

equatorial units are not necessarily strictly contemporaneous among themselves or with the
Shamvaian of Southern Africa. In Kasai (Congo),
the Dibaya metamorphic Complex and the "Charnockitic Cycle", which may represent two facies of one and the same cycle, have apparent
ages ranging from 2710 m.y. to 2470 m.y. The
younger ages probably represent the time of a
post-tectonic event, the syntectonic events being
older than 2530 m.y. The same time-range applies to the Zadinian (Lower Mayumbian) of Lower Congo and other ancient formations of Gabon
and Cameroun. The Kavirondian of East Africa
is limited at about 2550 m.y. and the Dodoman of
Tanzania is probably about the same age, with
post-tectonic phases possibly at about 2300 m.y.
and a syntectonic phase at about 2500 m.y.

E) Over 3000 m.y.

Only a few ages of this magnitude have been
obtained from Equatorial Africa. One extremely
old model age given by a galena in North Congo
yields a younger limit to the Bomu Complex at
circa 3480 m.y. In South Congo, a late stage of
the "Upper Luanyi Cycle" is 3270 m.y. or older.
In Kenya, a granite at Broderick Falls is at least
3150 ± 80 m.y. old but its relationship to the surrounding rocks is unknown and may remain so as
it is bounded by faults. The Nyanzian of East Africa may possibly belong to this group but the
evidence is by no means conclusive and it might
be somewhat younger.

17.2. *The Katangan Cycle*

Of recent years, awareness of the significance of this Cycle has been growing *. From an
important but essentially local feature, it has
grown to the status of a "Pan-African thermo-
tectonic episode" (Kennedy, 1964).

Table 17.1 summarizes what is known of the
geochronology of this Cycle in Equatorial Africa.
Only sparse indications exist as to the time of
the beginning of sedimentation; these are strongest in Katanga and it can be assumed provisionally that sedimentation was, at least locally, initiated shortly after 1250 m.y. (see p. 80). This
is to a certain extent supported by evidence from
Southern Angola (see p. 143) and from along the
eastern side of Lake Tanganyika (see p. 105). In
most areas, a rather significant event occurred

---

* See: Holmes and Cahen, 1957; Holmes, 1960; Cahen
  1961; Clifford, 1963; Cahen and Lepersonne, 1966;
  Clifford, in press; Kennedy, 1964; Vail, 1964.

at about 720 to 750 m.y. This appears to be an
important orogenic phase in Zambia which was
followed by sulphide mineralization, with a uranium mineralization in Katanga. In West Congo
and North Angola, Cameroun and North-East
Congo, sulphide mineralization occurred at approximately the same time; however, this mineralization is better substantiated in West Congo
than in the other areas. Some of this sulphide
mineralization has in later times been reworked
into its present position. The end of the major
episode(s) of folding appear(s) to be before about
620 m.y. in West Congo, Katanga and Zambia, in
the Mozambique Belt, and possibly also in East
Congo, although in Katanga some movement
seems to have continued till 570 m.y. and, in
other places, maybe even later. Distinctly post-
tectonic events occur from about 525 m.y. to
about 450 m.y. The present indications are thus
remarkably coherent considering the distances
involved, and it is obviously not expected that
similar events should occur at the same time all
over such a vast area.

Table 17.2 shows similar data for areas both
north and south of Equatorial Africa. As can be
seen, they are in keeping with results tabulated
in table 17.1. The most significant results come
from Hoggar (Picciotto et al., 1966) where the
sequence and chronology of events is very similar to that of the best known areas of Equatorial
Africa. For South-West Africa, only directly in-
terpretable data have been included in table
17.2; other results, all in the 450-560 m.y.
range, are to be found in Clifford [1963, 1965].

What is at present known of Equatorial and
Southern Africa during this orogeny is depicted
in fig. 17.2, which is not very different from
similar sketches already published (Clifford,
1963; Kennedy, 1964; Vail, 1964). Only one
amendment to previous syntheses appears worth
mentioning. Evidence has been presented (p. 108)
that the Congo Craton of Clifford [1963] and Kennedy [1964] should probably be sub-divided into
a Congo Craton proper (to the west of the Western Rift) and a Tanganyika Shield (to the east).
The area now occupied by large portions of the
Western Rift Valley was the site of a Katangan
age fold belt, which may have been of geosynclinal character.

In all the cratonic areas that have been distinguished with respect to the Katangan Orogeny,
the youngest post-tectonic manifestations (not
linked with either the Kibaran or Katangan Cycles) are about 1650 m.y. old; the folding of the
youngest dated belts forming part of these cratons is about 2100 m.y. old and practically all of

Table 17.1
The "Katangan" Cycle in Equatorial Africa.

m.y.	West Congo from Angola to Gabon	Cameroun	Katanga and Zambia	N. Eastern Congo and Western East Africa	Mozambiquian
450	Post-tectonic veins with sulphides and/or biotites; rejuvenation of biotites (455 to 510)		Post-tectonic sulphide veins; rejuvenation of biotites (460 to 490)		Rejuvenation of biotites, etc. (440–490)
500		Post-tectonic granites (500)			Post-tectonic pegmatites (480)
550	Post-tectonic pegmatites (510–525)	Later syntectonic granites (525 or earlier). Metamorphism (580 or earlier)	Post-tectonic facies of Hook Granite. Post-tectonic uranium mineralisation (520). Homogenisation (updoming?) of Nchanga Granite (570)		
600	Syntectonic intrusion (620 or earlier)-metamorphism (about 620 or earlier)		Uranium mineralisation in transcurrent faults near the end of orogeny (620)	Metamorphism of Bila-tion sediments (probably 630)	Post-tectonic pegmatites (600–650) Metamorphism, earlier than 650
650					Lava flow in Kisii Series (675)
700			Early episode of U mineralisation (> 670, prob. 720). Sulphide mineralisation about 720. Syntectonic Lusaka granite (about 740)		
750	Sulphide mineralisation (740)	Sulphide mineralisation (about 750)		Sulphide mineralisation (about 750)	
1050			Epigenetic lead mineralisation in Bushimay (about 1040)		
1150	Beginning of sedimentation approximately 1190 ± 90		Syngenetic lead mineralisation in Bushimay (about 1145)		
1250			Beginning of sedimentation shortly after 1250 ± 70		
1300				Beginning of Buanji sedimentation after Chimala granite (probably about 1300)	

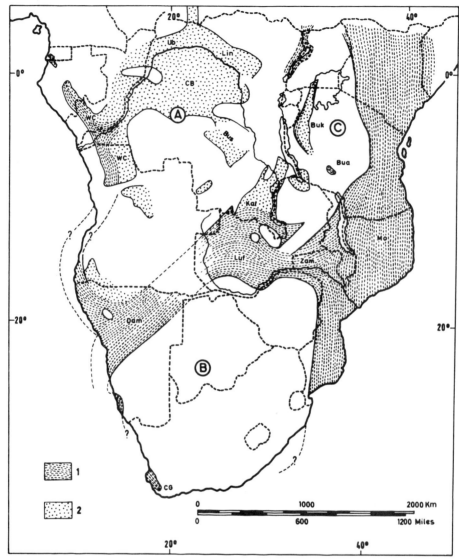

Fig. 17.2. The Katangan orogenic cycle in Equatorial and Southern Africa. 1: folded; 2: tabular. Stable areas: A: Congo Craton; B: Kalahari Craton; C: Tanganyika Shield. Kat-Luf: Katangan, folded during the Katangan or Lufilian Orogeny; Zam: Zambezi Belt; Mo: Mozambique Belt. This is probably a multicyclic belt, however at present only the final orogeny, circa 650 m.y. old can be shown; WC: West Congolian. This orogeny has set its mark on the older Zadinian and Mayumbian. In contrast with the Mozambiquian it is here possible to separate the three successive orogenics: Dam: Damaran; CG: Cape Granite; Bua: Buanji; Buk: Bukoban; Bus: Bushimay; Lin: Lindian; Ub: Ubangian; CB: tabular beds below the Palaeozoic and Mesozoic-Cainozoic covering of the Congo basin. These beds are linked to the Lindian on the one hand and to the West Congolian on the other.

Table 17.2

The "Katangan" Cycle in other parts of Africa.

m.y.	Hoggar (Picciotto et al., 1966)	Nigeria (Jacobson et al., 1963)	S.W. Africa (Burger et al., 1965; Clifford, 1963; Clifford, in press)	Madagascar (Delbos, 1964)
450	Rejuvenation of biotites (down to 460)	Rejuvenation of biotites (down to 480)	Rejuvenation of biotites (down to 460)	Rejuvenation of biotites (down to 460) Pegmatites and pyroxenites (480-500)
500	Post-tectonic pegmatite (510)	Pegmatites (510)	Post main orogeny pegmatite (500 ± 20)	
550	Post-tectonic rhyolite (550)	Older Granites, late syntectonic (540, or older)		"Sheet" granites (550)
600			Syntectonic pegmatites (600 ± 25)	Metamorphism (about 600)
650	Granites, some of which are considered syntectonic (650)	Sulphide mineralisation (older than 650)		
1200			Beginning of sedimentation circa 1190 ± 90	

the cratons were, to a large extent but not completely, already stabilized at circa 2500 m.y.

Ages and apparent ages in the same (450-650 m.y.) range occur also in Antarctica, where they cover large tracts of the continent (Picciotto and Coppez, 1963, 1964; Deutsch and Grögler, 1966), in India, where they are found in the Delhi and Malani Belts (Aswatharayana, 1964; Sarkar et al., 1964), and in the Northern Hemisphere.

17.3. *The Kibaran-Burundian-Karagwe-Ankolean Cycle*

This Cycle is also relatively well-defined. The correlation between the Kibaran, the Burundian and the Karagwe-Ankolean portions of the Belt has been established in the field and by geochronology. Outside the well-known south-west to north-east branch of this Belt, several tracts of rock appear to belong to the same cycle; these include the beds involved in the Irumide Belt of Zambia, the Ukingan, the Kingongolero Beds and possibly the Itiaso and Konse deposits of Tanzania. The evidence is still perhaps rather slender but appears to be unequivocal (p. 87). Those portions of Equatorial Africa which appear to have been deposited and deformed during this Cycle are depicted in fig. 17.3; the picture is possibly far from complete. To the south-west, the Kibaran and Irumide Belts disappear under the Kalahari sands whereas the Burundian-Karagwe-Ankolean appears to fade out towards the northwest. However, it is possible that the Belt, or part of it, emerges from under the Congo Basin as the Liki-Bembian of Ubangi but this is a guess based on similarity of lithology and of general strike and is not supported by age data. This hypothetical continuation is indicated by a question-mark in fig. 17.3. An area of the Central African Republic with apparent ages in the Kibaran range is also indicated by a question-mark.

The data pertaining to this Belt show it to have been initiated certainly after 2120 m.y. and probably after 1850 m.y.; an early metamorphic and granitic episode occurred at 1290 m.y. The main phase of the orogeny in Katanga took place shortly after; the post-tectonic aftermath was extremely long, and intrusive events are recorded from 1190 m.y. to 850 m.y.

In Southern Africa, Nicolaysen [1966] has recently shown the existence of a belt of metamorphism and intrusion, the events of which correspond in age to those of the Kibaran. This belt appears to transect the Continent from Namaqualand to Natal (fig. 17.2). The Namaqualand por-

179

Table 17.3

The "Kibaran" Cycle - comparison with other continents.

m.y.	Kibaran–Burundian–Karagwe–Ankolean	Grenville Province (Canada) (Leech et al., 1963; Krogh, 1964; Wanless et al., 1964) (recalculated with $^{87}Rb = 1.47 \times 10^{-11}y^{-1}$)	Scandinavia (Magnusson, 1965; Kulp and Neumann, 1961; Broch, 1964)
850	Rejuvenation of biotite (down to 830) Post-tectonic pegmatites and veins (850 to 1000)	Rejuvenation of biotite (down to 840)	Rejuvenation of biotite (down to 820) (S. Norway)
900	Post-tectonic tin granites (925 to 1020)		Pegmatites (S. Norway, 910)
950		Post-tectonic (Westport and Ridge) granites (950 ± 39)	
1000		Pegmatites (Cardiff, Wilberforce, 1020-1050) Metasomatism and metamorphism (Burleigh paragneiss, 1040 ± 33)	Pegmatites accompanying Bohus granite (Sweden, 1000)
1100	Post-tectonic pegmatites and veins (1100 to 1130)		Pegmatites (S. Norway, 1100)
1200	Late-syntectonic or post-tectonic granitisation (1190)	Pre-tectonic nepheline syenite (Blue Mountain, 1210 ± 41)	
1250	Syntectonic granitisation (1250 ± 70)		Metamorphism (Sweden, 1200-1300)
1300	Early or pre-tectonic granites (1290, 1300)		
	Veins and pegmatites in Ubendian Rusizian (1650-1700); pegmatites and post-tectonic granites in Bugunda–Toro–Kibalian, Rusizian–Ubendian and Luizian (1750-2050). Luizian metamorphism (2125)	Rejuvenation of biotites of Hudsonian Orogeny (down to 1580); Hudsonian Orogeny (Canada, 1800 or earlier), Penokean Orogeny (U.S.A., 1800 or earlier)	Rapakivi anorogenic granites (1650, Scandinavia). Svecofennian Orogeny (1750-1850, Scandinavia)

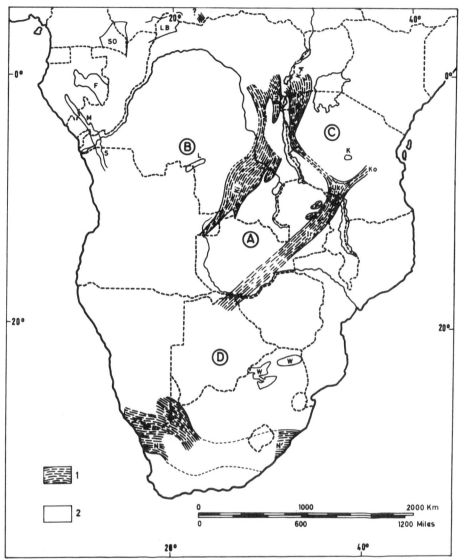

Fig. 17.3. The Kibaran Orogenic Cycle in Equatorial and Southern Africa (Southern Africa: after Nicolaysen, 1966).
1. More or less adequately dated belts of Kibaran age. Ki: Kibaran; Bu: Burundian; K-A: Karagwe-Ankolean; Ir: Irumide Belt; Uk: Ukingan; Ko: Konse; N-N': Namaqualand-Natal Belt (this belt of intrusion and metamorphism is mainly dated by pegmatites coeval with Kibaran pegmatites).
2. Tabular equivalents and possible equivalents, both folded and tabular. W: Waterberg (there is some geochronological evidence that the Waterberg is coeval with part of the Kibaran. It is certainly younger than the Bushveld Igneous Complex, circa 1950 m.y. old; LB: Liki Bembian (folded; SO: Sembe-Ouesso (folded); F: Francevillian (tabular); M and S: Mossouva and Sansikwa (tabular, subsequently folded with West Congolian; L: Lulua (folded). All these might correspond in age with the Kibaran.

181

tion, known of old (Holmes et al., 1950), may link up in some fashion with a south-westerly continuation of the Kibaran and Irumide Belts, though not exactly as those authors suggested, since the Damaran now appears to continue the Katangan and not the Kibaran (see, for instance, Clifford, 1965).

Taking the Continent as a whole, a pattern is beginning to emerge for the Kibaran "era", of belts of deformation and reactivation on the one hand and stable areas on the other. At present, four of these stable areas can be enumerated: (1) Katanga and North Zambia, (2) Kasai, (3) the Tanganyika Shield, and (4) Bechuanaland, Rhodesia and Transvaal. Of these, the first alone lost its stable characteristics during the Katangan Cycle which immediately followed the Kibaran, all the others remaining cratonic to the present day.

Ages in the same general time-range are known from the Grenville Province of Canada and its continuation in the U. S. A., from Scandinavia, India, and from Australia. Before commenting upon table 17.3 which compares data for the Kibaran, the Grenville Province and Scandinavia (which are at present the areas richest in age data of the time-range under consideration), it should be emphasized that the magnitude of these belts or provinces is such that there is little chance of similar events being the same age either in any one of these areas or in two or more, and no reason why they should be so. Attention has already been drawn (p. 91) to the fact that the apparent ages in a probable eastern branch of the Kibaran (in Tanzania, bordering the Ubendian) appear to show that corresponding events are somewhat earlier than in the main belt, although this needs further confirmation before arriving at far-reaching conclusions.

Precise age studies in the three areas compared in table 17.3 have only just begun but some significant facts can already be registered: (1) all the orogenic cycles which immediately preceded the one under consideration had post-tectonic manifestations approximately 1700 to 1800 m.y. old; these older cycles constitute the floor of the Kibaran geosyncline over much of its extension, as well as part of the floor of the Grenville (see Stockwell, 1964), and much of the floor of the "Dalslandian-Gothian" (Polkanov and Gerling, 1961; Magnusson, 1965), (2) the youngest post-tectonic event, denoted by rejuvenation of biotites, is approximately contemporaneous at about 830 m.y. in all three belts under consideration, (3) the post-tectonic history of all three areas is very long, 200 m.y. to 300 m.y.

It would be premature to draw further conclusions at present but what is already known leaves little doubt that in Equatorial Africa, North America and Southern Scandinavia, there are three portions of a major "world-wide" orogenic cycle, in the same way that the Circum-Pacific Orogeny and the Alpine and Himalayan Orogenies all belong to one major cycle as opposed, for instance, to the Hercynian Orogenic Cycle. This major orogenic cycle appears to be further represented in India where pegmatites in the Satpura Belt are dated at 950 and 1050 m.y., the best result being 955 ± 40 m.y. (Holmes, 1955; Aswathanarayana, 1964), and in Australia where apparent ages on pegmatite minerals from 940 m.y. to 1050 m.y. are also present (Wilson et al., 1961).

17.4. *Continental accretion*

The concept of marginal continental accretion appears to have been introduced by Suess (up to 1909) and has long seemed to be well supported by the structure of Europe and North America. In 1963, Engel reviewed the data for North America and concluded that marginal continental accretion had probably taken place. The same problem was discussed in 1963 by Michot on a more fundamental plane. Holmes [1965] has recently shown some inherent difficulties in this hypothesis.

We are only concerned here with an inquiry into the extent to which this concept is favoured or otherwise by the geochronology of Equatorial Africa. The picture is certainly not as simple as it appears to be in North America (fig. 17.4).

Ancient 'nuclei' at least 2500 m.y. old exist in various parts of Equatorial Africa. The Tanganyika Shield (ch. 5), the Kasai Shield (ch. 12), the Du Chaillu Shield (ch. 13) and the North Congo Shield (ch. 6) among others, can be mentioned as being more or less well defined by age determinations. Though the early stages of their evolution are often still shrouded in mystery, the history of some of these nuclei is known to go back to approximately 3500 m.y. so they cannot, by any means, be considered as stable entities throughout all their ancient history. These nuclei or 'shields' are taken here simply as a starting point to study the more recent evolution of this large part of the continent.

In one of the best-known shield areas, the Kasai Shield (ch. 12), the rocks older than 2500 m.y. are interrupted by a younger belt, the Luizian Belt, metamorphism of which took place

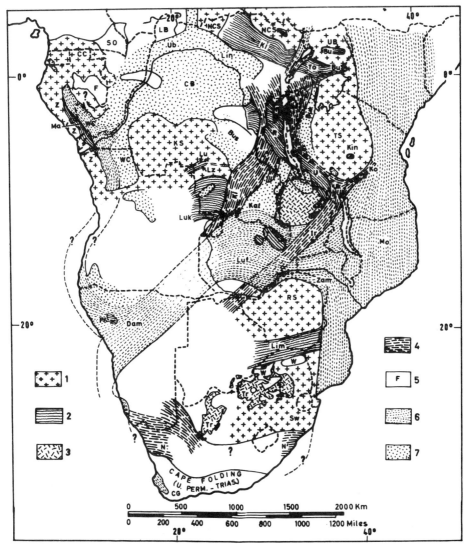

Fig. 17.4. Structural sketch of the Precambrian of Equatorial and Southern Africa (Southern Africa: Nicolaysen, 1966).
1. Areas folded and metamorphosed 2500-2600 m.y. ago or earlier. TS: Tanganyika Shield; RS: Rhodesian Shield; SW: Swazi Shield; KS: Kasai Shield; NCS: North Congo Shield; UB: Uganda Basement; Z: Zadinian; CC: Charnockitic Complex of N. Gabon and S. Cameroun.
2. Belts, folded and metamorphosed about 2100 to 1950 m.y. ago; post-tectonic events belonging to these belts can be as young as 1650 m.y. Ki-To-Bu: Kibali-Toro-Buganda Belt; U-R: Ubendian-Rusizian Belt; T: Tumbide orogeny (Lufubu schists); Lim: Limpopo or Messina Belt; Fr: Franzfontein granite; Ma: Mayumbian; Lz: Luizian; Luk: Lukoshian.
3. Beds, more or less tabular, corresponding in age, to 2. Transvaal and Griquatown "System", Bushveld Igneous Complex.
4. Belts, folded and metamorphosed about 1300 m.y. to 1100 m.y. Post-tectonic events belonging to these belts can be as young as 850 m.y. (see fig. 17.3).
5. Beds, more or less tabular or folded, which might correspond in age to 4 (see fig. 17.3).
6. Belts, folded and metamorphosed about 730 m.y. to 600 m.y. ago; post-tectonic events in these belts can be as young as 450 m.y. (see fig. 17.2).
7. Beds, more or less tabular, corresponding to 6 (see fig. 17.2).

183

circa 2120 m.y. ago. Caught up with and remobilized in this new Belt are various portions of the old shield area (Delhal and Ledent, 1965), so that the picture here is that of the establishment of a geosyncline on, and not adjacent to, a previously existing land mass.

At the moment it is not known whether, for instance, the Tanganyika Shield on the one hand, and the Kasai Shield on the other, were joined up together to form one land mass. At present they are separated by at least three fold belts or groups of fold belts: (1) the Ubendian-Rusizian and the Toro-Kibalian (both older than 1850 m.y. and probably more or less contemporaneous with the Luizian), (2) the Kibaran-Burundian-Karagwe-Ankolean, and (3) the Katangan. Some of these belts actually laid the pattern of the Western Rift Valley which is superimposed on them.

It is now known that the Kibaran Belt, for instance, girdles the present day outcrops of the Kasai Shield, which, including the younger Luizian Belt, acted as a foreland to the Kibaran Geosyncline. However, in the south-west portion of the Kibaran, geochronometric data (at present incomplete and therefore unpublished) indicate that in certain granitic masses formed in Kibaran anticlines, there exist remnants of more or less digested pre-Kibaran rocks. Similarly, in the Karagwe-Ankolean portions of the Belt in North Rwanda and Uganda, the arena granites, although intrusive into the Karagwe-Ankolean, are in part remobilized pre-Karagwe-Ankolean rocks (see p. 86). It is thus practically certain that the important Kibaran-Burundian-Karagwe-Ankolean Belt, although marginal to what is now exposed of the Kasai Shield including the more recent additions to it, most probably originated as a geosyncline cutting across previous shield areas which thus formed its floor.

However, both the western and eastern margins of Equatorial Africa appear to consist of relatively young belts, the youngest in this vast area. Along the Atlantic Ocean, the West Congolian Belt was active during the 750-450 m.y. time-range, as also was the Mozambiquian Belt which borders the coast of the Indian Ocean and probably includes the Seychelles Islands granite. So far as the Western Congo is concerned, the West Congolian Belt was built up in a geosyncline which succeeded at least two others, the earlier of which, trending east-north-east to north-east, was transected by the West Congo Belt of north-north-west trend, parallel to the present-day coast line. Nearly everywhere along the coast, the West Congolian Belt, although having had a profound influence on the oldest known of the pre-

existing belts, the Zadinian (Cahen and Lepersonne, 1966), is separated from the ocean depths by quite large stretches of relatively little-disturbed Zadinian beds with strikes directed more or less oceanwise (ch. 13). In the eastern margin of Equatorial Africa, we have seen that inliers of much older rocks are present in areas belonging to the domain of the Mozambique Belt (chs. 3 and 4). These marginal belts are linked by segments of the same age transecting the continent and cutting up previously existing shield areas.

Thus, not only is the pattern of fold belts much more intricate than in some other continents, so far as their distribution is known (see Holmes, 1961; Cahen, 1963; Nicolaysen, 1966). but in many cases, both inland and along the coast, it can be shown that the youngest belts have been built up in geosynclines resting on older floors which had previously undergone much the same type of evolution.

These geosynclines belong to the type called "segments orogéniques sur socle" (orogenic belts built up on a previously consolidated Sialic basement) by Michot (1963, 1965) as opposed to his "segments orogéniques fondamentaux" (fundamental orogenic belts). The latter alone produce true continental accretion, which is in effect, a transformation of an oceanic area into a continental area. Michot considers, however, that progressive cratonization of continents (in other words, recession of orogenic mobility) has, since the Phanerozoic and for some continents, during the Precambrian, proceeded centrifugally. What is known of the structure of Equatorial and Southern Africa during the Precambrian does not at present obviously favour even this form of zonation (fig. 17.4).

It is thus possible (and the age of the Seychelles granite favours this (see p. 32) that the pronounced general parallelism between the present coast lines of Equatorial Africa (and of part of Southern Africa) and the West Congolian and Mozambiquian portions of the youngest Precambrian Belt, may be due to fracturing along the general strike of the beds (continental drift) rather than to what, at first sight, would appear to be marginal continental accretion.

References

Aswathanarayana, U., 1964. Age determination of Rocks and Geochronology of India. Intern. Geol. Congr., 22nd, New Delhi, 1964.

Broch, O.A., 1964. Age determinations of Norwegian minerals up to March, 1964. Norg. Geol. Undersökelse, 228, 84-113.

Burger, A.J., O.Von Knorring and T.N.Clifford, 1965. Mineralogical and radiometric studies of monazite and sphene occurrences in the Namib Desert, South West Africa. Mineral. Mag. 35, 519-528.

Cahen, L., 1961. Review of Geochronological knowledge in Middle and Northern Africa, in: Geochronology of Rock Systems, Ann. New York Acad. Sci. 91, 535-566.

Cahen, L., 1963. Grands traits de l'agencement des éléments du soubassement de l'Afrique centrale. Esquisse tectonique au 1/5 000 000e.Ann.Soc.Géol. Belge, 1961-1962, 85, 183-195.

Cahen, L. and J.Lepersonne, 1966. Existence de trois orogénèses superposées dans le Précambrien du Bas-Congo. Compt. Rend. 262, 1181-1184.

Cahen, L. and J.Lepersonne, 1966. The Precambrian of Congo, Rwanda and Burundi, in: K.Rankama, Ed. The Precambrian, Vol. 3 (Interscience Publ. London and New York).

Clifford, T.N., 1963. The Damaran Episode of tectonothermal activity in South West Africa and its regional significance in Southern Africa. Univ. Leeds, Res. Inst. African Geol. 7th Ann. Rept. Sci. Results. p. 37-42.

Clifford, T.N., 1965. Structural units and tectonometallogenic provinces within the Congo and Kalahari Cratons of Southern Africa. Univ. Leeds, Res. Inst.African Geol. 9th Ann. Rept. Sci. Res. p. 30-33.

Clifford, T.N. (in press). The Damaran episode in the Upper Proterozoic-Lower Paleozoic structural history of Southern Africa. Geol. Soc. America.

Delbos, L., 1964. Signification et importance du cycle récent de 500 millions d'années dans le Précambrien de Madagascar. Ann.Fac.Sci.Univ. Clermont, no. 25, 39-52.

Delhal, J. et D.Ledent, in press. Quelques résultats géochronologiques relatifs aux formations du socle de la region de Luiza (Kasai). Bull. Soc. Belge Géol.

Deutsch, S. and N.Grögler, 1966. Isotopic age of Olympus Granite-Gneiss (Victoria Land, Antarctica). Earth Planet. Sci. Letters, 2, 82-84.

Holmes, A., 1951. The Sequence of Precambrian orogenic belts in south and central Africa. Intern. Geol. Congr. 18th, Great Britain 1948. Pt. 14, 254-269.

Holmes, A., 1955. Dating the Precambrian of Peninsular India and Ceylon. Proc. Geol. Assoc. Canada, 7, II, 81-106.

Holmes, A., 1960. A revised geological time-scale. Trans. Edinburgh Geol. Soc., 17, 183-216.

Holmes, A., 1965. Principles of Physical Geology (London-Edinburgh).

Holmes, A. and L.Cahen, 1955. African geochronology. Colonial Geol. Mineral Resources, 5, 3-39.

Holmes, A. and L.Cahen, 1957. Géochronologie africaine, 1956. Acad. Roy. Sci. Coloniales, Mém. Classe Sci. Nat. et Méd., V, no. 1, 169 pp.

Holmes, A., N.T.Leland and A.O.Nier, 1950. The age of uraninite from Gordonia, South Africa. Amer. J. Sci. 248, 81-94.

Jacobson, R.R.F., N.J.Snelling and J.F.Truswell, 1963. Age determinations in the Geology of Nigeria, with special reference to the older and younger granites. Overseas Geol. Mineral Resources 9, 168-182.

Kennedy, W.Q., 1964. The structural differentiation of Africa in the Pan-African (± 500 m.y.) tectonic episode. Univ. Leeds, Res. Inst. African Geol. 8th Ann. Rept. Sci. Results, p. 48-49.

Krogh, T.E., 1964. Strontium isotopic variation and whole rock isochron studies in the Grenville Province of Ontario. Ann. Rept. 1964. Dept. Geol. Geophys. p. 73-124.

Kulp, J.L. and H.Neumann, 1961. Some potassium-argon ages on rocks from the Norwegian Basement, in: Geochronology of Rock Systems. Ann. New York Acad. Sci. 91, 469-473.

Leech, G.B., J.A.Lowden, G.H.Stockwell and R.K. Wanless, 1963. Age determinations and Geological studies. Geol. Surv. Canada, Paper 63-17.

Magnusson, N.H., 1965. The Precambrian history of Sweden. Quart. J. Geol. Soc. London 121, 1-30.

Michot, P., 1963. La structure continentale. Acad. Roy. Belgique, Bull. Classe Sci., 5e sér., XLIX 12, 1337-1373.

Michot, P., 1965. Les orogènes fondamentaux (Die Grundorogene). Freiberger Forschungshefte, C 190, 1965, 49-62.

Nicolaysen, L.O., 1962. Stratigraphic interpretation of age measurements in Southern Africa, in: Petrologic Studies: A volume in honor of A.F.Buddington. (Geol. Soc. America) 569-598.

Nicolaysen, L.O., 1966. An extensive zone of 1000 million years old metamorphic and igneous rocks in Southern Africa. Coll. Intern. Géochronologie absolue, Nancy, 1965, Sci. Terre, Nancy.

Picciotto, E.E. et A.Coppez, 1963. Bibliographie des mesures d'âges absolus en Antarctique. Ann. Soc. géol. Belg., 1961-1962, 85, 163-308.

Picciotto, E.E. et A.Coppez, 1964. Bibliographie des mesures d âges absolues en Antarctique. (Addendum, Août, 1963). Ann. Soc. Géol. Belg., 1963-1964, 87, 115-128.

Picciotto, E., D.Ledent et C.G.Lay, 1966. Etude géochronologique de quelques roches du socle cristallophyllien du Hoggar (Sahara central). Coll. Intern. Géochronologie absolue, Nancy, 1965. Sci. Terre. Nancy.

Polkanov, A.A. and E.K.Gerling, 1961. The Precambrian Geochronology of the Baltic Shield, in: Geochronology of Rock Systems. Ann. New York, Acad. Sci. 91, 492-499.

Sarkar, S.N., A.A.Polkanov. E.K.Gerling and F.Y. Chukroy, 1964. Geochronology of the Precambrians of Peninsular India. A synopsis. Sci. Culture (Calcutta), 30, 527-537.

Stockwell, C.H., 1964. 4th Rept. Structural Provinces, Orogenies and time-classification of rocks of the Canadian Precambrian Shield. Geol. Surv. Canada Paper 64-17 (Pt. II).

Suess, E. up to 1909. Das Antlitz der Erde, 4 vols. (Vienna-Leipzig).

Vail, J.R., 1964. The Mozambique Belt of Eastern Africa. Univ. Leeds, Res. Inst. African Geol. 8th Ann. Rept. Sci. Results. 32-34.

Vail, J.R., 1965. Distribution of Radiogenic ages in Eastern Central Africa. Univ. Leeds, Res. Inst. African Geol. 9th Ann. Rept. Sci. Res. p. 27-29.

Wanless, R.K., R.D.Stevens, G.R.Lachance, R.Y.H. Rimsaite, C.H.Stockwell and H.Williams, 1964. Age determinations and Geological studies. Geol. Surv. Canada, Paper 64-17, Pts. I and II.

Wilson, A.F., W.Compston and P.M.Jeffery, 1961. Radioactive ages from the Pre-Cambrian Rocks of Australia, in: Geochronology of Rock Systems, Ann. New York Acad. Sci. 91, 514-520.

185

Editor's Comments
on Paper 16

16   CORLISS
The Origin of Metal-Bearing Submarine Hydrothermal Solutions

Paper 16 and the following groups of papers are concerned with the types of mineralization at median rises and mobile belts related to convecting mantle cell systems, and to the types of mineralization formed during the interval as one convecting system gives way to another.

Paper 16, by Corliss, is perhaps the most illuminating on ore-forming processes at median rises. Hydrothermal exhalations and volcanic emanations have often been cited but never specifically defined as sources of ore. Corliss showed that slowly cooled interior portions of mid-Atlantic ridge basalts are depleted in appropriate elements, and that the elements were probably mobilized by dissolution as chloride complexes in seawater, thus being available to form ore deposits. Ore formation is the process of concentrating metal into viable zones, and the process described by Corliss appears to be among the earliest in the ore-forming cycle. Fyfe (1974), also concerned with activity at the median rise, considered "that the cooling mechanism of shallow intrusive rocks is one involving massive seawater convective flow through the upper layers of ocean floor materials . . . circulating salt waters can perform pervasive chemical stripping and redeposition of elements (particularly transition metals) critical in the formation of ore-deposits."

16

Copyright © 1971 by the American Geophysical Union

Reprinted from *Jour. Geophys. Res.*, 76(33), 8128-8138 (1971)

The Origin of Metal-Bearing Submarine Hydrothermal Solutions

JOHN B. CORLISS[1]

Department of Oceanography, Oregon State University, Corvallis, Oregon 97330

Instrumental activation analyses for 16 major and trace elements in a suite of mid-Atlantic ridge basalts reveals that the slowly cooled interior portions of these submarine extrusions are depleted, relative to the quenched flow margins, in several elements that are enriched in pelagic sediments and manganese nodules (Mn, Fe, Co, the rare earth elements, and others). Many of these elements are excluded from the solid phases that crystallize from the melt, and thus are concentrated in residual liquids. Additional elements are mobilized during the deuteric alteration of early-formed olivine and the formation of immiscible sulfide liquids. It is suggested that these components of melt occupy accessible sites (e.g., intergranular boundaries) in the hot solid rock mass, and are mobilized by dissolution as chloride complexes in sea water introduced along contraction cracks that form during cooling and solidification. These solutions may be the metal-bearing 'hydrothermal exhalations' or 'volcanic emanations' that accompany submarine volcanism, which are often cited as a source of metals into the pelagic environment. Reasonable estimates of the amount of material involved suggest that a significant fraction of the mass of these elements that reside in pelagic sediments could have been supplied by this process. Recently described amorphous iron-manganese-silica material from the east Pacific rise may form by direct precipitation from these hydrothermal solutions following their introduction into sea water.

Submarine volcanic activity has often been cited as a source of metals in the pelagic environment. Murray [*Murray and Renard*, 1891] first suggested a relationship between modern pelagic iron and manganese deposits and volcanic eruptions. *Park* [1948] cited the pillow basalts, pelagic sediments, and manganese deposits on the Olympic Peninsula as evidence of such a correlation. *Hewett* [1963] documented the connection between continental Mn accumulations and volcanic activity, and suggested a similar relationship for pelagic Mn nodules and encrustations.

Several processes have been proposed for the transfer of elements from submarine basalt magmas into the deep marine environment [e.g., *Bostrom*, 1967]. One such mechanism is the long-term, low-temperature weathering (halmyrolysis) of volcanic rock exposed at the sea floor. Another is the mobilization of elements during the violent interaction of molten lavas with sea water to form palagonite tuffs [*Bonatti and Nayudu*, 1965]. A third mechanism, often

cited but never explicitly defined, is the action of 'hydrothermal exhalations' or 'volcanic emanations' that accompany submarine volcanic activity, which transport elements from the magma into sea water.

Bostrom and Peterson [1966, 1969] and *Bostrom* [1970] describe the enrichment of Fe, Mn, Cu, Cr, Ni, and Pb as metal oxide precipitates in pelagic sediments on the crest of the east Pacific rise. The zone of enrichment coincides with the zone of high heat flow over the crest of the rise. *Bender et al.* [1970] have determined accumulation rates in sediment cores from the east Pacific rise by the excess Th-230 method. They also measured the abundance of U, Th, Mn, Fe, and Cu in the cores. They find that the Mn accumulation rate is some 30 times greater at the crest compared to the flanks, and attribute the high rate to volcanic activity on the crest.

Bostrom and Peterson [1966] attribute these precipitates to 'ascending solutions, of deep seated origin, related to magmatic processes.' Several studies of ferromanganese nodules have cited these hypothetical solutions as sources for at least part of the metals found in these deposits [e.g., *Bonatti and Joensuu*, 1966; *Skornyakova et al.*, 1962; *Cronan*, 1967]. Layers of

[1] Present address: Department of Geology and Geophysics, Yale University, New Haven, Connecticut 06520.

metal-rich amorphous iron oxides in piston cores from near the east Pacific rise [*Sayles and Bischoff*, 1971] and in Deep Sea Drilling Project cores [*von der Borch and Rex*, 1970] and *von der Borch et al.* [1970] have also been attributed to the action of these hydrothermal solutions.

A recent review of the genesis of base metal ore deposits by *White* [1968] integrates experimental work and critical field observations that are particularly relevant to the present discussion if submarine 'volcanic emanations' are to be proposed as geologically significant sources of trace metals for marine environments. White has defined four critical aspects of the generation of base metal ore deposits, based on observations of two active metal-bearing hydrothermal systems (the Red Sea and Salton Sea areas) and on studies of fluid inclusions in ore and gangue minerals in three mining districts. To form such an ore deposit [*White*, 1968] the following must be true: (a) there must be a source of the ore constituents; (b) the ore constituents must dissolve in a hydrous phase (a Na-Ca-Cl brine in the cases he considered); (c) these solutions must migrate in directions controlled by local pressure gradients; and (d) the ore constituents must be selectively precipitated from these solutions in response to physical and chemical changes as the fluid moves into a new environment.

In this discussion we are concerned with the source of the metals and their dissolution and transport as 'hydrothermal exhalations.' A detailed study of the major and trace element distribution and petrography of a suite of basalts dredged from 22°N on the mid-Atlantic ridge samples has revealed evidence for both the mode of origin and the composition of such mineralizing solutions. The precipitation of these metals to form ore deposits requires an additional set of conditions; these are beyond the scope of the present paper.

PILLOW BASALTS AND HOLOCRYSTALLINE BASALTS FROM THE 22°N SUITE

Seven dredge hauls containing abundant basalt fragments were obtained in the course of a detailed survey of the mid-Atlantic ridge at 22°N on cruise 1965-1 of the R.V. *Thos. Washington*. Results of this survey are reported in *Melson et al.* [1966, 1968] and *van Andel*

and Bowin [1968]. Fifty-one samples of 31 rocks from this suite were analyzed by instrumental activation analysis by methods described by *Gordon et al.* [1968]. Details of the analyses reported here as well as additional work on these samples, which relates them to other suites of mid-Atlantic ridge basalt, are presented in *Corliss* [1970].

The most abundant rock types in the 22°N suite, as described by *Melson et al.* [1968], are pillow basalts and holocrystalline basalts, contrasted in Table 1. Thin sections of the pillow fragments reveal rare to common microphenocrysts of plagioclase or olivine, or both, set in a matrix ranging from variolitic glass at the outer pillow margins to a hyalopilitic ground mass largely crystallized to plagioclase and augite in the pillow centers, 20 to 30 cm from the glassy rims. In contrast, the massive, holocrystalline fragments are uniform in texture from margin to center; intersertal to subophitic with no glass. Fresh plagioclase laths and granular augite form the bulk of the rock, with minor amounts of olivine and saponite. The saponite was identified by *Melson et al.* [1968], who suggest it results from deuteric alteration of olivine. The petrography and chemistry of these rocks are discussed in more detail in *Corliss* [1970].

Melson et al. [1968] interpreted the holocrystalline rocks to be samples of the slowly cooled interiors of submarine basalt flows whose margins were quenched to form the pillows. The important distinctions in the present context are the presence of abundant glass and the lack of deuteric alteration effects in the pillows, contrasted with the absence of interstitial glass coupled with extensive deuteric alteration of the olivine in the holocrystalline rocks.

In addition to the textural differences, the high precision of the activation analysis data reveals significant chemical distinctions between the two rock types (Figure 1). The Na, Al, and Ca abundances from the unweathered cores of the holocrystalline fragments are identical to those in the pillows. In the holocrystalline rocks Cr is slightly higher, while Sc is slightly lower. On the other hand, Mn, Fe, Co, and REE and Hf (not shown), are significantly lower in the holocrystalline rocks. A similar relationship for Cu, Pb, and Ga is indicated by the results

TABLE 1. Contrast of Basalt Fragments Derived from Pillow Lavas with Those Derived from Massive Interiors of Flows in the 22°N Area (modified from *Melson et al.*, [1968])

Characteristics of:	Pillow	Massive
Glassy rinds	Abundant	Absent
Jointing	Radial	Columnar
Groundmass	Glassy	Holocrystalline
Texture	Hyalopilitic, intergranular	Subophitic, intersertal
Vesicles	Common	Rare
Deuteric alteration	Absent	Extensive olivine alteration

of emission spectrographic analysis reported by Thompson [*Melson et al.*, 1968]. I attribute these differences to mobilization and loss of some components of the initial magma from the interiors of the flows. In justifying this interpretation we must first explore other possible explanations for this difference.

It is possible that the holocrystalline rocks lose these elements during low temperature weathering after exposure by faulting. This would, of course, reduce their importance as sources of trace elements in the deep marine environment; only these few flows, whose interiors have been disrupted and exposed by faulting or slumping, would contribute these elements to sea water. Several lines of evidence suggest that this does not happen.

First, the dredged holocrystalline basalt fragments have altered zones or bands parallel to their broken surfaces. This alteration, visible in thin sections, is principally an oxidation of the Fe in the green saponite ('bowlingite') changing its color to reddish brown ('iddingsite'). In addition, analysis of these bands in one sample reveals that the Fe and K abundance increases sharply in these narrow bands, compared to the interiors of these holocrystalline fragments [*Corliss*, 1970]. The orientation of these bands parallel to the irregular surfaces of the broken fragments indicates that they formed after the rocks were broken and exposed to sea water, and the increase in Fe content outward suggests that this element, as well as K, was added to the rock from sea water rather than removed from it. The presence of significant amounts of saponite and related clay mineral phases, which have large ion-exchange capacities, presumably accounts for this addition of Fe from sea water to the holocrystalline rocks.

A similar observation has also been made by *R. A. Hart* [1970], who reports on a study of published dredged basalt analyses. He suggests that Fe and K, and other elements, are removed from sea water and added to these rocks when they are exposed at low temperatures over long periods of time.

Second, *Garlick and Dymond* [1970] have examined H_2O and ^{18}O values in a holocrystalline fragment with similar alteration banding. The outside and intermediate zones showed effects of hydration and exchange of oxygen with sea water while the center did not.

Third, *S. R. Hart's* [1969] data, as interpreted by *Corliss* [1970], indicate that Cs abundances in the centers of holocrystalline fragments are often very low, identical to those in the erupted magma, as indicated by the abundance in quenched glass. This suggests that the centers of these holocrystalline rocks are often unaffected by sea-water alteration at low temperature.

These three observations suggest that interiors of holocrystalline fragments with deuteric minerals which do not show visible effects of low-temperature alteration by sea water (such as oxidation of iron) have not been significantly affected by this process. Thus this type of alteration cannot be the cause of the depletions relative to pillow basalts shown in Figure 1.

Another possible cause for lower abundance of some metals in holocrystalline rocks relative to associated pillows would be enrichment of these elements in the glassy pillows by reaction with sea water. Two observations indicate this is not the case. First, nearly all of the pillow samples show no visible effects of alteration either in hand specimen or in thin section. A few pillow fragments do have traces of iron oxides on the exposed glassy surfaces, but the Fe content of interior samples of these rocks is identical to that of the completely unaltered pillow fragments. Second, comparison of samples from the pillow margins and interiors reveal higher and more variable K abundances in the interiors, where the glassy matrix is partially crystallized and sea water has presumably penetrated. Fe, on the other hand, is homogeneously distributed in the pillows [*Corliss*, 1970].

Another explanation for the chemical differences between pillows and holocrystalline rock is possible. It could be argued that the two compositions represent two distinct magma types present at 22°N, which have essentially identical Na, Al, Ca, and Sc contents, but differ in Fe, Mn, Co, and others (Figure 1). This explanation cannot be rigorously eliminated, but it requires an explanation of the observation that one of these magma types, which was found in six dredge hauls, only occurred as pillow basalt fragments, and the other, present in three dredge hauls, was found only as deuterically altered holocrystalline rocks, never as pillow fragments. Moreover, in two of the dredge hauls the two rock types occurred together. The suggestion that they are distinct magma types requires acceptance of a model significantly more complex than that proposed here.

As an alternative hypothesis, the favored one, to explain the chemical distinction between these two rock types, we may consider that some components of the magma are mobilized and lost during the slow crystallization and subsequent cooling of the flow interiors.

FRACTIONATION OF SUBMARINE BASALT FLOWS DURING COOLING

The composition of the magma when it reaches the sea floor is the end product of a series of geochemical processes, the most recent being partial melting in an upper mantle source region followed by transport to the surface accompanied by some degree of fractional crystallization. If we were to sample this lava as it arrives at the sea floor, and cool separate identical portions of it in closed containers (closed to gain or loss of matter) at various cooling rates, the texture and modal composition of each rock so formed would be different, but their bulk chemical compositions would necessarily still be the same. Very rapid quenching would produce homogeneous glass containing only those phenocrysts or microphenocrysts present in the liquid when sampled. In the more slowly cooled samples, the crystallizing solid phases would proceed to alter the composition of the residual liquid, markedly increasing the concentration of those elements rejected by the crystal lattices, the 'incompatible elements.' The residual liquids might differ considerably from the liquid from which the initial crystals formed, and could react with the solid phases that are now unstable in their presence. On further cooling in our hypothetical 'closed' system the residual components would form solid phases in the interstices of the previously formed crystalline mesh, and the composition of the total system would remain unchanged. I propose that the natural system is 'open' and that residual components of the erupted magma can be lost from the slowly cooling portions of the rock mass.

The distribution of radiogenic noble gases in pillow basalts provides strong evidence that when cooling is slow enough to allow crystallization, the rock is open to loss of mobile components. Radiogenic argon and helium, produced by the decay of ^{40}K, and U and Th

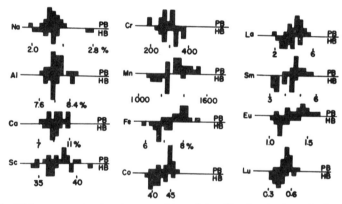

Fig. 1. Histograms contrasting the composition of the pillow basalts from the flow margins (PB) with the holocrystalline basalts from the flow interiors (HB). Abundances are ppm except where noted.

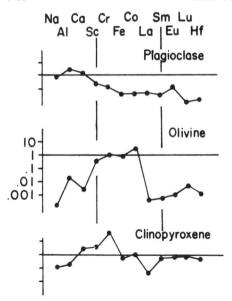

Fig. 2. Solid-liquid partition coefficients for 12 elements are shown in 3 major phases, which crystallize in the flow interiors. A partition coefficient is the ratio of the abundance of an element in the solid phase divided by its abundance in the melt.

isotopes in the upper mantle source of the magmas, are trapped by quenching in the glassy pillow margins, but have been lost from the pillow interiors where extensive crystallization has taken place [Dalrymple and Moore, 1968; Funkhouser et al., 1968; Dymond, 1970].

Evidence that slow crystallization will lead to the formation of residual liquids with high concentrations of several 'incompatible elements' can be seen in the partition coefficients presented in Figure 2. (The data plotted here are discussed in detail in Corliss [1970]). The three major phases that crystallize from oceanic tholeiite magmas at sea floor pressures are plagioclase, olivine, and clinopyroxene (subcalcic augite). In the present context, the important features of the partition coefficient data are that Cr and Sc tend to be located in the clinopyroxene lattice and Co in the olivine; that the REE and Hf are fractionated into the liquid; and that Fe enters into olivine while Na, Al, and Ca do not. Data on Mn from DeVore [1955] and Onuma et al. [1968], not

tabulated here, indicate that Mn is fractionated into both clinopyroxene and olivine. In addition, the separation of plagioclase, clinopyroxene, and olivine from a tholeiitic liquid tends to enrich the residual liquid in silica, as Yoder and Tilley [1962] have pointed out.

The deuteric alteration of olivine might be expected to mobilize additional elements. Data tabulated by Baker and Haggerty [1967] indicate that the green 'bowlingite', produced during deuteric alteration of olivine under non-oxidizing conditions, is depleted in Fe relative to the original olivine. It is reasonable to assume that other transition elements in the olivine lattice, which substitute for Fe^{2+} (Co and Mn), are also mobilized during this process.

An important set of observations relevant to these residual liquids, formed during the solidification of a tholeiitic magma, is reported by Peck et al. [1966] and Skinner and Peck [1969]. They document a detailed program of drilling and sampling during the cooling and crystallization of a tholeiitic basalt flow in Alae Lava Lake, Hawaii. The following observations are considered relevant:

1. The flow was 14 meters thick. The last interstitial melt solidified (at 980°C) 13 months after eruption. The entire flow cooled to 100°C in four years. (A submarine flow would cool faster; its boundaries are initially at 2°C and its upper surface is maintained at this temperature.)

2. As the melt solidified downward a zone of crystalline mesh and interstitial fluid formed between the solid rock above and the liquid below. The interstitial liquid in this zone was isolated from the bulk liquid below and could not mix with it by convection. These interstitial fluids were sampled several times by allowing them to flow into open drill holes. Analysis showed them to be depleted relative to the initial liquid, in Mg, Ca, Cr, and possibly Ni; and enriched in Fe, Ti, Na, K, P, F, Ba, Ga, Li, Y, and Yb, and in most samples, SiO_2 (Figure 3). Peck et al. [1966] note that this trend is distinct from the olivine fractionation trend of the Hawaiian tholeiites.

3. In some samples of the interstitial fluids, an immiscible sulfide-rich phase was found in the siliceous residual melt. The sulfur is apparently fractionated into the siliceous residual

liquids until it reaches a concentration adequate to form an immiscible phase. (The S content of the siliceous liquid from which the sulfides separated was 380 ± 20 ppm compared to ~ 100 ppm in the liquid below the zone of crystallization). Several elements are strongly fractionated into the sulfide phases. They contain 59% Fe, 5% Cu, 1% Ni, and presumably concentrate other trace metals such as Co, Pb, etc.

4. Deuteric alteration of olivine was observed at about 700°, corresponding to an increase in the ferric to ferrous iron ratio.

The formation of sulfide segregations has also been observed in dredged submarine pillow basalts from the mid-Atlantic ridge and Juan de Fuca ridge by *Moore and Calk* [1971]. They occur as 'spherules,' which form on vesicle walls during the quenching of the pillows, and larger 'globules,' which perhaps form during volatile fractionation prior to eruption. They contain abundant Fe, Cu, and Ni, which are strongly fractionated from the melt into the sulfide phase.

Another interesting set of observations relevant to this discussion of the origin of submarine hydrothermal metal-bearing solutions has been presented and interpreted by *Mackin and Ingerson* [1960]. They propose a model for the origin of magnetite mineralization in the

Iron Springs district in southern Utah. In this area, erosion has exposed Tertiary laccoliths of granodiorite porphyry surrounded by replacement ore bodies of magnetite and hematite in Jurassic limestone. They suggest that the source of the iron was biotite and hornblende, crystallized in the melt prior to intrusion. In the rapidly cooled margins of the intrusive body, these minerals are unaltered. In the interior, the biotite and hornblende phenocrysts 'were largely or completely destroyed by deuteric alteration, and the iron contained in them was released into the interstitial fluid of a slowly consolidating crystal mush.' They cite tension cracks produced in the semisolid crystalline mush by subsequent intrusion and uplift, as the pathways along which the fluids migrated into the adjacent limestone.

Mackin and Ingerson [1960] suggest that this 'deuteric release' mechanism is important in all ore deposits related to hypabyssal intrusives in which early formed mafic phenocrysts move to regions of lower pressure where they are unstable and are deuterically altered. This suggests a parallel to the olivine in oceanic tholeiite lavas. Olivine is commonly present as microphenocrysts in the erupted liquid (they appear in quenched pillows); these are unstable, and when cooled slowly they are altered.

The observations on the Alae Lava Lake and

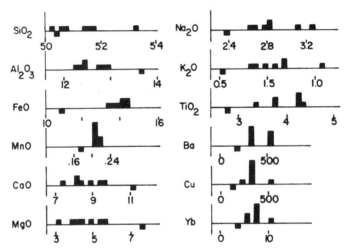

Fig. 3. Comparison of the composition of the initial liquid and residual liquids in the Alae Lava Lake tholeiite. The initial liquid composition is plotted below the lines, the compositions of the residual liquids above. (Data from *Peck et al.* [1966]).

on sulfide segregations in submarine pillows, and the model proposed for the mineralization of Iron Springs, coupled with the major and trace elements data on the 22°N pillows and holocrystalline basalts, suggest that during the cooling of the basalt flows, several elements, previously postulated as those transported by submarine 'hydrothermal solutions' or 'volcanic emanations,' are concentrated in residual phases during slow cooling and are subsequently lost from the flow interiors. The mechanisms that might account for this migration and loss must now be explored.

DISSOLUTION OF METALS AND THEIR TRANSPORT IN HYDROTHERMAL SOLUTIONS

The parallel joint surfaces of many of the dredged holocrystalline basalt fragments indicate that the flows in which they are formed are typical columnar basalts. The formation of contraction cracks has been observed in some detail in cooling lava lakes in Hawaii by *Peck and Minakami* [1968].

They conclude that the maximum temperature at which cracks initiate within the crust of the lava lake is probably about 900°C, or ~ 80°C below the solidus. Citing *Lachenbruch* [1962, p. 37], they point out that these tension cracks, once initiated, can propagate freely into a zone of no stress at higher temperatures. Through several lines of argument they argue that 'the maximum temperature at depths to which cracks will propagate near the center of the lava lakes must be close to 1000°C and perhaps as high as 1040°C,' into the zone where the residual melt has not solidified. In submarine flows, sea water is necessarily admitted into such fractures, and as the fracture systems develop, convective flow must occur.

Helgeson [1964], in a detailed discussion of hydrothermal ore deposition, supports the conclusion that 'most hydrothermal solutions are alkali chloride-rich electrolyte solutions containing predominant Na and Cl, lesser amounts of K and Ca, small amounts of SO_4, CO_2, HCO_3, Li, Rb, Cs, and other minor constituents . . .' (p. 80). This is, of course, a reasonable description of sea water. *White* [1968] suggests that intimate contact of a Na-Ca-Cl brine with solid phases that contain base metals, which are crystallizing or reconstituting at temperatures from 100° to 900°C, is 'the most favored circumstance for concentration of these metals

in the liquid phase,' probably as chloride complexes. *Krauskopf* [1967] presents experimental data which suggest that Fe and Mn would be readily dissolved by sea water which has been acidified by solution of the gases accompanying volcanic extrusions.

This model for the formation of metal-rich hydrothermal solutions requires that those components of the melt which are fractionated into residual solutions as discussed in the previous section, are available for dissolution in sea water when it is admitted by fracturing into their vicinity. This may occur initially at temperatures close to the solidus (~980°C) along the contraction fractures. In the bulk of the rock the interchange of matter between the sea-water phase and the residual magmatic phases must take place by migration along intergranular boundaries. As sea water dissolves the residual magmatic components, it also gains heat, and the temperature gradients so established drive a convective flow in the vertical fractures which brings fresh sea water to contact with the hot rocks and allows these solutions to enter the deep marine environment as submarine hot springs. These fluids are the hydrothermal solutions, or volcanic emanations, to which some submarine deposits are attributed.

This phenomenon has been directly observed on the submarine Banu Wuhu volcano in Indonesia [*Zelenov*, 1964]. The site of the 1919 eruption, as examined in 1963, was a submarine bank of dacite-andesite that ranged in depth from sea level to about 30 meters, from which jets of hot water streamed. The hot waters were of sea water composition with added Fe, Mn, and Si. 'Suspended iron and manganese hydroxides can be seen precipitating right under the water, . . . the rising jet starts becoming yellow and turbid about 1 meter above the bottom.'

These solutions precipitated 100 to 140 mg/liter of Fe and Mn hydroxides on cooling. Brown iron-manganese-silica sediment that is rich in trace metals covers the sea floor around these vents and is also carried away from the area in the current. It is reasonable that similar submarine hot springs occur in areas of volcanic activity along mid-ocean ridges.

GEOCHEMICAL BALANCES

If this model for the origin of the metal-bearing submarine hydrothermal solutions is correct, the proposed mechanism can contribute significant amounts of Fe, Mn, the REE, and presumably silica and other elements not measured, into the deep sea sedimentary environment. The magnitude of the effect can be estimated by comparing the loss from a given thickness of basalt underlying the ocean floor to the mass of these elements present in the overlying sea water, manganese nodules, and amorphous iron oxide beds. Table 2 presents the results of such calculations. The data show that basalt flows a few hundred meters thick which have generated hydrothermal mineralizing solutions according to the model presented here, may yield (1) amounts of Fe, Mn, Co, and the REE which are significantly greater than the amounts in the steady-state sea-water reservoir; (2) amounts of Mn, Co, and Cr which are of the same order of magnitude as the amounts in Mn nodules; (3) enough Ce to supply the anomalously high Ce content in manganese nodules although some excess of Sm (and the other REE) is indicated; and (4) a considerable excess of Fe over that present in Mn nodules to be accounted for. Evidence for the existence of this iron in pelagic sediments is found in data both on Pacific pelagic clays and recently discovered 'amorphous iron-oxide' beds.

Chester and Hughes [1969] have described results of selective leaching of samples of a pelagic clay core from the North Pacific. They interpret their results to indicate that there is 'excess iron over that required to form nodules' in the hydrogenous fraction of the core. *Goldberg and Arrhenius* [1958] also observed this 'excess' iron and suggested that it originates as 'free' colloidal iron, which is incorporated into the sediment and converted to goethite with time.

In the core studied by *Chester and Hughes* [1969], this 'excess' iron has an average concentration of ~900 ppm in five samples of the core, ranging from 600 to 1300 ppm. This concentration is too low to account for all of the excess iron produced by the model proposed here—enough excess iron for about 125 meters of sediment would be supplied by each meter of basalt—but it does suggest the existence of such excess iron.

More dramatic occurrences of non-nodular, presumably hydrogenous iron, have been discovered by the Deep Sea Drilling Project (DSDP), and in piston cores taken near the east Pacific rise described by *Sayles and Bischoff* [1971]. In DSDP cores from the western North Atlantic, abundant hematite is reported at sites 8 and 9A. The hematite colors the sediment brick red, and makes up 11.7% of the sediment in one sample. Siderite, a high-iron carbonate, is a significant component in several cores taken on leg III in the South Atlantic [*Rex,* 1970]. *Von der Borch and Rex* [1970] and *von der Borch et al.* [1970] report the occurrence, in several core holes in the Pacific, of thick sections of 'amorphous iron oxide precipitates.' At three sites on leg V, (37, 38, and 39) a 'basal amorphous iron oxide facies' immediately overlies basalt. In two holes (37 and 39) this facies consists of 5 to 6 meters of 'dusky yellowish-brown amorphous goethite mud.' In the third (38), it is made up of 9 meters of mixed amorphous iron oxide and calcified nannoplankton ooze overlain by 6 meters of '100% amorphous iron oxide sediment.' Overlying this basal facies at all three sites are 5- to 25-meter thick 'mixed amorphous iron oxide-detrital facies.' *Von der Borch et al* [1970] report the occurrence of amorphous iron-oxide

TABLE 2. Geochemical Budget of Elements Contributed to the Pelagic Environment by Submarine Volcanism

	Lost from Flow Interiors	Sea Water[1,2]	100-meter Equivalent[3]	Mn Crust[4]	100-meter Equivalent[5]	Amorphous Fe-Oxide Beds[6]	100-meter Equivalent
Fe	10,000 ppm	0.01 ppm	100,000 km	125,000 ppm	800 cm	160,000 ppm	625 cm
Mn	120 ppm	0.001 ppm	12,000 km	160,000 ppm	7.5 cm	50,000 ppm	24 cm
Co	4 ppm	0.0001 ppm	4,000 km	3,100 ppm	7.8 cm	160 ppm	250 cm
Cr	0	0.00005 ppm		10 ppm		10 ppm	
Sm	1 ppm	50×10^{-8} ppm	200,000 km	50 ppm	200 cm		
Ce	3 ppm	200×10^{-8} ppm	150,000 km	1,500 ppm	20 cm		

1. *Goldberg* [1965, pp. 164–165] (Cr, Sm, Ce).
2. *Brewer and Spencer* [1970] Pacific Deep Water (Fe, Mn, Co).
3. Depth of sea water containing amount of element lost by 100 meters of holocrystalline basalt.
4. *Cronan* [1967, p. 108].
5. Thickness of Mn crust containing same amount of element lost by 100 meters of holocrystalline basalt.
6. *Sayles and Bischoff* [1971].

194

layers interbedded with non-iron-rich calcareous nannoplankton ooze in the basal portions of holes 78 and 79 taken on DSDP leg X. These sediments are Eocene and Oligocene in age.

More recently, F. Sayles and J. L. Bischoff (unpublished manuscript, 1971) report on the occurrence of amorphous iron-manganese-silica beds cored east of the crest of the east Pacific rise at 9-10° South. This material is characterized by high concentrations of Ba, Cu, Pb, Zn, and Ni, and low $CaCO_3$. Three cores contained from ~ 1 to nearly 9 meters of this material, which they suggest 'is best explained by deposition of colloidal precipitates produced through the introduction of hot mineralized solutions into the ocean water.' They cite acoustic profiler (3.5 kHz) data to suggest that as much as 35 meters of this material may be present at one site.

SUMMARY AND CONCLUSIONS

It is observed that fine-grained holocrystalline basalts dredged along with glassy pillow fragments from a surveyed area on the MAR, are depleted relative to the pillows in Fe, Mn, Co, the REE, Cu, and Pb. The holocrystalline basalt fragments are assumed to be samples of the slowly cooling component of submarine extrusions. Pillow basalts are the rapidly quenched component of these flows, and their composition is close to that of the erupted liquid. These observations, along with information on the cooling behavior of tholeiitic lava masses in Hawaii and on hydrothermal mineralization on the continents, suggest a model for the origin of submarine hydrothermal solutions.

When basaltic rocks are erupted onto the sea floor, the margins of the extrusions are quenched to form pillows, which closely approximate the composition of the erupted liquid, including volatile components. The flow interiors are insulated by this shell of pillows, and cool slowly. Elements that do not readily enter the crystallizing phases are fractionated into residual silica-rich fluids, ultimately residing in residual solid phases in accessible sites along intergranular boundaries. Somewhat below solidus temperatures the early formed olivine is deuterically altered making additional elements accessible. Contraction fractures originate in the solidified shell at temperatures slightly below the solidus and propagate into the interior as it

cools, allowing volatile components to escape and admitting sea water into the hot crystallizing rock mass. This sea water is heated, dissolves escaping volatile components, and mobilizes metals in residual phases as chloride complexes. Convective flow in fracture systems should allow these hydrothermal solutions to emerge at the sea floor as submarine hot springs. Reasonable estimates of the amount of several elements mobilized in this way lead to the conclusion that a significant fraction of the mass of these elements that reside in pelagic deposits could have been supplied by this process.

Although it is presumed that the holocrystalline rocks studied here are from flow interiors near the sea-water interface, it is reasonable that the submarine ground water (sea water) could penetrate to considerable depths. Oxygen isotope data cited by *Muehlenbachs and Clayton* [1971] suggest that sea water is involved in the reactions producing greenstones from basalts beneath the mid-Atlantic ridge. *Taylor* [1970] cites oxygen isotope data on diorite and granodiorite intrusions in the western Cascades in Oregon to suggest that 'much so-called deuteric hydrothermal activity is probably caused by heated ground waters rather than by H_2O released during magmatic crystallization.' He suggests that a volume of water about equal to the volume of rock has been flushed through these intrusive stocks by convective circulation of ground water during much of the cooling and crystallization history.

The amorphous iron oxide deposits near the east Pacific rise and in DSDP cores and the 'excess' hydrogeneous iron in pelagic clays presumably form when these iron-rich solutions emerge from submarine hot spring vents and mix with sea water, which cools and oxidizes then precipitates a ferric-hydroxide floc which may settle in topographic lows near the vents or be widely dispersed, depending on bottom current conditions. *Dasch et al.* [1971], who reported data on the amorphous iron-manganese-silica sediments described by Sayles and Bischoff (1971), find that the isotopic composition of the abundant lead in this material is typical of mid-ocean ridge basalt lead. The strontium isotopic composition is predominantly that of sea water, with a small volcanogenic component. *Bender et al.* [1971] have found that the REE abundance patterns and the isotopic composition of Sr and U in an east Pacific rise core indicate that these elements are predominantly derived from sea water while the Pb isotopic composition indicates that the Pb is of local volcanic origin. These observations combined with the well-known tendency of many elements to co-precipitate with ferric hydroxide suggest that the mechanism proposed here can concentrate elements not only by mobilization of the residual components of submarine volcanic extrusions, but also by the extraction of trace constituents from sea water.

The model proposed here implies that we can determine the contribution of submarine volcanism to sea water for many elements by comparing the quenched pillow basalts with slowly cooled holocrystalline rocks from the same extrusions. In addition to inherited radiogenic helium and argon and the elements studied in this work, it is likely that other elements present in the magma and trapped on quenching are released to sea water from the slowly cooling portions of the flow. These might include other trace metals, silica, halogens, sulfur, rare gases, and nitrogen. Further studies of these rocks should provide data on the rate of supply of these elements to the hydrosphere from the upper mantle.

Acknowledgments. Sincere thanks are due Tj. H. van Andel, G. G. Goles, G. R. Heath, and J. R. Dymond for encouragement and critical discussions in the course of this work.

This work was done as part of a Ph.D. dissertation at Scripps Institution of Oceanography, University of California, San Diego. Principal support was provided by an American Chemical Society, Petroleum Research Fund¹ grant to Tj. H. van Andel. Additional financial assistance was provided by the National Science Foundation and the National Aeronautics and Space Agency.

REFERENCES

Baker, I., and S. E. Haggerty, The alteration of olivine in basaltic and associated lavas, *Contrib. Mineral. Petrol., 16,* 258–273, 1967.
Bender, M., W. Broecker, V. Gornitz, and U. Middel, Accumulation rate of manganese and related elements in sediments from the East Pacific Rise (abstract), *Eos Trans. AGU, 51,* 327, 1970.
Bender, M., W. Broecker, V. Gornitz, and U. Middel, R. Kaye, S. S. Sun, and P. Biscaye, Geochemistry of three cores from the east Pacific rise, *Earth. Planet. Sci. Lett.,* in press, 1971.
Bonatti, E., and O. Joensuu, Deep sea iron deposit

from the South Pacific, *Science, 154,* 643–645, 1966.
Bonatti, E., and Y. R. Nayudu, The origin of manganese nodules on the ocean floor, *Amer. J. Sci., 263,* 17–39, 1965.
Bostrom, K., The problem of excess manganese in pelagic sediments, in *Researches in Geochemistry,* edited by P. Abelson, p. 421–452, John Wiley, New York, 1967.
Bostrom, K., Submarine volcanism as a source for iron, *Earth Planet. Sci. Lett., 9,* 348–354, 1970.
Bostrom, K., and M. N. A. Peterson, Precipitates from hydrothermal exhalations of the East Pacific Rise, *Econ. Geol., 61,* 1258–1265, 1966.
Brewer, P. G., and D. W. Spencer, Trace element intercalibration study, *Tech. Rep. 70-62,* Woods Hole Oceanographic Institution, 1970.
Chester, R., and M. J. Hughes, The trace element geochemistry of a North Pacific pelagic clay core, *Deep Sea Res., 16,* 639–654, 1969.
Corliss, J. B., Mid-ocean ridge basalts: 1. The origin of submarine hydrothermal solutions, 2. Regional diversity along the Mid-Atlantic Ridge, Ph.D. dissertation, University of California, San Diego, 1970.
Cronan, D. S., The geochemistry of some manganese nodules and associated pelagic deposits, Ph.D. thesis, University of London, London, England, pp. 108–125, 1967.
Dalrymple, G. D., and J. C. Moore, Argon 40: Excess in submarine pillow basalts from Kilaula Volcano, Hawaii, *Science, 161,* 1132–1135, 1968.
Dasch, E. J., J. R. Dymond, and G. R. Heath, Isotopic analysis of metalliferous sediment from the East Pacific Rise (abstract), *Geol. Soc. Amer. Abstr. Programs,* in press, 1971.
DeVore, G., Crystal growth and the distribution of elements, *J. Geol. 63,* 471, 1955.
Dymond, J., Excess argon in submarine basalt pillows, *Geol. Soc. Amer. Bull., 81,* 1229–1232, 1970.
Funkhouser, J. G., E. D. Fisher, and E. Bonatti, Excess argon in deep-sea rocks, *Earth Planet. Sci. Lett., 5,* 95–100, 1968.
Garlick, G. D., and J. Dymond, Oxygen isotope exchange between volcanic materials and ocean water, *Geol. Soc. Amer. Bull., 81,* 2137–2142, 1970.
Goldberg, E. D., and G. Arrhenius, Chemistry of Pacific pelagic sediments, *Geochim. Cosmochim. Acta, 13,* 153–212, 1959.
Gordon, G. E., K. Randle, G. G. Goles, J. B. Corliss, M. H. Beeson, and S. S. Oxley, Instrumental activation analysis of standard rocks with high resolution x-ray detectors, *Geochim. Cosmochim. Acta, 32,* 364–396, 1968.
Hart, R. A., Chemical exchange between sea water and deep-ocean basalts, *Earth Planet. Sci. Lett., 9,* 269–279, 1970.
Hart, S. R., K, Rb, Cs contents with K/Rb, K/Cs ratios of fresh and altered submarine basalts, *Earth Planet. Sci. Lett., 6,* 295–303, 1969.
Helgeson, H. C., *Complexing and Hydro-*

thermal Ore Deposition, Macmillan, New York, 1964.

Hewett, D. F., M. Fleischer, and N. Conklin, Deposits of the manganese oxides-supplement, *Econ. Geol.*, *58*, 1-50, 1963.

Krauskopf, K. B., Separation of manganese from iron in sedimentary processes, *Geochim. Cosmochim. Acta*, *12*, 61-84, 1967.

Lachenbruch, A. H., Mechanics of thermal contraction cracks and ice-wedge polygons in permafrost, *Geol. Soc. Amer. Spec. Pap. 70*, 1962.

Mackin, J., and E. Ingerson, A hypothesis for the origin of ore-forming fluid, Geol. Surv. Res., 1960, *U.S. Geol. Surv. Prof. Pap. 400-B*, B1-B2, 1960.

Melson, W. G., and Tj. H. van Andel, Metamorphism on the Mid-Atlantic Ridge, 22° N latitude, *Mar. Geol.*, *4*, 165-186, 1966.

Melson, W. G., G. Thompson, and Tj. H. van Andel, Volcanism and metamorphism in the Mid-Atlantic Ridge, 22° N latitude, *J. Geophys. Res.*, *73*, 5925-5941, 1968.

Moore, J. G., and L. Calk, Sulfide spherules in vesicles of dredged pillow basalt, *Amer. Mineral.*, *56*, 476-488, 1971.

Muehlenbachs, K., and R. N. Clayton, The oxygen isotope geochemistry of submarine greenstones (abstract), *Eos Trans. AGU*, *52*, 365, 1971.

Murray, J., and A. F. Renard, Deepsea deposits, report of the scientific results of H. M. S. Challenger, 1873-1876, 372-378, 1891.

Onuma, N., H. Higuchi, H. Wakita, and H. Nagasawa, Trace element partition between two pyroxenes and the host lava, *Earth Planet. Sci. Lett.*, *5*, 47-51, 1968.

Park, C. F., The spilite and manganese problems of the Olympic Peninsula, Washington, *Amer. J. Sci.*, *244*, 305-323, 1948.

Peck, D. L., T. L. Wright, and J. G. Moore, Crystallization of tholeiitic basalt in Alae Lava Lake, Hawaii, *Bull. Volcanol.*, *29*, 629-656, 1966.

Peck, D. L., and T. Minakami, The formation of columnar joints in the upper part of the Kilauean Lava Lakes, Hawaii, *Geol. Soc. Amer. Bull.*, *79*, 1151-1166, 1968.

Rex, R. W., X-ray mineralogy studies: Leg 2 Deep Sea Drilling Project. 1970. in *Initial Reports of the Deep Sea Drilling Project, vol. 3*, edited by M. N. A. Peterson, U.S. Government Printing Office, Washington, 509-582, 1970*b*.

Skinner, J. J., and D. L. Peck, An immiscible sulfide melt from Hawaii, *Econ. Geol. Mono. 4*, 310-322, 1969.

Skornyakova, N. S., P. F. Andruschchenko, and L. S. Fomina, The chemical composition of iron-manganese concentrations of the Pacific Ocean, *Okeanologiya Akad. Sci.*, *2*, 2, 1962.

Taylor, H. P., Oxygen isotope evidence for large-scale interaction between meteoric ground waters and Tertiary Diorite inclusions, Western Cascade Range Oregon (abstract), *Eos Trans. AGU*, *51*, 453, 1970.

van Andel, Tj. H., and C. O. Bowin, Mid-Atlantic ridge between 22° and 23°N latitude and the tectonics of mid-ocean rises, *J. Geophys. Res.*, *73*, 1279-1293, 1968.

von der Borch, C. C., and R. W. Rex, Amorphous iron oxide precipitates in sediments cored during leg 5, Deep Sea Drilling Project. 1970, in *Initial Reports of the Deep Sea Drilling Project, vol. 5.*, edited by D. A. McManus, U.S. Government Printing Office, Washington, 1970.

von der Borch, C. C., W. D. Nesteroff, and J. Galehouse. 1970, Iron rich sediments cored during Leg VIII of the Deep Sea Drilling Project, in *Initial Reports of the Deep Sea Drilling Project, vol. 8*, edited by J. Tracey, U.S. Government Printing Office, Washington, 1971.

White, D. E., Environments of generation of some base-metal ore deposits, *Econ. Geol.*, *63*, 301-335, 1968.

Yoder, H. S., Jr., and C. E. Tilley, Origin of basalt magmas: an experimental study of natural and synthetic rock systems, *J. Petrol. 3*, 342, 1962.

Zelenov, K. K., Iron and manganese in exhalations of the submarine Banu Wuhu Volcano, Indonesia, *Doklady Akademii Nauk SSR*, Engl. Transl., *155*, 94-96, 1964.

(Received June 18, 1971;
revised August 27, 1971.)

Editor's Comments
on Papers 17 Through 20

17   UPADHYAY and STRONG
       Geological Setting of the Betts Cove Copper Deposits, Newfound-
       land: An Example of Ophiolite Sulfide Mineralization

18   NALDRETT
       Nickel Sulphide Deposits—Their Classification and Genesis, with
       Special Emphasis on Deposits of Volcanic Association

19   GOODWIN and RIDLER
       The Abitibi Orogenic Belt

20   BOLDY
       Excerpts from *Geological Observations on the Delbridge Massive*
       Sulphide Deposit

This group of papers is concerned with initial-stage (geosynclinal) metallogeny. Upadhyay and Strong (Paper 17) are among the many authors who have compared their work with that on the Troodos Massif in Cyprus. Most workers in Cyprus, however, have been concerned with the various types of mineralization, whereas Upadhyay and Strong were primarily concerned with the setting of the ore in the ophiolite environment. Descriptions of Kuroko deposits similarly deal largely with mineralization rather than setting, and no benchmark paper is known that fits the theme of this volume (but see Lambert and Sato, 1974).

Since the discovery of the Kambalda nickel deposits in Western Australia, interest in nickel deposits in mobile belts has increased markedly, and many papers have been written on various types of deposit and ultramafic hosts. Paper 18, by Naldrett, is recent but may prove to be the benchmark in classifying both hosts and deposits; the latter include not only the nickel of Naldrett's title but copper, which in the cupreous pyrite of the ophiolites has no associated nickel. Hutchinson and Searle (1971) show pentlandite and pyrrhotite associated with the ultramafics at the base of the Troodos Complex 20,000 feet below the pillow lava series and cupreous pyrite.

I would not make the assumption, as Naldrett did, that the sulfur in nickel-rich sulfides is derived from a primary mantle source. The Lynn

Lake intrusive, as a case in point, is post-Wasekwan, pre-Sickle, and so early stage, and not a class 1–(i) sill contemporaneous with eugeosynclinal vulcanism. Early-stage diapiric differentiated intrusives, of which the Lynn Lake intrusive is considered here to be a squeezed off part, appear to have the ability (not necessarily unique) of ingesting metal from the crust and either retaining it in disseminated form or regurgitating it in concentrations. At Lynn Lake I suspect the nickel silicate from the mantle combined with copper sulfide from the felsic volcanic host to the intrusive to give nickel–copper sulfide ores. The Lynn Lake body is in accord with the concept that older bodies (1700 b.y.) are more likely hosts for nickel sulfides than younger ones, for Bilibin showed the Phanerozoic of the USSR to have diapiric intrusives with no more mafic fraction than gabbro. Emslie and Moore (1961) gave an excellent discussion on the origin of Lynn Lake.

Goodwin has been developing the theme of volcanic cyclicity since his 1961 paper. Ridler, prepared to propound on exhalites at any opportunity, is his co-author in Paper 19, their most comprehensive work. Although this paper covers a broader field than eugeosynclines, vulcanism is the prime interest of both authors. The eugeosynclinal phenomena were perhaps best developed during the Superior era (2800–2350 m.y.) in the Canadian Shield.

Goodwin reiterated his idea that crustal floor spreading is a valid model for this orogenic belt and yet noted substantial north–south shortening. To me, the two are incompatible, and I see no reason to consider the Abitibi orogenic belt as differing in nature from other orogenic belts at subduction zones. The later extract from Plumstead (Paper 28) is concerned with biogenic deposition of gold and uranium in the Witwatersrand, which I relate to an interval. Biogenic activity, on the evidence of graphite bands, may well have aided base metal deposition, at Sherridon, Manitoba, for example, in the geosyncline.

Paper 20, by Boldy, is the essence of the work of the Conzinc school. In contrast, the leaders of the school were not writers, although King's review (1973) counters that argument. The difference may be that Boldy used figures; his Figures 9 and 11 are particularly apt.

Haddon King is the undisputed leader of the Conzinc school, whose ideas on the sedimentary origin of ore spread from Broken Hill, Australia, to many areas. King was encouraged in his ideas by Garlick and Brummer when he visited them in Northern Rhodesia; Brummer gives all credit in leadership to Garlick and also suggested to me that I look up the much earlier work of Schneiderhohn. The Canadian branch of the Conzinc school, Moss, Edwards, Gilmour, and Spence, were the authors of the now generally accepted volcanogenic concept of the origin of ores, first at Noranda and subsequently for perhaps all the major copper-zinc deposits of Superior Province. The works of Pereira, the European representative of the school, appear twice in the present volume (Papers 13 and 26), and, also from Europe, Oftedahl is given much credit by Edwards. Stanton (1955) was also early in the field, both in Australia and Canada, and he has been a major influence.

17

Geological Setting of the Betts Cove Copper Deposits, Newfoundland: An Example of Ophiolite Sulfide Mineralization

H. D. Upadhyay and D. F. Strong

Abstract

The Betts Cove copper deposits occur in one of the eight Lower Ordovician ophiolite complexes of Newfoundland. The deposits consist essentially of pyrite, chalcopyrite and minor sphalerite, displaying both syngenetic (e.g., sedimentary slump folds) and epigenetic (e.g., fracture-filling and replacement) features. Although localization of the sulfides appears to have been controlled by fault zones, careful geological mapping shows that they are fundamentally related to the ophiolite stratigraphy. Furthermore, these deposits are localized near the base of the volcanic sequence, indicating formation in the very earliest stages of volcanism. Such features are common to other ophiolites in Newfoundland, as well as those of Cyprus, Greece and Turkey. The Betts Cove and other Newfoundland deposits are thus taken to represent typical "Cyprus-type" volcanic exhalative mineralization, produced at an oceanic ridge during the earliest stages of Cambro-Ordovician sea-floor spreading. At a later stage the Betts Cove sulfides underwent remobilization and emplacement along fault zones.

Introduction

It is now recognized that most of the mafic-ultramafic complexes ("ophiolites") of Newfoundland (Fig. 1) probably represent the oceanic crust-mantle assemblages produced during Cambro-Ordovician sea-floor spreading (Bird and Dewey, 1970; Church and Stevens, 1971; Dewey and Bird, 1971; Upadhyay et al., 1971; Williams, 1971; Strong, 1972). In this context the tectonostratigraphic setting and genesis of the base metal deposits of central and western Newfoundland need to be re-examined with regard to volcanic setting (e.g., Strong and Peters, in press; Upadhyay and Smitheringale, 1972), but especially in the light of information available on deposits associated with well-known ophiolites such as Troodos (Hutchinson and Searle, 1971; Constantinou and Govett, 1972).

Although the Newfoundland ophiolites have long been known to contain sporadic copper mineralization (Snelgrove, 1928; Williams, 1963), no systematic pattern for this mineralization has yet been established, and it is especially noteworthy that no relation between the Bay of Islands deposits in Western Newfoundland (York Harbour, Crabbs Brook) and those of Notre Dame Bay (Little Bay, Whalesback, Betts Cove) has yet been recognized. The recent advances in the understanding of the Newfoundland ophiolites have revealed some general patterns, and we take the Betts Cove area as a representative case.

Geology of the Betts Cove Area

The Betts Cove area, located on the eastern part of the Burlington Peninsula, is underlain by the Lower Ordovician Snooks Arm Group (Fig. 1), a belt of rocks stretching over nearly 10 miles from Tilt Cove in the northeast to Betts Cove in the southwest (Snelgrove, 1931; Neale, 1957). The Snooks Arm Group consists of a basal ophiolite suite (Upadhyay et al., 1971) that comprises, from bottom to top, ultramafic rocks, gabbro, sheeted dikes, pillow lavas and volcaniclastic sediments; this sequence is conformably overlain by a sequence of pillow lava, sediments and pillow lava (Figs. 2, 3D). The ophiolite complex is most completely preserved in Betts Cove area; elsewhere a fault contact exists between the ultramafic assemblage and the pillow lava. The basal ultramafic member of the Snooks Arm Group is separated by a fault from the older Nippers Harbour Group and younger quartz-feldspar porphyries to the west. The ultramafic rocks are discontinuous and strongly sheared in the vicinity of the fault but are elsewhere identified as pyroxenite, harzburgite and dunite. These lithologies are regularly interlayered on all scales from several millimeters to 1.5 meters, and show pronounced cumulus textures.

The upper contact of the ultramafic member is extremely variable in character but normally consists of a complex transition zone of interlayered pyroxenite and gabbro which grades upward into gabbro.

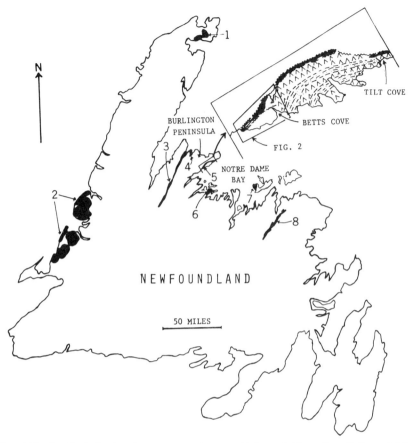

Fig. 1. Map of Newfoundland showing the occurrence of some ophiolites. 1. Hare Bay; 2. Bay of Islands; 3. Baie Verte; 4. Ming's Bight; 5. Betts Cove; 6. Pilleys Island; 7. Moretons Harbour; 8. Gander River. Inset shows the con figuration of the Snooks Arm Group.

The gabbro member is regionally discontinuous and cut by abundant basic dikes which increase in number toward the top, grading into the sheeted dike unit which consists of 100 percent dikes. Where gabbro is absent the sheeted dikes die out rapidly downward into the ultramafic member.

The overlying pillow lavas are generally faulted against the sheeted complex, but in some cases there is a rapid but continuous upward decrease in number of dikes to pillow lavas. The pillow lavas also contain pillow breccias, aquagene tuffs and abundant interstitial red chert. The sedimentary units include immature quartz-free volcaniclastic sediments, pyroxene-andesite tuff and agglomerate, graywacke, minor argillite, and red and green chert.

More detailed descriptions are given by Upadhyay et al. (1971).

Sulfide Mineralization

Betts Cove and Tilt Cove are the two localities within the Snooks Arm Group where copper has been mined. The Betts Cove deposits were discovered in 1860 and mined between 1875 and 1885 yielding 130,000 tons of hand-picked ore averaging about 10% copper, along with 2,450 tons of pyrite (Douglas et al., 1940). The Tilt Cove deposits, on the other end of the ophiolite belt (Fig. 1, inset) were discovered in 1857 and were mined during two periods, from 1864 to 1917 and from 1957 to 1967 yielding several million tons of copper ore. Apart from these two main areas, there are numerous sulfide showings within the volcanic rocks of the Snooks Arm Group between Betts Cove and Tilt Cove.

In the Betts Cove area (Fig. 2) sulfide mineralization occurs at a number of places within the sheeted

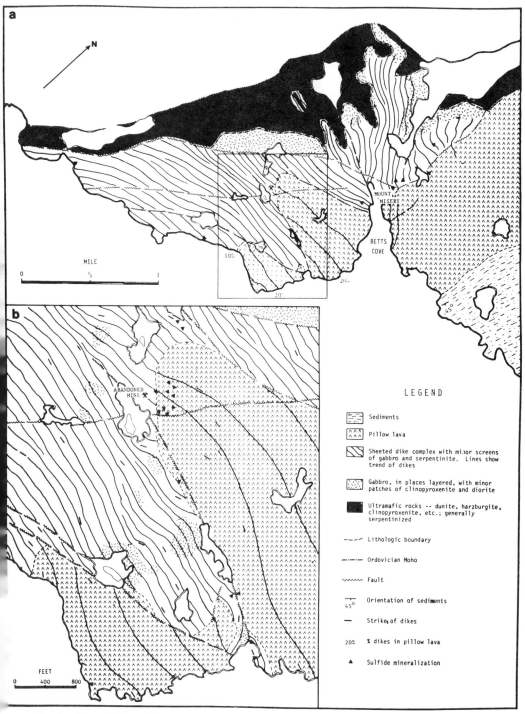

FIG. 2a. Generalized geological map of Betts Cove ophiolite complex (modified after Upadhyay et al., 1971).
2b. Detailed map of the Betts Cove mine area.

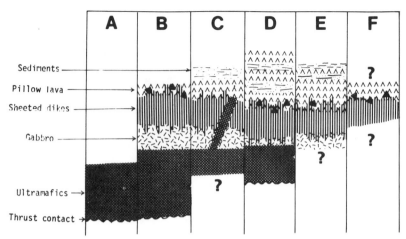

FIG. 3. Schematic cross sections and probable correlations of some Newfoundland ophiolites. Solid triangles indicate sulfide mineralization. A. Hare Bay (Williams, 1971); B. Bay of Islands (R. W. Hutchinson, pers. comm.; Williams and Malpas, 1972); C. Ming's Bight (R. Norman, pers. comm.); D. Betts Cove (Upadhyay et al., 1971); E. Pilleys Island; F. Moretons Harbour (Strong, 1972).

dikes and pillow lavas; sporadic pyrite and minor chalcopyrite are particularly common in sheeted dikes. Pyrite is sparsely disseminated throughout the gabbro, sheeted dikes, and pillow lavas. Detailed mapping of the Betts Cove mine and Mount Misery areas shows that the dominant concentration of sulfides occurs along or near the contact between sheeted dikes and overlying pillow lavas. This stratigraphic control contradicts the suggestion of earlier workers that the nearby dioritic-gabbroic (Snelgrove, 1931; Douglas et al., 1940) and granitic (Baird, 1951) intrusions caused an epigenetic hydrothermal mineralization localized along pre-existing faults. Futhermore, since the gabbro rocks occur as screens within the sheeted dikes (i.e., they do not intrude the dikes), they could not have been responsible for any epigenetic mineralization.

The following evidence is taken as indicating that the chloritized fault zones which carry many of the sulfide concentrations postdate the mineralization, i.e., they did not act as "controls" for sulfide deposition. The sulfides occur in all forms ranging from banded massive lenses to disseminated zones. The largest sulfide bodies occur within chloritized fault zones, most of which lie close to the sheeted dikes/ pillow lava contact. Ore samples from such parts are strongly foliated (Fig. 4a).

Mineralization in pillow lavas shows maximum concentration in the spaces between pillows although it is disseminated through them. The sulfide-rich rims are now foliated, producing an augen structure around the massive pillows. The concentration of

sulfides between pillows is taken as a primary feature which was further accentuated by deformation, with the interstices of the pillows being more susceptible to shearing and remobilization than the interiors of the massive pillows.

Massive and banded sulfides are best seen in material from the abandoned mine dump, where numerous blocks show sedimentary lamination and slump folds (Fig. 4b) which are indicative of initial sedimentary deposition of sulfides. These were later modified by deformation and remobilization as suggested by veins and stringers of sulfides within the associated silicate minerals.

Mineralogy

Mineralographic studies of the Betts Cove sulfides show that pyrite is by far the most abundant sulfide mineral, followed by chalcopyrite and local concentrations of sphalerite; native iron (Deutsch and Rao, 1972) occurs elsewhere in the Betts Cove pillow lava. Although there appears to be a definite paragenetic sequence, with pyrite being the earliest phase to crystallize generally as dispersed euhedral crystals in a sphalerite-chalcopyrite matrix (Fig. 4c), pyrite, sphalerite, and less commonly. chalcopyrite form individual laminae with sharp contacts that are fairly consistent in thickness. In the case of pyrite, for instance, crystals show a constant size within any particular lamina. Chalcopyrite occurs as disseminated stringers and also interstitial to euhedral pyrite crystals, but most commonly fills fractures within highly brecciated pyrite grains (Fig. 4d). Sphalerite

FIG. 4a. Foliation developed in a massive sulfide specimen. Light material is pyrite, dark is chlorite. 4b. Sedimentary slump fold in a massive sulfide specimen. Light material is pyrite, dark is sphalerite. 4c. Primary euhedral pyrite crystals and associated chalcopyrite in a silicate matrix (reflected, plane polarized light). 4d. Fractured and granulated pyrite crystals with remobilized chalcopyrite "healing" the fractures (reflected, plane polarized light). 4e. Strongly foliated pyrite, with minor chalcopyrite, in a silicate matrix (reflected, plane polarized light). 4f. Foliation shown by sulfides and silicate gangue. Note the deformation lamellae in quartz, on the lower left part of the photo (transmitted light, crossed nicols).

In 4c, d, e the light material is pyrite, medium is chalcopyrite and the dark is silicate gangue. Note the gradation from undeformed (c), through moderately deformed (d), to highly deformed (e) sulfides.

appears last in the sequence surrounding and filling fractures in both chalcopyrite and pyrite.

The hand specimens of undeformed rocks show more clearly that these sulfides are interbedded in alternating layers rich in one particular mineral (Fig. 4b), indicating repeated precipitation of each phase. The paragenetic sequence described above is

interpreted as reflecting a relative ease of remobilization, with brittle deformation of pyrite and more plastic deformation of chalcopyrite and sphalerite— an interpretation supported by the experiments of Gill (1969) and the observations of Hutchinson (1965) and Suffel et al. (1971).

There exists a complete gradation from unfoliated

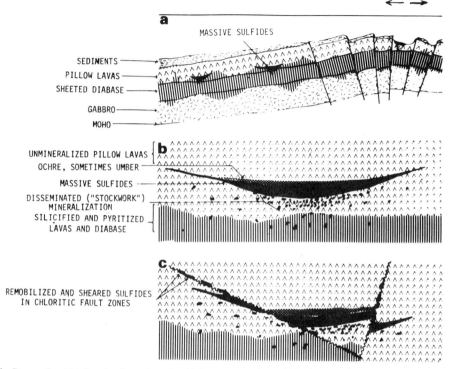

FIG. 5. Proposed model for the formation of ophiolite-type massive sulfide deposits (modified after Hutchinson and Searle, 1971; Constantinou and Govett, 1972). a. Massive and "stockwork" mineralization produced by volcanic exhalative activity in an oceanic ridge-type environment during sea-floor spreading. Note the position of the sulfides at the base of the pillow lava sequence. b. Enlarged schematic cross section of a typical massive sulfide deposit showing sequence from "stockwork" mineralization to ochre. c. Deformed and remobilized equivalent of *b*, showing the emplacement of massive sulfides in chloritic fault zones, a feature common in Betts Cove and some other Newfoundland ophiolites.

to highly foliated ore (Figs. 4c, d, e). This feature can be seen on any scale, from microscopic to outcrop size. Silicates, particularly quartz, associated with sulfides are foliated in the same direction as deformed sulfides (Fig. 4f).

Discussion

It has been suggested (Upadhyay et al., 1971) that the Betts Cove ophiolites represent oceanic crust comparable to other ophiolites such as the Troodos (Gass, 1968; Moores and Vine, 1971), Oman (Reinhardt, 1969), and the Bay of Islands (Williams, 1971; Williams and Malpas, 1972) ophiolite complexes. It can be shown not only that the mineralization is similar in each area but also that it occupies a similar stratigraphic position in each ophiolite sequence. Throughout the Bay of Islands complex the mineralization appears to be at the contact between sheeted dikes and pillow lavas, just as at Betts Cove (Figs. 2, 3). This stratigraphic control is also recognized for all major deposits of Cyprus (G.

Constantinou, pers. comm.). Furthermore, detailed studies of the York Harbour deposit in the Bay of Islands complex show it to be very closely comparable in all respects to the Troodos deposits (R. W. Hutchinson, pers. comm.). These similarities of overall geological environment, stratigraphic control sedimentary features, and pyrite-chalcopyrite-sphalerite mineralogy lead us to adopt an hypothesis comparable to those of Hutchinson and Searle (1971) or Constantinou and Govett (1972) for the Troodos deposits, although the latter ascribe much importance to secondary submarine leaching. We suggest the following model (Fig. 5) for the formation of Betts Cove deposits:

(1) Lower Ordovician sea-floor spreading resulting in production of the ophiolite (oceanic crust) complex (Fig. 5a), accompanied in the very earliest stages by volcanic exhalative mineralization to produce typical massive sulfides underlain by disseminated "stockwork" mineralization (Fig. 5b).

(2) Faulting and shearing at a later stage resulted in the remobilization of the sulfides and their location in chloritic fault zones (Fig. 5c), producing the sequence of textures demonstrated in Figures 4c–4e.

The mineralization resulted primarily from syngenetic precipitation of sulfides from upwelling geothermal brines during hiatuses in volcanism in an oceanic ridge-type environment, with the less important disseminated or "stockwork" mineralization produced by replacement and cavity-filling in the underlying rocks during ascent of the volcanic exhalations. All evidence on mineralization in Betts Cove and the Bay of Islands ophiolites conforms to this interpretation, and we suggest that mineral exploration in these and other such areas can be most successful if emphasis is placed on understanding of the ophiolite stratigraphy.

Acknowledgments

This study was financed by Geological Survey of Canada Grant 37–70 to D. F. Strong and National Research Council of Canada Grants A-7975 to D. F. Strong and A-5540 to E. R. W. Neale. We have benefited greatly from discussions with R. W. Hutchinson and H. R. Peters, and from the discussions and guidance (to D. F. Strong) in Cyprus of A. Panayiotou and G. Constantinou. E. R. W. Neale and W. G. Smitheringale kindly read the manuscript and offered constructive suggestions. We acknowledge these with gratitude.

DEPARTMENT OF GEOLOGY
MEMORIAL UNIVERSITY OF NEWFOUNDLAND
ST. JOHN'S, NEWFOUNDLAND, CANADA

H. D. U. PRESENT ADDRESS:
DEPARTMENT OF GEOLOGY
ST. FRANCIS XAVIER UNIVERSITY
ANTIGONISH, NOVA SCOTIA, CANADA
July 5, August 10, 1972

REFERENCES

Baird, D. M., 1951, The geology of the Burlington Peninsula, Newfoundland: Canada Geol. Survey Paper 51-21.

Bird, J. M., and Dewey, J. F., 1970, Lithosphere plate-continental margin tectonics and the evolution of Appalachian orogen: Geol. Soc. America Bull., v. 81, p. 1031–1060.

Church, W. R., and Stevens, R. K., 1971, Early Paleozoic ophiolite complexes of the Newfoundland Appalachians as mantle-oceanic crust sequences: Jour. Geophys. Research, v. 76, p. 1460–1466.

Contantinou, G., and Govett, G. J. S., 1972, Genesis of sulfide deposits, ochre, and umber of Cyprus: Inst. Mining Metallurgy Trans., sec. B, v. 81, p. B34–46.

Deutsch, E. R., and Rao, K. V., 1972, Preliminary magnetic study of Newfoundland ophiolites, showing the presence of native iron [abs.]: Am. Geophys. Union Trans., v. 53, p. 734.

Dewey, J. F., and Bird, J. M., 1971, The origin and emplacement of the ophiolite suite; Appalachian ophiolites in Newfoundland: Jour. Geophys. Research, v. 76, p. 3179–3206.

Douglas, G. V., Williams, D., and Rove, O. N., 1940, Copper deposits of Newfoundland; Newfoundland Geol. Survey Bull. 20.

Gass, I. G., 1968, Is the Troodos Massif of Cyprus a fragment of Mesozoic ocean floor?: Nature, v. 220, p. 39–42.

Gill, J. E., 1969, Experimental deformation and annealing of sulfides and interpretation of ore textures: ECON. GEOL., v. 64, p. 500–508.

Hutchinson, R. W., 1965, Genesis of Canadian massive sulfides reconsidered by comparison to Cyprus deposits: Canadian Inst. Mining Metallurgy Trans., v. 68, p. 286–300.

—— and Searle, D. R., 1971, Stratabound pyrite deposits in Cyprus and relations to other sulfide ores: Soc. Mining Geologists Japan, Spec. Issue 3, p. 198–205.

Moores, E. M., and Vine, F. J., 1971, The Troodos Massif Cyprus, and other ophiolites as oceanic crust; evaluation and implications: Royal Soc. [London] Philos. Trans., v. 268, p. 443–466.

Neale, E. R. W., 1957, Ambiguous intrusive relationship of the Betts Cove-Tilt Cove serpentinite belt, Newfoundland: Geol. Assoc. Canada Proc., v. 9, p. 95–107.

Reinhardt, B. M., 1969, On the genesis and emplacement of ophiolites in Oman Mountains geosyncline: Schweizer. mineralog. petrog. Mitt., v. 49, p. 1–30.

Snelgrove, A. K., 1928, The geology of the central mineral belt of Newfoundland: Canadian Mining Metall. Bull. 197, p. 1057–1127.

—— 1931, Geology and ore deposits of Betts Cove-Tilt Cove area, Newfoundland: Canadian Mining Metall. Bull. 228, p. 477–519.

Strong, D. F., 1972, Sheeted diabases of central Newfoundland: new evidence for Ordovician seafloor spreading: Nature, v. 235, p. 102–104.

—— and Peters, H. R., in press, The importance of volcanic setting for base metal exploration in central Newfoundland: Canadian Inst. Mining Metallurgy Trans.

Suffel, G. S., Hutchinson, R. W., and Ridler, R. H., 1971, Metamorphism of massive sulfides at Manitouwadge, Ontario, Canada: Soc. Mining Geologists Japan, Spec. Issue 3, p. 235–240.

Upadhyay, H. D., Dewey, J. F., and Neale, E. R. W., 1971, The Betts Cove ophiolite complex, Newfoundland: Appalachian oceanic crust and mantle: Geol. Assoc. Canada Proc., v. 24, p. 27–34.

Upadhyay, H. D., and Smitheringale, W. G., 1972, The Gullbridge copper deposit, Newfoundland: volcanogenic sulfides in cordierite–anthophyllite rocks: Canadian Jour. Earth Sci., v. 9, p. 1061–1073.

Williams, H., 1963, Relationship between base metal mineralization and volcanic rocks in northeastern Newfoundland: Canadian Mining Jour., v. 84, p. 39–42.

—— 1971, Mafic-ultramafic complexes in western Newfoundland Appalachians and the evidence of their transportation: a review and interim report: Geol. Assoc. Canada Proc., v. 24, p. 9–26.

—— and Malpas, J. G., 1972, Sheeted dikes and brecciated dike rocks within transported igneous complexes, Bay of Islands, western Newfoundland: Canadian Jour. Earth Sci., v. 9, p. 1216–1219.

18

Reprinted from *Canadian Mining Metallurg. Bull.*, **66**(739), 45–63 (1973)

Nickel Sulphide Deposits — Their Classification and Genesis, With Special Emphasis on Deposits of Volcanic Association

A. J. NALDRETT, Professor,
Department of Geology,
University of Toronto,
Toronto, Ont.

ABSTRACT

Nickel sulphide deposits may be classified conveniently in relation to the types of ultramafic and mafic bodies with which they are associated. In this paper, nine major classes of ultramafic bodies are recognised. Broadly, they are divided into those of orogenic and non-orogenic association. The orogenic group is comprised of (1) syn-volcanic bodies (subdivided into class 1-(i), sills, and class 1-(ii), lenses), (2) alpine bodies, (3) Alaskan-type zoned intrusions and (4) bodies associated with major crustal sutures. The non-orogenic group consists of (5) large stratiformly layered complexes, (6) large sills and sheets that are the intrusive equivalents of flood basalts, (7) medium-sized sub-volcanic intrusions, (8) small sub-volcanic sheets and dykes, and (9) alkalic ultramafic rocks in ring complexes and kimberlite pipes.

Estimates of the millions of tons of nickel metal associated as sulphide nickel deposits with each of the different classes of bodies, including past production and present reserves, are as follows (E denotes economic, M marginal or non-economic deposits): 1-(i), E > 0.86; 1-(ii), E = 2.89, M = 3.48; 2, Nil; 3, Nil; 4, E = 2.39, M = 0.76; 5, E = 9.92, M = 0.52; 6, Nil; 7, E >> 1.0; 8, M = 0.21; 9, Nil. The importance of the contribution to the world's supply of nickel made by deposits associated with class 1-(ii) lenses has only recently been recognised, and, as ore reserves in the Western Australian and other new nickel camps grow, this class may eventually rival the Sudbury Camp (class 5) as being the pre-eminent nickel producer in the western world.

Available experimental data suggest that below about 80 km in oceanic regions, sulphides will be liquid and may tend to settle downwards, leaving a portion of the mantle

ANTHONY J. NALDRETT was born in London, England and studied geology at the University of Cambridge (B.A. 1956). He emigrated to Canada in 1957 and worked for Falconbridge Nickel Mines Ltd. until 1959, when he left to work with Professor J. E. Hawley at Queen's (M.Sc. 1961; Ph.D. 1964) on problems related to nickel deposits. After spending three years at the Geophysical Laboratory in Washington, D.C., working with Dr. G. Kullerud, he returned to the University of Toronto. He now holds the rank of Professor and is involved in teaching economic geology. Dr. Naldrett has recently returned from a sabbatical year in Western Australia, where he worked with CSIRO on problems associated with the Eastern Goldfields nickel camp.

PAPER SUBMITTED: in December of 1972; revised manuscript received on May 25, 1973.

KEYWORDS: Ore deposits, Nickel sulphides, Ultramafic bodies, Volcanic rocks, Ore genesis, Mineral exploration, Copper-nickel ratios, Abitibi belt, Alexo mine.

CIM TRANSACTIONS: Vol. LXXVI, pp. 183-201, 1973.

immediately below this depth depleted in sulphur. At depths of about 300 km, the settling sulphides may become solid again and tend to concentrate. Many ore-bearing magmas appear to have been derived from depths of around 300 km. Experimental data also suggest that if a portion of the mantle undergoes partial melting, the sulphides will also melt and be removed, probably in solution in the silicate melt, leaving a refractory residuum depleted both in sulphur and in low-melting-point silicate constituents. Recent studies indicate that alpine ultramafic bodies are fragments of this refractory residuum, thus accounting for the absence of important associated nickel sulphide deposits. Alkalic magmas may be generated in the portion of the mantle from which liquid sulphides have settled, thus accounting for the lack of significant deposits associated with these rocks. All of the upper mantle is thought to have undergone melting, with the removal of picritic or tholeiitic partial melts several times in the last 3.0 x 10⁹ yrs. The progressive depletion in the concentration of sulphur in the mantle produced in this way may account for the observation, made in this paper, that the majority of nickel sulphide ores are older than 1.7 x 10⁹ yrs.

Copper-nickel ratios of nickel sulphide deposits vary with the class of host rock, high ratios reflecting a more mafic, less ultramafic host.

The nickel-rich, syn-volcanic class 1(ii) lenses form an integral part of the sub-volcanic - volcanic areas in which they occur, some lenses being demonstrably extrusive. A volcanic model for ore deposition is important to bear in mind when exploring for deposits of this type.

INTRODUCTION

THE AIM OF THIS PAPER is to present a classification of nickel sulphide deposits that may be helpful, both in understanding their genesis and in guiding exploration. In this context, genesis means more than just the manner of concentration and emplacement of the sulphides. In most cases, the evidence is fairly clear that nickel sulphide deposits have been introduced as immiscible sulphide-oxide liquids; thus the manner of their emplacement is much less obscure than the reason why they occur in a certain area associated with certain rocks. The approach taken in this paper is to classify nickel deposits in terms of the particular classes of ultramafic or mafic bodies with which they are associated, pointing out classes of bodies devoid of important deposits, to discuss the possible behaviour of sulphur in the mantle and the importance of this in relation to the generation of the different classes of ultramafic/mafic rocks, and then to discuss features of one particular class of volcanic association that may have a bearing on the localization of ore.

CLASSIFICATION OF BODIES OF ULTRAMAFIC AND RELATED MAFIC ROCKS

Among recent attempts to classify ultramafic and related mafic bodies, Wyllie (1967) has stressed the size, form and chemical composition of the bodies and Thayer (1971) has emphasized the distinction among those bodies that crystallized *in situ* (authigenic). those that were emplaced as solids (allogenic) and

TABLE 1 — Classification of Ultramafic and Associated Mafic Bodies

Class	Examples	Tectonic Environment	Other Remarks
A. BODIES ASSOCIATED WITH OROGENIC BELTS			
1. Rocks contemporaneous with eugeosynclinal volcanism			
(i) Gravity-differentiated sills and complexes			
(a) Ultramafic-rich sub-type	Abitibi Belt, Canada (Naldrett & Mason, 1968; McRae, 1969).	Intruded within a eugeosynclinal sequence. Crystallization preceded major folding.	Ratio of mafic to ultramafic rocks highly variable, possibly due to periodic "decanting" of residual magma during crystallization.
	Barberton Mountainland, S. Africa (Viljoen & Viljoen, 1970).		
	E. Goldfields, Australia (Williams & Hallberg, 1972)		
(b) Anorthositic sub-type	Doré Lake Complex, Quebec (Allard, 1970)	Although some of this sub-type are conformable and have crystal-lized before folding, others are transgressive and their early in-trusion is less certain.	
	Bell River Complex, Quebec (Sharpe, 1965)		
	Kamiscotia Complex, Timmins, Ontario.		
(ii) Ultramafic lenses	Abitibi Orogen (Naldrett & Mason, 1968; Pyke, Naldrett & Eckstrand; 1972).	Closely associated with volcanism; largely conformable with volcanic stratigraphy. No post-consolidation movement has occurred during folding.	Lenses emplaced as a suspension of olivine crystals in a picritic liquid. Stratiform successions of thin lenses in which marked assymetry in quenching, lack of cross-cutting features and unusual Pahoe-hoe forms indicate extrusion in some cases (see text).
	N.W. Ontario (Watkinson & Irvine, 1964).		
	Cape Smith - Wakeham Bay belt and Labrador Geosyncline, (Fahrig, 1962; Wilson et al., 1969).		
	Barberton Mountainland, E. Trans-vaal (Viljoen & Viljoen, 1969).		
	Eastern Goldfields, Australia (Nesbitt, 1971; McCall, 1971; McCall & Leishman, 1971; Hallberg & Williams, 1972)		
2. Alpine-type bodies			
(i) Large obducted sheets	New Caledonia (Guillon, 1969)	Restricted to Phanerozoic con-suming plate margins: (i) as sheets thrust over sediments and volcanics forming continental or island-arc crustal sequences; (ii) as thrust wedges in flysch sedi-ments; (iii) as recognizable units or chaotic blocks associated with mélanges; (iv) as possible diapiric bodies exploiting suture zones.	Interpreted as fragments of oceanic lithosphere, the lower ultramafic portion being largely material that crystallized within the mantle and an upper basaltic-gabbroic portion representing layers 2 & 3 of the oceanic crust. See text for more complete explana-tion.
	Papua (Davies, 1968)		
(ii) Ophiolite complexes	Vourinos (Moores, 1969)		
	Troodos (Gass, 1967, 1968)		
	Bay of Islands (Smith, 1958; Dewey & Bird, 1971; Irvine & Findlay, 1972)		
(iii) Deformed ophiolite complexes and chaotic blocks in mélange terranes	Canyon Mountain, Oregon (Thayer, 1963)		
	Twin Sisters, Washington (Ragan, 1967)		
	Vermont Serpentinites (Jahns, 1967)		
	Iran (Gansser, 1969)		
	Franciscan Series, California (Hamilton, 1969; Hsu, 1968)		
(iv) Possible diapirs	Mt. Albert, Quebec (MacGregor, 1962)		

Continued

those that were emplaced partly as solid and partly as liquid (polygenic). The classification presented in Table 1 (and in a less updated form by Naldrett and Gasparrini, 1971) stresses the tectonic setting into which bodies were emplaced, while at the same time drawing on criteria that both Wyllie and Thayer have used.

An initial distinction is made between those bodies that were emplaced in an orogenic tectonic environment and those that were emplaced in a non-orogenic environment*. Let us first consider the orogenic en-

vironment. At present the majority of ultramafic bodies appear to fall into three major classes. The first of these covers bodies that were emplaced during the initial, volcanic stage of an orogenic belt, before the onset of major folding. Two major types are recognized — gravity-differentiated sills (class 1(i) a & b) and lenses showing little or no differentiation (class 1-(ii)). The sills fall into two sub-types — (a) those rich in peridotite and pyroxenite and containing normal gabbros and/or norites and (b) those with less ultramafic rocks and containing much anorthosite,

(continental or oceanic), a plate undergoing incipient rifting or one undergoing the major rifting that eventually gives rise to the development of new oceanic crust at an accreting plate margin. The terms are applied to the Precambrian without necessarily implying that plate movement was occurring during this period.

3. Alaskan-type zoned intrusions	Intrusions of Alaska and British Columbia, including: Duke Island (Irvine, 1959); Union Bay (Ruckmick & Noble, 1959); Tulameen (Findlay, 1969) Intrusions of the Urals.	Intruded along orogenic belts either after or during the late stages of the main folding and metamorphism, but before the emplacement of granite batholiths.	The marked zoning characteristic of this class is largely due to multiple intrusions of successive magmas, each successively more ultramafic than the preceding intrusion and tending to intrude through the core of the preceding intrusion.
4. Intrusions associated with major crustal sutures	Intrusions of the Manitoba nickel belt (Wilson & Brisbin, 1961; Coats, 1966; Quirk *et al.*, 1970). Intrusions of the English River (Ontario) gneissic belt.	Intruded along major faults in crustal sutures ("mobile belts") characterized by complex folding, high-grade metamorphism, and distinctive gravity anomalies and crustal thicknesses.	Little work has been done on ultramafic bodies of this class, and it still remains to be determined whether the majority of the ultramafic bodies were emplaced during deformation or whether they were emplaced in rocks that subsequently underwent intense deformation. Bodies of the English River gneissic belt are probably truly representative of the class, although many of those in Manitoba may be stratigraphically controlled.
B. BODIES OCCURRING IN MORE STABLE AREAS			
5. Large stratiformly layered complexes	Bushveld (Wager & Brown, 1967) Stillwater (Jackson, 1961; Hess, 1960) Muskox (Irvine & Smith, 1967) Duluth (Taylor, 1964; Phinney, 1970) Kiglapait (Morse, 1961) Sudbury (Naldrett *et al.*, 1970)	Stable or relatively stable cratons	Have differentiated *in situ* as a result of gravitational crystal settling. Removal of residual magma and influxes of fresh magma have occurred on occasions.
6. Large sills and sheets that are the intrusive equivalent of flood basalts	Palisades Sill, New Jersey (F. Walker, 1940; K. Walker, 1969) Insizwa-Ingeli Intrusion, S. Africa (Maske, 1966) Dufek Intrusion, Antarctica (Ford & Boyd, 1968)	Generally occur in cratonic areas, in which block faulting, epeirogenic uplift and extrusion of flood basalts is occurring.	Have differentiated *in situ* as a result of gravitational crystal settling. Chemically these intrusions appear to be the intrusive equivalents of the flood basalts with which they are closely associated.
7. Medium-sized intrusions associated with non-orogenic volcanism	Skaergaard (Wager & Brown, 1968) Rhum (Wager & Brown, 1968) Intrusions of Noril'sk-Talnakh area (Godlevskii, 1959; Zolotuchin & Vasilev, 1967).	Located in non-orogenic areas.	Have differentiated *in situ* as a result of gravitational crystal settling. In some cases, the intrusions appear to have been differentiation chambers for associated volcanic rocks.
8. Small sheets and intrusions associated with non-orogenic volcanism	This class includes a wide variety of intrusions ranging from tholeiitic to alkalic in composition associated with volcanic activity as exemplified by the Permo-Carboniferous volcanism of the Lowland valley of Scotland and by the Thulean Tertiary igneous province.		
9. Alkalic ultrabasic rocks in ring complexes and kimberlite pipes	Numerous	Cratons	

TABLE 2 — Major Sulphide Nickel Camps of the World

Nickel Camp	Host Rocks	Production plus Reserves (in 10^6 tons of nickel metal)*	Average Cu/(Cu + Ni)	Age (in 10^9 yrs)	Important References
WESTERN AUSTRALIA					
Eastern Goldfields	Class 1-(ii) lenses	E 1.84 M 3.00	0.062 0.036	>2.667 ± 0.027 Ref. Turek (1966) Arriens (1971)	Woodall & Travis (1969) Hancock *et al.* (1971) McCall (1971) Purvis, Nesbitt & Hallberg (1972)
	Class 1-(i) complex sill	E 0.03	0.260		
CANADA					
Abitibi plus Shebandowan and Rankin Inlet	Class 1-(ii) lenses	E 0.2	0.226	2.75 — 2.80 Ref. Krogh & Davis (1971)	See Naldrett & Gasparrini (1971) for references
Lynn Lake and Dumbarton mine	Class 1-(i) sills	E 0.25	0.380	Probably ∼ 2.75	Ruttan (1955) Milligan (1960) Emslie & Moore (1961) Karup-Moller & Brummer (1971)
Cape Smith - Wakeham Bay	Class 1-(ii) lenses	E 0.65 M 0.07	0.252	Probably older than 1.7-1.8. Ref. Dimroth (1970)	Wilson *et al.* (1969) Kilburn *et al.* (1969)
Thompson - Wabowden belt	Class 4	E 2.0	0.064	Probably older than 1.7-1.8 and possibly Archean	Zurbrigg (1963) Kilburn *et al.* (1969) Quirke *et al.* (1970) Coats and Brummer (1971)
Gordon Lake	Class 4	E 0.02	0.296	2.75 - 2.80. Ref. Krogh & Davis (1971)	Scoates (1970)
Sudbury	Class 5	E 9.7	0.468	Previously given as 1.704 (Fairbairn *et al.*), but more probably 1.9-2.0 (Souch *et al.*, 1969, Gibbins & McNutt, 1972)	See Hawley (1962) for early references. More recent ones bearing on the ores include: Naldrett & Kullerud (1967); Cowan (1968); Souch *et al.* (1969); Naldrett *et al.* (1970, 1972)
Great Lakes Nickel	Class 8	M 0.21		Probably 1.1	
Lac Renzy	Class unknown. Occurs in a differentiated sill in Grenville amphibolites and paragneisses.	E 0.014	0.507	Not known	
Giant Nickel	Class unknown. Complex zoned intrusion with peridotite pipes occurring in pyroxenite and hornblendite.	E 0.012		Eocene	Aho (1956) Muir (1971) Muir & Naldrett (1973)
SOUTHERN AFRICA					
Empress Mine (Rhodesia)	Probable class 1-(i) sill	E 0.29	0.328	Between 2.9 & 3.3. Refs. Nicolaysen (1962); Bliss & Stidolph (1969); Stowe (1971)	Sharpe (1964) Eales (1964)
Trojan, Madziwa, Inyati & Shangani (Rhodesia)	Class 1-(ii)	E uncertain, but supposedly about 0.2	0.048		Le Roux (1964)
Pikwe & Selibe (Botswanna)	Class 4? Sill-like bodies of amphibolite in the Limpopo metamorphic belt	E 0.39	0.511	>1.94 ± 0.06 Ref. Nicolaysen (1962)	
Bushveld Igneous Complex (South Africa)	Class 5. Ore occurs in pegmatitic pipes & in the Merensky reef	E 0.22 (The Merensky ore is mined largely because of its Pt content)	0.344	1.95 ± 0.15 Ref. Nicolaysen (1958)	Liebenberg (1970)
UNITED STATES					
Stillwater Complex	Class 5	M uncertain, but much in excess of 0.32	0.500	2.75. Ref. Nunes & Tilton (1971)	
Duluth Complex	Class 5	M uncertain, but much in excess of 0.2	0.750	1.115 ± 0.015. Ref. Silver & Green (1963)	

Continued

anorthositic gabbro and commonly bands rich in ilmenite or titaniferous magnetite. The sub-type (a) sills are clearly pre-folding as are some examples of sub-type (b), although other examples of this latter sub-type (e.g. the Kamiscotia intrusion) are highly discordant and their early age is less well established.

The lens-like bodies (discussed in greater detail later) are very common in Archean greenstone belts [forming up to 30 per cent of the 11,000 ft of the Komati Formation in the Barberton Mountainland (Viljoen and Viljoen, 1969, 1970)], moderately common in the Proterozoic [for example, the Cape Smith - Wakeham Bay belt of the Ungava Peninsula] and relatively rare in the Phanerozoic [Baffin Bay (Clarke, 1970), the Rambler area, Newfoundland (Gale, 1973), and possibly the serpentinites of the D'Urville Island - Mossburn belt, New Zealand (according to Grindley, 1958)].

The second class covers "alpine-type" ultramafic bodies; these will be discussed in more detail later in this paper. The third class covers a distinctive group of concentrically zoned ultramafic bodies reported from the cordillera of Alaska and British Columbia and also in the Urals, and referred to as Alaskan-type intrusions. They appear to have been intruded late in the orogenic cycle, after most of the folding had ceased. Irvine (1967a) and Findlay (1969) point out that they form a suite with a distinctive and somewhat alkalic petrography and chemistry.

A fourth orogenic class has been created to cover syn- or late-orogenic, fault-controlled ultramafic bodies that are associated with major crustal sutures or "mobile belts". These sutures are marked by some or all of the following features: major fault zones, belts of contrasting metamorphic grade within or on either side of the suture, positive and sometimes parallel negative gravity anomalies, and unusual seismic profiles for the base of the crust. When first proposed (Naldrett and Gasparrini, 1971), this class was envisaged as being typified by the ultramafic bodies of the Thompson-Wabowden belt in Manitoba. It has since become clear that, although some of the bodies along this belt may be fault-controlled (Coats and Brummer, 1971), others are distinctly strata-bound (Quirke *et al.*, 1970) and therefore possibly pre-tectonic. The post-deformation age of all ultramafic bodies in belts of this type and therefore the importance of this class of fault-controlled bodies should be regarded as uncertain until more information is available about the Thompson and other belts; it is possible that most of the important examples of this class may be shown to belong in Class 1.

Turning to bodies found in non-orogenic areas, the classification draws heavily on the breakdown proposed by Wyllie. Class 5 covers the large, stratiform, layered complexes, including the Bushveld, Muskox, Stillwater, Duluth and Sudbury intrusions. Class 6 covers the distinctive, very large igneous sheets such as the Dufek intrusion in Antarctica [6-7 km thick, more than 8000 sq. km in area (Ford, 1970)] and the Insizwa complex in South Africa [Maske, 1966]. Class 7 includes medium-sized intrusions such as the Skaergaard body and Class 8 covers many smaller intrusions with varied form and a variety of magmatic affiliations. Both of these last two classes are closely related to volcanism. Finally, alkalic ultramafic rocks related to carbonatites and the various types of kimberlites are grouped in Class 9.

NICKEL SULPHIDE DEPOSITS IN RELATION TO CLASSES OF ULTRAMAFIC BODIES

Some of the classes outlined above are hosts to numerous important nickel sulphide deposits and others are hosts to very few or none at all. This is documented in Table 2 and Figure 1. In the table, the individual deposits or nickel camps throughout the world are grouped according to the class of ultramafic-mafic body with which they are associated. The ore reserves and production figures are the best available to the author. In some cases they are published figures from the literature, the press or company reports; in others they are no more than educated guesses. The figures for the Russian camps are particularly poor and are merely tokens of the amounts present (the Noril'sk - Talnakh camp is one of the largest in the world, although this is not apparent from Table 2).

Figure 1 provides a summary of the data in Table 2. The vertical axis shows the reserves plus production in millions of tons of nickel metal for deposits associated with different classes of host rocks subdivided according to geography. Figures above the horizontal line are for economic deposits and those below the line are for marginal or non-economic deposits. The characterization as economic or non-economic is based on the arbitrary decision of the author as to whether the deposit would be mineable if situated near Sudbury in eastern Ontario. Ultramafic bodies of class 1 have been subdivided into gravity-differentiated sills (class 1-(i)) and ultramafic lenses (class 1-(ii)). Classes 5 and 6 and then 7 and 8 have been grouped together in the figure.

The absolute predominance of the Sudbury Camp, comprising the Canadian contribution to the column headed classes 5 and 6, is very clear from the figure.

U.S.S.R.					
Pechenga	Class 1-(i) sills	Unknown, but 1940 figure was E 0.32	0.321	1.72 - 1.78 Ref. Polkanov & Gerling (1960)	Haapala (1969) Gorbunov (1959) E. N. Eliseev (1959) N. A. Eliseev *et al.* (1961)
Noril'sk-Talnakh	Class 7	Unknown, but greatly in excess of 1.0	0.714	Mid-Triassic. Ref. Godlevskii (1959)	Godlevskii (1959) Smirnov (1966) Zolotuchin & Vasilev (1967)
Monchegorsk	Class 7	Reputedly 0.064	0.286	Paleozoic	Haapala (1969)
FINLAND					
Kotalahti	Class unknown	E∶0.08		Exact age unknown, but >1.75-1.85. Ref. Eskola (1963)	Haapala (1969)

*E refers to economic deposits; M refers to marginal and non-economic deposits

212

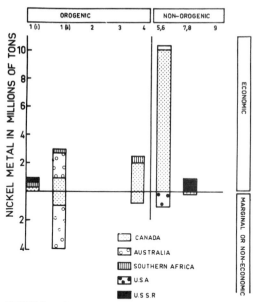

FIGURE 1 — Present reserves plus past production of sulphide nickel as a function of the class of ultramafic body with which the deposit is associated. See Table 1 for a description of the different classes.

The non-economic deposits of this column are those at the base of the Stillwater and Duluth complexes. The column headed Classes 7 and 8 is more important than it appears, because it contains the Noril'sk - Talnakh deposits. However, if Sudbury is discounted, ultramafic bodies of the orogenic division are more important than those of the non-orogenic division. Amongst these, the ultramafic lenses of class 1-(ii) stand out in importance, primarily due to the recent discoveries in Western Australia, but also due to discoveries in Rhodesia and Canada. The sills of class 1-(i) seem less important, although the figures for the Pechenga Camp falling in this class are based on 1940 reserves proved by INCO (Boldt and Queneau, 1967) and obviously bear little relation to the production over the intervening 30 years plus the present reserves. Amongst the sills, the ultramafic, sub-type (a) bodies are the important ones from the point of view of nickel, no significant deposits having been found with the sub-type (b) anorthositic bodies. The major contribution to the class 4 deposits comes from the Thompson-Wabowden belt; it has already been stressed that this class is not well established and it is possible that the deposits should be placed in class 1.

Classes 2 and 3, the alpine and Alaskan-type bodies, are conspicuously poor hosts for nickel sulphide deposits (as pointed out previously by Thayer, 1971; Naldrett and Gasparrini, 1971; Irvine and Findlay, 1972). The intrusion containing the Giant Nickel deposit (Aho, 1956; Muir, 1971) has been described as being Alaskan-type (Taylor, 1967), but it is distinctly tholeiitic in character and has a more irregular zoning pattern than other Alaskan bodies. Alpine bodies may contain millerite - cobalt - nickel arsenide deposits (Pollock, 1959), but these are very small and of a totally different character than the dominantly sulphide-bearing deposits being discussed here. Ultramafic rocks of alkalic association (class 9) are also unlikely hosts for nickel sulphides.

ABSENCE OF DEPOSITS ASSOCIATED WITH CERTAIN CLASSES OF ULTRAMAFIC BODIES

The majority of nickel sulphide deposits are regarded as being the result of the emplacement of a sulphide magma, generaliy in association with a host silicate magma (see, for example, Souch et al., 1969; Naldrett and Kullerud, 1967; Hudson, 1972). In order to develop a reasonable nickel-rich sulphide magma, it is necessary for the sulphide phase to form at an early stage, and equilibrate with the silicate magma before this has been severely depleted in nickel by the crystallization and removal of olivine (see Wager, Vincent and Smales, 1957, for a convincing argument on this requirement). If we make the assumption that the sulphur is derived from a primary mantle source, the early saturation implies that the silicate magmas must have been particularly rich in sulphur at the time of their generation. This section is concerned with current ideas on the genesis of the different classes of ultramafic bodies and how these ideas may govern the concentration of sulphur within the bodies.

Present-Day Views About the Origin of Alpine Ultramafic Bodies

Before the reasons for the absence of major nickel sulphide deposits in alpine ultramafic bodies can be discussed, it is necessary to define what is meant by the term "alpine". In the past, the term has been used loosely to include any ultramafic body that has been involved in alpine-style folding. Thayer (1960) drew attention to many points distinguishing alpine bodies from stratiform intrusions, some of the distinctions indicating that alpine bodies are more than merely tectonically deformed stratiform bodies. Green (1964) pointed to evidence suggesting that some alpine ultramafic rocks have crystallized at high pressure and has suggested that these bodies may represent complete mobilization of portions of the upper mantle. Thayer (1967) emphasized the close association of gabbros with many alpine peridotites and suggested that the association is due to differentiation by gravitational crystal settling in the upper mantle, followed by emplacement higher in the crust as a crystal mush. Hopson (1964), Challis (1965) and McTaggart (1971) challenged this view, interpreting some alpine bodies as tectonically deformed high-level differentiates.

An important advance in understanding alpine ultramafic rocks has come from a better appreciation of ophiolite complexes. Although the term ophiolite has also been used loosely in the past, being applied to any close association of sodic pillowed basalts, ultramafic rocks and cherts, most authorities (Moores and Vine, 1971; Dewey and Bird, 1971; Church, 1972) would now apply it strictly to define a particular stratigraphic succession of ultramafic, mafic and volcanic rocks, capped in some cases by chert. This is illustrated by the idealized section shown in Figure 2, which is adapted from Moores' (1969) reconstruction of the Vourinos complex. The basal portion, which is the largest part of the complex, consists of a sequence of highly folded, banded harzburgites and dunites, cut by dunitic, pyroxenitic and gabbroic dykes. The banding shows no relation to the over-all outlines of the body, nor to the contact with the overlying rocks. Podiform chromite deposits occur toward the top of the harzburgitic portion. A zone of pyroxenite overlies the harzburgites and is in turn overlain by gabbro, sheeted, massive and then pillowed diabase and finally chert. The contacts of the pyroxenite and higher units are roughly parallel to one another and do not show the folding that has affected the banded harzburgite.

Ophiolite complexes showing many of the essentials of the stratigraphy described above are found throughout the Alpine-Himalayan chain, in Japan, in the Franciscan series of the North American cordillera, in the Ural orogen, and in the Quebec and Newfoundland portions of the Appalachian belt. Commonly the complexes are associated with flysch-type sediments. In many cases the complexes are disrupted by faulting, ultramafic and mafic portions may become detached from one another and, *in extremis*, all that may remain may be a chaotic mélange of ultramafic, gabbroic and other blocks in a matrix of flysch. It seems likely that the majority of ultramafic bodies formerly classed as alpine are fragments of ophiolite complexes that have been tectonically deformed to varying degrees, and in this paper I use this restricted definition for alpine bodies.

Large slices of ultramafic rocks showing much of the ophiolite stratigraphy have been thrust (Fig. 3) over Mesozoic and Tertiary sediments and volcanics in New Caledonia (Guillon, 1969; Avias, 1967) and Papua (Davies, 1968). The magnitude of these slices and the way in which their underlying thrusts dip toward the Pacific Ocean leave little doubt but that they are obducted slices of oceanic lithosphere. Ophiolite complexes are also interpreted as oceanic lithosphere (Coleman, 1971; Moores and Vine, 1971), some authorities regarding the contact between the harzburgites and the overlying rocks as the Moho (Dewey and Bird, 1971). Although the emplacement of ophiolite blocks is not well understood, the association of many of them with fossil subduction zones is well established, and it seems likely that most of them represent fragments of either the down-going or over-riding plate that have been thrust into the soft sediments collecting in the trench marking the subduction zone or squeezed up through major fractures in more consolidated strata adjacent to the trench.

In their work on the Bay of Islands ophiolite complex, Irvine and Findlay (1972) have demonstrated that the lowermost 10,000 ft of harzburgite equilibrated at high pressure. Plagioclase is absent in this zone; instead, alumina is present both in pyroxene and in chrome spinel. Olivine and hypersthene appear to have crystallized in equilibrium together and show no reaction relation. The fabric of the lowermost zone is largely metamorphic and little evidence of cumulus textures is present. In contrast, the overlying feldspathic peridotite, pyroxenite and gabbro have crystallized at much lower pressures and show textures suggestive of igneous accumulation. Pointing to the lack of cryptic variation in the basal harzburgites (the olivine composition varies by less than 2 mole per cent forsterite throughout the whole of the sequence), they argue that if these were igneous cumulates they would have required an impossibly large thickness of magma from which to crystallize (possibly 100,000 ft). The overlying feldspathic rocks show a cryptic variation analogous to most layered intrusions and are probably the result of gravitational differentiation of a tholeiitic magma at relatively low pressure. Irvine and Findlay conclude that the high-pressure harzburgites represent refractory solid material remaining behind after a section of the mantle has been melted at depth and the partial melt has been removed. Following this melting stage, the refractory solid residuum made its way up to the top of the mantle without undergoing further melting and tholeiitic magma was placed on top of it, perhaps at a spreading ridge.

The evidence used by Irvine and Findlay to support their arguments is common to most of the ophiolite complexes that have been studied in detail. In this paper, therefore, alpine ultramafic rocks are interpreted as portions of the upper mantle that have been emplaced tectonically in the solid state. They are regarded as the refractory, solid residuum left behind in the mantle after partial melting has occurred and a ultramafic or tholeiitic melt has been removed.

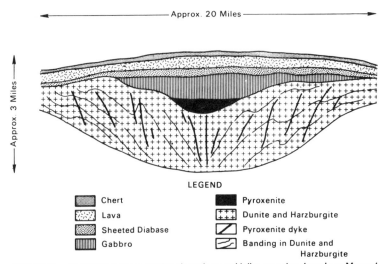

Approx. 20 Miles

Approx. 3 Miles

LEGEND

Chert

Lava

Sheeted Diabase

Gabbro

Pyroxenite

Dunite and Harzburgite

Pyroxenite dyke

Banding in Dunite and Harzburgite

FIGURE 2 — An idealized cross section through an ophiolite complex, based on Moores' (1969) reconstruction of the Vourinos complex

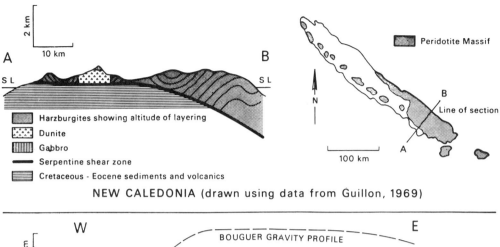

NEW CALEDONIA (drawn using data from Guillon, 1969)

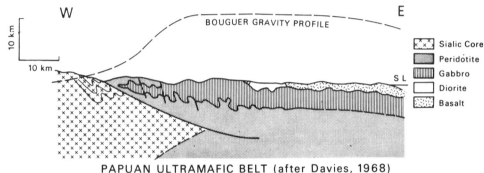

PAPUAN ULTRAMAFIC BELT (after Davies, 1968)

FIGURE 3 — Illustrations of the New Caledonian and Papuan ultramafic massifs after Guillon (1969) and Davies (1968).

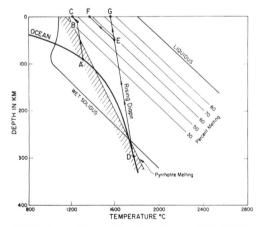

FIGURE 4 — A P-T projection illustrating the field of melting for pyrolite containing 0.1 wt. % H₂O based on the data of Wyllie (1971). The contours indicate the percentage of melting at varying depths and temperatures, after Green (1972). The line labelled "pyrrhotite" is from Sharpe (1969). The diagonally shaded area indicates the probable zone of complete melting of iron-nickel sulphides to sulphide-oxide liquids in the mantle. The line labelled "ocean" is the oceanic geotherm of Ringwood (1969).

The Physical State of Sulphides in the Mantle, Their Possible Distribution and Their Behaviour During Partial Melting

Figure 4 illustrates the melting interval of pyrolite* containing 0.1 wt. per cent water at the pressures found in the crust and upper portion of the mantle. The interval is contoured in terms of the percentage of melting that will be experienced by the pyrolite under any combination of depth and temperature. The likely range in liquidus temperatures for sulphides in the mantle is also shown. The upper limit is based on the data of Sharp (1969) for the maximum melting temperature of pyrrhotite. The diagonal shading illustrates the area in which liquidus temperatures for sulphide-rich liquids in the mantle are most likely to fall, bearing in mind that these liquids will (i) contain small amounts of oxygen and (ii) contain nickel and perhaps copper, all of which will tend to lower their liquidus temperatures from that of the maximum value for pure pyrrhotite (Naldrett, 1969).

A curve showing the present-day oceanic geotherm is also superimposed on the figure.

*Pyrolite is an imaginary rock composed of 3 parts peridotite and 1 part tholeiitic basalt that is often (Green and Ringwood, 1967) taken to represent primitive mantle.

It can be seen that the geotherm cuts across the expected field of melting for sulphides, so that at depths of more than 70-80 km much of the sulphide will probably be molten. A silicate melt will also exist in this zone, although in the presence of only small amounts of water ($\sim$0.1 wt.%) the percentage of such a melt will be small (1-3%)*. It is likely that when this zone is disturbed in response to the build-up of tectonic stresses, the dense sulphides will tend to percolate downwards, thus leaving this portion of the mantle depleted in sulphur. The depth at which sulphides will start to melt is probably much greater under continental areas than under the oceans. However, the predictions outlined above are based on estimates of present-day isotherms. It is likely that geothermal gradients were steeper at earlier stages of the earth's history (Hart *et al.*, 1970), reducing the depth necessary for sulphide melting and sulphide depletion to occur. The depth to which sulphides will remain liquid and therefore the depth to which they are likely to settle is less certain. If one extrapolates Sharp's data on pyrrhotite to greater pressures than the 65 kb to which they were determined, the difference in slope between the melting of pyrrhotite (2.1°C per km of depth) and a reasonable average for the geotherm between 200 km and 700 km (0.67°C per km of depth; Wyllie, 1971, p. 31) suggests that the geotherm may drop back below the field of melting of sulphides at a depth of around 300 km, effectively damming the settling process and possibly creating a sulphide-enriched zone at this depth.

The mechanism by which silicate magmas are commonly regarded as forming in the mantle (Ringwood, 1969) is for some gravitational instability to develop (perhaps triggered by a tectonic process, by water vapour or by magma rising from the descending slab of a subduction zone, or perhaps as a result of normal convective overturn), with the result that hot material from depth moves up into a lower pressure regime, following an adiabatic cooling curve. The gradient of this curve, dt/dP, is much less than the geotherm, so that the material from depth is hotter than its surroundings and, if it rises high enough, suffers appreciable partial melting.

The paths marked AB and DEG in Figure 4 represent adiabats that will be followed by pyrolite diapirs rising from points A and D on the oceanic geotherm. In the case of the diapir rising from A, it is assumed that the partial melt separates from the remaining solid portion of the diapir (the "refractory residuum") after about 40 per cent melting at point B and then rises non-adiabatically, at its own liquidus temperature, to reach the surface at point C. Two alternatives are shown for the diapir originating at D. In one case the molten portion is assumed to separate after about 55 per cent melting at point E, rising non-adiabatically after this to reach the surface at point F. In the other, the diapir is assumed to rise adiabatically all the way to the surface without separation of melt from the refractory residuum. In this extreme case, the diapir will have suffered as much as 75 per cent melting.

In simple cases similar to those just discussed, the composition of the partial melt will depend on the depth at which separation of liquid from refractory residuum occurs and on the degree of partial melting

suffered by the diapir. In general, where the degree of partial melting exceeds 35 per cent, it becomes the more important variable, more extreme melts having more ultramafic compositions.

Although there are certain major exceptions, it is a general rule that nickel sulphide deposits occur in association with rocks that were emplaced as ultramafic magmas (c.f. the section following this). It can be seen from Figure 4 that the generation of an ultramafic magma (50-70 per cent partial melting) requires a diapir to originate at a much greater depth (300 km and greater), and to rise adiabatically over a much greater distance before separation of partial melt and refractory residuum, than the generation of a basaltic magma (30-40 per cent partial melting). Green (1972) has already pointed this out. It is interesting to note that the minimum depths from which ultramafic magmas are likely to originate correspond roughly with the minimum depths at which sulphides may tend to concentrate in the mantle.

It can also be seen from Figure 4 that if a diapir rises adiabatically far enough to undergo appreciable partial melting, the sulphides within it will also be molten. A number of possibilities are open to the liquid sulphides. If the degree of partial melting is high enough and the concentration of sulphide in the zone of melting low enough, they may all dissolve in the silicate melt and be carried upwards in solution. Alternatively, if excess sulphide is present, some will dissolve in the silicate melt and the excess may either:

(1) be squeezed out along with the silicate melt and be carried upwards as fine droplets suspended in the melt;

(2) sink out of the zone of melting;

(3) stay where it is.

With regard to the possibility of solution, Haughton and Skinner (1972) report that the solubility of sulphur in a silicate melt is strongly dependent on temperature. Raising the temperature of a basaltic melt from 1100 to 1200°C causes a five-fold increase in sulphur solubility.

H. Shima (personal communication) has found that at 1450°C picritic (ultramafic) melts*, analogous to those formed on the partial melting of pyrolite, dissolve about 0.16 wt. per cent S. At a depth of 25 km, these melts would be in equilibrium with harzburgite at about 1500°C — and at 50 km at about 1550°C. If we consider the effect of temperature alone and assume that an effect similar to that observed by Haughton and Skinner would apply to the melts studied by Shima, the maximum sulphur solubility of picritic melts at the 25-50-km depth (i.e. at about the depth represented by point E in Fig. 4) would be two to five times the values at the surface, i.e. 0.3 to 0.8 wt. per cent. On separating from the source diapir and rising to the surface non-adiabatically along path E-F, the drop in temperature would be sufficient to cause the greater part of the sulphur to come out of solution as fine droplets of liquid sulphide.

Skinner *et al.* (1972) have drawn attention to the possible importance of pressure on the solubility of sulphur in silicate melts, but state that "pressure is apparently not an important variable throughout the crustal pressure range except in-so-far as it affects the H_2O contents and therefore the bulk compositions

---

*The presence of silicate melt in this zone has been suggested as the explanation for the low seismic velocities that are experienced here (Kushiro *et al.*, 1968; Lambert and Wyllie, 1970).

---

*The reason that picritic rather than tholeiitic basaltic melts are discussed at this stage will become apparent in a later section.

of magmas". The effect of pressure at sub-crustal depths is not known. Another unknown factor is the abundance of sulphides in the mantle. It is clear then that we have no way of telling whether any partial melt produced in the mantle is likely to be saturated or unsaturated in sulphides. What does seem likely is that the saturation sulphur solubility may be of the order of several tenths of a per cent or more and that this sulphur may tend to segregate as an immiscible sulphide liquid as the silicate melt rises and cools after separation from its source diapir, provided that the melt was nearly saturated or at least relatively rich in sulphur in the first place.

As stated above, if the abundance of sulphur in any portion of the mantle exceeds the amount that can dissolve in the partial melt, alternatives 1, 2 and 3 become possibilities. In view of the tendency for the silicate fraction of partial melting to become concentrated and separated from the refractory residuum, moving upward as a discrete body of magma, it seems most unlikely that the even less viscous liquid sulphides will stay in position. Rather, it seems probable that they will be carried away from the zone of melting as fine droplets, either to move upwards with the silicate melt toward the surface or to concentrate and percolate downwards through the mantle.

To turn now to the implications of this discussion with regard to the distribution of nickel sulphide deposits in rocks, it can be seen that it is not in the examples of the refractory residuum (i.e. in the allochthonous, tectonically emplaced, alpine-type ultramafic bodies) that one should look for significant sulphide concentrations, but in the liquid products of partial melting; that is, in the autochthonous mafic or ultramafic bodies that were emplaced as magmas.

One further point should be borne in mind. There is still much discussion as to the exact P-T-PH₂O regime under which different classes of magma are generated in the mantle (c.f. Wyllie, 1971, for a good summary of the discussion). It is possible that magmatypes that appear to be poor hosts for concentrations of nickel sulphides (e.g. akalic magmas, Alaskan-type ultramafic magmas) were generated at depths between 80 and 300 km where the mantle may be depleted in sulphide.

NICKEL SULPHIDE MINERALIZATION THROUGHOUT TIME AND POSSIBLE DEPLETION OF THE MANTLE IN SULPHUR

In Figure 5, nickel sulphide reserves and production are shown as a function of the age of the host rocks. It is seen that the Archean and early Proterozoic were the periods when the majority of ore was emplaced[1]. The only major deposits significantly younger than 1.7 x 10⁹ yrs are those in the Duluth complex (1.115 x 10⁹ yrs, Silver and Green, 1963) and in the Noril'sk area (middle Triassic). The reason for the old age of nickel deposits is not clear[2], although a possible explanation is that the mantle has become

depleted in sulphur with time. The logic behind this is as follows.

The importance of the initial sulphur concentration in determining whether a given silicate magma is or is not likely to give rise to a nickel sulphide deposit has been emphasized in the preceding section. If our present theories of plate tectonics are extended into the Archean, they require the mantle to have undergone partial melting several times in the last 3 x 10⁹ years. Regardless of the correctness of this extension, the mantle must have been subjected to considerable partial melting throughout this period of time. As has been shown above, the available evidence on sulphur solubilities suggests that if an area of the mantle is subjected to partial melting, the sulphide present will be concentrated in the partial melt and much of it removed, leaving a residue depleted in sulphur. It is likely that much of the mantle will become severely depleted in sulphur as time proceeds. The concept of the mantle becoming depleted in trace elements with time has been suggested by Hart et al. (1970) on the basis of comparative studies of Archean and Phanerozoic volcanic rocks.

Subsequent partial melts derived from a depleted mantle will have much lower sulphur contents and, on rising to the surface and cooling, will become saturated in sulphur at a much later stage in their history than the first melt. They will therefore be less likely to give rise to nickel sulphide orebodies, thus accounting for the rarity of such ores in younger rocks.

A key assumption behind this argument is that the sulphur involved in any deposit is juvenile and is derived entirely from the mantle. If, on the other hand, an initially sulphur-poor basic melt assimilates a large amount of sulphur from the wall rocks of its magma conduit as it rises toward the surface, it may also become saturated in sulphur and give rise to magmatic nickel sulphide deposits. In this regard it is interesting that the sulphides of the two young camps (Noril'sk and Duluth), which constitute the major exceptions to the generalization about the old age of nickel deposits, contain significant amounts of sulphur derived from crustal rather than mantle sources. Godlevski and Grinenko (1963) point out that the average S³²/S³⁴ ratio of 15 sulphide samples from Noril'sk is 22.02 (equivalent to a δS³⁴ value of 9.52‰), and attribute the heavy sulphur to contamination by Devonian anhydrite beds through which the host rocks have been intruded, coupled with a reduction of the sulphate to sulphide by carbon from coal measures through which they have also been intruded. Study of some ore zones near the base of the Duluth complex has shown that these contain heavy sulphur (δS³⁴ = +13 to +15‰), that the sulphides are intimately mixed with graphite and that the host rocks contain many partially digested aluminous inclusions (P. W. Mainwaring, personal communication). In one area, 40 miles north of Duluth, Minnesota, the sulphur isotope data are consistent, with as much as 75 per cent of the sulphur being derived from the underlying sulphide- and graphite-bearing Virginia slate.

---

[1]The age of the Sudbury Nickel Irruptive is usually quoted as ±1.704 × 10⁹ yrs (Fairbairn et al., 1968), but zircon ages of 1.9 × 10⁹ yrs (reported by Souch et al., 1969) agree with recent Sr-Rb measurements (Gibbins and McNutt, 1972), suggesting that the age is more likely to be 1.9 to 2.0 × 10⁹ yrs.
[2]Laznicka (1973) has suggested that nickel sulphide deposits form only at great depths and that deposits in younger rocks have yet to be uncovered by erosion. The impact hypothesis for the triggering of the Sudbury Nickel Irruptive would require it to be emplaced under

little more than 5000 ft of Onaping formation. As is discussed later in this paper, the West Australian nickel ores are associated with volcanism. The style and grade of metamorphism of the greenstone belts in which they occur show little evidence of these rocks having suffered deep burial. The Noril'sk ores are associated with sub-volcanic intrusions. These three camps account for about three-quarters of the world's supplies of sulphide nickel and all are shallow rather than deep-seated phenomena. Laznicka's suggestion as it relates to nickel deposits therefore appears invalid.

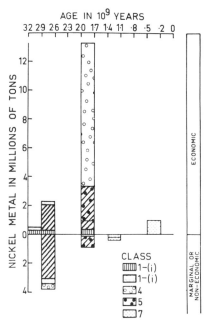

FIGURE 5 — Present reserves plus past production of sulphide nickel as a function of the age of the host rocks. The shading refers to the class of ultramafic body that is host to the ore, with the exception that the pattern of diagonal lines superimposed on the stippling indicates that the age shown for the portion of the ore so patterned is a minimum and not an absolute age.

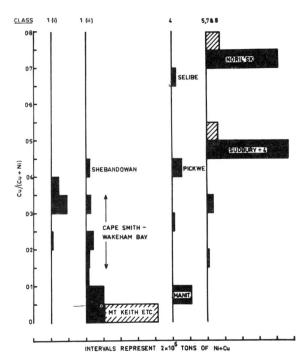

FIGURE 6 — $Cu/(Cu+Ni)$ ratios of nickel sulphide deposits as a function of the class of the associated ultramafic/mafic rock. The size of the bar is proportional to the total content of $Ni+Cu$ falling within each division, except in the case of Sudbury, where the bar is ¼ of its true size. Economic deposits are indicated by a solid bar, marginal or non-economic deposits by diagonal cross-hatching.

COPPER - NICKEL RATIOS

The average $Cu/(Cu+Ni)$ ratios of all camps for which data are available are listed in Table 2. Figure 6 illustrates the same ratio as a function of the class of host rock. Where detailed breakdowns of ratios for different mines within a single camp were available, these are shown in the figure to provide an idea of the spread in ratios. However, in many cases this has not been possible and the average for each of these camps as a whole has been used.

The correlation between the class of host rock and the $Cu/(Cu+Ni)$ ratio of the ore is very apparent from the figure. The tendency for the ratio to increase as the host rock varies from peridotite to norite has been known for a long time (Wilson and Anderson, 1959). The low ratios associated with class 1-(ii) bodies reflect the primitive magnesium-rich nature of the magmas giving rise to these bodies. There are certain exceptions to this rule, however. The Cape Smith - Wakeham Bay deposits are associated with very magnesian rocks (Wilson *et al.*, 1969), as is the Shebandowan deposit (Watkinson and Irvine, 1964), but the former has a Cu/(Cu + Ni) ratio of 0.252 and the latter a ratio of 0.4, contrasting with a typical value for the class as exemplified by the figure of 0.062 for the economic Western Australian deposits.

The higher $Cu/(Cu+Ni)$ ratios of the class 1-(i) sills suggests that the magma for these may be less primitive than that responsible for the class 1-(ii) lenses. In Canada, analytical data on closely associated sills and lenses support this view. Naldrett and Mason

(1968) have demonstrated that the Mg/Fe ratio of the most magnesian rocks of the lenses is higher than that for similar rocks from a sill in the Dundonald-Clergue area of the Abitibi belt, and T. N. Irvine (personal communication) has reached the same conclusion on the basis of over 1000 analyses of ultramafic rocks from throughout the Superior province.

In general, both large and medium-sized intrusions occurring in non-orogenic regions contain sulphide ores in which copper is either nearly as concentrated or even more concentrated than nickel. Again, this reflects the dominantly mafic rather than ultramafic character of these bodies. It is worth stressing, however, that despite the over-all mafic composition of the intrusions, the ore deposits at Noril'sk and in the Duluth, Stillwater and Bushveld complexes are associated with olivine-bearing rocks. Even at Sudbury, long noted for the association of nickel sulphides with norite, the orebodies are closely related to a series of ultramafic intrusions. These may have been derived from a zone of ultramafic cumulates that crystallized from the magma at the same time and in the same place as the sulphides segregated and were then emplaced, together with the sulphides, into the present sites of the orebodies (Naldrett *et al.*, 1972).

Deposits occurring in mobile belts (class 4) show a very wide range of $Cu/(Cu+Ni)$ ratios, reflecting the varying petrology of their host rocks (Manitoba, peridotite-dunite; Pikwe and Selibe, amphibolite-pyroxenite) and again emphasizing the possible artificiality of this class. The similarity of the Manitoba ratios with those for the West Australian camp is consistent with the view expressed above that the former camp may represent a highly metamorphosed variant of the class 1 deposits.

ULTRAMAFIC LAVAS AND RELATED ROCKS

Now that the importance of silicate magmas in nickel sulphide mineralization has been stressed, one type of ultramafic magmatic activity that seems particularly favourable will be discussed; this covers the ultramafic lenses appearing as class 1-(ii) in Table 1.

Bodies of this type are found in the Archean portions of the Canadian (Naldrett and Mason, 1968; Naldrett, 1972), South African (Viljoen, 1969), Rhodesian and West Australian (Nesbitt, 1972; McCall, 1972; McCall and Leishman, 1972) shields, and in the Proterozoic Cape Smith - Wakeham Bay belt of the Ungava Peninsula (Wilson et al., 1969).

Typically, the lenses are conformable and range from a few hundred feet to several miles in length and a few tens to several thousands of feet in thickness. They have an original mineralogy consisting of olivine, pyroxene (largely augite) and accessory

chrome spinel. Simple lenses consist of an inner core rich in olivine (perhaps 75 modal per cent), with pyroxene increasing at the expense of olivine toward the margins. The marginal zones sometimes show an unusual texture composed of skeletal olivine and pyroxene, formally described as "bird-track" or "quench" texture. Complex lenses consist of a sequence of alternating olivine-rich and olivine-poor sheets, the latter showing the skeletal texture.

A particularly well exposed example of a complex lens has been described by Pyke, Naldrett and Eckstrand (1972) from Munro twp., Ontario. Over 60 thin units are exposed on the outcrop (Fig. 7). The units may be very short, pinching out after only a few feet, or may be much longer, extending 600 feet across the whole width of the outcrop. A typical unit (Fig. 8) consists of a thin basal chill zone, a few inches thick, overlain by a zone (the B zone) composed of equant olivine crystals set in a pyroxenitic matrix. The overlying zone (the A zone) is in marked contrast and consists of large, book-like aggregates of bladed, skeletal olivine crystals. The material between the olivine crystals in both zones consists of skeletal py-

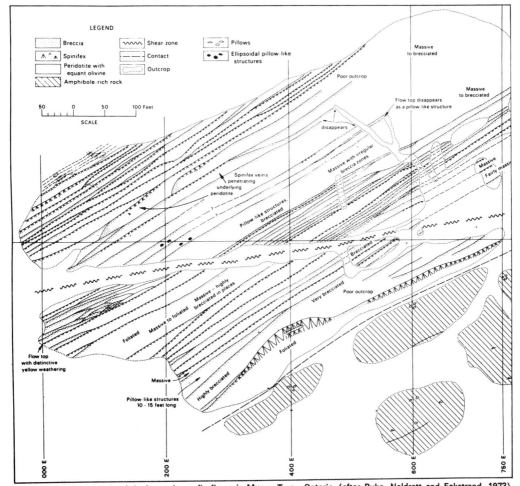

FIGURE 7 — An outcrop of Archean ultramafic flows in Munro Twp., Ontario (after Pyke, Naldrett and Eckstrand, 1973).

roxene and chromite set in a brown chloritic matrix interpreted as alteration after glass. The orientation of the blades in the A_2 zone is sub-perpendicular to the plane of the units and the size of the books decreases progressively upwards until they grade into a fine chill zone at the upper contact (A_1 zone). Because of its similarity to a species of Australian grass known as spinifex, the texture of the A_2 zone is referred to as spinifex texture.

The textures of the B zone are fairly typical of all cumulus peridotites (with the exception of the skeletal nature of the material interstitial to the olivine crystals), but the spinifex textures resemble those in slags, and suggest that a silica-poor liquid, with a low viscosity and accordingly high rate of internal diffusion for its chemical constituents, has undergone rapid cooling in a strong thermal gradient. There is no evidence of compositional variation within the spinifex zone and thin-section study of many samples has shown that no phenocrysts are present. Analyses of this material are therefore representative of the liquid portion of the units and indicate that this has the composition of olivine-rich picrite.

Pyke *et al.* interpret the units at Munro as lava flows and regard them as the result of the repeated extrusion of a picritic liquid carrying up to 40 per cent olivine phenocrysts in suspension. The liquid quenched at its upper contact with sea water and, because of its low viscosity, the phenocrysts settled rapidly to form the cumulus B zone, leaving an overlying zone of liquid depleted in phenocrysts. A strong thermal gradient was established through this liquid, with isotherms parallel to the upper contact, and the spinifex texture was formed by blades of olivine crystals growing rapidly downwards, sub-perpendicular to the isotherms.

Before leaving this subject, it should be stressed that the presence of spinifex texture is not proof of extrusion; it merely indicates rapid cooling of an olivine-rich liquid and therefore emplacement of this liquid in a cool, supracrustal environment, not necessarily on surface. The proof of extrusion at Munro lies in the multiple nature of the units, their thinness, the much greater degree of chilling at the top than at the bottom, the fact that they are never observed to cut one another and also the complex, almost Pahoehoe-like nature of some of the units described by Pyke *et al.* Elsewhere, where field relations are not so clear, one may be looking at spinifex rock developed at the margins of a feeder to a surface flow, or at the margins of a shallow intrusive lens.

Distribution of Class 1-(ii) Lenses in the Abitibi Belt

The locations of all known areas of spinifex rocks in the Abitibi belt are indicated in Figure 9. As the significance of the texture has only recently been recognized, there are probably many more areas where it is exposed than are shown in the figure and it is unwise to make too many generalizations at this stage. However, two types of occurrence can be distinguished provisionally. The more common of these covers the area west and south of Timmins, Ontario. Here the lenses containing the spinifex rocks are found within a typical Archean sequence of tholeiitic to high-alumina basalts and andesites with significant intercalations of dacitic and rhyodacitic rocks. Major layered sills are absent. Magnetite iron formation is common, often closely associated with the ultramafic rocks.

The second type of occurrence (shown by stippling in Fig. 8) has only been identified in a belt extending east-southeast from south of Cochrane, Ontario toward the Quebec border. Here the spinifex rocks occur in close association with class 1-(i) differentiated sills. Many of the basalts of the area are very magnesian, some containing up to 18 per cent MgO. Preliminary studies suggest that there is a continuous gradation from normal basalt to the ultramafic magma forming the spinifex zones. Acidic rocks are apparently restricted to very thin (50-200 ft) pyroclastic horizons and oxide iron formation is absent. Chloritic horizons, rich in graphite and sulphide, are present and may represent intravolcanic sediments. The "primitive" nature of the igneous rocks in this belt raises the possibility that the belt represents a fragment of Archean oceanic floor. Nickel sulphide deposits occur with ultramafic bodies of both types of occurrence (Texmont and Langmuir mines with the first type; Alexo mine with the second type). It remains to be seen, however, whether the suggested distinction between the two types is a valid one and, if so, whether there is any difference in the number and perhaps size of the deposits associated with each of the two types.

SULPHIDE DEPOSITS ASSOCIATED WITH ULTRAMAFIC LENSES

Sulphur Saturation in Ore-Bearing Magmas

The importance of the class 1-(ii) lenses as hosts for nickel sulphides has already been stressed. In the majority of these cases (see Naldrett and Gasparrini, 1971), sulphides occur toward the base of their host rocks, suggesting gravitational settling of a sulphide liquid as the mode of concentration. The question now arises as to whether, at the time of emplacement, the magmas of the lenses were saturated in sulphur and carrying the greater part of their sulphides as immiscible droplets, or whether they were unsaturated, with all of the sulphides in solution.

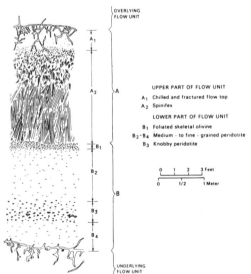

FIGURE 8 — A section through a typical spinifex-bearing flow unit (after Pyke, Naldrett and Eckstrand, 1973).

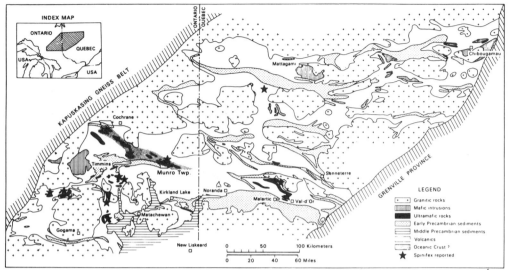

FIGURE 9 — Reported occurrences of spinifex-bearing ultramafic rocks in that portion of the Superior province known as the Abitibi orogenic belt.

The rapid crystallization of the liquid portion of the lenses to produce skeletal textures has been demonstrated above. It is thought unlikely that most of the sulphides now forming the bottom accumulations were in solution when this rapid crystallization started, as they would have been unable to sink quickly enough, and would have been trapped before concentrating at the bottom. It is therefore concluded that, regardless of whether or not they were saturated in sulphur when they were generated, the magmas of those lenses that contain basal sulphide concentrations were saturated and carrying sulphides as immiscible droplets as they rose toward the surface*. Once the magmas emerged from their conduits and spread out as horizontal sheets (either intrusive or extrusive), that is, once near-vertical flow was replaced by near-horizontal flow, the sulphides started to settle toward the bottom of the moving bodies of magma.

Typical Sections Through Ore Deposits

Typical sections through deposits associated with class 1-(ii) lenses appear in Figure 10, in which a section through a portion of the Alexo mine in Ontario is compared with Woodall and Travis's (1969) section through a portion of the Lunnon shoot, Kambalda. Features in common to both sections are: (1) massive sulphides occur at the base of the ore zone (the banded ore at Lunnon consists of massive sulphides in which individual minerals have become concentrated in thin bands); (2) the massive sulphides are in sharp contact with the overlying disseminated ore, which consists of a continuous network of sulphides in peridotite (net-texture); (3) this continuous network is itself in sharp contact with an overlying, very much weaker, discontinuous dissemination which grades upward into peridotite with a very low sulphur content (0.03 wt. % S at Alexo). The sharp

FIGURE 10 — Comparison of typical sections through two ore deposits associated with Archean, class 1-(ii), ultramafic lenses. The Alexo mine is 25 miles east-northeast of Timmins, Ontario and the Lunnon shoot is near Kambalda, Western Australia.

*This conclusion that ore-bearing magmas are saturated in sulphur at the time of extrusion gives rise to a possible method for discriminating ultramafic rocks that are possible hosts for ore from those that are unlikely. The spinifex rocks are regarded as representing the liquid portion of the ultramafic magmas, free of phenocrysts. If the rocks are not too altered the sulphur content of the spinifex rock may well represent the original sulphur content of this liquid. If the sulphur content of a spinifex zone of a particular unit is high (equal to or in excess of the saturation value), the chances are reasonable that a sulphide accumulation may be present somewhere at the base of this unit. If it is low, well below the saturation value, the chances are much poorer. Experimental work is currently in progress to determine the saturation sulphur contents of silicate liquids analogous to spinifex rocks.

221

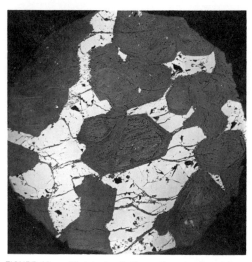

FIGURE 11 — "Net texture". Serpentine pseudomorphous after euhedral olivines encircled by pyrrhotite (greyish white) and pentlandite (white). The sulphides occupy a position in the rock fabric analogous to that of pyroxene in fresh peridotite (15x, plane polarized light.)

FIGURE 12 — A sample of Alexo ore showing the sharp upward termination of net-textured ore with weakly mineralized peridotite. The sample measures 10 cm from top to bottom.

contacts between massive and net-textured ore, and between net-textured ore and weakly mineralized peridotite, define very smooth surfaces which are parallel to contacts between stratigraphic units in the country rocks and therefore seem to represent the horizontal at the time the ore deposits were forming.

The Lunnon shoot differs from Alexo in that the massive ore of the former is banded with numerous fine "schlieren" of pentlandite, occurring parallel to the contacts, and very commonly a band of pyrite occurring close to the upper contact with the disseminated peridotite. No similar banding has been observed at Alexo. Ewers and Hudson (1972) interpret both the pentlandite and pyrite banding as secondary features that have formed as the ore cooled. The Lunnon shoot also differs from Alexo in that, although in the former the massive ore directly overlies basalt, in the latter a thin band of pyroxene-rich peridotite separates the ore from the underlying basalt.

Net texture, as seen in Figure 11, is the result of sulphides enclosing olivine crystals which, despite their serpentinization, look remarkably euhedral. The sharp contact between the net-textured ore and the discontinuous dissemination is illustrated in Figure 12. In this sample, which is from Alexo, the change occurs over ½ inch. The continuity of the sulphide net in the net-textured ore and the sharp break with overlying discontinuous dissemination has been confirmed by resistivity measurements on core from a number of Western Australian ore deposits, including the Lunnon shoot (W. E. Ewers, personal communication).

"Billiard Ball" Model for Sulphide Mineralization

The observations described above can be explained with reference to a very simple model. Consider a large beaker partly full of billiard balls; imagine that this is then filled with water and that the balls are denser than the water so that they remain at the bottom. The situation will be as shown in Figure 13(a). Now suppose that mercury is poured into the beaker

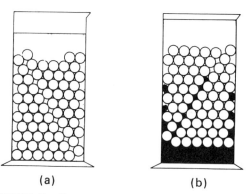

(a)   (b)

FIGURE 13 — Illustration for the "billiard ball" model.

(Fig. 13b). The mercury, shown in black, will sink to the bottom and the balls will tend to float on top of this. The weight of the overlying balls will force the lowest ones down into the mercury to an extent necessary to provide flotation for those above the mercury. If the beaker is then frozen before all the mercury has had a chance to percolate to the bottom, the end product may be as shown in Figure 13b. It is not difficult to see the analogy between the massive mercury and massive sulphides, between the overlying zone of balls immersed in mercury and net-textured ore, between the water enclosing the billiard balls and the picritic liquid enclosing olivine and, lastly, between the zone free of billiard balls at the top of the beaker and the spinifex zone*.

―――――
*The explanation presented here differs markedly from that of sulphurization suggested seven years ago by myself in this journal for the Alexo mine (Naldrett, 1966). Two main lines of evidence have led me to change my mind about Alexo (although not necessarily about the
'(Continued overleaf)

1. Sub-surface magma reservoir
    (Ungava Situation ?)

2. Thicker olivine-rich portions of
    flows, closer to feeder

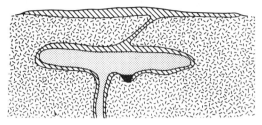
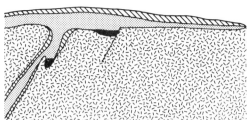

3. Pool of lava overflowing intermittently
    to give rise to a series of
    'spinifex capped' flows at edges

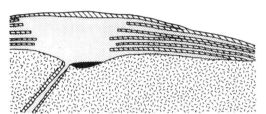

LEGEND

Country rocks

Rocks with > 50% olivine

Rocks with < 50% olivine (spinifex ?)

Sulfides

FIGURE 14 — Speculations on some possible relationships between sulphide deposits associated with class 1-(ii) lenses and their host rocks.

Some Speculations as to the Location of Sulphide Deposits

This model, the concept of ultramafic flows and the conclusion that the flow magmas were carrying droplets of immiscible sulphide during their extrusion have been used to speculate on possible relationships between sulphide concentrations and their host rocks. It should be borne in mind that the model is one of three phases — a dense sulphide liquid, moderately dense olivine crystals and a slightly less dense picritic liquid. These will be trying to settle out from one another under the influence of gravity, with the sulphides tending to settle most quickly, followed by the olivine crystals and with the picritic liquid moving farthest afield.

Drawing 1 in Figure 14 illustrates a possible sub-surface settling chamber, in which sulphides and olivine crystals tend to remain behind and the picritic liquid continues to the surface to extrude as flows. The geology of the Ungava (Cape Smith - Wakeham Bay) nickel belt corresponds in many respects to this model. Drawing 2 illustrates a flow in which the picritic liquid has moved farthest from the source, so that the direction of the source is indicated by an

over-all decrease in the ratio of spinifex (A_2 zone) material to massive (B zone) peridotite. Because they would tend to settle most rapidly, concentrations of sulphides are more likely to be found close to the source than far away from it. Sulphide concentrations may also occur in structures along the margin of the feeder dyke to the flow and zones of disseminated sulphides may occur within the feeder itself (possibly analogous to the Dumont nickel prospect near Amos, Quebec). Drawing 3 illustrates how a lava lake may develop, fed either from below (as shown) or perhaps laterally from off the section of the drawing. The lake is envisaged as overflowing its banks from time to time to give rise to a series of peripheral, spinifex-capped flows. It may act as a settling pond for sulphides.

CONCLUSIONS

In this paper, nickel sulphide deposits are classified in terms of the bodies of ultramafic rocks with which they are associated. It is argued that an understanding of the genesis and emplacement of these rocks offers the best approach to understanding their associated ore deposits, and in discriminating between likely ore-bearing environments and unlikely ones.

The following points are regarded as particularly important:

1. On the basis of the distribution of existing deposits, alpine, Alaskan-type and alkalic ultramafic bodies appear to be poor bets as exploration targets. The apparent lack of deposits in alpine peridotites may be due to the fact that these rocks are portions of the mantle emplaced in the solid state and that sulphides originally present in them were removed dur-

validity of sulphurization in some other cases). The first is the realization that the zoned sequence from (pyrrhotite + pentlandite) through pentlandite to heazlewoodite that extends away from the ore into the host peridotite at Alexo is more readily explained in terms of nickel released during serpentinization reacting with sulphides already present in the rock than by my sulphur diffusion model. The second is the inability of the hypothesis of replacement of pyroxene by sulphide, which I proposed in my earlier paper to explain net-texture, to account for the sharp upper boundary of the zone of net-textured ore.

ing partial melting of the mantle. The lack of deposits in alkalic ultramafic rocks may be due to these magmas having been generated at a level within the mantle at which sulphides are normally liquid and out of which, therefore, the sulphides have settled.

2. Again on the basis of existing deposits, Archean and early Proterozoic ($>1.7 \times 10^9$ yrs old) ultramafic bodies are more likely hosts for nickel sulphides than younger bodies, unless an external source of sulphur has been readily available to be assimilated by the younger magmas. This relation may be due to several periods of partial melting having depleted the present mantle in sulphur with respect to the mantle in early Precambrian times.

3. Lens-like ultramafic bodies occurring in Archean and early Proterozoic greenstone belts form one of the most important classes of ultramafic rocks in terms of their potential as hosts for nickel ores.

4. A possible extrusive origin for both rocks and ore should be borne in mind when exploring the lens-like ultramafic bodies.

ACKNOWLEDGMENTS

I am grateful to the following people for criticism of the manuscript, although they are not responsible for any deficiencies that may be present: Professor G. M. Anderson and U. Kretchmar of the University of Toronto, and W. E. Ewers and Drs. J. A. Hallberg and D. R. Hudson of CSIRO, Perth, Australia.

F. Jurgeneit and Miss Wendy Mather drafted the figures and B. O'Donovan took care of photography. The research was supported by National Research Council grant No. A4244.

REFERENCES

Aho, A. E., 1956. Geology and genesis of ultrabasic nickel-copper pyrrhotite deposits in the Pacific Nickel property, southwestern British Columbia. Econ. Geol., 51, 444-481.

Allard, G. O., 1970. The Doré Lake complex of Chibougamau, Quebec — a metamorphosed Bushveld-type layered complex. Symposium on the Bushveld and other layered intrusions, Geol. Soc. S. Africa, Spec. Publ. 1, 477-491.

Arriens, P. A., 1971. The Archean geochronology of Australia. Geol. Soc. Australia, Spec. Publ. 3, 11-24.

Avias, J., 1967. Overthrust structure of the main ultrabasic New Caledonian massives. Tectonophysics, 4, 531-541.

Bliss, N. W., and Stidolph, P. A., 1969. The Rhodesian Precambrian basement complex: a review. Trans. Geol. Soc. South Africa.

Boldt, J. R., Jr., and Queneau, P., (ed.), 1967. The winning of nickel: its geology, mining and extractive metallurgy. Longmans, Toronto.

Brett, P. R., and Bell, P. M., 1968. Melting relations in the Fe-rich portion of the system Fe-FeS at 30 kb pressure. Carnegie Inst. of Washington, Year Book 67, 198-199.

Challis, G. A., 1965. The origin of New Zealand ultramafic intrusions. Jour. Petrol., 6, 322-364.

Church, W. R., 1972. Ophiolite: its definition, origin as oceanic crust, mode of emplacement in orogenic belts, with special reference to the Appalachians. Publications of the Earth Physics Branch, Dept. Energy, Mines & Resources, Canada, 42, No. 3, 71-86.

Clarke, D. B., 1970. Tertiary basalts of Baffin Bay: possible primary magma from the mantle. Contr. Mineral. Petrol., 25, 203-224.

Coats, C. J. A., 1966. Serpentinized ultramafic rocks of the Manitoba nickel belt. Unpub. Ph.D. thesis, Univ. of Manitoba.

Coats, C. J. A., and Brummer, J. J., 1971. Geology of the Manibridge nickel deposit, Wabowden, Manitoba. Geol. Assoc. Canada, Spec. Pub. No. 9, 155-165.

Coleman, R. G., 1971. Plate tectonic emplacement of upper mantle peridotites along continental edges. Jour. Geophys. Res., 76, 1212-1222.

Davies, H. L., 1968. Papuan ultramafic belt. Proc. 23rd Int. Geol. Congress, 1, 209-220.

Dewey, John F., and Bird, John M., 1971. Origin and emplacement of the ophiolite suite: Appalachian ophiolites in Newfoundland. Jour. Geophys. Res., 76, No. 14, 3179-3206.

Dimroth, E., 1970. Evolution of the Labrador geosyncline. Bull. Geol. Soc. Amer., 81, 2717-2742.

Eales, H. V., 1964. Mineralogy and petrology of the Empress nickel-copper deposit, Southern Rhodesia. Trans. Geol. Soc. South Africa, 67, 173-200.

Eliseev, E. N., 1959. The geochemistry of important sulphide copper-nickel provinces of the U.S.S.R.: Problems of Geochemistry, No. 1. Ivan Franko State University of Lvov.

Eliseev, N. A., Corbusov, G. J., Eliseev, E. N., Maslenikov, V. A., and Utkin, K. N., 1961. Ultrabasic and basic intrusives of Pechenga. Academy Nauk, Moscow-Leningrad.

Emslie, R. F., and Moore, J. M., Jr., 1961. Geological studies of the area between Lynn Lake and Fraser Lake. Manitoba Mines Branch Pub. 59-4.

Eskola, P., 1963. The Precambrian of Finland, in The Precambrian, Vol. 1; ed. K. Rankama, p. 145-264, John Wiley, New York.

Ewers, W. E., and Hudson, D. R., 1972. An interpretive study of a nickel-iron sulphide ore intersection, Lunnon Shoot, Kambalda, Western Australia. Econ. Geol. (in press).

Fairbairn, H. W., Faure, G., Pinson, W. H., and Hurley, P. M., 1968. Rb-Sr whole-rock age of the Sudbury lopolith and basin sediments. Can. Jour. Earth Sci., 5, 707-714.

Fahrig, W. F., 1962. Petrology and geochemistry of the Griffis Lake ultrabasic sill of the central Labrador trough, Quebec. Geol. Surv. Canada, Bull. 77, 39 p.

Findlay, D. C., 1969. Origin of the Tulameen ultramafic-gabbro complex, southern British Columbia. Can. Jour. Earth Sci., 6, 399-426.

Ford, A. B., 1970. The development of the layered series and the capping granophyre of the Duflek intrusions of Antarctica. Symposium on the Bushveld and other layered intrusions, Geol. Soc. S. Africa, Spec. Pub. 1, 492-510.

Ford, A. B., and Boyd, W. W., Jr., 1968. The Duflek intrusion, a major stratiform gabbroic body in the Pensacola mountains, Antarctica. Proc. 23rd Int. Geol. Congr., 2, 213-228

Gale, G. H., 1973. Palaeozoic basaltic komatite and ocean-floor type basalts from northeastern Newfoundland. Earth and Plan. Sci. Letters, 18, 22-28.

Gansser, A., 1959. Ausseralpine Ophiolit probleme. Econ. Geol. Helv., 52, 659-680.

Gass, I. G., 1968. Is the Troodos massif of Cyprus a fragment of Mesozoic ocean floor? Nature, 220, 39.

Gass, I. G., 1967. The ultrabasic volcanic assemblage of the Troodos massif, Cyprus. Ultramafic and related rocks (ed. P. J. Wyllie), 121-134, New York, John Wiley and Sons.

Gibbins, W. A., and McNutt, R. H., 1972. Rubidium-strontium studies of the Sudbury Irruptive (Abstract). Program, Geol. Soc. Amer.

Godleveskii, M. N., 1959. Traps and ore-bearing intrusions of the Noril'sk region. Gosgeoltekhizdat, Moscow, p. 69.

Godleveskii, M. N., and Grinenko, L. N., 1963. Some data on the isotopic composition of sulphur in the sulphides of the Noril'sk deposit. Geochemistry, No. 1, 35-41.

Gorbunov, G. I., 1959. Principles of distribution of sulfide copper-nickel deposits in the region of Pechenga (Kola peninsula). Geologiya Rudnykh Mestorozhdeniy No. 1.

Green, D. H., 1972. Archean greenstone belts may include terrestrial equivalents of lunar maria? Earth Planet. Sci. Letters, 15, 263-270.

Green, D. H., 1964. The petrogenesis of high temperature alpine-type peridotite at the Lizard, Cornwall. Jour. Petrol., 5, 134-188.

Green, D. H., and Ringwood, A. E., 1967. The genesis of basaltic magma. Contrib. Mineral. Petrol., 15, 103-190.

Grindley, G. W., 1958. The geology of the Eglington valley, Southland. New Zealand Geol. Surv., Bull. No. 58.

Guillon, Jean-Hughes, 1969. Données nouvelles sur la composition et la structure du grand massif péridotitique du sud de la Jouvelle, Caledonie. Cah. ORSTOM, sér. Géol., Vol. I, No. 1, 7-25.

Haapala, P. S., 1969. Fennoscandian nickel deposits. Econ. Geol., Mono. 4, 262-275.

Hallberg, J. A., and Williams, D. A. C., 1972. Archean mafic and ultramafic rock associations in the Eastern Goldfields region, Western Australia. Earth Planet. Sci. Letters, *15*, 191-200.

Hamilton, W., 1969. Mesozoic California and the underflow of Pacific mantle. Bull. Geol. Soc. Amer., *80*, 2409-2430.

Hancock, W., Ramsden, A. R., Taylor, G. F., and Wilmshurst, J. R., 1971. Some ultramafic rocks of the Spargoville area, Western Australia. Geol. Soc. Australia, Spec. Pub. 3, 269-281.

Hart, S. R., Brooks, C., Krogh, T. E., Davis, G. L., and Nava, D., 1970. Ancient and modern volcanic rocks: a trace element model. Earth Planet. Sci. Letters, *10*, 17-28.

Haughton, D. R., and Skinner, B. J., 1972. Chemistry of sulfur in melts. Vol. of Abstracts, 24th Int. Geol. Congress, p. 420.

Hawley, J. E., 1962. The Sudbury ores: their mineralogy and origin. Canadian Mineral., 7, pt. 1, 207 p.

Hess, H. H., 1960. Stillwater igneous complex, Montana. Geol. Soc. Amer., Mem. 80, 230 p.

Hopson, C. A., 1964. The crystalline rocks of Howard and Montgomery Counties, *in* The geology of Howard and Montgomery Counties, Baltimore, Maryland Geol. Survey, 27-215.

Hsu, K. J., 1968. Principles of mélanges and their bearing on the Franciscan-Knoxville paradox. Bull. Geol. Soc. Amer., *79*, 1063-1074.

Hudson, D. R., 1972. Evaluation of genetic models for Australian sulfide nickel deposits. Proc., Aust. Inst. Mining and Metall., 59-69.

Irvine, T. N., 1967a. The Duke Island ultramafic complex, southeastern Alaska. Ultramafic and related rocks (ed. P. J. Wyllie), 84-96, New York, John Wiley & Sons.

Irvine, T. N., 1959. The ultramafic complex and related rocks of Duke Island, S.E. Alaska. Unpub. Ph.D. thesis, California Inst. of Technology.

Irvine, T. N., and Findlay, T. C., 1972. Alpine peridotite with particular reference to the Bay of Islands igneous complex. Publ. of the Earth Physics Branch, Dept. of Energy, Mines & Resources, Canada, Vol. 42, No. 3, 97-128.

Irvine, T. N., and Smith, C. H., 1967. The ultramafic rocks of the Muskox intrusion, Northwest Territories, Canada. Ultramafic and related rocks (ed. P. J. Wyllie), 38-49, New York, John Wiley & Sons.

Jackson, E. D., 1961. Primary textures and mineral associations in the ultramafic zone of the Stillwater complex, Montana. U.S. Geol. Surv., Profess. Paper, *358*, 106 p.

Jahns, R. H., 1967. Serpentinites of the Roxbury district, Vermont. Ultramafic and related rocks (ed. P. J. Wyllie), 137-159, New York, John Wiley & Sons.

Karup-Moller and Brummer, J. J., 1971. Geology and sulfide deposits of the Bird River claim group, southeastern Manitoba. Geol. Assoc. Canada, Spec. Pub. 9, 143-154.

Kilburn, L. C., Wilson, H. D. B., Graham, A. R., Oguro, Y., Coats, C. J. A., and Scoates, R. F. J., 1969. Nickel sulfide ores related to ultrabasic intrusions in Canada. Econ. Geol., Mono. 4, 276-293.

Krogh, T. E., and Davis, G. L., 1971. Zircon U-Pb ages of Archean meta-volcanic rocks in the Canadian Shield. Carnegie Inst. of Washington, Year Book 70, 241-242.

Kushiro, I., Syons, Y., and Akimoto, S., 1968. Melting of a peridotite nodule at high pressures and high water pressures. Jour. Geophys. Research, *73*, 6023-6029.

Lambert, I. B., and Wyllie, P. J., 1970. Melting in the deep crust and upper mantle and the nature of the low velocity layer. Phys. Earth Planet. Interiors, *3*, 316-322.

Laznicka, P., 1973. Development of nonferrous metal deposits in geological time. Can. Jour. Earth Sci., *10*, 18-25.

Le Roex, H. D., 1964. Nickel deposit on the Trojan claims, Bindura district, Southern Rhodesia, *in* The geology of some ore deposits in southern Africa (ed. S. H. Haughton). Geol. Soc. South Africa, Vol. 2, 509-521.

Liebenberg, L., 1968. The sulfides in the layered sequence of the Bushveld igneous complex. Unpub. Ph.D. thesis, University of Pretoria.

Liebenberg, L., 1970. The sulfides in the layered sequence of the Bushveld igneous complex. Geol. Soc. S. Africa, Special Pub. No. 1, 108-207.

MacGregor, I. D., 1962. Geology, petrology and geochemistry of the Mount Albert and associated ultramafic bodies of central Gaspé, Quebec. Unpublished M.Sc. thesis, Queen's University.

MacKenzie, D. B., 1960. High temperature alpine-type peridotite from Venezuela. Bull. Geol. Soc. Amer., *71*, 303-318.

Maske, S., 1966. The petrography of the Ingeli mountain range. Annals of the Univ. of Stellenbosch, *41*, Series A, No. 1, 109 p.

McCall, G. J. H., 1971. Some ultrabasic and basic igneous rock occurrences in the Archean of Western Australia. Geol. Soc. Australia, Spec. Pub. No. 3, 429-442.

McCall, G. J. H., and Leishman, J., 1971. Clues to the origin of Archean eugeosynclinal peridotites and the nature of serpentinization. Geol. Soc. Australia, Spec. Pub. No. 3, 281-300.

McRae, N. D., 1969. Ultramafic intrusions of the Abitibi area, Ontario. Can. Jour. Earth Sci., *6*, 281-304.

McTaggart, K, C., 1971. On the origin of ultramafic rocks. Bull. Geol. Soc. Amer., *82*, 23-42.

Morse, S. A., 1961. The geology of the Kiglapait layered intrusion, coast of Labrador, Canada. Unpublished Ph.D. thesis, McGill University.

Milligan, G. C., 1960. Geology of the Lynn Lake district. Manitoba Mines Branch, Pub. 57-1.

Moores, E. M., 1969. Petrology and structure of the Vourinos ophiolitic complex of northern Greece. Geol. Soc. Amer., Spec. Paper 118, 74 p.

Moores, E. M., and Vine, F. J., 1971. The Troodos massif, Cyprus, and other ophiolites as oceanic crust — evaluation and implications. Phil. Trans. Roy. Soc., London, Series A, 443-466.

Muir, J. E., 1971. A study of the '4600' orebody at the Giant Nickel Mine, Hope, B.C. Unpub. M.Sc. thesis, University of Toronto.

Muir, J. E., and Naldrett, A. J., 1972. A natural occurrence of two-phase chromium-bearing spinels. Canadian Mineral. (in press).

Naldrett, A. J., 1972. Archean ultramafic rocks. Publ. of the Earth Physics Branch, Ottawa, Vol. 42, No. 3, 141-151.

Naldrett, A. J., 1969. A portion of the system Fe-S-O between 900 and 1080°C and its application to ore magmas. Jour. Petrol., *10*, 171-201.

Naldrett, A. J., Bray, J. G., Gasparrini, E. L., Podolsky, T., and Rucklidge, J. C., 1970. Cryptic variation and the petrology of the Sudbury Nickel Irruptive. Econ. Geol., *65*, 122-155.

Naldrett, A. J., and Gasparrini, E. L., 1971. Archean nickel sulfide deposits in Canada: their classification, geological setting and genesis with some suggestions as to exploration. Geol. Soc. Australia, Spec. Pub. No. 3, 201-226.

Naldrett, A. J., Hewins, R. H., and Greenman, L., 1972. The main Irruptive and the sub-layer at Sudbury, Ontario. Proc. 24th Int. Geol. Congress, Sect. 4, 206-214.

Naldrett, A. J., and Kullerud, G., 1967. A study of the Strathcona mine and its bearing on the origin of the nickel-copper ores of the Sudbury district, Ontario. Jour. Petrol., *8*, 453-531.

Naldrett, A. J., and Mason, G. D., 1968. Contrasting Archean ultramafic igneous bodies in Dundonald and Clergue townships, Ontario. Can. Jour. Earth Sci., *5*, 111-143.

Nesbitt, R. W., 1971. Skeletal crystal forms in the ultramafic rocks of the Yilgarn block, Western Australia: evidence for an Archean ultramafic liquid. Geol. Soc. Australia, Spec. Pub. No. 3, 331-350.

Nicolaysen, L. O., 1962. Stratigraphic interpretation of age measurements in South Africa; *in* Petrologic Studies: a volume in honour of A. F. Buddington. Geol. Soc. Amer., 569-698.

Nicolaysen, L. O., de Villiers, J. W. L., Burger, A. J., and Strelow, F. W. E., 1958. New measurements relating to the absolute age of the Transvaal system and of the Bushveld igneous complex. Trans. Geol. Soc. S. Africa, *61*, 137-163.

Nunes, P. D., and Tilton, G. R., 1971. Uranium-lead ages of minerals from the Stillwater igneous complex and associated rocks, Montana. Bull. Geol. Soc. Amer., *82*, 2231-2250.

Phinney, W. C., 1970. Chemical relations between Keewanawan lavas and the Duluth complex, Minnesota. Bull. Geol. Soc. Amer., *81*, 2487-2496.

Polkanov, A. A., and Gerling, Z. K., 1960. Precambrian geochronology of the Baltic Shield, Mezhdwnar. Geol. Kongr., XXI Sess., Dokl. Sov. Geologov. Problema 3.

225

Pollock, D. W., 1959. Sulfide paragenesis in the Eastern Metals deposit, Montmagny county, Quebec. Econ. Geol., *54*, 234-247.

Purvis, A. C., Nesbitt, R. W., and Hallberg, J. A., 1972. The geology of part of the Carr Boyd complex and its associated nickel mineralization, Western Australia. Econ. Geol., *67*, 1093-1113.

Pyke, D. R., Naldrett, A. J., and Eckstrand, O. R., 1973. Archean ultramafic flows in Munro township, Ontario. Bull. Geol. Soc. Amer., *84*, 955-978.

Quirke, T. T., Jr., Cranstone, D. A., Bell, C. K., and Coats, C. J. A., 1970. Geology of the Moak - Setting Lakes area, Manitoba (Manitoba Nickel Belt); Geol. Assoc - Min. Assoc. Canada, Guidebook, Field trip No. 1, 23rd Ann. Meeting, Winnipeg.

Ragan, D. M., 1967. The Twin Sisters dunite, Washington, *in* Ultramafic and related rocks (ed. P. J. Wyllie), 160-166, New York, John Wiley & Sons.

Ringwood, A. E., 1969. Composition and evolution of the upper mantle. Geophys. Monograph 13, The Earth's crust and upper mantle (ed. Pembroke J. Hart), 1-17.

Ruckmick, J. C., and Noble, J. A., 1959. Origin of the ultramafic complex at Union Bay, southeastern Alaska. Bull. Geol. Soc. Amer., *70*, 981-1018.

Ruttan, G. D., 1955. Geology of Lynn Lake. CIM Bulletin, *48*, 339-348.

Sasaki, A., 1968. Sulfur isotope study of the Muskox intrusion, District of Mackenzie. Geol. Surv. Canada, Paper 68-46.

Scoates, R. F. J., 1970. The Gordon Lake nickel deposit. Unpublished Ph.D. thesis, University of Manitoba.

Sharp, W. E., 1969. Melting curves of sphalerite, galena and pyrrhotite and the decomposition of pyrite between 30 and 65 kilobars. Journ. Geophys. Research, *74*, 1645-1652.

Sharpe, J. I., 1965. Field relations of Matagami sulfide masses bearing on their disposition in time and space. CIM Bulletin, *58*, 951-961; CIM Trans., *68*, 265-278.

Sharpe, J. W. N., 1964. The Empress nickel-copper deposit, Southern Rhodesia, *in* The geology of some ore deposits in southern Africa (ed. S. H. Haughton). Geol. Soc. South Africa, Vol. 2, 497-508.

Silver, L. T., and Green, J. C., 1963. Zircon ages for middle Keewanawan rocks of the Lake Superior region (Abstract). Trans. Am. Geophys. Union, *44*, p. 107.

Skinner, B. J., Fernandez, L. A., and Althaus, E., 1972. Primary sulfide phases precipitated from magmas: their nature and causes of precipitation (Abstract). Canadian Mineral., *11*, 580-581.

Smirnov, M. F., 1966. The structure of Noril'sk nickel-bearing intrusions and the genetic types of their sulfide ores. All-Union Scientific Research Inst. of Mineral Raw Materials (VIMS), Moscow, 60 p.

Smith, C. H., 1958. Bay of Islands igneous complex, western Newfoundland. Geol. Surv. Canada, Mem. 290.

Souch, B. E., Podolsky, T., and geological staff of The International Nickel Co. of Canada, Ltd., 1969. The sulfide ores of Sudbury: their particular relation to a distinctive inclusion-bearing facies of the Nickel Irruptive. Econ. Geol., Mono. 4, 252-261.

Stowe, C. W., 1971. Summary of the tectonic development of the Rhodesian Archean craton. Geol. Soc. Australia, Spec. Pub. No. 3, 377-384.

Taylor, H. P., Jr., 1967. The zoned ultramafic complexes of southeastern Alaska, *in* Ultramafic and related rocks (ed. P. J. Wyllie), New York, John Wiley and Sons.

Taylor, R. B., 1964. Geology of the Duluth gabbro complex near Duluth, Minnesota. Minn. Geol. Surv.. Bull. 44, 63 p.

Thayer, T. P., 1971. Authigenic, polygenic and allogenic ultramafic and gabbroic rocks as hosts for magmatic ore deposits. Geol. Soc. Australia, Spec. Pub. No. 3, 239-252.

Thayer, T. P., 1967. Chemical and structural relations of ultramafic and feldspathic rocks in alpine intrusive complexes, *in* Ultramafic and related rocks (ed. P. J. Wyllie), 222-239, New York, John Wiley and Sons.

Thayer, T. P., 1963. The Canyon Mountain complex, Oregon, and the alpine mafic magma stem. U.S. Geol. Surv., Profess. Paper 475-C, C82-C85.

Thayer, T. P., 1960. Some critical differences between alpine-type and stratiform peridotite-gabbro complexes. Rep. 21st Int. Geol. Congress, pt. XIII, 247-259.

Turek, A., 1966. Rb-Sr isotopic studies in the Kalgoorlie-Norseman area, Western Australia. Unpub. Ph.D. thesis, Aust. Nat. University.

Viljoen, M. J., and Viljoen, R. P., 1969. Evidence of the existence of a mobile extrusive peridotitic lava from the Komati formation of the Onverwacht group. Geol. Soc. S. Africa, Spec. Pub. No. 2, 55-85.

Viljoen, R. P., and Viljoen, M. J., 1970. The geology and the geochemistry of the layered ultramafic bodies of the Kaapmuiden area, Barberton Mountain Land. Symp. on the Bushveld and other layered intrusions, Geol. Soc. S. Africa, Spec. Pub. No. 1, 661-688.

Wager, L. R., and Brown, G. M., 1968. Layered igneous rocks. London, Oliver and Boyd.

Wager, L. R., Vincent, E. A., and Smales, A. A., 1957. Sulfides in the Skaergaard intrusion, East Greenland. Econ. Geol., *52*, 855-903.

Walker, F., 1940. Differentiation of the Palisades diabase, New Jersey. Bull. Geol. Soc. Amer., *51*, 1059-1106.

Walker, K. R., 1969. The Palisades sill, New Jersey, a reinvestigation. Geol. Soc. Amer., Spec. Pap. 111.

Watkinson, D. H., and Irvine, T. N., 1964. Peridotitic intrusions near Quetico and Shebandowan, northwestern Ontario: a contribution to the petrology and geochemistry of ultramafic rocks. Can. Jour. Earth Sci., *1*, 63-98.

Wilson, H. D. B., and Brisbin, W. C., 1961. Regional structure of the Thompson - Moak Lake nickel belt. CIM Bulletin, *54*, 815-822; CIM Trans. *64*, 470-477.

Williams, D. A. C., and Hallberg, J. A., 1972. Archean layered intrusions of the Eastern Goldfields region, Western Australia. Contr. Min. Petrol. (in press).

Wilson, H. D. B., Kilburn, L. C., Graham, A. R., and Rambal, K., 1969. Geochemistry of some Canadian nickeliferous ultrabasic intrusions. Econ. Geol., Mono. 4, 294-309.

Woodall, R., and Travis, G. A., 1969. The Kambalda nickel deposits, Western Australia. Publ. 9th. Commonwealth Mining & Metallurgical Congress, Paper 26, p. 17.

Wyllie, P. J., 1971. The dynamic earth. John Wiley & Sons Inc., New York.

Wyllie, P. J., 1971. Experimental limits for melting in the earth's crust and upper mantle. Geophys. Monograph Series, *14*, 279-301.

Wyllie, P. J., 1967. Petrography and petrology, *in* Ultramafic and related rocks (ed. P. J. Wyllie), 1-6, New York, John Wiley & Sons.

Zolutuchin, V. V., and Vasilev, U. R., 1967. Particular characteristics of formation of some trap intrusives of the southeast Siberian platform. Nauka, Moscow, 231 p.

Zurbrigg, H. F., 1963. Thompson Mine geology. CIM Bulletin, *56*, 451-460; CIM Trans., *66*, 227-236.

19

Reprinted from *Symp. Basins Geosynclines Canadian Shield,* A. J. Baer, ed.,
Geol. Surv. Canada, 1970, pp. 1–24

THE ABITIBI OROGENIC BELT

A. M. Goodwin
University of Toronto, Toronto, Ontario

and

R. H. Ridler
University of Western Ontario, London, Ontario

Abstract

This east-trending tectonic unit, some 500 by 150 miles in dimensions,
is the largest single continuous Archean greenstone belt in the Canadian
Shield. It is a characteristic Archean orogenic composite belt featuring mafic
to felsic volcanics with coeval intrusions, volcanic sediments, both clastic
and chemical including banded iron formation, and several large granitic batholiths. Metallogenic patterns conform to lithic distributions. Low- to medium-
rank greenschist facies prevail. Supracrustal rocks have been isoclinally
folded about east-trending, undulating axes resulting in substantial lithofacies compression. The belt is bounded north and south by granitic-
metasedimentary crystalline terrains; it is truncated abruptly east and west
by Grenville and Kapuskasing crystalline rocks, both products of younger
Precambrian events. Thus the present belt represents only part of an original
Archean tectonic entity.

A number of well-recognized mafic to felsic volcanic centres with
intercalated volcaniclastics, iron formation and igneous intrusions are present. Despite common features each centre constitutes a semi-independent lithic
assemblage of limited stratigraphic continuity. The centres may be confidently
ascribed to processes of igneous differentiation and eruption largely by way of
central vents. Three main stages of igneous eruption are recognized in the
belt. The total time span was probably several hundred thousand years. Major
geochronological problems remain.

Recognized mafic to felsic volcanic centres and accompanying clastics
with iron formation are concentrated in two main east-trending bands, each 50
miles wide, which respectively cross the northern and southern parts of the
belt. In contrast, the intervening or median part of the belt, also 50 miles
wide, is underlain, as known, by uniform tholeiitic basalt, fine-grained clastics, and major granitic batholiths.

On this basis, the Abitibi belt is viewed as a remnant of a bilaterally symmetrical intra-tectonic orogen rather than a conventional asymmetrical
continental-oceanic tectonic interface (e.g. island arc). Crustal thinning
and other orogenic activities may be attributed either to conventional geosynclinal downsinking of thin, supple Archean crust or to spreading apart of
Archean crustal blocks in the manner of present-day ocean floor spreading.

INTRODUCTION

Abitibi belt, 475 miles long by 125 miles wide, is the largest continuous Archean (older than 2,500 million years) greenstone belt in the
Canadian Shield. Located in the southeastern part of the Superior tectonic
province this east-trending belt is truncated east and west by northeast-
trending crystalline rocks of Grenville province and Kapuskasing subprovince
respectively. The present belt is part only of an originally longer Archean
orogen which in Archean time included now-crystalline Grenville rocks immediately on the east and, likely, orogenic rocks of Wawa belt to the west.

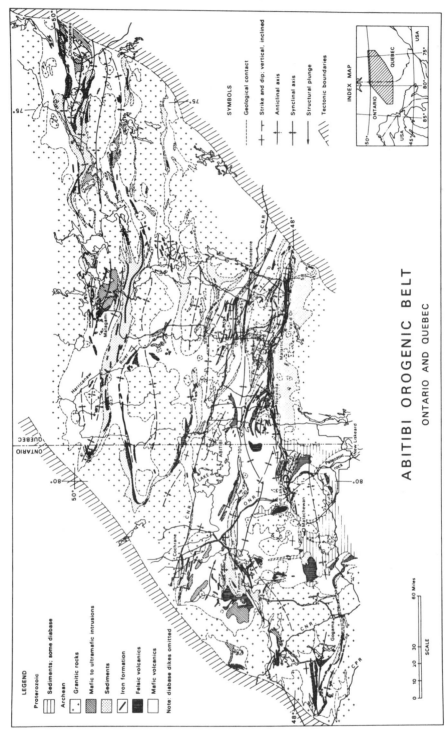

Figure 1. Geologic map of Abitibi orogenic belt. Omitted from the map for cartographic reasons are diabase dikes and some thin pyritiferous carbonaceous (± chert) zones in the Noranda-Val d'Or-Senneterre area. The latter are shown in Figure 2 as sulphide facies iron formation.

Abitibi belt has been geologically studied and explored for more than sixty years. Complete regional air-photo, aeromagnetic and gravity as well as local E.M. surveys are available. Many parts have been geologically mapped in detail. Sustained mineral exploration and mine development over the years have provided a wealth of data there being at least one hundred and fifty producers (past and present) including Au-Ag, Cu-Zn-Au-Ag, Ni-Cu, Fe, Mo-Bi, Li and asbestos deposits. However, extensive drift-covered tracts especially north of the main CN railway obscure many key geologic relationships. Despite the attention received, the major geologic problems including those of stratigraphic correlation and tectonic construction remain largely unresolved. Therefore additional studies, regional and detailed, direct and indirect, will be required for many decades to come.

In common with most greenstone belts of the Canadian Shield the geologic record of this belt doubtless includes a long and complex history of Archean events. So far, age-dating has not penetrated significantly the radiometric barrier imposed by the Kenoran event (at 2,500 million years) (Wanless *et al.*, 1968; see also Roscoe, 1965). This paper presents a brief integrated statement on relationships pertaining to tectonic evolution of this important primitive crustal unit. A principal purpose is to stimulate similar analysis of other Archean orogens in this and other Precambrian Shields of the world.

GENERAL GEOLOGY

The Abitibi lithic assemblage (Fig. 1) is characteristic of many Archean greenstone belts of the Canadian Shield. Older supracrustal and igneous rocks, now substantially deformed and of common greenschist facies, have been intruded by syntectonic to post-orogenic felsic to mafic intrusions ranging up to those of large batholithic dimensions. Numerous younger Precambrian diabase dikes, mostly of the Matchewan swarm, transect. Several tongues of flat-lying Proterozoic rocks and one of Paleozoic rocks protrude from the south. Pleistocene glaciation resulted in both clean-scouring of thick, steeply inclined Archean stratigraphic sections and accumulation of thick glacio-fluvial and lacustrine deposits which obscure bedrock relations particularly in northern parts of the region.

Volcanic Rocks

A variety of volcanic rocks is present. Mafic to felsic flows and pyroclastics of calc-alkaline chemical affinity, all representative of the orogenic suite (Baragar and Goodwin, 1969) greatly predominate. Some alkaline volcanic rocks are present at Kirkland Lake. Basalt flows and associated gabbroic intrusions are common in the lower parts of volcanic assemblages. Andesitic flows and pyroclastics are intercalated with basalt and characteristically increase in proportion upwards. Felsic rocks of dacite to rhyolite composition are generally present in upper stratigraphic parts. Most volcanic assemblages in the region display a generalized mafic to felsic compositional sequence of this type. In some areas the sequence is repeated in whole or in part.

Mafic volcanic rocks comprising basalt and andesite in that order of abundance, are the predominant supracrustal rocks of the region. They are broadly distributed particularly in the central and westen parts. Much remains to be learned about their petrochemistry. Thick mafic (i.e. basalt-andesite) accumulations exceeding 40,000 feet thick are present in Blake River group west of Noranda, and in Skead group south of Kirkland Lake. Similar thicknesses of mafic rocks may be present elsewhere in the region. In the Blake River group a lower zone of low-Al_2O_3 mafic volcanics, approximately 20,000 feet thick, is conformably overlain by equivalent thicknesses of high-Al_2O_3 mafic effusives (Baragar, 1968; Baragar and Goodwin, 1969). Similar low- and high-Al_2O_3 basalt and andesite are present in Swayze area to the west (Goodwin, 1968). In the Skead group south of Kirkland Lake a thick mafic assemblage approximately 28,000 feet thick which includes basanite lava flows exhibits significant upward increase in K_2O content (Ridler, 1970). Volcanic piles north and south of Timmins include tholeiitic basalt and andesite. Preliminary results of a reconnaissance geochemical cross-section of the volcanic rocks between Cadillac, Malartic area in the south and Matagami area in the north indicate a preponderance of basaltic rocks except for the felsic rocks shown in Figure 1 (J. Descarreaux, pers. comm.). Sparse chemical analyses of mafic volcanic rocks in the Matagami area suggest the presence of normal tholeiitic pillow basalt and andesite. In the Chibougamau area to the northeast, low-K_2O tholeiites including basaltic rocks with very low K_2O, high Na_2O and intermediate $K/_{Rb}$ ratios are present (Gunn, 1969).

Mafic lava flows are commonly 50 to 100 feet thick but flows up to 800 feet thick are known. Pillow forms, palagonite, variolites, amygdules and hyaloclastites are widespread and abundant thereby indicating prevailing subaqueous accumulation of the mafic lava. Primary textures are, on the whole, remarkably well preserved (Moorhouse, 1970).

Overlying the great thicknesses of mafic volcanic rocks are numerous concentrations of felsic volcanic rocks, each concentration characteristically representing a felsic eruptive centre (Fig. 1). Principal concentrations occur at Val d'Or, Noranda, Kirkland Lake, Timmins, Matagami and Chibougamau. Rocks of dacitic composition are most common in the felsic concentrations. Rhyolite is present locally as, for example, at Noranda, Timmins, Swayze, Joutel and Matagami but is absent in other centres, such as, for example, Kirkland Lake. Pyroclastic forms are very common in felsic concentrations of Swayze, Val d'Or, Kirkland Lake, Timmins, Matagami and Chibougamau areas. Massive rhyolite lava flows intercalated with andesite flows are common at Noranda. Subaqueous ash-flows recently identified in Noranda area (R.S. Fiske, pers. comm.) may be widespread in other volcanic piles of the region. Trachytic flows and pyroclastics are present at Kirkland Lake.

In addition to specific felsic concentrations, the regional volcanic pile contains many thin discontinuous felsic tuff zones, some notably spherulitic, which are attributed to widespread wind and/or water distribution. Such zones are potentially valuable marker-horizons particularly where intercalated within thick uniform mafic flow accumulations.

Although mineralogically altered, felsic volcanic rocks exhibit a wide range of internal textures including spherules, phenocrysts, vesicles, perlite, pumice and shards, all commonly in excellent state of preservation (Moorhouse, 1970).

Sediments

Clastic zones range from thin discontinuous units to broad regional, east-trending belts of which the principal are situated south of Noranda-Val d'Or, east of Kirkland Lake, near Timmins, northeast of Lake Abitibi, south and southwest of Matagami and west of Chibougamau (Fig. 1). Sediments characteristically occupy higher stratigraphic positions. Most sediments are structurally conformable with associated layered rocks. However, local unconformities have been established at Kirkland Lake and Timmins.

Sediments of the "poured-in" turbidite association predominate thereby suggesting rapid accumulation in tectonically unstable environments. Two principal facies are identical: volcanogenic and flyschoid. The Timiskaming facies, a variation, is present locally. Rocks in the volcanogenic facies, ranging up to 10,000 feet thick include greywacke, shale, lithic sandstone, conglomerate, breccia and iron formation. Identifiable clastic components compare closely in composition with nearby volcanic rocks and have obviously been derived in large part from them by rapid mechanical erosion and subaqueous

deposition in nearby troughs and basins. Chaotic textures and polymictic
unsorted materials are distinctive features. This facies which is characteristic of strongly active tectonic zones, typically displays intricate soft-
sediment deformational structures. Graded bedding and abrupt facies changes
are common. Iron formation is present. Carbonaceous and pyritiferous zones
are intercalated with finer grained clastic phases.

A prominent belt of more regularly bedded flyschoid facies (Pontiac
group) lies south of Noranda-Malartic-Val d'Or. It comprises rhythmically bedded greywacke-argillite sequences of uniform construction and composition which
lack marked lateral facies changes. According to J. Holubec (in preparation)
the facies is 7,000 to 8,000 feet thick and was developed in a comparatively
stable tectonic environment in proximity to a stable crustal domain to the south.

A variant facies, the Timiskaming, is present in narrow, east-trending
zones near Kirkland Lake-Noranda and Timmins. This facies is characterized by
coarse-grained detritus including conglomerate, abrupt lateral facies changes
and erosional unconformities. It occupies narrow zones of pronounced tectonic
instability which have been attributed to the local development of steep slopes
(J. Holubec). At Kirkland Lake the facies is associated with a unique assemblage of trachytic flows and pyroclastics.

Iron formation is widely distributed (Figs. 1 and 2). Principal bands
occur in Swayze, south of Timmins and Kirkland Lake-Val d'Or areas in the south
and within a broad east-trending belt south of Matagami in the north. Local
bands are present at Chibougamau, in the Barraute-Quevillon area of the central-
eastern part and elsewhere. The iron formations are important stratigraphic
elements and the closest approach to marker horizons in the regional assemblage.
Their continuity across broad parts of the region demonstrates the presence of
large extensive basins of deposition in Abitibi time.

Iron formations of the region are classified as oxide, carbonate or
sulphide facies depending on the predominant iron mineral. The three facies
are transitional across the region (Fig. 2). Oxide (magnetite) facies is common in Swayze, south Timmins, Kirkland Lake and Val d'Or areas in the south and
in the principal belt south of Matagami in the north. Generally, oxide facies
is transitional eastward and northeastward through narrow carbonate (siderite-
ankerite-dolomite) zones to sulphide (pyrite-pyrrhotite) facies iron formation.
These facies transitions are considered to define original shelf-to-basin
bathymetry. For example in the Timmins area, magnetitic iron formation in the
south is transitional northeastward through a narrow zone of sideritic iron
formation to thin cherty pyritiferous iron formation. Magnetitic iron formation at Kirkland Lake is transitional eastward to siliceous carbonate iron
formation as at Larder Lake and farther to thin pyritiferous zones at Noranda
(Ridler, 1970). Similarly, magnetitic iron formation near Val d'Or is transitional northeastward to thin chert-pyrite-carbonaceous zones representing sulphide facies iron formation. Similar transitions are present elsewhere
(Fig. 2).

Mafic Intrusions

Numerous mafic sheets, sills and dikes occur in the mafic lava accumulations. They are readily confused with diabasic phases of thick flows and
commonly represent intrusive phases of the volcanism. They are also intrusive
into felsic volcanic phases as at Noranda.

Differentiated sheets and sills up to several thousand feet thick typically comprise basal zones of peridotite, usually serpentinized, transitional
upwards through pyroxenitic or mafic gabbro or norite, to gabbro and quartz
gabbro and, locally, to granophyre (micropegmatite). In Chibougamau-Opemiska
area, differentiated mafic sills include peridotite-pyroxenite, gabbro-
pyroxenite and diorite-quartz diorite transitions. The Chibougamau anorthosite, 30 miles in diameter, comprises coarse-grained greenish white assemblages
composed of saussuritized plagioclase and chloritized ferromagnesian minerals.

231

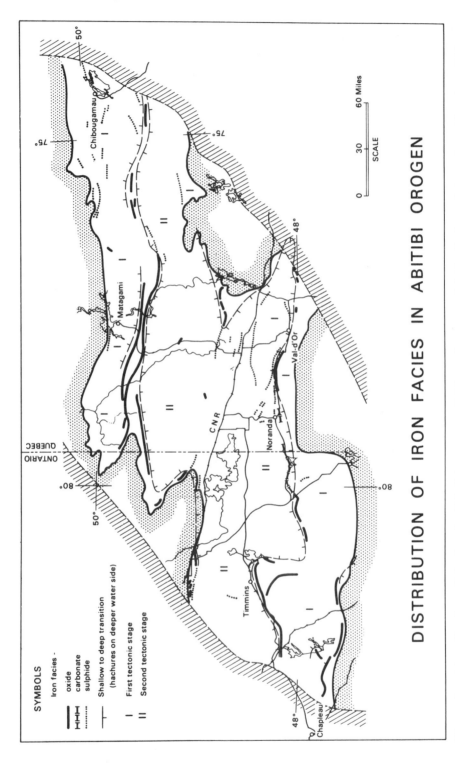

DISTRIBUTION OF IRON FACIES IN ABITIBI OROGEN

Figure 2. Distribution of facies in Abitibi orogen. Oxide-carbonate-sulphide facies of iron formation transitions delineate original shelf to basin slopes within the orogen. Tectonic stages refer to postulated episodic progressions in construction of the orogen. Presumed younger volcanic-sedimentary rocks are present towards the centre of the orogen.

The Bell River complex at Matagami, at least 15,000 feet thick, comprises strongly and rythmically banded norite, anorthosite, pyroxenite and associated gneissic cataclastic rocks. Another differentiated complex occurs at Kamiskotia Lake north of Timmins.

Granitic Rocks

A great variety of granitic rocks is present in the region but only a brief statement is made on the subject. Foliated to massive granitic rocks commonly underlie the margins of the region. They are associated with more or less metavolcanic and metasedimentary rocks. Where intrusive relations are apparent the granitic rocks have intruded the volcanic and sedimentary rocks and hence post-date them in terms of last emplacement. Large granitic complexes along the north and south boundaries may include primitive crystalline basement. By its nature the demonstration of the reality of such sialic basement is very difficult if indeed possible. The granitic rocks consist in large part of weakly to strongly foliated granodioritic gneiss of possible metasedimentary origin, together with varying proportions of schists, gneiss, migmatites, and undoubted igneous plutons. Based on geochemical reconnaissance of similar crystalline terrain in Ungava region of northern Quebec the average chemical composition of these crystalline rocks approaches that of common granodiorite (Eade, *et al.*, 1966).

In addition granitic plutons are widely distributed within the orogenic belt itself. At least nine major plutons each 40 to 100 miles in diameter are present, most in the northern part of the region. Additional smaller plutons enclosed within volcanic rocks are closely associated with felsic volcanic rocks e.g. Noranda, Malartic, Kirkland Lake and Chibougamau, of which they may represent coeval intrusive equivalents. The Round Lake batholith south of Kirkland Lake contains a southern part of leucrocratic quartz diorite and a northern, presumably younger part of hornblende granite. The Otto stock on the northeast is composed of syenite, nepheline syenite and locally, pegmatitic phases. To the east, in the Malartic area, the Bourlamaque batholith comprises a core of quartz diorite and a border of albite granite. The LaCorne-La Motte group of plutons in the north Malartic area, approximately 40 miles long by 16 miles wide, comprises respectively hornblende-, biotite-, and muscovite-granite phases, the last named associated with spodumene-bearing pegmatites and molybdenite mineralization. To the west the Taschereau, Palmarolle and Flavrian plutons comprise cores of pink granodiorite and borders of grey quartz diorite. Many other types of plutons are present in the belt of which the above provide some example.

Radiometric ages of various intrusions in the Kirkland Lake area, for example, indicate that plutonic activity has occurred there repeatedly from Archean to Mesozoic time. Particular examples are the Round Lake batholith (2,400 million years), the Otto stock (1,700 million years) and the Kimberlite dikes in the Upper Canada Mine which are ascribed to a Jurassic-Cretaceous age (Ridler, 1970). Clearly there are major problems concerning the times and mechanisms of igneous emplacement as well as the origin of sialic material contained in the plutons.

Lithic Proportions

A planimeter survey of the geologic map (Fig. 1) including only those rocks within the orogenic belt (Fig. 3) provides an assessment of lithic composition. Areas of cover rocks of Proterozoic and Paleozoic ages totalling 1,268 sq. mi. have been excluded. The belt has been divided into southern and northern parts with the main CN railway as the dividing line.

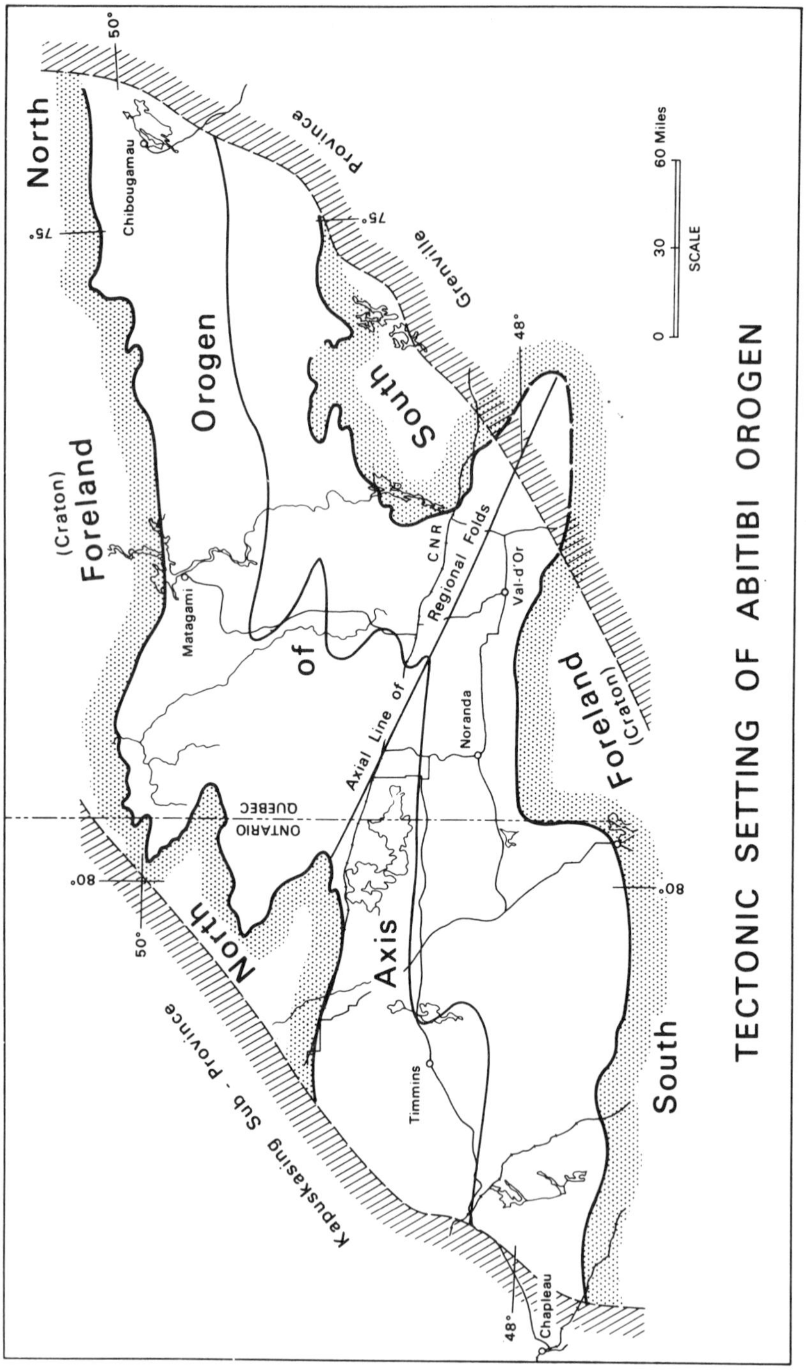

Figure 3.   Tectonic setting of Abitibi orogen. Boundaries of the forelands are defined on the basis of preponderant crystalline granitic rocks. S-shaped form of the orogen is thought to be due to a structural deformation.

Abitibi Orogen

	Southern Part	Northern Part	Total
Area	15,948 sq. mi.	20,811 sq. mi.	36,759 sq. mi.
Granitic rocks	26.2 per cent	37.0 per cent	32.3 per cent
Mafic volcanics	47.5 "	44.2 "	45.6 "
Felsic volcanics	5.5 "	2.2 "	3.6 "
Sediments	18.8 "	13.8 "	16.0 "
Mafic intrusions	2.0 "	2.8 "	2.5 "
	100.0 per cent	100.0 per cent	100.0 per cent

Recalculating to Archean supracrustal rocks only (i.e. excluding granitic rocks and mafic intrusions) the following lithic proportions are indicated to be present in the belt:

	Southern Part	Northern Part	Total
Mafic volcanics	66.2 per cent	73.5 per cent	70.0 per cent
Felsic volcanics	7.6 "	3.6 "	5.5 "
Sediments	26.2 "	22.9 "	24.5 "
	100.0 per cent	100.0 per cent	100.0 per cent

Thus the indicated ratio mafic volcanics: felsic volcanics is 13:1. Stated otherwise, felsic volcanic rocks (dacite, rhyodacite, rhyolite and minor trachyte) form an indicated 5.5 per cent of the regional volcanic assemblage. The figures must be viewed with caution considering the extensive drift cover in the northern parts particularly. Past experience suggests that further field studies in the region will reveal additional bands of felsic volcanic rocks, and result in reclassification of some sediments as felsic volcanic rocks. In the meantime the figure of 5.5 per cent felsic volcanics is offered as a working minimum, the true figure possibly approaching 7 to 10 per cent felsic volcanic rocks.

The northern part of the region includes substantially higher proportions of internal granitic rocks in the form of large batholiths and less felsic volcanic rocks. Despite the note of caution expressed in the previous paragraph the stated proportions are considered to reflect a genuine difference in lithic proportions between the northern and southern parts of the region. Major problems remain concerning the origin of the sialic material and the time and mechanism of emplacement of these large northern batholiths.

Relatively more sediments are present in the southern part of the orogen (i.e. 18.8 vs 13.8 per cent). This reflects inclusion within the orogen of Pontiac sediments - the main flyschoid facies - situated south of Noranda-Malartic-Val d'Or (Fig. 1). Present indications are and future studies may prove that this facies was derived from a craton to the south and should more properly be included in the southern foreland (Fig. 3). However in order to be consistent in defining the boundaries of the forelands i.e. preponderant granitic rocks, the Pontiac facies has been included in the orogenic belt for present purposes.

TECTONIC SETTING

The tectonic setting and principal parameters of the Abitibi orogen are illustrated in Figure 3. The orogen lies between two northeast-trending boundaries, the Grenville province on the east and the Kapuskasing subprovince

235

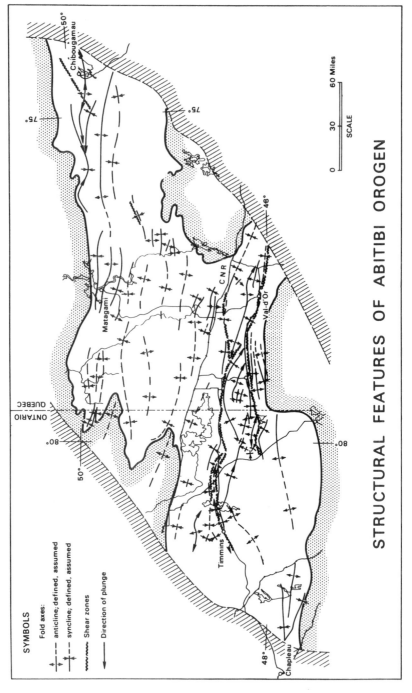

Figure 4. Structural features of Abitibi orogen. Additional folds and shear-zones may be present in drift-covered parts, especially in the north.

on the west. The orogen proper lies between two postulated cratonic forelands, the boundaries of which are defined on the basis of preponderant granitic rocks. Foreland boundaries are in fact gradational from predominantly supra-crustal rocks within the orogen to predominantly crystalline complexes without. Also the foreland complexes locally include substantial masses of metavolcanic and metasedimentary rocks. This is particularly so in the north foreland as well as in the western part of the south foreland. The axis of the orogen has been placed midway between the forelands with due regard to internal lithic distribution.

Accordingly the Abitibi orogen constitutes a generally east-trending tectonic unit approximately 500 miles long and 60 to 100 miles wide. Litho-logically it features orogenic supracrustal rocks, both volcanics and sedi-ments, together with prominent granitic plutons. It is intracratonically con-tained between predominantly granitic forelands. The present S-shaped form reflects structural deformation.

STRUCTURE

The first-order structure of the region is an S-shaped regional fold which is delineated by the boundary surface of the orogen (Fig. 3). Thus the north boundary surface extends from Chibougamau vicinity on the east to Bradburn township north of Timmins on the west, and the south boundary surface from Marceau township situated south of Doda Lake on the east to the vicinity of Chapleau on the west. Two conspicuous second order S-folds which form part of the regional fold are present on both boundary-surfaces. Those on the north boundary surface protrude eastward in the area north of Lake Abitibi near the Ontario-Quebec interprovincial boundary. Those on the south boundary surface lie respectively northeast of Val d'Or and southwest of Noranda the latter indicated by the conspicuous southward trend of rock units south of Larder Lake. The axial line of the regional folds trends northwesterly (Fig. 3).

Folding of the compositional layering within the orogen is widespread (Fig. 4). The folds have east-trending axial lines, are characteristically isoclinal, doubly plunging, and commonly bifurcate along strike. Those to the north of the CNR line commonly plunge to the west e.g. Chibougamau and Matagami whereas those to the south commonly plunge to the east e.g. Noranda, Malartic, Kamiskotia. On a regional scale granitic plutons are preferentially distributed along anticlines and supracrustal rocks along synclines. On a local scale synclines commonly bifurcate to enclose an anticline which includes central plutons e.g. Noranda, Malartic, Matagami and Chibougamau. In the west Malartic area, volcanic (Malartic group) and sedimentary (Kewagama) rocks have been folded into a west-facing anticline; these rocks underlie younger vol-canic rocks situated to the west and north.

Several prominent, steeply inclined, east-trending, discontinuous shear-zones of undetermined displacement have been identified in the southern part of the region e.g. Porcupine-Destor and Larder Lake "breaks". They follow lithofacies boundaries for the most part, including sedimentary-volcanic inter-faces. Other regional shear zones of this type may be present particularly in drift-covered areas to the north. Recent field studies of the Larder Lake "break" cast doubt on its continuity and structural significance (Ridler, 1970). This suggests that a re-examination of the entire problem of the regional "breaks" is in order.

Comparison with modern volcanoes in island arc settings (e.g. Solo zone in East Java, *see* Van Bemmelen, 1949, Fig. 16, p. 26) suggests that the present elliptical outline of the Abitibi volcanic complexes (Fig. 5; *see* page 12) is due to foreshortening in a north-south direction of originally cir-cular volcanic complexes. The present axial ratios of the volcanic complexes average 5:1. This implies that originally circular complexes have been fore-shortened by at least 50 per cent. If the complexes were also foreshortened

237

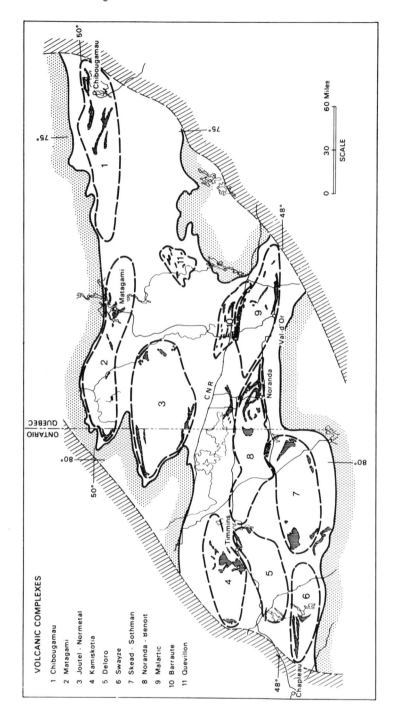

Figure 5. Distribution of volcanic complexes in Abitibi orogen. Felsic volcanic rocks shown by vertically lined pattern. Approximate boundaries of volcanic complexes including cogenetic intrusions and sediments is shown by heavy dashed lines. Present elliptical shape of presumably originally circular volcanic complexes is attributed to compression folding.

in an east-west direction as is suggested by the presence of doubly-plunging east-trending fold axes then the amount of north-south foreshortening may have substantially exceeded 50 per cent.

The present elliptical shape of the volcanic complexes together with closely spaced isoclinal folds suggests that the mechanism of deformation was compression folding. Such deformation would also produce substantial vertical extension of the rocks. Thus Archean orogenic assemblages originally 35,000 feet thick (10 km) may have been increased in thickness to approach that of the existing continental crust (40 km). More definite statements on the mechanics and state of deformation would require additional structural studies in the field.

VOLCANIC COMPLEXES

The regional stratigraphy is dominated by the presence of semi-independent ellipsoidal volcanic-sedimentary domains each of which is termed a volcanic complex. Nine major and two minor volcanic complexes have been delineated in the region (Fig. 5) each with a mafic to felsic volcanic sequence, associated intrusions and sediments.

The best known domain is that of the Blake River group herein called the Noranda-Benoit complex (number 8 in Fig. 5). The lowermost identifiable strata of this group which are of prevailing mafic composition extend from Nighthawk Lake situated 15 miles east of Timmins, 1) eastward to north Noranda area, thence southeastward to Malartic, and 2) southeastward to Kirkland Lake, thence eastward to Malartic, the two boundary lines thus outlining the complex. An identifiable mafic to felsic volcanic sequence with associated intrusions and sediments is contained within this boundary. The single felsic volcanic centre lies at Noranda in the eastern part of the complex. Although lateral extensions of Blake River rocks may fall outside this boundary the bulk of the preserved assemblage appears to lie within it.

Similarly, other volcanic complexes have been delineated with the felsic volcanic concentrations of the region serving in their identification (Fig. 5). Some boundaries have been tentatively defined only e.g. terminations of Skead-Sothman (#7), Matagami (#2), Chibougamau (#1) and Kamiskotia (#4) complexes. Barraute complex (#10) may be a structural extension of Malartic complex (#9). Quevillon complex (#11) is of uncertain definition as are the mutual boundary areas of Deloro (#5), Swayze (#6) and Skead-Sothman (#7) complexes. Other as yet unidentified complexes may be present in the region. Modifications of existing boundaries are to be expected on the basis of future work.

Despite these problems of boundary definition it is suggested that the pattern of complexes illustrated in Figure 5 accurately reflects the style and format prevailing during construction of the Abitibi orogen - a constructional style dominated by development of numerous semi-independent mafic to felsic volcanic piles with coeval intrusions and sediments. Analogy with recent and modern volcanic piles including those of linear tectonic (especially island arc) association suggests that the Abitibi volcanic complexes originally had circular outlines, each apparently in the order of 60 to 100 miles diameter. Their present elliptical outlines are attributed to structural deformation.

The defined volcanic complexes of the region are concentrated along the northern and southern boundaries of the orogen. Thus comparison of Figures 3 and 5 shows that volcanic complexes 1 to 4 lie north of the axis of the orogen and close to the northern boundary, whereas complexes 6 to 11 lie south of the axis of the orogen and close to the southern boundary. This pattern may reflect original linear distribution of strato-volcanic complexes along tectonic structures such as geanticlinal uplifts essentially parallel to the margins of the developing orogen.

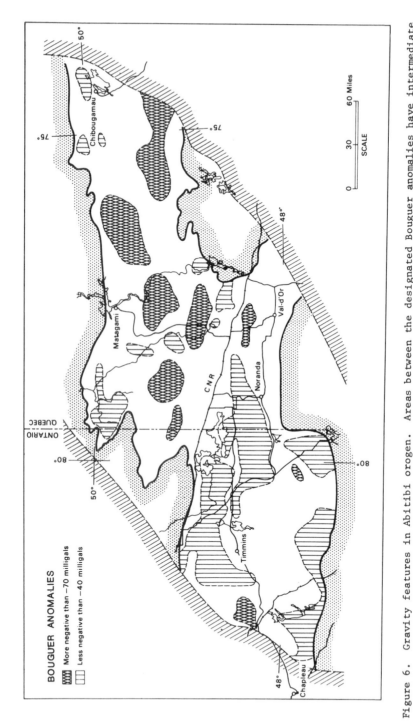

Figure 6.   Gravity features in Abitibi orogen. Areas between the designated Bouguer anomalies have intermediate gravity expressions of −70 to −40 milligals. Bouguer anomalies more negative than −70 milligals correspond to granitic batholiths. Anomalies less negative than −40 milligals correspond to large areas of mafic igneous rocks.

GRAVITY

Gravity and geologic features are closely related in Abitibi orogen. This may be seen by comparing the Bouguer anomaly map (Fig. 6, which is based on Gravity Map of Canada GMC 67-1) and the geologic map (Fig. 1). The close correspondence of gravity and geologic features indicates that the regional gravity measurements summarized in the gravity map substantially reflect surface and near-surface crustal features rather than deep-crustal or mantle features.

Six principal high negative (less than - 70 milligals) Bouguer anomalies north of the CN railway coincide remarkably well with the main granite batholiths of the region. By demonstrating the presence in the crust of substantial bodies of low density material, they indicate that the batholiths have deep roots. Three high negative anomalies south of the CN railway coincide with smaller granitic batholiths (near Val d'Or, Kirkland Lake and Timmins). In contrast several granitic batholiths within the orogen do not have comparable high negative gravity expressions e.g. Matachewan, Timmins, Gogama areas. Such batholiths may terminate at shallow depths or alternatively, contain igneous material of intermediate or assorted densities.

The principal low negative (greater than - 40 milligals) Bouguer anomalies coincide with 1) major mafic volcanic-intrusive assemblages and 2) flat-lying Proterozoic diabase along the southwestern margin. For example, thick, extensive mafic assemblages are reflected by major low negative anomalies northwest of Noranda (Blake River group mainly) and north of Timmins. Four local mafic assemblages are reflected by correspondingly small, low negative gravity anomalies northwest of Senneterre, between granitic batholiths near Joutel-Poirier, west of Matagami, and near Chibougamau.

All other parts of the orogen, which yield an intermediate gravity expression (between -70 and -40 milligals) are underlain apparently by mixed lithic assemblages which collectively lack an anomalous gravity expression.

Gravity relationships clearly demonstrate that the part of the orogen north of the main CN railway contains less total mass (i.e. more low-density rock) than the southern part. This is attributed mainly to the presence of numerous granite batholiths with deep roots and proportionately less mafic igneous rock in the north. In addition some undisclosed deeper crustal influence may be reflected.

It is particularly significant that high negative Bouguer anomalies (less negative than -70 milligals) coincide with internal granite batholiths only and not with the regional granitic complexes of the northern and southern forelands as defined in Figure 3. This provides a ready method of differentiating true granitic batholiths from the mixed lithic predominantly granitic assemblages of the forelands. In this connection two high negative Bouguer anomalies, each presumably an expression of a granite batholith, lie outside the orogen; and within the bordering crystalline complexes. One lies in north foreland rocks, 50 miles north of Matagami and the other in Grenville province rocks 50 miles southeast of Chibougamau (*see* Gravity Map of Canada). With the exception of these two high negative areas the forelands have intermediate gravity expressions suggesting the presence of mixed crystalline complexes of assorted gneissic-migmatitic-plutonic composition.

METALLOGENESIS

Type and Distribution of Mineralization

The region contains several thousand known mineral occurrences including 150 producers (past and present) of which 57 lie in Ontario and 93 in

241

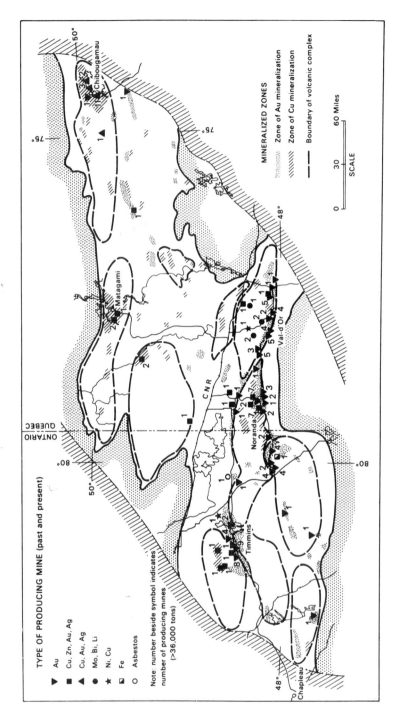

Figure 7.   Metallogenic relations in Abitibi orogen.   Distributions of main Au and Cu mineralized zones and of producing mines (greater than 36,000 tons; past or present) are shown relative to volcanic complexes.

Quebec. Most occurrences are either gold-bearing quartz veins or disseminated to massive sulphide (Cu-Zn-Au-Ag) concentrations. Others include magnetitic iron formation, asbestos, Ni-Cu, Zn-Ag-Pb, Mo-Bi and Li.

The main mineralized zones based on known occurrences are shown in Figure 7. A principal east-trending Au zone extends intermittently from Sothman-Matachewan area on the west through Kirkland Lake, Larder Lake, Noranda and Malartic to Val d'Or area on the east. A second well-established Au zone extends eastward from Timmins vicinity through Duparquet thence curving southeastward to join the first zone at Malartic. Smaller discontinuous Au zones are widely distributed in the region, for example, in the Swayze, south Timmins, Benoit and north Malartic areas; north of the CN railway in the Doda-Opawica area, and southwest of Chibougamau and Quevillon areas. Principal zones of Cu mineralization, on the basis of producers and known occurrences, lie in north Timmins, Noranda, Malartic-Val d'Or, Joutel, Matagami and Chibougamau areas. Smaller zones lie west and south of Matagami, west of Chibougamau and in the Barraute-Quevillon area. Several other small isolated mineralized patches without obvious pattern are present in the region as illustrated in Figure 7.

Lithic Relationships

Most Au deposits lie in intermediate to felsic volcanic rocks, their intrusive equivalents or associated clastic or exhalative sediments. Included are those at Swayze, Matachewan, Kirkland Lake, Timmins, Cadillac, Malartic and Val d'Or. Sulphide deposits of the. Cu-Au and Cu-Zn-Ag types are associated with intermediate to felsic volcanic complexes at Noranda, Joutel, Timmins and Matagami. Mo-Bi and Li deposits are associated with pegmatitic rocks in north Malartic area. A single iron producer lies in banded iron formation enclosed in felsic tuff and mafic lavas south of Kirkland Lake. Asbestos deposits are associated with differentiated ultramafic complexes north of Matheson and west of Timmins. Ni-Cu deposits lie in mafic to ultramafic intrusions near Cochrane, south of Timmins and north of Malartic. Cu-Ag-Au deposits are associated with gabbroic, noritic and anorthositic intrusions of volcanic association at Chibougamau (Duquette, in press) and Cu-Zn-Ag deposits with anorthositic intrusions at Matagami (Sharpe, 1965).

Relation to Volcanic Complexes

As illustrated in Figure 7 most mineralized zones including all but three of the 150 producers lie within or marginal to the designated volcanic complexes. Principal Au zones lie at the margins of complexes and have traditionally been interpreted as being primarily associated with "shear-zones" e.g. Porcupine-Destor and Larder Lake "breaks". For example, one Au zone extends from Matachewan through Kirkland Lake at the northern margin of Skead-Sothman complex (#7 in Fig. 5) eastward along the southern margins of the Noranda-Benoit (#8) and Malartic (#9) complexes; a second Au zone extends from the mutual boundary of Kamiskotia (#4) and Deloro (#5) complexes eastward along the northern and northeastern margins of the Noranda-Benoit (#8) complex. Other, leaner, Au-bearing zones are present in the western parts of Swayze (#6) Skead-Sothman (#7) and Deloro (#5) complexes. In addition, several small Au zones without obvious relation to volcanic complexes occur in the area southwest of Chibougamau and north of the main CN line.

Most Cu-bearing sulphide zones lie in the volcanic complexes directly associated with felsic volcanic rocks. They are most common in the eastern parts of the following six volcanic complexes: Chibougamau (#1), Matagami (#2), Joutel-Normetal (#3), Kamiskotia (#4), Noranda-Benoit (#8) and Malartic (#9). Cu-Au and Cu-Zn-Ag producers are present in Matagami (#2), Joutel-

243

Normetal (#3), Kamiskotia (#4), Noranda-Benoit (#8) and Malartic (#9) complexes whereas Cu-Au-Ag producers are present in Chibougamau (#1) complex. Other small local zones of Cu mineralization occur in Barraute (#10) and Quevillon (#11) complexes, in the central and western parts of Matagami (#2) and Chibougamau (#1) complexes and elsewhere north of the main CN railway as illustrated in Figure 7.

Mo-Bi and Li occurrences are associated with pegmatitic phases of felsic intrusions in the centre of Malartic (#9) complex. Magnetitic iron formation is associated with volcanic tuff and mafic lavas in northeastern Skead-Sothman (#7) complex. Ni-Cu producers lie in ultramafic intrusions in or near Malartic (#9) and Noranda-Benoit (#8) complexes.

Lithic associations common to sulphide deposits are felsic volcanic concentrations reflecting central vent eruptions, sulphide facies iron-formation reflecting reducing environments, and absence or scarcity of coarse clastics reflecting offshore, deeper water sites of accumulation. Such deeper water environments may be attributed in part to tectonic collapse including cauldron subsidence following felsic volcanic discharge. Conversely some lithic associations are inimical to sulphide occurrences: e.g. oxide facies iron formation and clastic sediments, especially conglomerate. This association reflects topographically higher, shallow water, partially oxidizing shelf-environments. Accordingly, most mineralized sulphide zones in the region are directly associated with felsic volcanic concentrations having sulphide facies iron formation nearby e.g. Noranda, Malartic, Matagami and Chibougamau. On the other hand, many Au occurrences are directly associated with carbonate facies iron formation either at felsic volcanic-intrusive centres or at the margins of volcanic complexes.

Stratigraphic Relations

Cu-Zn-Au-Ag concentrations have the most direct stratigraphic relationship. Many lie at specific volcanic contacts, commonly felsic-mafic transition in upper parts of stratigraphic successions e.g. Kamiskotia, Kidd Creek Noranda, Matagami, Joutel and Malartic areas. Cu-Au-Ag deposits lie along fractures at the northern margin of a volcanic-enclosed, anorthositic complex at Chibougamau. Some Ni-Cu deposits lie in ultramafic intrusions either within stratigraphically lower mafic volcanic rocks or upper felsic volcanic rocks e.g. Marbridge and Alexo mines.

Au deposits have variable degrees of stratigraphic associations. One family of deposits has endogenous relations to felsic alkaline intrusions in the form of subvolcanic sills, discordant plugs or fairly large stocks. The gold occurs disseminated in irregular zones often associated with pyrite or concentrated into various types of vein structures. Although the plutons are believed to be cogenetic volcanic equivalents, and therefore Archean in age, it is possible that post-Archean igneous events have been accompanied by gold mineralization. Examples are the gold mines of Kirkland Lake, Granada, Yonge-Davidson, Eldrich, Powell-Rouyn and Sullivan mines. Flow and pyroclastic equivalents of the above class of plutons may be enriched in gold, either disseminated or in vein form. Examples are Dome, Hollinger and McIntyre mines of the Timmins area and certain parts of Upper Canada mine east of Kirkland Lake. Gold deposits intimately associated with carbonate facies iron formation include Omega and Kerr Addison (Ridler, 1970); those with interflow deposits include Wasamac mine west of Noranda and Dome deposit of Timmins area. Gold deposits of possible placer derivation in whole or in part include Pamour mine, lying in quartzite east of Timmins and McWatters mines in conglomerate south of Noranda, each situated in the upper part of the local stratigraphic succession. In addition, many of the gold deposits lie in dilatant shear and fracture zones into which the precious metal together with accompanying quartz appears to have been concentrated from the foregoing original stratigraphic sites during brittle deformation. This has partly obscured primary stratigraphic distribution and has produced a distribution pattern which has been tradition- ally related to a regional fracture system.

Finally, magnetitic iron formations are enclosed in volcanic rock in upper stratigraphic parts of local successions e.g. Adams mine at Kirkland Lake, whereas asbestos deposits lie in differentiated mafic sills and complexes towards the base of local stratigraphic successions, e.g. Munro mine north of Matheson.

TECTONIC INTERPRETATION

It has been established that Archean rocks of the Abitibi orogen have been structurally deformed to a considerable degree. Relations indicate that component lithofacies have been tightly folded about east-trending axes, resul- ting in substantial north-south shortening. Regional faults may be present. Deformations of this type have combined to obscure original tectonic relation- ships. Despite these reservations and uncertainties which are common to Archean greenstone belts in general, a tectonic reconstruction based on avail- able data is offered for consideration as a working hypothesis.

The nature, distribution and composition of volcanic and sedimentary rocks points clearly to the presence in Archean time of a mobile belt or tec- tonically unstable, linear basin complex. Archean mafic to felsic volcanic assemblages have their modern analogues in orogenically active, thin-crustal environments such as those of island arcs which are characteristic of continent-ocean interfaces. Archean sediments, notably greywacke, conglomerate and breccia, are of the high-energy, "poured-in" turbidite association. Thus tectonic mobility is a clearly indicated characteristic of the orogen. Further, the present location of the east-trending orogenic belt between adjoining pre- dominantly granitic terrains suggests that the orogen developed between two sialic forelands. The presence of flyschoid clastics (Pontiac group) in Noranda-Val d'Or area derived from a granitic provenance to the south (Holubec in preparation), reinforces this interpretation of the presence of a sialic foreland south of Noranda during Abitibi time. Therefore, we deduce that Abitibi orogen developed intracratonically upon thin mobile crust, between adjoining forelands of predominantly sialic composition. The apparent absence of flyschoid facies along the northern margin of the orogen and along the west- ern part of the southern margin, together with the presence of some volcanic and sedimentary rocks within the adjoining predominantly granitic terrains sug- gest that comparatively subdued sialic forelands of mixed lithic composition existed in those parts during Abitibi time.

Two elements of regional stratigraphy are used to reconstruct tectonic development of the orogenic assemblage, namely facies transitions of iron and relative ages, as known, of specific volcanic complexes.

The main shelf-to-basin transitions in the orogen, based on iron for- mation facies-relationships, extend east-west across the northern and southern parts of the region. In addition other local transitions are present. The transitions, in effect, separate common shelf-type oxide facies iron formation, present along the northern and southern parts of the orogen, from deeper water or basinal sulphide facies iron formation towards the centre of the orogen. These relationships, in brief, point to the presence of tectonic basin(s). Successive basins may have been present during progressive development of the orogen. The local presence of sulphide facies iron formation near Chibougamau and Matagami may be due to cauldron subsidence or to some other form of local tectonic collapse.

Detailed stratigraphic studies in Kirkland Lake area (Ridler, 1970) indicate that an older volcanic assemblage (Skead group) mainly to the south of Kirkland Lake proper underlies younger Blake River volcanic rocks to the north (*see* Figs. 1 and 5); that older Skead volcanics constituted a stable shelf

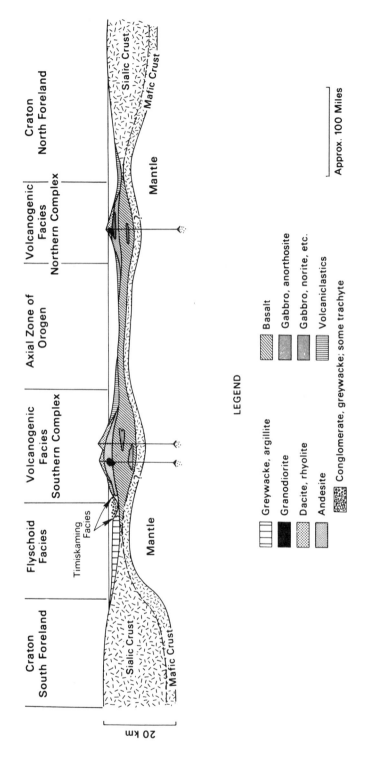

Figure 8. Hypothetical tectonic reconstruction of Abitibi orogen. Vertical cross-section. Length of the orogen is approximately 500 miles. Distance between southern and northern volcanogenic facies may have been in the order of 200 miles. Vertical scale is approximate only. The line of the section is intended to cross Noranda in the south and Matagami in the north. The smaller size and relief of the northern sialic foreland is indicated by the absence of flyschoid facies along its southern margin. The nature and thickness of mafic crust underlying the orogen is conjectural. In the absence of direct evidence of sialic contribution to the volcanic and sedimentary assemblages of the orogen (except for the flyschoid facies, i.e. Pontiac, at the south margin), no subjacent sialic crust is shown as being present, at least during the early development stages of the orogen.

246

relative to the Blake River basin of mafic volcanic accumulation to the north; and futhermore that the shelf-to-basin transition is marked by oxide- and carbonate-facies iron formation. In the Timmins area the east-trending iron formation lies between underlying Deloro volcanics to the south and what is generally interpreted as younger Tisdale volcanics to the north. Malartic volcanic and sedimentary (Kewagama) rocks stratigraphically underlie volcanic rocks to the west (Blake River) and to the north.

Therefore it is suggested that the main shelf-to-basin transitions of the region, marked by iron formation, delineate volcanic assemblages of slightly different ages. The younger assemblages are interpreted as occupying the deeper water side, towards the basin and hence mainly towards the axis of the orogen. Accordingly, the following volcanic complexes and rock groups appear to possess some degree of contemporaneity: in the south, Deloro (#5 in Fig. 5), Swayze (#6), Skead-Sothman (#7) and Malartic (#9) complexes and Pontiac sediments; in the north, Matagami (#2) and Chibougamau (#1) complexes. Somewhat younger and bearing a similar degree of contemporaneity, on the same basis, are Kamiskotia (#4), Noranda-Benoit (#8) and Joutel-Normetal (#3) complexes together with associated belts of sediments. Thus the postulated sequence of events is one of episodic progression in tectonic development from orogenic margins to the axis of the orogen. This working hypothesis requires detailed testing. In this regard radiometric dating, isotopic and paleomagnetic studies, and detailed stratigraphic studies may be fruitful.

Most of the volcanic complexes with their felsic concentrations lie close to the boundaries of the orogen. All felsic concentrations lie within 40 miles of a foreland boundary and all but two (Kamiskotia and Joutel) lie within 25 miles. Thus the main bulk of defined felsic volcanic rock is situated near the margins of the orogen. Because most mineral deposits (excluding asbestos and Ni) are closely associated with felsic rocks, it follows that metallogenic patterns bear a similarly close spatial relationship to foreland boundaries. The two exceptions referred to above are in 1) the Timmins area where major Au and Cu-Zn deposits lie 40 and 30 miles respectively from the northern foreland boundary and 2) the Joutel Cu-Zn deposit approximately 40 miles east of a folded spur of the same boundary. However, rock assemblages in both of these areas are highly deformed and extensively drift-covered so that their true stratigraphic positions relative to original foreland boundaries are difficult to interpret.

Within the limitations imposed by the existing state of deformation, drift-cover and knowledge of stratigraphic relations the main tectonic features of the orogen have been reconstructed and presented diagrammatically in Figure 8. Thus the orogen is considered to have developed intracratonically between two sialic forelands. Accumulation of mafic to felsic volcanic complexes proceeded, presumably under deep-fracture control, in proximity to sialic forelands i.e. at thick-thin crustal interfaces. Igneous differentiation produced effusive products ranging from basalt through andesite to rhyolite, in that general order of abundance. The role of sialic contamination in development of felsic differentiates has not been adequately tested but may have been slight. Within the orogen and remote from sialic forelands, tholeiitic basalt was apparently the predominant volcanic effusive, to the exclusion of felsic differentiates. This apparent restriction of felsic differentiates to the vicinity of sialic forelands may reflect, in addition to deep-fracture control at thick-thin crustal interfaces, a more advanced stage of igneous differentiation in high-level magma chambers at topographically higher, near-shore sites. Conversely, the central part of the orogen with predominant tholeiitic basalt and fine-grained sediments may reflect deeper water basinal sites at lower elevation resting upon thinner mafic crust.

Construction of volcanic complexes apparently proceeded in epidsodic progression, the axis of successive volcanic accumulation shifting away from forelands and towards the axis of the orogen. Accordingly younger assemblages appear to lie towards the axial zone. This distribution pattern may reflect progressive tectonic spreading of sialic forelands possibly a manifestation of crustal floor spreading in Archean time. Accordingly the apparent preference of Abitibi volcanic complexes for the margins of the forelands may correspond to the world-wide distribution of currently active volcanoes which are mainly in areas where the moving sea-floor turns down under the continents (Heirtzler, p. 9).

As previously stated, metallogenic patterns are intricately related to volcanic patterns in the orogen. Many deposits, e.g. asbestos, Cu-Ni, Mo-Bi, Li, Cu-Zn-Au-Ag and iron are directly enclosed in original lithic hosts either intrusive or extrusive. Other deposits, e.g. Au-Ag, have been involved in structural deformation with more or less migration of components to favourable sites. On a regional scale, there is little doubt that mineralization represents an integral part of the tectonic evolution of the orogen.

CONCLUSIONS

The east-trending Abitibi orogen, 500 miles long by 60 to 100 miles broad, is bounded east and west by northeast-trending crystalline rocks of the Grenville province and Kapuskasing subprovince respectively. The orogen is intracratonically contained between northern and southern predominantly granitic forelands. Its present S-shaped form reflects severe structural deformation, dominated internally by tight isoclinal east-trending folds. The original orogen constituted a much longer and in all likelihood, broader mobile belt.

Based on planimeter survey of the geologic map, the present orogen is 37,000 square miles in area. It is underlain by 32.3 per cent granitic rocks, 45.7 per cent mafic volcanics, 3.6 per cent felsic volcanics, 16.0 per cent sediments and 2.5 per cent mafic intrusions. The proportion of granitic rocks, in the form of large batholiths, is higher in the northern parts of the orogen, a fact supported by regional gravity anomalies.

Archean supracrustal rocks, of predominantly orogenic type, are products of tectonically unstable, thin-crustal, mobile environments. Mafic to felsic volcanic sequences of predominant calc-alkaline affinity correspond in significant degree to those of modern island arcs situated at continent-ocean interfaces. Accompanying clastic sediments including volcanogenic and flyschoid facies have the essential characteristics of turbidite associations including greywacke-conglomerate assemblages, graded bedding, soft rock slump structures and other evidence of rapid, high energy, "poured-in" accumulation histories.

Iron formations, widely distributed in the regional assemblage, comprise oxide, carbonate and sulphide facies. The facies are arranged across the orogen in shelf to basin transitions corresponding to original basin bathymetry. Regional stratigraphy is dominated by the presence of numerous semi-independent, elliptical volcanic masses, each with a mafic to felsic extrusive sequence, cogenetic intrusions and sediments. Of eleven such volcanic complexes within the orogen four lie close to the northern foreland and six close to the southern foreland.

Mineral occurrences and ore deposits, mainly Au-Ag, Cu-Zn-Ag, Cu-Au-Ag, Ni-Cu, Mo-Bi-Li, asbestos and iron, preferentially lie within volcanic complexes, where they are commonly associated with felsic rocks. Cu-Zn-Au-Ag deposits favour deeper water lithic associations marked by felsic volcanic concentrations, sulphide facies iron formation and scarcity or absence of coarse clastics. Many Au-Ag occurrences are associated with shear zones in shelf facies with felsic concentrations, clastic sediments and carbonate facies iron formation. Asbestos and Ni deposits, are directly·associated with mafic intrusions.

Abitibi orogen is considered to have been developed intracratonically between spreading sialic forelands. Igneous differentiation leading to development of mafic to felsic volcanic sequences occurred within the orogen under presumed deep-fracture control in proximity to sialic forelands which represented the thick-thin crustal interfaces of Abitibi time. The role of sialic contamination in development of felsic differentiates is uncertain but may have been minimal. Accumulation of supracrustal assemblages apparently proceeded in episodic progression, with the younger rocks being distributed towards the axis of the orogen.

Tectonic development may be attributed to either, 1) intracratonic ocean-floor spreading in response to mantle convection in the manner of modern plate tectonics, or, 2) withdrawal of deep-crustal support with consequent downsinking and progressive filling of the resulting mobile belt or basin complex in the manner of conventional geosynclinal development. Analogy with modern crustal architecture lends some support to the concept of Archean ocean-floor spreading. However a great deal of additional work would be required to test the hypothesis.

ACKNOWLEDGMENTS

We have drawn freely upon the following published maps in compiling the geologic map and preparing other illustrations of the Abitibi orogenic belt: Ontario Department of Mines - Timmins Kirkland Lake Sheet (Map 2046), Coral Rapids-Cochrane Sheet (Map 2161), Chapleau-Foleyet Sheet (Map 2116); Department of Natural Resources, Quebec-Noranda, Matagami, Val d'Or and Chibougamau area (Maps 1600-iii, iv and v); Geological Survey of Canada - Mineral Deposits of Canada (Map 1252A); and Observatories Branch of Department of Energy, Mines and Resources, Ottawa - Gravity Map of Canada (Map GMC 67-1).

We gratefully acknowledge the value of discussions on specific topics with G. Duquette, P. Sauvé, W.M. Schwerdtner, A.O. Sergiades, J I. Sharpe and M. Latulippe.

All illustrations were prepared by F. Jurgeneit, University of Toronto.

REFERENCES

Baragar, W.R.A.
    1968:   Major element geochemistry of the Noranda volcanic belt, Quebec-Ontario; *Can. J. Earth Sci.*, vol. 5, pp. 733-790.

Baragar, W.R.A., and Goodwin, A.M.
    1969:   Andesites and Archean volcanism of the Canadian Shield; *Proc. of the Andesite Conf., Intern. Upper Mantle Project*, Sc. Rept. 16, pp. 121-142.

Duquette, G.
 in press: Archean stratigraphy and ore relationships in the Chibougamau district; Ministère Richesses Naturelles, Quebec.

Eade, K.E., Fahrig, W.F., and Maxwell, J.A.
    1966:   Composition of crystalline rocks and fractionating effects of regional metamorphism; *Nature*, vol. 211, No. 5055, pp. 1245-1249.

Goodwin, A.M.
    1965:   Mineralized volcanic complexes in the Porcupine-Kirkland Lake-Noranda region, Canada; *Econ. Geol.*, vol. 60, No. 5, pp. 955-971.

    1968:   Archean protocontinental growth and early crustal history of the Canadian Shield; *XXIII Intern. Geol. Cong. Proc.*, vol. 1, pp. 69-89.

Goodwin, A.M., and Shklanka, R.
  1967:   Archean volcano-tectonic basins: form and pattern; *Can. J. Earth Sci.*, vol. 4, pp. 777-795.

Gunn, B.M.
  1969:   Geochemistry of Archaean metavolcanics, Chibougamau; *in* Report on Volcanology, *Can. Geophys. Bull.*, vol. 21.

Heirtzler, J.R.
  1968:   Sea-floor spreading; *Sci. Am.*, vol. 219, No. 6, pp. 60-70.

Holubec, J.
  in prep.: Lithostratigraphy, structure and deep crustal relations of Archean rocks, Rouyn-Noranda area, Quebec.

Moorhouse, W.W.
  1970:   Atlas of textures in Archaean and modern volcanic rocks; *Geol. Assoc. Can.*, SP 8.

Ridler, R.H.
  1970:   Relationship of mineralization to volcanic stratigraphy in the Kirkland-Larder Lakes area, Ontario; *Proc. Geol. Assoc. Can.*, vol. 21, pp. 33-42.

Roscoe, S.M.
  1965:   Geochemical and isotopic studies, Noranda and Matagami areas; *Can. Inst. Mining Met.,Trans.*, vol. LXVIII, pp. 279-285.

Sharpe, J.I.
  1965:   Field relations of Matagami sulphide masses bearing on their disposition in time and space; *Can. Inst. Mining Met., Trans.*, vol. LXVIII, pp. 265-278.

Wanless, R.K., Stevens, R.D., Lachance, G.R., and Edmonds, C.M.
  1968:   Age determinations and geological studies; *Geol. Surv. Can.*, Paper 67-2, Pt. A.

Van Bemmelen, R.W.
  1949:   The geology of Indonesia; Government Printing Office, The Hague, 732 pp.

20

Reprinted from *Canadian Mining Metallurg. Bull.*, 1045, 1054 (Sept. 1968)

Geological Observations on the Delbridge Massive Sulphide Deposit

J. Boldy

Kennco Explorations (Canada) Ltd.

ABSTRACT

This presentation is based on observations made on a recently discovered massive sulphide deposit in the Noranda district of Northwestern Quebec. The essential components of the ore deposit are described and several self-explanatory illustrations are provided.

The Delbridge deposit occurs within a complex assemblage of porphyritic siliceous lavas, pyroclastics and intrusives of early Precambrian age. The key geological feature is the presence of a lenticular rhyolite breccia pile intercalated within the stratigraphically upper section of the porphyritic siliceous volcanic sequence. The breccia pile overlies a former eruptive center.

The massive sulphide deposit is strata-bound and is located on the southern flank of the breccia pile. The massive sulphides are confined to a chert (sinter) unit which itself is localized in its distribution, occurring at the interface between a footwall chlorite alteration zone and the ore-bearing breccia horizon intercalated within the pile.

Pyrite, sphalerite and chalcopyrite are the principal sulphides present in the ore. Minor amounts of galena, tetrahedrite and native silver also occur. Of particular interest is the close association of the massive pyrite-sphalerite mineralization to palaeogeographic depositional features exhibited by the chert unit — suggestive that metallization occurred during a lull in the explosive activity of the Delbridge breccia.

Although the deposit is of modest proportions, it nevertheless exhibits certain key features common to other massive sulphide deposits, in addition to features not known elsewhere in the district. It is distinguished by its high zinc to copper ratio and the high silver content of the ore.

At Delbridge, the relationship of a massive sulphide deposit to a specific center of explosive siliceous volcanism is now documented. This deposit is offered as an example of metallization of a flank fissure which was the site of solfataric activity and within which the various metals were rhythmically precipitated. Preliminary observations suggest that the deposit is of volcanic exhalative origin, active during Precambrian time.

JULIAN BOLDY is a graduate of Trinity College, Dublin, where he obtained his M.A. and M.Sc. degrees in geology. He immigrated to Canada in 1956 and, for the following twelve years, served as an exploration geologist for the Falconbridge Group of companies.

From 1957 to 1962, Mr. Boldy was on the exploration staff of Giant Yellowknife Mines Ltd. Between 1962 and 1967, he was based at Noranda, Que., where he was involved in the discovery and development of the Delbridge deposit. In 1966, he was appointed resident geologist for Falconbridge at Noranda.

Early in 1968, he joined Kennco Explorations (Canada) Limited, and is now working out of their Toronto office.

THE PAPER WAS SUBMITTED: in March of 1968. It will appear in the Transactions of the Institute.

KEYWORDS IN THIS PAPER: Delbridge deposit, Massive sulphides, Strata-bound sulphides, Sulphide mineralization, Mine geology, Breccia pile, Sphalerite, Chalcopyrite, Paragenesis, Chert.

INTRODUCTION

THE NORANDA MINING DISTRICT of Northwestern Quebec is one of the greatest metal-producing centers in the Canadian Precambrian Shield. Of particular geological interest is the occurrence of a number of massive sulphide deposits containing appreciable amounts of copper, zinc, silver and gold within an Archean volcanic environment (*Figure 1*). Since production commenced in 1927, a total of 82 million tons of ore have been extracted from twelve mines and have yielded 1,800,000 tons of copper, 700,000 tons of zinc, 17,-000,000 ounces of silver and 10,000,000 ounces of gold. Approximately 80 per cent of this production has come from the Horne-Quemont complex.

[*Editor's Note:* Material has been omitted at this point.]

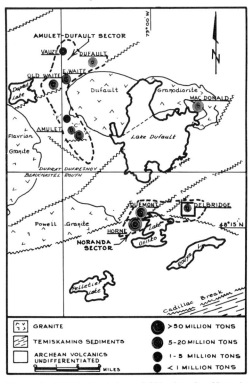

Figure 1.—Principal massive sulphide deposits, Noranda district.

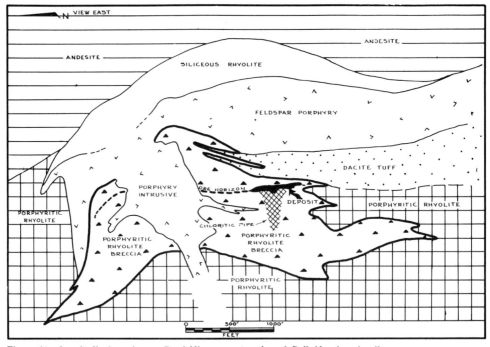

Figure 11.—Longitudinal section — Pre-folding reconstruction of Delbridge breccia pile.

CONCLUSIONS

In all probability, the elongated aspect of the Delbridge ore deposit represents an infilling of an original trough or valley, which was the site of a flank fissure along which solfataric activity occurred, and within which the various metals were rhythmically precipitated with the chert. Preliminary observations suggest that the orebody is of volcanic exhalative origin and was emplaced in a relatively shallow but rapidly changing environment, located on the flanks of a rhyolite eruptive center which was active during Precambrian times and subsequently tilted to its present position during folding.

ACKNOWLEDGMENTS

The writer would like to thank the management of Delbridge Mines Limited, in addition to Mr. G. P. Mitchell, exploration vice-president for Falconbridge, for permission to publish this paper. I would also like to thank Dr. W. G. Robinson and the late Dr. A. S. Dadson for suggestions and criticisms of the text, and for the provocative discussions we had on numerous occasions. Thanks are also due to Mr. U. Jarvi for his underground geological mapping, Mr. R. Buchan for his mineralogical examination of drill-core samples and Mr. A. Brewer for drafting the illustrations.

Furthermore, I am indebted to the various geologists — past and present — whose ideas have consciously or sub-consciously affected me during my study of this fascinating mining district.

SELECTED BIBLIOGRAPHY

Boldy, J. (1967), Delbridge Mines Ltd., *C.I.M. Centennial Field Excursion Volume*, Northwestern Quebec and Northern Ontario, pp. 58-61.
Buchan, R. (1967), mineraligical examination of D'Eldona samples, Falconbridge company report #536.
Dresser, J. A., and Dennis, T. C. (1944), *Geology of Quebec*, Vol. II, Descriptive Geology; Quebec Department of Mines, Report No. 20, pp. 74-113.
Dresser, J. A., and Dennis, T. C. (1949), *Geology of Quebec*, Vol. III, Economic Geology; Quebec Department of Mines, Report No. 20, pp. 337-383.
Dugas, J. (1966), "The Relationship of Mineralization to Precambrian Stratigraphy in the Rouyn-Noranda Area, Quebec," *Precambrian Symposium, Geol. Assoc. Canada*, pp. 43-54.
Gilmour, P. (1965), "The Origin of the Massive Sulphide Mineralization in the Noranda District, Northwestern Quebec, *Geol. Assoc. Canada*, vol. 16, pp. 63-80.
Hodge, H. J. (1967), Horne Mine, Noranda Mines Ltd., *C.I.M. Centennial Field Excursion Volume*, Northwestern Quebec and Northern Ontario, pp. 40-45.
Purdie, J. J. (1967), Lake Dufault Mines Ltd., *Centennial Field Excursion Volume*, Northwestern Quebec and Northern Ontario, pp. 52-57.
Roscoe, S. M. (1965), "Geochemical and Isotopic Studies, Noranda and Mattagami Areas," *C.I.M. Bulletin*, Vol. 68, No. 641, pp. 965-971.
Sharpe, J. I. (1967), "Metallographic Portrait of the Noranda Area," *C.I.M. Centennial Field Excursion Volume*, Northwestern Quebec and Northern Ontario, pp. 62-63.
Spence, C. D. (1967), "The Noranda Area," *C.I.M. Centennial Field Excursion Volume*, Northwestern Quebec and Northern Ontario, pp. 36-39.
Weeks, R. (1967), Quemont Mining Corporation Ltd., *C.I.M. Centennial Field Excursion Volume*, Northwestern Quebec and Northern Ontario, pp. 46-51.
Wilson, H. D. B. (1967), Presidential Address — "Volcanism and Ore Deposits in the Canadian Archean," *Geol. Assoc. Canada*, Vol. 18, pp. 11-30.
Wilson, M. E. (1941), "Noranda District, Quebec," *Geol. Assoc. Canada*, Mem. 229.

Editor's Comments
on Papers 21 Through 24

This group of papers is concerned with early-stage metallogeny, although firm correlation is not established in any of the papers. Early-stage diapiric intrusives (and perhaps others) appear to have the ability to ingest metal from the country rock and either retain it in disseminated form or regurgitate it as concentrations. The application to Naldrett's theme is described in the previous section. The phenomenon, although undoubtedly appreciated by older generations of geologists, was novel to me. Equally novel was the application of a way of thinking evident in literature from the USSR, the consideration of the history of a particular metal, from magmatic source to crust, probably in a volcano, concentration by an intrusive, and ultimately erosion and hence dispersion or concentration in placer form.

I first noted variations in rocks of magmatic origin in diabase dykes at the Ontario–Manitoba border. On the evidence of aeromagnetic maps (N.T.S. 53K), the dykes vary in iron content according to the iron content of the belts of country rock they cross, being consistently about 100 gammas above background.

Aeromagnetic maps, geochemical maps for iron, show remarkable variations in the iron contents of suites of igneous rocks from one area to another. In the area from Noranda westward across the Ontario–Quebec border, a full suite of mafic to felsic extrusive to intrusive rocks shows no magnetic relief except over zoned quartz diorite–granodiorite diapiric intrusives. The adjoining area to the north (south of Lake Abitibi) has strong magnetic striping related to the extrusives and enhanced values related to diapiric syenite. I suspect the diapirs in the two areas had the same magmatic origin. I wonder, too, how other trace elements compare and how such geochemical signatures relate to ore. In Paper 21, Maucher adds tungsten to the list of metals that are ingested and thrown out, in this case with ore potential.

Zoning is characteristic of early-stage diapirs, and it was Lowell's appreciation of the zoning of the San Manuel deposit (Paper 22) that led to his discovery of the faulted-off Kalamazoo body and provided the guide to similar features in most of the porphyry deposits. J. M. Guilbert of the University of Arizona is his co-worker in most of these studies. Because the Laramide age of many porphyry deposits in the cordillera is that of the Cretaceous early stage of the Hellenides, for a long time I assumed that the Laramide was early stage. From the work of Scheibner and Zonenshain et al. (Papers 9 and 10), it is clear that this is not necessarily so, and the cordillera must be interpreted on the basis of rock sequences and not by equating with time elsewhere.

Alan Clark and his graduate students at Queen's, Kingston, Ontario, have shown the migration of igneous activity in the Andes of Chile and Argentina, where porphyry copper is of prime interest. Although the Laramide and early stage may dominate in porphyry coppers, they are clearly not restrictive. The essence of the work of this school appears in Paper 23, and fuller documentation is keenly awaited. In his talks on the topic, Clark notes the contemporaneity of vulcanism, plutonism, and metallization in six episodes, about 190, 95 to 105, 60, 40, 20 and 10 m.y. ago but is emphatic that these episodes refer strictly to the Copiapo transect of the Chile–Argentine Andean cordillera. Other transects show other episodes. Clark also remarks on the lack of geosynclinal development.

Legends on some metallogenic maps seem to me unduly concerned with form. In Paper 24, Gorzhevskii and Kozerenko show lucidly how form is related to local condition. Form is in fact secondary to the petrochemical nature of the host.

21

Reprinted from the *24th Intern. Geol. Congr.*, Sec. 4, 83–87 (1972)

Time- and Stratabound Ore Deposits
and the Evolution of the Earth

ALBERT MAUCHER,
F.R. Germany

ABSTRACT

The study of stratabound deposits of stibnite ($\pm$ cinnabarite) and of scheelite in different parts of the world shows that these deposits, like many other stratabound ore deposits, are coeval with their host rocks. The ore minerals here are not extraneous curiosities, but are "rock-forming minerals" in special facies of rock series having cogenetic lithology. The ores are incorporated in special time-rock units, not only on a regional but on a world-wide scale. The bulk of the scheelite deposits especially shows this timebound character in relation to metavolcanic rocks of Ordovician to Silurian age.

A large part of the "younger" tungsten deposits, associated with granites, derived their tungsten content from these older stratabound scheelite deposits during granitisation processes. The bulk of some other metals has been supplied to the earth's crust in specific geologic time periods, mostly as a result of volcanic activity. These metals do not always accumulate to economic deposits, but they may just have formed geochemical enrichments, which occasionally can be used as a time indicator. The reasons why specific metals are bound to specific geologic time periods are discussed, especially with regard to the evolution of the earth's mantle and crust and to evolutionary trends in ore deposition. The differences between deposits created by metal supplies from the mantle and those created by self-sufficiency of the crust are emphasized.

MANY STRATABOUND DEPOSITS — especially the massive sulfide ores in volcanic-sedimentary complexes and the lead-zinc-ores in carbonate rocks — are coeval with their host rocks and result from the same sedimentary and diagenetic processes. The arguments for the cogenetic development of the ore and rock minerals in these deposits are also valuable for stratabound deposits of stibnite, cinnabar and scheelite which our institute has been dealing with in the last decade. During exploration of new stibnite reserves in the mining region of Ballao-Villasalto (Sardegna, Italy), an apparently tectonically controlled ore vein system, we realized for the first time the stratabound character of the bulk of the ore (Angermeier, 1964). The ore shows two generations of typical patterns which occur in the same unit of Early Paleozoic (Ordovician-Silurian) age. This unit is made up of graphite schists, twice alternating with layers of porphyroids (metatuffs) and limestone.

The first generation of mineralization forms massive to banded streaks and lenses of mostly fine-grained stibnite without gangue minerals but with irregularly dispersed strings and patches of scheelite. It is restricted to one bed in the graphite-sericite schist just above the upper porphyroid layer. Based on tectonic

---

Authors' addresses are given at the back of this book.

measurements, proving that ore and host rock have undergone the same tectonic deformations, of which the oldest are of Variscan age, the ores must be pre-Variscan. Any genetic relationship to the partially syn-, partially post-tectonic Variscan granites of South Sardegna is herewith excluded. Locally, sedimentary fabrics (rhythmic bedding, glide folds, etc.) are still recognizable in parts of the massive and the banded ore beds.

The second generation forms small, partially brecciated, short fissure veins without great economic importance. The fissures, controlled by the Variscan tectonics, occur in the whole rock unit and are mainly filled with calcite or quartz, depending on the lithology of the transected rocks. They contain stibnite only near the stratabound ores, the ore content being restricted to the small rock unit consisting of the ore stratum, the porphyroid layer and the limestone just below. No ore has ever been found in the veins above the ore stratum or underneath the lower porphyroid horizon. The veins are not continuous feeder veins leading to an ore source in the depth, but metamorphic veins owing their mineral content to leaching of constituents of the nearby rocks into the veins. Sometimes scheelite is also found in these veins very near the primary ore, never having been mobilized over a great distance.

The stratabound character of the stibnite, its paragenesis with scheelite and the special unit of alternating submarine volcanic and sedimentary rocks induced us to look for similar or corresponding deposits. A remarkable number of stratabound stibnite deposits, all of them bound to very similar volcanic-sedimentary series of stratified rocks, occur in Austria, Czechoslovakia, Germany, Greece, Italy, Spain and Turkey. Besides the great congruency of all these deposits, there are some peculiarities and differences in the rock-facies and the paragenesis of the ore minerals in some of them. For example, not all contain scheelite, the pyrite content is very variable, many are just stibnite deposits, some have alternating strata of stibnite and cinnabar, and others contain cinnabar as the only ore mineral. Our research work is still going on, but we have already published some of our results (Maucher, 1965; Höll, 1966; 1970 a; Saupe, 1967; Maucher, and Saupé, 1967; Fellerer, 1968; Lausen, 1969; Schulz, 1969). One of the most surprising results is that all these deposits are not only strata-bound to very similar rock units related to metavolcanics, but that most of these ore-bearing units are of Lower Paleozoic age, ranging between Upper Cambrian and Upper Silurian. The ore deposits are not only bound to a stratum but to a time-rock unit — they are timebound. After having recognized that in a special region the ore content of a metamorphic rock unit can be used for dating the rocks and that some rock units contain "geochemical-stratigraphical marker horizons" (Maucher and Höll, 1968) with Sb, Hg, W (Mo) contents considerably augmented above their clarke, we started to examine the metal contents of special metamorphic rock units of the Alps of debatable, presumably Lower Paleozoic, age. Because the Sb and Hg content of rocks is very easily mobilized and dispersed during metamorphism, we concentrated our interest on the relatively immobile tungsten content, especially the content of scheelite. Our studies were started in the region of the scheelite-magnesite deposit, Tux, in Austria (Höll and Maucher, 1967) and have been extended through the Austrian Alps over a range of more than 500 km from Grisons, Switzerland in the west to Styria, Austria in the east. Over this distance we found more than 50 unknown scheelite occurrences in the rock units in question (Höll, 1970 b, 1971) and new ones may still be found. In some occurrences, the scheelite content is very low and finely dispersed in the basic metavolcanics or metasediments; in others it forms strings and impregnations of economic value. The metamorphic scheelite-bearing rocks are very similar to those described as "tactites". However, our ore-bearing rock units are not contact metamorphic and most of them are far from granites or granite contacts; the overwhelming majority do not contain any carbonate rocks or minerals.

Scheelite is a primary rock mineral in the volcanics and sediments, changing its mineralogical behaviour according to the grade of metamorphism, thus forming very small grains in the greenschist facies, but porphyroblasts of 10- to 20-mm size in rocks of the amphibolite facies.

This dependence is evident in the whole region and very well observable in the most important of the newly found deposits, the deposit of Felbertal near Mittersill (Federal state of Salzburg, Austria), which is well exposed because of much drilling and underground exploration work. There the ore-bearing unit consists of a metamorphic rock sequence of 150 to 250 m thickness, depending on the configuration and the relief of the former marine basin in which the sediments and volcanics had been deposited. The contact of the unit with the underlying schists (graphite, biotite and muscovite schists with lydites and limestones) is sharp. No tungsten has been found in these underlying rocks. The ore-bearing unit itself is characterized by rocks rich in amphiboles. There are two cyclic repetitions of dark, coarse-grained amphibolites rich in pargasitic-edenitic amphiboles with Cr contents, which grade into hornblende schists, hornblende prasinites and biotite-albite schists of the lower greenschist facies. The main ore bed is bound to the upper part of the lower "hornblende series", containing two very fine grained horizons rich in silica or banded tuffite-quartzite horizons. The ore consists of impregnations, strings and patches of scheelite (with little tungstenite) of varying grain size according to the metamorphic facies of the ore-bearing rock beds. The higher metamorphic facies contain coarser grained zones and, besides the porphyroblastic scheelite, small fissures with scheelite and quartz also occur.

There are great similarities between this deposit and the wellknown scheelite deposit of Sandong, South Korea (So, 1968), where the ore is stratabound, time-bound, genetically related to metavolcanic layers interbedded with metapelitic strata of the same age and restricted to six very extensive but relatively thin beds. These beds differ from the normal metapelitic beds of the Myobong formation in their characteristically high amphibole content; the scheelite content varies with the amphibole content. The main ore bed still shows typical sedimentary fabrics. Small metamorphic fissure veins extending only a few decimeters into the capping or underlying bed, containing quartz and scheelite, transect the rocks but are only ore-bearing adjacent to the six ore-bearing horizons.

The ore beds and fissures contain sporadically very small amounts of pyrrhotite, chalcopyrite, magnetite, ilmenite, hematite, molybdenite, bismuth, bismuthinite and joseite (Bi_4Te_2S).

In the Felbertal region, the scheelite-bearing unit occasionally also contains other ore minerals worth mentioning: pyrrhotite, magnetite, hematite, ilmenite (mostly metamorphosed to rutile or anatase), pentlandite, chalcopyrite, tetrahedrite, sphalerite, molybdenite, tungstenite, bismuth, bismuthinite, silver and gold rich in silver (electrum), pyrite, and rare arsenopyrite and cassiterite; crystals of beryl (emerald or aquamarine) have also been found. Because Be contents are also mentioned from the very similar stratabound scheelite deposits of the Pyrénés (Guitard, and Lafitte, 1958; Fonteilles, and Machairas, 1968), a systematic geochemical investigation of the ore-bearing series and the nearby rocks by spectral analysis and neutron activation is being carried out. The contents of Be, W, Mo, Bi, Ag and Au are typically augmented above their clarke only in the ore unit. The relationship of these elements to the metavolcanics (especially to the very basic ones) is evident.

257

Not all the scheelite deposits of the Austrian Alps lie in rocks of the greenschist or amphibolite facies. Some of them are found in the regions of higher metamorphism, in corresponding rock units, as for instance the gold-pyrite-scheelite deposit of Schellgaden, in the so-called "central-gneisses" which are rich in amphibolites, hornblende gneisses and migmatites. The ores occur in banded lenses rich in quartz and sulfides, parallel to the schistosity. In other parts of the granite-gneisses, scheelite is found only in fissure veins together with quartz, but these veins also occur just in special units of the partially granitized rocks.

From all our stratigraphical, lithological, geochemical and tectonic investigations, we can conclude: The scheelite occurrences in the crystalline rocks of the Alps are stratabound and timebound to a rock series of Lower Paleozoic age. Tungsten, gold, silver, bismuth and beryllium are genetically related to basic volcanism, especially to its rest solutions rich in silica. All the ore-bearing strata show the same tectonic and metamorphic influences as their geologic frame. According to the grade of metamorphism, ranging from the greenschist facies to partially anatectic granite-gneisses, scheelite forms respectively impregnations, strings and patches in greenschists, porphyroblasts combined with pegmatoid lenses or fissure veins with quartz in amphibolites, bands, lenses and veins with quartz (with or without gold) in gneisses. The accompanying elements, Au, Ag, Bi, Cu, Be, etc., are partially enriched with the scheelite; they form, in places, separate metamorphic hydrothermal enrichments. Complete anatexis and granitization would lead to a granitic magma and magmatic ore deposits of tungsten, gold, silver, copper, bismuth and beryllium, etc. during differentiation, thus possibly even creating genuine contact-metasomatic scheelite tactites.

The observations and conclusions about the scheelite occurrences in the Alps are also valuable for other distinctive types of mineralization accompanying granites and for the relation of mineral deposits to regional geology in many other regions of the world. They may answer some questions raised by (Park, 1964) and (Krauskopf, 1971), concerning the scheelite deposits in the proximity of the Nevada batholith, and give some hints about the source of ore metals. Our observations show that scheelite is not restricted to granite marble contacts, that scheelite has been enriched in a pre-intrusive phase and that it is not necessary to drill into deep regions above the asthenosphere to find layers containing tungsten, gold, silver, bismuth, copper, beryllium, etc. having concentrations a few tens of times greater than the average for the earth's crust. Layers of this kind are also to be found in the older metamorphic rock series which contain stratabound dispersed metal contents related to volcanic activity in the proximity of the famous Variscan granites of the Erzgebirge, the Harz mountains, Cornwall and Spain. The granites are granitization products of these pre-existing metal-bearing rocks and any granite magma is only capable of forming ore deposits if the pre-existing rock series contained the appropriate metals. Insofar as deep-seated volcanism is the source of ore metals in the earth's crust, the heterogeneous distribution of ore deposits depends on the natural tectonic and magmatic laws of distribution of volcanic activity in time and space. The fact that the volcanic activity does supply the same elements in volcanic belts all over the world during the same volcanic period, but that there are maxima of special elements at special times, leads to the conclusion that the metals are derived from laterally homogeneous regions of the mantle, but that these regions during the course of time show distinct differences of special metal contents due to differentiation processes in the mantle. This may, besides the different stages in the development of the earth's crust and their tectonic history, also be one of the causes of evolutionary trends in ore deposition (Pereira and Dixon, 1965).

258

REFERENCES

Angermeier, H.-O., 1964. Die Antimonit-Scheelit-Lagerstätten des Gerrei (Südostsardinien, Italien) und ihr geologischer Rahmen. Inaug.-Diss., 62 p., Univ. Munich.

Fellerer, R., 1968. Geologische und lagerstättenkundliche Untersuchungen zwischen Passo Cereda und Forcella Aurine (südliche Pala-Gruppe-Norditalien). Inaug.-Diss., 78 p, Univ. Munich.

Fonteilles, M., and Machairas, G., 1968. Eléments d'une description pétrographique et métallogénique du gisement de scheelite de Salau (Ariège). Bull. du Bureau de Recherches géologiques et Minières, Section II, 3, p. 63-85.

Guitard, G., and Lafitte, P., 1958. Les cacaires métamorphiques et es skarns du Pic de Costabonne (Pyrénées-Orientales). Science de la Terre, 6, p. 59-137, Nancy.

Höll, R., 1966. Genese und Altersstellung von Vorkommen der Sb - W - Hg - Formation in der Türkei und auf Chios/Griechenland. Bayer. Akad. Wiss. math. - naturw. Kl. Abh, n. F. 127, 118 p.

Höll, R., and Maucher, A., 1967. Genese und Alter der Scheelit-Magnesit-Lagerstätte Tux. Sitzungsber. Bayer. Akad. Wiss. math.-naturw. Kl, p. 1-11.

Höll, R., 1970a. Die Zinnober-Vorkommen im Gebiet der Turracher Höhe (Nock-Gebiet/Österreich) und das Alter der Eisenhut-Schieferserie. N. Jb. Geol. Paläont. Mh., p. 201-204.

Höll, R., 1970b. Scheelitprospektion und Scheelitvorkommen im Bundesland Salzburg/Österreich. Chemie der Erde, p. 185-203.

Höll, R., 1971. Scheelitvorkommen in Österreich. Erzmetall, 24, p. 273-382.

Krauskopf, K. B., 1971. The source of ore metals. Geochim. Cosmochim. Acta, 35, p. 643-659.

Lahusen, L., 1969. Die schicht- und zeitgebundenen Antimonit-Scheelit-Vorkommen und Zinnoberverezungen der Kreuzeck- und Goldeckgruppe in Kärnten und Osttirol, Österreich. Inaug. - Diss., 139 p., Univ. Munich.

Maucher, A., 1965. Die Antimon-Wolfram-Quecksilber-Formation und ihre Beziehungen zu Magmatismus und Geotektonik. Freiberger Forschungshefte, C 186, p. 173-188.

Maucher, A., und Saupé, F., 1967. Sedimentärer Pyrit aus der Zinnober-Lagerstätte Almadén. Miner. Deposita, 2, p. 312-317.

Maucher, A., und Höll, R., 1968. Die Bedeutung geochemisch-stratigraphischer Bezughorizonte für die Altersstellung der Antimonit-lagerstätte von Schlaining im Burgerland, Österreich. Miner. Deposita, 3, p. 272-285.

Park, Ch. F., Jr., 1964. Is geology field work obsolete? Econ. Geol., 59, p. 527-537.

Pereira, J., and Dixon, C. J., 1965. Evolutionary trends in ore deposition. Trans. Inst. Mining and Metall., 74, p. 505-527.

Saupé, F., 1967. Notre préliminaire concernant a genèse du gisement de mercure d'Almaden, Province de Ciudad Real, Espagne, Miner. Deposita, 2, p. 36-33.

Schulz, O., 1969. Schicht- und zeitgebundene paläozoische Zinnober-Vererzung in Stockenboi (Kärnten). Sitzungsber. Bayer. Akad. Wiss. math. - naturw. Kl., p. 113-139.

So, Chil-Sup., 1968. Die Scheelit-Lagerstätte Sangdong. Inaug. - Diss., 71 p., Univ. Munich.

22

Geology of the Kalamazoo Orebody, San Manuel District, Arizona

J. DAVID LOWELL

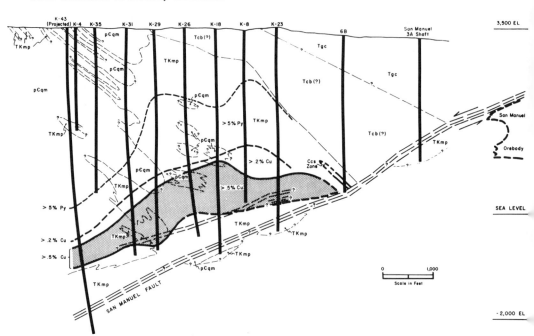

FIG. 5. Longitudinal section of Kalamazoo orebody looking northwest.

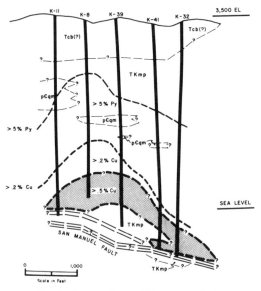

FIG. 6. Cross section of Kalamazoo orebody looking northeast.

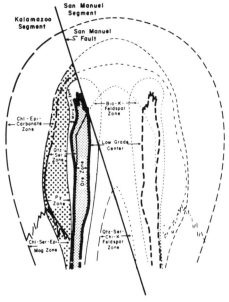

FIG. 7. Schematic section of San Manuel-Kalamazoo orebody

23

Reprinted from *Canadian Mining Metallurg. Bull.*, 65(719), 37 (1972)

THE EVOLUTION OF A METALLOGENIC PROVINCE AT A CONSUMING PLATE MARGIN: THE ANDES BETWEEN LATITUDES 26° AND 29° SOUTH

A. H. Clark and M. Zentilli

Queen's University, Kingston, Ontario

Significant metallic mineralization throughout the Mesozoic-Cenozoic Andean Province of northern Chile and northwestern Argentina has consistently been related to the emplacement of granitic plutons or to andesitic-to-rhyolitic volcanism. Between latitudes $26°$ and $29°$S epigenetic, hydrothermal deposits of copper and other metals formed immediately following the intrusion of felsic stocks of Upper Jurassic (140-155 m.y.), Middle Cretaceous (90-120 m.y.), Lower Paleocene (60-67 m.y.), Upper Eocene (39-44 m.y.), Oligocene-Miocene (22-23 m.y.) and Miocene-Pliocene (5-12 m.y.) age.

The Cenozoic intrusions, associated with essentially contemporaneous volcanics, exhibit local prophyritic facies, and mineralization was concentrated close to the intrusive-extrusive interface. Ore deposits of the porphyry copper clan occur in all post-Mesozoic intrusive provinces, but major centres have so far been recognized only in Upper Eocene and Miocene-Pliocene stocks. Disseminated copper and copper-silver deposits, stratabound but not strictly syngenetic, have developed in marine and continental andesites of Middle Jurassic to Cretaceous age, and in continental Tertiary rhyolites. Large bodies of magnetite, of apparent contact metasomatic origin, are, with few exceptions, restricted to sequences of andesite intruded by Middle Cretaceous plutons.

From the initiation of the Andean orogen, in Triassic-Jurassic times, to the mid-Tertiary, the locus of epizonal magmatism and associated mineralization has migrated east-southeastward at a gradaully increasing rate (from ca. 0.06 to 0.10 cm/yr), to form a series of discrete long-itudinal metallogenetic sub-provinces, each of which evolved within a specific metallogenetic phase. This systematic progression was abruptly terminated in the Oligocene, and, by the Miocene, calc-alkaline volcanism was taking place simultaneously in several areas across a 250-km-wide belt to the east of the Eocene sub-province. Subsequently from the Pilocene to the Recent, volcanism has been focused at an intermediate position to form the strato-volcanoes of the High Andes. Porphyry copper mineralization associated with this magmatic activity in Argentina has possibly shown a parallel westward recession since the Miocene.

The emplacement of high-level igneous rocks and hydrothermal ore deposits in this region is considered to have resulted from the generation of magmas through subduction of oceanic lithosphere beneath continental South America. The structural framework of a Benioff zone may have been imposed in the Jurassic, but the present general configuration of the continental margin probably derives from a major re-orientation of plate motions in the Miocene.

The pre-Mesozoic basement, in general sparsely mineralized, appears to have contributed little to the development of the igneous rocks and mineralization of the overprinted Andean mobile belt at this latitude. The recurrent copper specialization of this metallogenetic province reflects a consistency in the source materials and processes of sub-crustal magma generation.

24

Reprinted from *Intern. Union Geol. Sci.*, Ser. A, No. 2, 161–165 (1970)

On the Depth Problem of Postmagmatic Deposits

(Abstract)

D. I. Gorzhevskii & V. N. Kozerenko

U.S.S.R.

Several different methods have been suggested for the determination of the depth of formation of postmagmatic deposits: (1) geological (tectonostratigraphic); (2) determination of the depth of formation of magmatic rocks, with which the deposits are associated; (3) structural; and (4) mineralogical, including the use of gaseous-liquid inclusions in the minerals. A number of important theoretical studies has been devoted to the discussion of the physico-chemical side of this problem.

Geological methods can be applied only to deposits located in well studied areas, where stratified rock masses of sedimentary and sedimentary-volcanic deposits are developed. Furthermore, each of the methods is not definitive in certain aspects and further development of the geologic, mineralogic-structural methods of depth determination is necessary. Special mention should be made of the necessity of geobarometric research and determination of the size of pre-ore cavities and voids.

A tabular summary of the methods of determining the depth of postmagmatic ore deposits is presented. The summary presents the following depth zones: (1) near surface — less than 0.5 km, (2) subvolcanic — 0.5 to 1.0 km- (3) hypabyssal — 1 to 2 km, (4) lesser mesoabyssal — 2 to 3.5 km, (5) greater mesoabyssal — 3.5 to 6 km and (6) abyssal — 6 to 8 km. The geologic, mineralogic and structural characteristics that are presented for each depth zone are: (1) skarnization, (2) propylitization, (3) argillization, silicification and greisenization, (4) sulfides, (5) oxides and sulfates, (6) ore structures, (7) ore textures and (8) main types of enclosing structures. Examples of deposits in each depth zone are also given.

Table 1. Main features of depth facies in postmagmatic deposits

Depth facies of the top parts of ore bodies. Assumed depth of incipience (in km)	Near-surface <0.5	Subvolcanic 0.5 1.0
I. Main types of hydrothermal-metasomatic processes		
1. Skarnization (formation of calcareous-silicate skarns)	Not characteristic	Sometimes found in the composition of garnet. Ion Ca^{2+} is predominant
2. Propylitization and green-rock alteration	Propylitization near the surface (plagioclase, zeolite, chlorite, calcite)	Subvolcanic propylitization (plagioclase, adularia, chlorite, calcite)
3. Argillization, formation of secondary quartzites, greisenization	Solfatara argillization, alunitization	Argillization and formation of secondary quartzites
II. Peculiarities of the mineralogical composition of the ores		
1. Sulphides	In ore bodies sulphides play a subordinate role. Occurrences of cinnabar, realgar, stibnite, pyrargyrite and proustite	Often found as hypogene-melnikovite and marcasite, wurtzite, sulfosalts
2. Oxides and sulfates	Widely developed; magnomagnetite is characterized by very subordinate amount of ferrous iron	Occurrences of barite and hypogene gypsum. Hematite abundant. Magnomagnetite with predominance of ferrous iron over magnesium
III. Specific features in the structure and texture of ores		
1. Ore structures	Disseminated ores are predominant	Often colloform-eutaxitic structures
2. Ore texture		Characteristic are gel, crypto-crystalline and perlitic textures

(table continues)

263

Table 1. (cont.)

Hypabyssal 1–2	Mesoabyssal		Abyssal 6–8
	lesser depth 2–3.5	greater depth 3.5–6	
Characteristic. Ions Mn²⁺ and Fe²⁺ are often present in the garnet	Slightly characteristic. Ions Fe²⁺ and Mg²⁺ often present in garnets	Not characteristic	Not characteristic
Hypabyssal propylitization: albite, epidote, chlorite, calcite, quartz, pyrite	Green-rock alteration: albite, zoisite, clinozoisite	?	?
Formation of secondary quartzites	Greisenization	?	?
Sulfides in ore bodies are often predominant			
Often predominance of pyrite, occurrences of hypogene marcasite, grey copper ores	Along with pyrite and grey copper ores pyrrhotite and arsenopyrite are appearing		
Occurrences of barite and hematite. Magnomagnetite has a minimum magnesia content in respect to ferrous iron	Hypogene sulfates are not characteristic. Along with hematite magnetite is appearing		
Along with colloform-eutaxitic structures also drusy and geode are occurring	Massive structures are predominant		Eutaxitic structures widely predominant
Characteristic are fine-grained and metacolloidal structures, as well as emulsion structures of the decomposition of solid solutions	Predominance of granular structures; also reticulate and lattice structures of the decomposition of solid solutions	Characteristic are crystalline-granular structures and granular structures of the decomposition of solid solutions	

264

Table 1. (cont.)

Depth facies of the top parts of ore bodies. Assumed depth of incipience (in km)	Near-surface <0.5	Subvolcanic 0.5 1.0
IV. Main types of enclosing fracture structures	Non-persistent ruptures	Ruptures and pinnate joints. Reticulate zones of stock-works
Examples of deposits and areas	Deposits of native sulphur on Kamchatka, in Japan, New Zealand, Chile; alunites in Transcarpathia and Lesser Caucasus. Sassolin in Italy and Tibet. Mercury deposits in Transcarpathia	Gold-ore deposits of the Carpathians, Cordillera, Balei in Transbaikal region. Tin ore deposits of Malyi Khingan, Sinancha in Sikhote Alin, Oruro in Bolivia

Hypabyssal 1 2	Mesoabyssal		Abyssal 6 8
	lesser depth 2 3.5	greater depth 3.5 6	
Along with ruptures, shear fractures are appearing with breccia and leatherstone	Characteristic are shear fractures filled with leatherstone. More rarely ruptures and pinnate joints are present	Characteristic are ground shear fracture cleavages	Flow cleavage zones, more rarely fault zones
Some copper-pyrite deposits of the Southern Urals (Bliava, Sibai), Turiinsk mines in the Urals; copper-porphyry deposits in Central Kazakhstan. Some polymetallic deposits of Karamazar (Altyn-Topkan) and of Rudnyi Altai (Zmeinogorsk, Nikolaevsk, Zavodinsk). Tin ore deposits of Sikhote-Alin (Lifudzin, Khrustal'noe) and of Bolivia (Uncia). Antimony-mercury deposits of the Tien Shan (Khaidarkan, Chauvai)	Polymetallic deposits of Rudnyi Altai (Belousovsk, Berezovsk,) Karamazar, Chalaga. Tin ore deposits of Cornwall and the Erzgebirge. Tungsten-bearing skarns of Middle Asia (Langar, Koitash)	Tin ore deposits of Kalba(Chudsk), copper-pyrrhotite deposits of the Altai (Vavilonsk, Karchiginsk)	Copper, gold-ore and polymetallic deposits of old crystalline shields (Flin-Flon, Sulivan, Sherrit-Gordon, Hornin Canada; Broken Hill and Mount-Isa in Australia; Sovetskii mine in the U.S.S.R.)

265

Editor's Comments
on Paper 25

25   HOSKING
Excerpts from *Aspects of the Geology of the Tin-Fields of South-East Asia*

Tin is one of the main products associated with the intermediate (main orogenic) stage, together with tungsten, lithium, beryllium, and the rare earths. A relationship of tungsten to source beds was described by Maucher in Paper 21. The major source of tin in Texasgulf's Kidd Creek deposit, plus known geochemical provinces of tin associated with vulcanism, draws attention to comparable possibilities of source bed tin.

It is interesting to compare Hosking (Paper 25) and Zonenshain et al. (Paper 9). Zonenshain et al. showed two sequences: (1) Lower Mesozoic (T–J1) with eugeosyncline in central Kalimantan, backtrough at the edge of the Malay Peninsula, and granites of South Malaya and the tin islands, and (2) Late Cretaceous–Early Paleogene with eugeosyncline in Arakan, batholiths of Burma, and the Malayan tin-bearing granitic belt. Hosking showed tin associated with granites of Triassic and Late Cretaceous age, and noted that it seems possible, in some instances, that the Triassic granite magma assimilated earlier lodes associated with the Carboniferous granite. He found it instructive to consider Bilibin's views concerning the sequence of events. Hosking noted that much of the tin province of southeast Asia is inadequately mapped.

As I compare the details of the granite belts in Malaya (Paper 25, Figure 4) with the distribution of belts in southeast Asia as a whole, perhaps numbering as many as a dozen (Figure 7), I cannot but wonder at the outcome of detailed studies such as those by Scheibner and his colleagues in New South Wales. In this conclusion I join with Hosking. The theme was also developed by Burton (1970). Recurrent remobilization seems probable, a topic that Stoll (1965) also discussed in describing the metallogenic provinces of the Andes.

25

Reprinted from *2nd Tech. Conf. Tin, Intern. Tin Council,* London, 1–4, 20–41
(1970)

ASPECTS OF THE GEOLOGY OF THE TIN-FIELDS OF SOUTH-EAST ASIA
By

K. F.G. Hosking, A.C.S.M., D.Sc., M.I M.M., F.G.S.

Synopsis

The tin-belt of south-east Asia is over 1,800 miles in length. Its geology is complex and much more work has to be done before it is reasonably well understood and before the factors governing the nature and disposition of the various types of tin deposit found there can be stated with any real degree of confidence.

Much that is found in this paper is conjectural, which is not surprising when, for example, it is realised that published accounts of detailed studies of the mineralogic character of south-east Asian hard-rock tin deposits are virtually non-existent. Nevertheless, the paper will serve a useful purpose if it stimulates people to make good the defects.

An attempt has been made to indicate the variations of the primary and secondary tin mineralisation theme encountered in what is the most important tin province in the world, to comment on certain questions of the genesis of the deposits and on the factors which appear to have determined the distribution of these tin deposits.

The question of the relationship between the tin deposits and granite rocks is considered in some detail and the main conclusion that emerges is that Bilibin's views concerning mineralisation in mobile belts is applicable to the one under review and that in West Malaysia these views have already suggested further target areas.

Methods used in south-east Asia for the exploration and evaluation of tin deposits are briefly reviewed and, in view of the present marked interest in the offshore tin potential, a comprehensive survey in the target areas in the province under consideration is presented.

Finally the writer recommends that an international attack should be made on those problems of south-sast Asian geology which when solved are likely to facilitate the search for further tin deposits not only in south-east Asia but elsewhere in the world.

Introduction

The south-east Asian metallogenetic province is the most important tin producer in the world. The major part extends from North Burma via Peninsular Burma and Thailand and West Malaysia to the 'Tin Islands' of Indonesia (of which the most important are Banka, Billiton and Singkep) covering, in all, a distance of more than 1,800 miles. Appendages to the main belt occur in south-central Thailand and Laos: these are comparatively unimportant and will not figure largely in this paper.

267

The post-World War II literature contains a number of important papers, etc., (albeit, a limited number) which deal with one aspect or another of the tin deposits of Thailand, W. Malaysia and Indonesia, and the works of the following are particularly relevant:— Brown *et al.* (1953); Aranyakanon (1961); The Department of Mineral Resources of Thailand (1967); Kaewbaidhoon and Aranyakanon (1961); Ingham and Bradford (1960); Bradford (1961); Singh and Bean (1967); Krol (1960); van Overeem (1960); Adam (1960); Osberger (1967 (A), 1967 (B), 1968); Osberger and Romanowitz (1967). During this period little of consequence seems to have been written about the tin deposits of the other countries of the province* and one has to refer to Jones (1925) for adequate accounts in English of some of the deposits there that were investigated during the first quarter of this century, and to supplement this reading, in the case of the Burmese deposits, by reference to the excellent paper on the Mawchi mine by Hobson (1940) and to the more general accounts of Coggin-Brown (1924).

Short of writing a book it would not be possible to give a comprehensive account of this tin province whose length is more than 18 times that of the tin province of south-west England, and such a book would be far from satisfactory because much of the region is inadequately mapped and so a great deal of its complex geology is unknown or imperfectly understood. Similarly there is little point in repeating descriptions of deposits which have appeared recently in the literature (see, for example, Kaewbaidhoon and Aranyakanon (1961); Ingham and Bradford (1967); Bradford (1961) and Adam (1960).

It seems, therefore that the writer can best serve the interests of the Congress by high-lighting certain general geological features of the various types of deposit which collectively, in his opinion, may facilitate an evaluation of its tin potential.

The general geological setting

Apart, possibly, from the eastern appendages noted above, the province is the result of a complex series of events which occurred within a north-south-trending mobile belt whose age is unknown, but in which sediments were accumulating, and migmatites were locally developing in the early Proterozoic:**. The end-product of the sedimentological, igneous, metamorphic, tectonic and ore depositional events within the belt is a complex of sediments, in which calcareous members are prominent, with which are intercalated volcanics which, in West Malaysia, at least, are dominantly acid and intermediate types: the earlier components have been

---

* The best recent general account of the south-east Asian deposits is to be found in 'Tin ore resources of Asia and Australia' published by the U.N., in N. York, in 1964.

** Khoo (personal communication) believes, and has cogent reasons for believing, that in the north of Burma this geosyncline swings to the west and continues into Tibet.

subject for the most part to considerable regional metamorphism. Emplaced within the earlier sedimentary-volcanic accumulations are granites of varying ages which collectively form what may be provisionally termed an elongate composite batholith whose original surface possessed a strong topography and which extends throughout the length of the tin province. These granites often impressed remarkably little thermal metamorphism on the host rock and those that so behaved were, in Hutchison's view (private communication), emplaced in the Mesozone at depths varying from 5 to 9 miles and where the temperature of the invaded rocks approximated to that of the granitic magma.

Whilst pegmatite/aplite and other types of granitic veins are commonly associated with the granite there is a general paucity of porphyitic microgranitic dykes which in south-west England are termed 'elvans' and which there are almost constant companions of the tin lodes.

Tin and certain other metalliferous deposits are spatially closely related to some of these granitic intrusives and, in at least one instance, to granitic effusives; however, it is convenient to discuss the tin/granite relationships in some detail later. In Tertiary times much of the tin province was a mountainous belt and was part of the continental mass known as Sunda Land.

Prolonged denudation up to the Pleistocene resulted in planation, to base level, of the marginal areas of the northern and central portions of the tin belt and the virtually complete peneplanation of the southern portion now largely occupied by the islands of Singkep, Banka and Billiton and the adjacent sea areas.

It is not known when the primary tin deposits (the last of which were probably deposited not later than the Cretaceous) were uncovered, but the evidence suggests that generally it was not (surprisingly) until the Pleistocene; Osberger (1967 (B)) gives cogent reasons for believing that the formation of the Indonesian stanniferous placers commenced in the Middle to Upper Pleistocene and that some of them did not complete their development until the Holocene. Doubtless, however, some cassiterite was released somewhat earlier and, indeed, Zwierzycki (1934) records that a little cassiterite occurs in a Tertiary conglomerate *c*. 18 km west of Palembang (Sumatra). There is, however, as yet no strong reason for thinking that the known onshore Tertiary deposits or any that may occur offshore are likely to contain economically interesting concentrations of tin, and yet it is difficult for the writer to believe that nowhere, in the tin belt under review, during Tertiary times, were there primary stanniferous deposits sufficiently well-exposed to yield, at that time, by weathering appreciable concentrations of cassiterite to the overburden and drainage systems. Perhaps some of the accumulations of tin ore in the Pleistocene Older Alluvium of western Malaysia and in contemporaneous stanniferous placers elsewhere in the belt owe their presence, in part, to the reworking of Tertiary stanniferous detritus.

Major fluctuations in the Pleistocene sea-level, together with isostatic adjustments and tilting are probably the most important factors which determined the development of the stanniferous placers but it is convenient to defer further discussion of these topics until later.

Throughout its history this tin belt has been subjected to faulting and, although this aspect of its geology has not been entirely neglected, in fact, far too little is known about it to discuss it from a regional point of view. Faulting certainly played a major rôle in determining the original disposition of the primary tin deposits, and may be responsible, to no small extent, for the present broad form of the land masses of the tin belt. Certain aspects of this topic will be discussed in more detail later.

A 'recent' event, the significance of which will be apparent later, is the out-pouring of basalt in the Kuantan district of Pahang (West Malaysia) *c.* 1.8 million years ago*.

[*Editor's Note:* Material has been omitted at this point.]

* The writer is grateful to J.D Bignell of the Department of Geology and Mineralogy, Oxford, who is at present Leverhulme Research Scholar at the Department of Geology, University of Malaya, for this information and, indeed, for much more data re radiogenic ages of granitic rocks within the tin belt together with numerous discussions on this topic.

Aspects of the relationship between the primary tin deposits and the granitic rocks

In this section the writer proposes to deal with certain temporal, spatial and mineralogical/chemical relationships between the primary tin deposits of south-east Asia and the granitic rocks.

Temporal relationships

When one considers the dimensions of the granites of the tin belt it seems unrealistic to believe that they were formed during a geologically short time span. Indeed, many who have worked on the granites of Thailand and Burma have long held that two granites of distinctly ages existed there, that the two were mineralogically distinct and that tin deposits were for the most part only associated with one of them. Thus Chhibber (1934, p. 338) notes that "two types of granite have been recognised in the Mergui district (of Burma). One, a coarse porphyritic granite devoid of tin; the other, a biotite-granite which is tin-bearing. The former is the older of the two". Brown *et al.* (1953, pp. 45-46) note "there is considera-evidence to indicate that the granitic intrusive rocks of Thailand are of two distinct geologic ages and slightly different from one another mineralo-gically". They go on to say that the older intrusions are commonly hornblende-biotite granites and that in these "apparently the mineralisation of tin has been rare or absent", whilst the younger intrusions, with which "almost all of the tin and tungsten deposits of Thailand are associated" are of binary granite. Tentatively they consider the older granite to be of Triassic age and the younger to be of Late Cretaceous age: the latter are being accepted in part because Scrivenor (1931, pp. 19-23) had suggested that the Malayan granites were of Late Cretaceous or Early Tertiary age. More recently some Thai geologists have been of the opinion that the earlier granite may, locally, be more strongly tin-bearing than was originally thought (see Work and problems on tin in Thailand, Paper 25 of 'A technical Conference on tin', London 1967).

271

During the past few years radiogenic dates (potassium/argon and rubidium/strontium) of many samples of granite from south-east Asia, but particularly from West Malaysia, have been obtained. Unfortunately not all of them have yet been published and it has been stressed by Snelling, Bignell and Harding (1968), who have been much involved in this work, that it is premature to be dogmatic concerning the ages of intrusion of the granites. They are also careful to point out that potassium/argon dates may well be dates of events other than the intrusion of granite. Figure indicates radiogenic dates of Malaysian granites that have been abstracted from an unpublished map by Hutchison who, unfortunately, has not differentiated between K/A and Rb/Sr dates (Figure 4).

According to Bignell (personal communication) there is some evidence for believing that in West Malaysia there are locally granites of Carboniferous, Triassic, and Upper Cretaceous/Tertiary ages.

The granites of Bakri (Johore) and Kedah Peak, with which are associated cassiterite/tantalite-columbite-bearing pegmatites, are tentatively thought to be of Carboniferous age, as is the granite at Pahang Consolidated mine. Other granites of indicated Carboniferous age occur locally elsewhere in north-east Malaya and in the Main Range. Many samples of the Main Range granite are probably of Triassic age, and with the Triassic granite are associated the numerous vien swarms, pegmatites and occasional lodes and the major stanniferous placer deposits which have been derived from such deposits by weathering processes. Finally, the few isolated granitic masses of indicated Upper Cretaceous/Lower Tertiary age, of which the best known is Mount Ophir (Johore), are virtually devoid of associated mineralisation, although Bignell (personal communication) noted a small vein of galena in one of these late granites near Mount Ophir.

It is of additional interest to note that two samples of granite from near Ranong (Thailand) yielded Lower Cretaceous ages.

It is instructive to consider these data in the light of Bilibin's (1968) views concerning the sequence of events in mobile belts, and although what follows is based on West Malaysian data, the results might apply, in a general way, to the whole of the tin belt.

According to Bilibin (1968) meaningful primary tin deposits are first formed in what he terms the middle stages of development of a folded zone, that is, when the geosyncline is being converted into a folded belt and when syntectonic batholithic intrusions of acid and ultra-acid potassic granites are emplaced. These tin deposits, which are generally believed to be genetically related to the granites, are largely pegmatites, often containing tantalite/columbite in association with the cassiterite, or hydrothermal veins and lodes of varying structural and mineralogical complexity. During the late stage of development of the folded zones tin-sulphide deposits, associated with acid-intermediate volcanics, may also develop. Such tin deposits are usually xenothermal in character.

It may be tentatively concluded that in West Malaysia a geosyncline was converted into a folded zone during the Variscan orogeny and that the tin deposits of Bakri, Kedah Peak and Pahang Consolidated mine were emp-

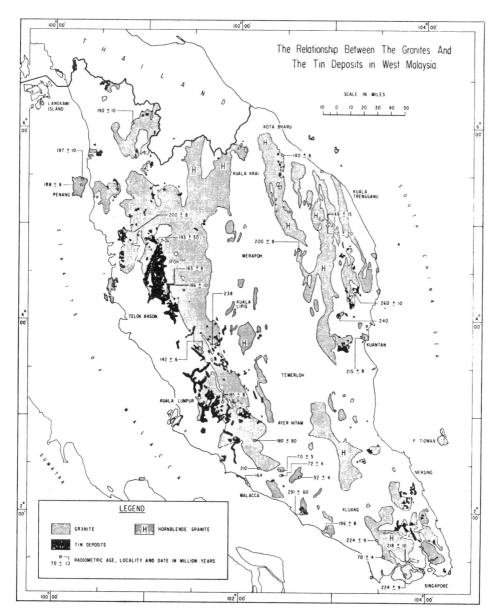

Figure 4

laced during the middle stages of this event immediately after the Lower Carboniferous granites had been emplaced. The xenothermal Manson lode, which is associated with volcanics, was formed during the late stages, in Permian times.

The geosyncline was then rejuvenated and further folding, sediment-ation, igneous activity, etc., occurred, and in the middle phases of this major geologic event the majority of the granite that is now uncovered was emplaced; that was during Triassic times. Associated with this late kine-matic mesozone granite was the development of a multitude of stanniferous pegmatites, veins and lodes, from which later the cassiterite-bearing placers largely developed. It seems possible that, in some instances, the Triassic granitic magma assimilated earlier lodes associated with the Carboniferous granite and that the tin was subsequently forced out and deposited largely as cassiterite in fractures, etc., in the early-consolidated apical portions of cusps and in the immediately neighbouring country rock. With progressive consolidation of the batholith what might be termed "normal Triassic granite cassiterite" was emplaced in the vicinity of the high-spots of the granitic mass. On occasion this mineralisation was superimposed on the "metal front" mineralisation and sometimes it was not. When the former occurred, exceedingly tin-rich vein swarms, such as never occur in south-west England, were developed. On the other hand "Carboniferous" vein swarms and "Triassic" vein swarms that do not contain a double dose of tin, are of about the same grade as those of south-west England.

In West Malaysia there is no evidence of post-Triassic tin miner-alisation associated with acid volcanics; perhaps this is because such effusives are not much in evidence there. On the other hand, it could be that their place has been taken by the few small Upper Cretaceous-Lower Tertiary post-orogenic granites. If this is so then perhaps there may be some pri-mary tin mineralisation associated with them: this, however, has not yet been demonstrated and considering the extent to which tin deposits have been sought in this country it might with good reason be argued that any such mineralisation which may be associated with these granites will have little or no economic significance. Nevertheless interesting concentration of tin could be hidden in as xenothernal type deposits.

That Bilibin's 'theory' is applicable to West Malaysia receives fur-ther support from the fact that he considers the last major igneous event that occurs within mobile belts is the outpouring of basalt and, as noted earlier the radiogenic of the Kuantan basalt is 1.8 m.y.

In the light of Bilibin's views the discovery of the xenothermal nature of the Sungei Ketubong deposit suggests that further deposits of this type, which may contain economically interesting concentrations of tin and several other metals, should be sought in the vicinity of those numerous acid-intermediate volcanics of post-Carboniferous/pre-Triassic age which extend down the centre of West Malaysia, virtually from one end to the other.

It seems to the writer that the dating of the granites and acid-inter-madiate effusives throughout the tin belt far (as it as is possible) should receive a high degree of priority and that, in view of what has been written above, it should at least be considered whether or not interpretation of the geology of the extra-Malaysian countries of the tin belt in the light of Bilibin's views might facilitate the search for tin (and other) deposits.

Spatial relationships

Much of what the writer has to say that is new concerning the spatial relationships between the tin deposits and the granites has already been said in the previous section: certain of the other important spatial relationships he has already noted elsewhere (Hosking 1967) and hence, so far as these are concerned, only the following brief recapitalation is necessary here:—

 i. In West Malaysia and in Peninsular Thailand the strongest tin mineralisation occurs where these regions are widest, that is, where the emplacement of granite was strongest.

 ii. Lode/vein swarms tend to develop in the vicinity of the apices of steep-sided granite cusps, and the strongest tin mineralisation in an area of limited extent often centres round those cusps which penetrated to the greatest heights above a given datum. (This is well seen at Phuket where the major tin mineralisation is centred around the highest granite cusps: these occupy the central third of the island).

 iii. Major lodes may be associated with the apices of prominent granitic cusps (as at Mawchi) or may develop along the flanks of a granitic ridge (as at Pahang Consolidated).

There are, however, certain other important spatial relationships which must be noted, particularly as they have hitherto received scant attention.

In the tin belt generally there is a paucity or almost a complete lack of what may be broadly termed post-granite/pre-lode porphyry dykes of igneous origin which in south-west England, for example, are the almost constant companions of the tin lodes. Hutchison states that "granite porphyry, quartz porphyry, and felsite occur commonly throughout West Malaysia. It is, however, very doubtful if any of these so-called porphyries are of plutonic, or even of igneous origin. All are foliated, or sheared, or have a flow structure. — — the majority can now confidently be considered as the products of metamorphism ...."*

It is abundantly clear then that the emplacement of porphyry dykes is not a necessary adjunct to the development of tin lodes. However, the fact that swarms of structurally complex lodes which have strike and dip dimensions of scores of Cornish lodes and of those of Pahang Consolidated are very rare in the province under consideration suggests that if the regional forces are inadequate to produce long-strike fissures to accommodate porphyry dykes they are generally inadequate to produce them during the immediately

---

*The above has been extracted from a yet unpublished account of plutonic activity in West Malaysia which is to constitute a chapter of a book, now nearing completion, on the geology of the country.

succeeding period of hypothermal mineralisation. However, that Cornish-type lodes occur at Pahang Consolidated, where dykes generally are sparce and where pre-lode dykes are probably (although not absolutely certainly) absent, indicates that the dimensions and broad structural characteristics of primary tin deposits cannot be predicted with certainty by simply considering whether pre-lode dykes are present or absent. Clearly, the structural features and general dimensions of the lodes in a given area are largely dependent on the nature of the faulting to which the area was subject before and during the period of mineralisation. It seems likely, in the writer's view that lodes and veins in the tin belt commonly developed along normal faults originating from tension fractures generated between pairs of transcurrent faults, as Adam (1960) so ably demonstrated at Klappa Kampit and as Garnett (1961) showed to be the case locally in south-west England. The strike lengths of such lodes and veins would then be determined, to a large extent, by the horizontal distance between the transcurrent faults.

It is not entirely out of place to note here that, although fault phenomena of the tin belt have not been neglected, a vast amount of work still needs to be done before it is known with some reasonable degree of certainty how faults controlled primary mineralisation and to what extent the original tin distribution pattern has been modified by faulting. There is little doubt that large post-primary tin mineralisation wrench faults exist in the province and recently Stauffer (1969) has, as a result of a map study, pointed out that there appears to be a major one, trending approximately W.N.W.—E.S.E., which transects West Malaysia and intersects the Main Range granite a little to the north of Kuala Lumpur. This fault seems to have effected a lateral displacement of the granite and associated primary tin deposits of at least 14 miles. Appreciation of the effects of this indicated fault serves to focus interest on the tin potential of certain areas to the north of it.

Further study of the wrench faults may also facilitate the search for submarine tin placers. Thus, if as the writer suspects, the stanniferous islands of Samui and Phangan, off the east coast of Peninsular Thailand, were originally to the south of the island of Phuket, and owe their present position to displacement along a N.E.—S.W.—trending wrench fault, then there is little point in searching for offshore tin deposits immediately to the south of Phuket or to the north of the other two islands mentioned.

Whilst individual lodes and veins show little or no suggestion of primary zoning, there are some indications of regional zoning. Thus, the metal distribution map of the Kinta Valley (which is inset in the Malayan mineral distribution map published by the Geological Survey in 1966 tungsten occurrences are very close to the granite contact whilst those of lead/zinc lie in the 'sedimentary' rocks at some distance from the contact (Figure 5). It may, of course, be eventually demonstrated that this distribution pattern is the product of several distinct phases of mineralisation, each of which is separated by considerable intervals of geologic time.

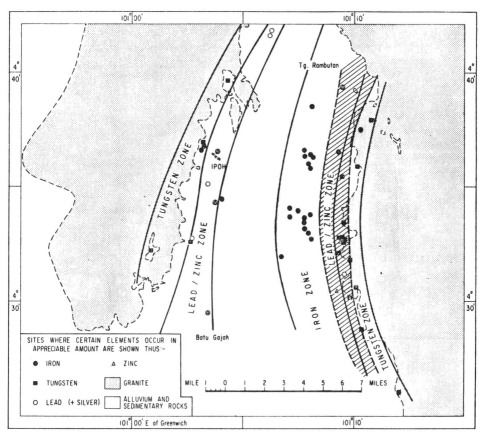

METAL ZONES IN THE KINTA VALLEY
(BASED ON DATA TAKEN FROM THE MINERAL DISTRIBUTION MAP (1966)
OF THE GEOLOGICAL SURVEY OF MALAYSIA
(K.F.G. HOSKING, 1969)

Figure 5

277

Chemical/mineralogical relationships

Although the problem of chemical/mineralogical relationships between the granites and the tin deposits continues to be investigated in south-east Asia there is little that the writer can add to that which he has already written on the subject (see Hosking, 1965 and 1967). He would, however, stress that the radiogenic ages of the granites must be taken into account when interpreting the results of any mineralogic/chemical studies relating to the problem under discussion. He would also make a plea for workers in this field to note precisely the natural environment of each sample they study. It is important to know if the sample was taken from the top of a cusp or from far down the flank of a granite ridge: it is important to know how near is the nearest mineralisation and the nearest granite contact.

A study of the minor and trace element content of the various heavy minerals in the granites may well yield the most useful results. As noted earlier, samples of ilmenite from Malaysian tin deposits are distinctly stanniferous and can contain 1,500 ppm. tin: perhaps some contain much more. Zircon from a similar environment may be strongly stanniferous. Perhaps ilmenites and zircons from granitic areas, of limited extent, with contain primary tin deposits may be distinctly more stanniferous than those occurring in granite, say, a mile from such tin deposits. It is hoped that work in progress in the Department of Geology of the University of Malaya will answer this and similar questions. Clearly the results of this work might be such that they would advise that trace element analysis of selected heavy minerals collected during the reconnaissance sampling of submarine sediments might facilitate the delineation of stanniferous areas worthy of more detailed investigation.

The hornblende-rich "granites" of West Malaysia and Thailand pose some interesting problems: most of them lack associated tin deposits and the geochemical question that requires an answer is: Why is this so?

It may be, as Sullivan (1948) suggested long ago, that any tin in the magmas from which these rocks developed was incorporated in the hornblendes. Clearly, a study of the tin content of hornblende is worth under-taking it for no other reason than that the findings would be valuable when interpreting the results of the analysis of stream sediments for tin when carrying out geochemical exploration in areas where hornblende granites might occur or are known to occur.

Secondary tin deposits

In view of the recent excellent series of papers on this subject in which Osberger was either the sole or part author and because the present writer has recently written about this topic at some lenght,* the content of this section can, with justification, be severely limited.

Although many excellent papers concerning the genesis of the Indonesian placers have been written (for example, Krol (1960), van Overeem (1960) and Osberger (1968)) there are really no outstanding

* Much of what appears in this section has been taken with little alteration from the writer's paper "The off shore tin deposits of South-east Asia" (in the press).

papers on the genesis of the tin placers found elsewhere in the belt, and there is still a great deal of uncertainty concerning the development of stanniferous placers through south-east Asia as a whole. It seems, however, difficult to escape from the conclusion that the major fluctuations in the sea level in the Pleistocene, due to the removal of sea-water as ice and the subsequent return of water to the oceans as the ice melted, played a major part by inducing rejuvenation of the rivers, accompanied by rapid valley erosion, followed by transgression of the land by the sea that re-worked already developed onshore placers. Possibly during and after this time the land/sea relationship was further modified (although perhaps, only slightly) by isostatic adjustments and local tilting.

There is little doubt that during the Pleistocene eluvial cassiterite accumulated on valley slopes, hillsides, etc., where primary deposits were subject to weathering, and some of the cassiterite so liberated migrated under gravity and with the aid of run-off water to form base-of-slope or colluvial deposits* or into the rivers where it tended to become concentrated as a result of running water preferentially removing associated lower specific gravity components of roughly the same size as the cassiterite grains. Of course, depending essentially on the velocity of the water and the size of the cassiterite grains, little or much of the tin-ore was transported downstream. However, probably largely because of the brittle nature of cassiterite, grains of the latter are rapidly reduced to exceedingly small particles in running water and this, plus the progressive dilution of the sediments by tin-barren debris as the distance from the primary cassiterite source increases, results in the fact that economically interesting tin alluvials are unlikely to be found at distances greater than about 8 km from the primary source (Emery and Noakes, 1968).

Locally the rivers ran along granite/invaded rock contacts where they encountered primary mineralisation, and elsewhere they crossed stanniferous vein areas. Conditions appear to have been such that, as these rivers deepened their valleys in an endeavour to become graded, large masses of quartz together with cassiterite were released from the veins only to become stranded whilst the more readily transportable components were carried downstream. These essentially residual deposits, termed kaksa in Indonesia and Karang in Malaysia, are the richest of all the onshore developed placers. During kaksa development the finer cassiterite was, of course, transported downstream, but on occasion this also was eventually induced to settle and to accumulate to such an extent that an economically important deposit developed.

Cassiterite from primary sources near the coast was transported by rivers to the sea and, although the larger grains tended to remain in the river valleys, some of the finer fractions were transported by longshore currents and deposited in elongate zones below low-water level parallel to

*Tropical weathering and heavy rain cause frequent land-slides on the steep valley-sides of south-east Asia and these spectacular events were probably one of the major means by which cassiterite was transported from the slopes into the bottoms of the valleys.

the coast. Only when a beach was backed or underlain by a primary tin source did cassiterite accumulate in useful amounts in the littoral zone.

Transgression of the land by the sea caused the reworking and upgrading of some of the kaksa deposits and the local deposition of low-grade stanniferous marine deposits.

As the valleys became progressively more nearly graded the quantity of cassiterite added to the valley alluvium from the reworking of upstream placers and from primary sources on the slopes of the valleys naturally decreased, so that barren alluvium commonly overlies the tin placers. However, there is reason for believing that the barren superficial alluvium barren from the miner's point of view overlying kaksa-type deposits probably contains appreciably greater concentrations of tin than those occurring in such alluvium elsewhere — a fact that might well facilitate exploration of offshore deposits of this type.

There is no doubt that the offshore tin placers of south-east Asia that are of major economic importance are the eluvial, colluvial and valley-type deposits that have been reworked by the sea, together with those deposits of that cassiterite which was originally transported by rivers and concentrated in valleys before being reworked and drowned. All other possible types are either very rare or are of low grade. Types of stanniferous placer encountered in south-east Asia are shown in Figure 6.

The distribution of stanniferous placers in south-east Asia (Figure 7.)

Onshore cassiterite placers capable of being worked in the immediate or forseeable future will continue to be located in south-east Asia. They will be located in areas previously set aside for rice-growing or for the production of timber or occupied by rubber plantations, etc. They will also be found in areas which, although prospected in the past, were not developed, largely because of the grade of the ground, and they will be located under deep alluvial cover which may be stanniferous and already worked, or barren and perhaps covered with mangrove swamp. Vast areas that fall into this last category exist between the tin fields fringing the west side of the Main Range of West Malaysia and the sea. Locally there may well be tin beneath the thick alluvial cover, but its discovery will depend on a much better understanding of the overall geology of the country and particularly of those factors which determined the original and present disposition of the primary tin deposits. It will also depend on improved boring and other exploration techniques. The exploitation of any deposits of this type which may be found will depend, of course, on tonnage, grade and the price of tin, and probably the availability of improved methods of working deposits in such an environment.

Improved mining and beneficiation methods coupled with a satisfactory price for tin concentrates and possibly increased sales of accessory minerals such as monazite, zircon, ilmenite and xenotime will probably permit tailings to be re-worked for many years.

Although there is no doubt that economically viable onshore tin mining is becoming progressively more difficult to find in south-east Asia,

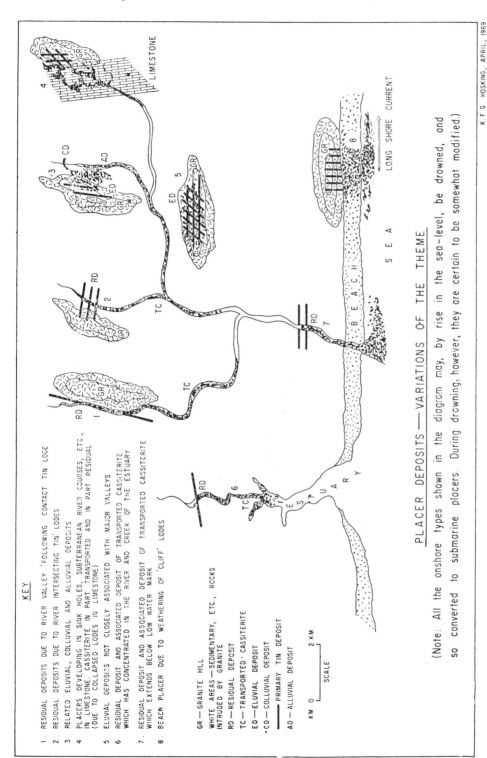

KEY

1. RESIDUAL DEPOSITS DUE TO RIVER VALLEY 'FOLLOWING' CONTACT TIN LODE
2. RESIDUAL DEPOSITS DUE TO RIVER INTERSECTING TIN LODES
3. RELATED ELUVIAL, COLLUVIAL AND ALLUVIAL DEPOSITS
4. PLACERS DEVELOPING IN SINK HOLES, SUBTERRANEAN RIVER COURSES, ETC, IN LIMESTONE. CASSITERITE IN PART TRANSPORTED AND IN PART RESIDUAL (DUE TO COLLAPSED LODES IN LIMESTONE)
5. ELUVIAL DEPOSITS NOT CLOSELY ASSOCIATED WITH MAJOR VALLEYS
6. RESIDUAL DEPOSIT AND ASSOCIATED DEPOSIT OF TRANSPORTED CASSITERITE WHICH HAS CONCENTRATED IN THE RIVER AND CREEK OF THE ESTUARY
7. RESIDUAL DEPOSIT AND ASSOCIATED DEPOSIT OF TRANSPORTED CASSITERITE WHICH EXTENDS BELOW LOW WATER MARK
8. BEACH PLACER DUE TO WEATHERING OF 'CLIFF' LODES

GR — GRANITE HILL
WHITE AREAS — SEDIMENTARY, ETC, ROCKS INTRUDED BY GRANITE
RD — RESIDUAL DEPOSIT
TC — TRANSPORTED · CASSITERITE
ED — ELUVIAL DEPOSIT
·CD — COLLUVIAL DEPOSIT
——— PRIMARY TIN DEPOSIT
AD — ALLUVIAL DEPOSIT

KM 0   2 KM
    SCALE

PLACER DEPOSITS — VARIATIONS OF THE THEME

(Note: All the onshore types shown in the diagram may, by rise in the sea-level, be drowned, and so converted to submarine placers. During drowning, however, they are certain to be somewhat modified.)

K. F. G. HOSKING, APRIL, 1969

Figure 6

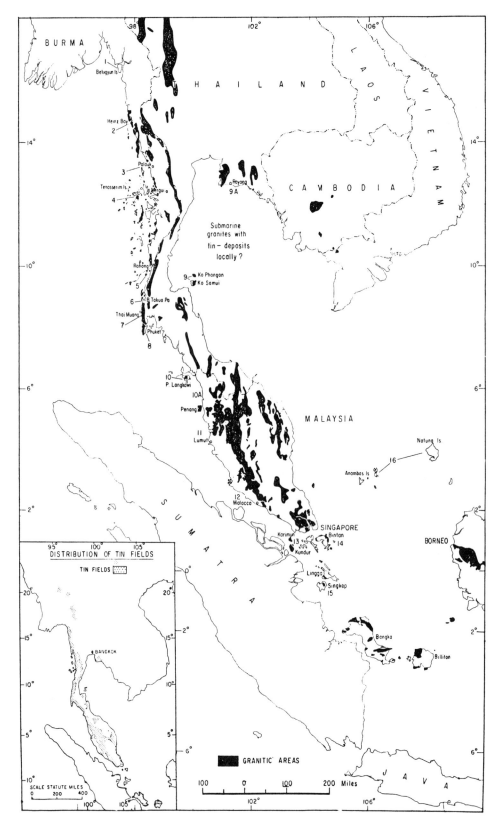

Figure 7 Map of the tin fields of south-east Asia.

Key to Figure 7

1. *BELUGYUN ISLAND.* STANNIFEROUS PLACERS. FROM HERE TO THE THAI BORDER AT RANONG SCORES OFF GOOD OFF-SHORE TARGETS EXIST.

2. *HEINZE BASIN.* SnO_2 DREDGED FROM TIDEWAYS.

3. *SPIDER ISLAND. (AT MOUTH OF PALAUK RIVER)* SnO_2 (AND WOLFRAMITE) RECOVERED FROM BEACH SANDS.

4. *TENASSERIM DELTA AND LAMPA AND NEIGHBOURING ISLANDS.* SnO_2 RECOVERED FROM THESE LOCALITIES.

5. *RANONG AND COAST TO SOUTH.* ONSHORE PLACERS LOCALLY EXTEND TO COAST. PRIMARY TUNGSTEN DEPOSIT AT COAST (AT KAU CHAI). A GOOD TARGET AREA.

6. *TAKUAPA.* SUCTION DREDGE WORKING OFFSHORE.

7. *THAI MUANG.* SnO_2 IN BEACH SANDS. OFFSHORE POTENTIAL?

8. *PHUKET.* DREDGES OPERATING OFF E. COAST. SOME BAYS ON W. COAST KNOWN TO CONTAIN IMPORTANT CONCENTRATION OF SnO_2. RECENT ILLICIT OFFSHORE MINING OFF W. COAST.

9. *KO PHANGAN AND KO SAMUI.* POOR TIN MINERALISATION ON ISLANDS. SnO_2 OCCURS OFFSHORE BUT POTENTIAL UNCERTAIN.

9A. *RAYONG.* SnO_2 IN BEACHSANDS AND OFFSHORE.

10. *LANGKAWI ISLANDS.* 2% Sn IN ANDRADITES: "TRACE" SnO_2 IN BEACH SANDS. POSSIBLY AN OFFSHORE POTENTIAL EXISTS BETWEEN MAINLAND.

(10A.) AND ISLANDS W. OF KEDAH PEAK. SnO_2 IN BEACH SANDS.

11. *LUMUT-DINDINGS* SnO_2 IN BEACH SANDS AND OFFSHORE AT TANJONG HANTU AND PULAU KATAK. SITE OF RECENT ILLICIT OFFSHORE MINING.

12. *MALACCA.* SnO_2 IN MAINLAND BEACH SANDS AND OFFSHORE, FROM S. LINGGI TO S. UDANG, ALSO IN BEACH SANDS OF PULAU BESAR (AN ISLAND). WITH (11) THE MOST PROMISING MALAYSIAN OFFSHORE AREA.

13. *KARIMUN AND KUNDUR* TIN-BEARING ISLANDS BUT MINERALISATION NOT STRONG. POSSIBLY AN OFFSHORE POTENTIAL EXISTS.

14. *BINTAN.* SMALL PLACERS ON ISLAND. POSSIBLY AN OFFSHORE POTENTIAL EXISTS.

15. *"THE TIN ISLANDS"* MARKED OFFSHORE MINING AND EXPLORATION ACTIVITY.

16. *ANAMBAS AND NATUNA ISLANDS.* CASSITERITE REPORTED, BUT OUTSIDE TIN BELT PROPER. IT IS UNLIKELY THAT AN OFFSHORE POTENTIAL EXISTS.

it would be the greatest mistake to think, as some have, that the end of this province as a great producer of tin from onshore placer deposits is in sight.

Nevertheless, the offshore tin potential of south-east Asia is now receiving for greater attention than ever before. This interest is due to the increasing difficulty of finding sufficiently large onshore deposits whose characteristics are such that they can satisfactory yield profits when conventional mining and beneficiation methods are used. It is also due to the fact that there are good geologic reasons for thinking that submarine stanniferous deposits of merit, that have not yet been exploited, exist off the coasts of south-east Asia, and to the fact that rich submarine deposits are at present being exploited there. Another favourable feature of off-shore placer mining lies in the fact that a dredge can cheaply and quickly be moved from one deposit to another: this is in marked contrast to the state of affairs onshore.

Large offshore tin deposits have long been known and exploited in the seas fringing the major "Tin Islands" of Indonesia, and recently mining companies have indicated their wish to search for further deposits in the shallow waters between these islands and also in the vicinity of some of the islands to the north, as some of the latter, for example, Kundur, Karimun and Bintan, contain onshore tin deposits. Interest has also been shown in the shallow sea areas fringing the Anambas and Natuna groups of islands in the China Sea, as a little cassiterite has been recorded on the land there (Osberger, 1967 (A), p. 97). However, as these islands are far to the east of the main tin belt it would be surprising if offshore tin deposits of any merit were found there.

In West Malaysia much more interest is being shown in the off-shore areas margining the west coast than the east one, and the reasons for this are clear. Generally speaking there is a west tin belt and an east one, each paralleling its associated coastline, and each extending throughout the length of the country. The west belt is vastly richer in tin than the other, and locally in Malacca and Perak cassiterite has been recovered from modern beach sands and other coastal deposits. Recently a fleet of small boats, converted to primitive suction dredges, has been carrying out illegal mining in the sea off Tanjong Mengkudu in the Dindings District of Perak (Straits Times, Dec. 25th, 1968, p. 7): this is surely a further indication that this offshore area is worth prospecting.

A little cassiterite is known to occur in the beaches of some small islands just off the mainland and due west of Kedah Peak. Finally, although only traces of cassiterite occur on the beaches of the Langkawi group of islands, stanniferous andradite garnets and stannite occur there although probably in limited amount. Thus it is clear that agents capable of promoting the development of tin-bearing minerals had locally invaded the region, and so, perhaps somewhere between these islands and the mainland drowned deposits of cassiterite may occur.

Along the east coast there are no onshore tin deposits known to the writer that fringe the sea, nor, indeed, are there any known to him which are within 8 km of the sea. Therefore obvious offshore targets do not exist there; but it must not be forgotten that the locations of a number of the offshore Indonesian deposits were not indicated by geologic studies of the adjacent islands.

Off the west coast of Peninsular Thailand cassiterite is recovered to the east of Phuket and a Takuapa, but there is a number of other offshore areas in which tin is known to occur in amounts that almost certainly would support dredging operations. In particular, Kamera Bay, Bang Thao Bay and Haad Surin, all on the west coast of Phuket, are known to contain stanniferous placers, and perhaps up to 1,000 miners were recently recovering cassiterite illegally from the last-mentioned place. The beach sands of Thye Maung have long been known to be stanniferous and rumour has it that this place has also been the site of illegal offshore mining.

The estuary separating Thailand from Burma contains stanniferous alluvium which might well be of economic importance, whilst the offshore strip bordering the land to the south of Ranong has long been regarded as a good prospect in view of the fact that there locally mineralised granite is not far removed from the coast, and possibly a granite ridge underlies the coastal strip at no great depth (a fact suggested by the presence of wolframite veins at Kau Chai). In addition, useful onshore tin placers virtually reach the coast in the Ranong area and in the Bankajkrud river valley.

The east coast offers no immediately attractive offshore targets except those in the waters fringing the weakly tin-mineralised islands of Samui and Phangan. These offshore areas have been subject to a degree of investigation but the results are not available for publication.

At Rayong, at the north of the Gulf of Thailand, along the shoreline fine tin is very widespread and is believed to owe its existence to the erosion of granite wash which serves as bedrock, and is exposed along the shore. In accumulation areas which contain dark minerals or heavy minerals tin is generally found in an unusually large quantity. Those areas extending beyond the shoreline towards the sea should be future prospecting grounds. (The Department of Mineral Resources of Thailand, 1967, p. 5). One is also led to wonder if there are granites and associated tin deposits under the Gulf of Thailand that constitute a connecting link between the east tin belt of Malaya and the tin fields to the north of the Gulf. If this drowned link does exist, then, judging by its onshore counterparts, any stanniferous placers associated with it are likely to be sporadically disposed and probably of minor economic importance.

Much of the offshore belt fringing the west coast of Burma, from Moulmein in the north to Victoria Point in the extreme south, is worthy of intensive exploration for tin placers. In support of this contention sufficient is it to note that cassiterite has been recovered from the tideway of the Heinze Basin (EAFE, 1964, p. 28), from the estuarine reaches of the Tenasserim River and from the beach sands on Spider Island at the mouth of the Palauk River (Jones, 1925, p.229). In addition, a tin/tungten belt traverses Lampi and a number of islands of the Mergui Archipelago. Cassiterite has also been won from Belugyun Island, near Moulmein (Jones, 1925, p. 219). These examples are surely ample indication that a great and largely untapped tin potential lies in these Burmese waters.

Aspects of the search for further tin deposits in south-east Asia

The types of exploration target are four-fold, namely:-

i. hard-rock deposits that must be mined largely by underground methods,

ii. hard-rock deposits that may be mined by open-cast methods,

iii. onshore placers, and

iv. offshore placers.

Hard-rock deposits that must be mined largely by underground methods have not yet been generally sought for in a really determined manner. This is probably due to the fact that most mining companies prefer, for obvious reasons, to concentrate on placers and other deposits that can be worked by open-cast methods. As stated earlier, it is quite unknown whether any, a few, or many major deposits of this type yet remain to be found in the belt, but the chances of finding such deposits would seem to be best in north-east Malaya (where Pahang Consolidated mine occurs). Eventually, also, areas in the Kledang Range which hold strong lodes must be drilled in order to discover if these lodes persist in depth and if, so, what are their characteristics in depth.

Hard-rock tin deposits that may be mined by open-cast methods are plentiful but some are small and many are of low grade. The larger of these should generally receive greater attention than they have in the past as some of them may be capable of being mined profitably provided the operations are on a large tonnage basis and the mills are designed and run by competent people.

The search for hard-rock deposits falling in the first category generally involves in eastern Malaya, at any rate, the selection of likely areas on geologic and historic grounds. Such areas may then be subject to geochemical surveys, involving the analysis of stream sediments and of soil samples from spurs and ridges. At the same time a photogeological interpretation of the area is made (provided air-photographs are available) and this is followed by routine geological mapping of the area. Should the deposits be known to be sulphide-rich (as is the Manson lode described earlier) then the geochemical investigations are supplemented by appropriate geophysical ones. Targets thus delineated are drilled and appropriate portions of the core are analyzed or converted to thin and polished sections for examination under the microscope in order not only to identify the species present but also to discover, at an early stage, any mineral beneficiation problems which the ore may present.

Of course, if one were investigating the deeper zones of an area that had already been mined superficially, and such as occur, for example, in the Kledang Range, then the exploration programme may be essentially a diamond-drilling one. The search for onshore placer deposits varies somewhat according to whether one is searching for a workable deposit in an area that has not before been investigated, or to only a very limited extent, or is looking for a worthwhile lateral extension of an area currently being worked, or for deep tin deposits within known stanniferous ground. The search is also guided by the kind of mining operation one wishes to establish: a small block of undulating ground may be adequate for a palong mine but a large expanse of flat ground is required for a large dredge.

In new areas, in West Malaysia, the Geological Survey advocates, at least in some instances, a rapid reconnaissance survey involving essentially semi-quantitative determinations of the tin (and possibly gold) content of superficial stream sediments and bank samples by panning known volumes of material: generally, however, this seemingly sensible technique has been little practised.

A study of air-photographs of a new area is, of course, worthwhile, not only because it reveals (albeit, with varying degrees of certainty) the distribution of the major lithologic units and the disposition of any granite contacts (for which tin deposits seem to have a penchant) but it also serves to establish whether suitable dredging areas are likely to occur there and it facilitates the choice of the best means of gaining access into it.

Limited experimental resistivity studies have been carried out over alluvial areas in West Malaysia, primarily with a view to determining the bedrock profile, and, in particular, to locate any deep troughs which might exist at limestone/granite contacts. The work has met with a degree of success (W.K. Lee: personal communication), but it must be left to the investigators to discuss this and other aspects of their investigations.

Commonly, areas which for geological and/or other reasons are thought to be reasonable exploration targets are investigated solely by obtaining samples, by drilling with a Banka or mechanical drill (depending on the expected depth of the alluvium, the nature of the ground accessibility, etc.), from points on a grid. The initial grid in West Malaysia may be a 32-chain one but in areas of promise the grid may be progressively decreased until the bores are only 2 chains apart. Now the more progressive companies employ statistical methods to determine the minimum number of bores necessary for the adequate evaluation of a given deposit (Broadhurst and Batzer, 1964).

The cassiterite content of the samples is generally obtained by unsophisticated methods and separation of the cassiterite from gangue by panning is common to all of them.

There is little point in discussing offshore exploration techniques as these have been dealt with recently in a most comprehensive series of papers by Osberger (1967 (A), 1967 (B), 1968) and Osberger and Romanowitz (1967) and have just been reviewed by the present writer (Hosking (1969): in press).

Recommendations and Conclusions

To date many great tin deposits have been discovered in south-east Asia by the employment of simple and comparatively cheap exploration techniques.   Now new fields are much more difficult to find and effective search for these must employ more sophisticated and more expensive methods of search and of evaluation, and the selection of target areas should be essentially a product of the geological interpretation of the region.   Unfortunately, the geology of the tin belt contains far too many unknowns so that whether a selected target area proves to be a good one or a bad one is, to a greater extent than is desirable, a matter of chance.   In south-east Asia the exploration geologists is often in a position similar to an archaeologist who is trying to reconstruct a pot when great pieces of it are missing : both need more pieces before their reconstructions are likely to be correct. In order that the geologist should have the necessary pieces, it is desireable that work on the following topics should be vastly accelerated or initiated, as the case may be :

   i.   Many more radiogenic ages of granitic rocks, and of the primary mineral deposits when possible, of the tin belt should be established, and temporal relationships between tin deposits and granitic rocks should be investigated concurrently.

   ii.   Chemical/mineralogical studies aimed at throwing further light on the relationships between granitic rocks and primary tin deposits should be intensified.   Investigations should involve not only studies of the composition of granitic rocks and some of their components close to and removed from mineralisation but also variations in composition encountered in each of these environments.   The possibility that compositional variations may be related to the original topography of the granites should be examined.   These are only some of the topics that should be tackled.   Variations in the habits of accessory species in the granite and, in the case of zircons, variations in the distribution pattern of radioactive centres are further fields of research likely to prove fruitful.

   iii.   Many more structural studies of the tin belt are desperately needed and, in particular, the relationships between the distribution of primary and secondary tin deposits and faults need to be clarified.

   iv.   The geology of the tin belt should be interpreted in the light of Bilibin's views as this may reveal further areas where tin and other deposits of economic importance might be found.

   v.   Detailed mineralogical investigations of all available hard-rock tin deposits should be carried out with the aid of both simple and sophisticated techniques.   Apart from throwing light on the genesis of such deposits and yielding data re the amenability of the material to the various mineral beneficiation processes, it may well, on occasion, indicate the presence of hitherto unknown components of economic importance.   Similar studies on the heavy components of stanniferous placers are also not without merit.

vi.   Determined efforts should be made to establish and interpret
Pleistocene and Recent geological events throughout the belt so
that an overall and reliable history of the genesis, etc., of the
stanniferous placers of south-east Asia can be established.   In
this connection there is clearly a need for the dating of a vastly
greater number of samples of organic material found in the
placers than has yet been done.

For such a programme to be carried out in an adequate way a
considerable number of pieces of sophisticated and expensive equipment have
to be available and a considerable number of people would need to participate in the field and laboratory work.

Clearly such a programme would have the greatest chance of success
if members of all the countries in the tin belt were active participants and
if, in addition, technical assistance could be obtained from other countries
who had a special interest in tin, particularly when the assistance was of
a type that could not be obtained in south-east Asia.   At present radiogenic
dating cannot be done in south-east Asia nor can anyone in the region un-
dertake electron probe studies of minerals: assistance in such work from
countries outside the belt would be necessary for the programme to succeed.

If such an international programme were contemplated it is the view
of the writer that a centre should be established in south-east Asia from
which the work could be directed and the results processed and despatched
to interested parties.

It is obviously not appropriate to develop the theme further here.
The writer believes the idea is worthy of consideration not only because
the economic geology of the tin belt is still a riddle wrapped in a mystery
inside an enigma despite certain naive utterances, from time to time, to the
contrary, but also because the results which might accrue from such a programme would be likely to facilitate the search for tin and other metal-
bearing deposits elsewhere in the world.

Acknowledgements

It is impossible to acknowledge individually all those who have
contributed data, thoughts and ideas to this paper, but the writer would
like to single out a few of them for 'special mention'.   He particularly
wants to acknowledge the general assistance and encouragement given to
him by the Chief Inspector of Mines of Malaya, Inche Mohammed Salleh.
Mr. Santokh Singh, Deputy Director of the Geological Survey of Malaysia,
has spared no efforts in his endeavour to provide the data which the writer
was constantly requesting.   He owes debts of gratitude to Mr. Henry Hod-
ding, General Manager of Pahang Consolidated Co. Ltd., and to Inche Hal-
lam Rasip, Director of Eastern Mining and Metals Sdn. Bhd., for allowing
him access to their properties, supplying him with a great deal of information and permitting him to publish certain relevant parts of it.   His
colleagues, Dr. C.S. Hutchison and Mr. J.H. Leow, have facilitated the writer's

289

studies concerning the granites and mineral deposits respectively; to these two gentlemen he extends his thanks. Finally he wishes to express his gratitude to Inche Srinivass and Che Haidar for the great deal of time and effort they spent translating the writer's Heath Robinson drawings into those which grace this paper, and to Inche Jaafar for dealing in his usual efficient way with the photographic aspects of this work.

References

ADAM, J.W.H.: On the geology of the primary tin deposits in the sedimentary formation of Billiton. Geologie en Mijnbouw, *39*, 405—426, 1960.

ALEXANDER, J.B.: Geology and mineral resources of the Bentong area, Pahang. Geol. Surv. W. Malaysia, District Memoir 8, 1968.

ALEXANDER, J.B. and FLINTER, B.H.: A note on varlamoffite and associated minerals from the Batang Padang district, Perak, Malaya, Malaysia. Min. Mag., *35*, 622—627, 1965.

ARANYAKANON, P.: The cassiterite deposit of Haad Som Pan, Ranong Province, Thailand. Rep. Invest. no. 4, Roy. Dept. Mines, Bangkok, Thailand, 1961.

BILIBIN, Yu. A.: Metallogenic provinces and metallogenic epochs. Trans. by E.A. Alexandrov. Queens College Press, Flushing, N. York, 1968.

BRADFORD, E.F.: The occurrence of tin and tungsten in Malaya. Proc. 9th Pacific Science Congress, *12*, 378—398, 1960.

BROADHURST, J.K. and BATZER, D.J.: Valuation of alluvial tin deposits in Malaya with special reference to exploitation by dredging. Paper 5 Symposium on opencast mining, quarrying and alluvial mining. Instn. Min. Metall., London, 1954.

BROWN, G.F. *et al.*: Geologic reconnaissance of the mineral deposits of Thailand. Geol. Surv. Memoir no. 1, Roy. Dept. Mines, Bangkok. Thailand, 1953.

BYRDE, E.W.: Discussion. Trans. Instn. Min. Metall., Lond., *29*, 369, 1920.

CHHIBBER, H.L.: The geology of Burma. McMillan and Co., Lond., 1934.

COGGIN-BROWN, J.: A geographical classification of the mineral deposits of Burma. Records, Geol. Surv. India, *56*, pt. 1, 65-103, 1924.

EMERY, K.O. and NOAKES, L.C.: Economic placer deposits of the Continental Shelf. pp. 95-111 of Tech. Bull. 1 of ECAFE Committee for co-ordination of joint prospecting for mineral resources in Asian offshore areas. Printed by Geol. Surv. Japan, 1968.

FITCH, F.H.: The geology and mineral resources of the neighbourhood of Kuantan, Pahang. Memoir no. 6 (new series), Geol. Surv. Dept. Federation of Malaya, 1952.

GARNETT, R.H.T.: Structural control of mineralisation in south-west England Mining Mag., *105*, 329-337, 1961.

GONI, J.: Contribution a l'étude de la localisation et da la distribution des éléments en traces dans les minéraux et les roches granitiques.   No. 45, Mémoires du B.R.G.M., Paris, 1966.

GRIFFITH, S.V.: Tin and wolfram in Burma. Mining Mag., 212-215, 1956.

HOSKING, K.F.G.: The search for tin.   Mining Mag., *113*, 261-273, 308-383, 448-461, 1965.

------------------: The relationship between primary deposits and granitic rocks. pp. 269-306 of 'A technical conference on tin', London 1967.

HOBSON, V.G.: The development of the mineral deposit at Mawchi as determined by its geology and genesis.   Trans. Mining, Geol. Metall. Inst. India, *36*, pt. 1, 35-78, 1940.

INGHAM, F.T. and BRADFORD, E.F.: Geology and mineral resources of the Kinta Valley, Perak, Federation of Malaya. Geol. Surv. District Memoir 9, 1960.

JONES, M.P.: Some impressions of the tin mining industry of Southeast Asia. Trans. Section A, Instn. Min. Metall., *76*, A1-A13, 1967.

JONES, W.R.: Tinfields of the world. London, 1925.

KAEWBAIDHOON, S. and ARANYAKANON, P.: Tin and tungsten deposits of Thailand. Proc. 9th Pacific Science Congress, *12*, 400-404, 1961.

KROL, G.L.: Theories on the genesis of the kaksa.   Geologie en Mijnbouw, *39*, 437-443, 1960.

OSBERGER, R.: Prospecting tin placers in Indonesia. Mining Mag., *117*, 97-103, 1967 (A).

------------------: Dating Indonesia cassiterite placers. Mining Mag., *117*, 260-264, 1967 (B).

------------------: Billiton tin placers : types, occurrences, and how they were formed. World Mining, 34-40, June 1968.

OSBERGER, R. and ROMANOWITZ, C.M.: How the offshore Indonesian tin placers are explored and sampled. World Mining, Nov., 1967 (Reprint, 7 pages).

ROE, F.W.: The geology and mineral resources of the Fraser's Hill area. Selangor, Perak and Pahang. Federation of Malaya Geol. Surv. Memoir no. 5, 1951.

SCRIVENOR, J.B.: The geology of Malayan ore-deposits. MacMillan and Co., London, 1928.

------------------: The geology of Malaya. MacMillan and Co., London, 1931.

SNELLING, N.J., BIGNELL. J.D. and HARDING, R.R.: Ages of Malayan granites. Geologie en Mijnbouw, *47*, 358-359, 1968.

STAUFFER, P.H.:   The Kuala Lumpur fault zone; a proposed major strike-slip fault across Malaya.   Geol. Soc. Malaysia newsletter no. 15, 2-4, Nov. 1968.

SULLIVAN, C.J.: Ore and granitisation. Econ. Geol., *43*, 471-498, 1948.

SINGH, SANTOKH, D. and BEAN, J.H.:   Some general aspects of tin minerals in Malaysia, pp. 459-478 or 'A technical conference on tin', London, 1967.

van OVEREEM, A.J.A.:   The geology of the cassiterite placers of Billiton (Indonesia).   Geologie en Mijnbouw, *39*, 444-457, 1960.

WILLBOURN, E.S.:   The Beatrice mine, Selibin, F.M.S. Mining Mag., *45*, 338-341, 1931, and *46*, 20-24, 1932.

SWIERZYCKI, J.:   Enkele nieuwere geologische waarnemingen op de tinei-landen en op Sumatra betreffende het tinvraggstuk.   De Mijning, *14*, 171-176, 1953, Bandung.

Editor's Comments
on Paper 26

26   PEREIRA
Further Reflections on Ore Genesis and Exploration

Paper 26 is concerned with late-stage metallogeny. Pereira is the English link in the Conzinc school, led by Haddon King in Australia and followed by Frank Moss in Canada. Much of Pereira's work has been concerned with the final products of the cycle "under altogether less spectacular tectonic and volcanic conditions . . . . Their setting is predominantly calcareous or dolomitic. Rammelsberg in W. Germany, Meggen in Westphalia, and Mount Isa in Australia are typical examples." Pereira and his co-author Dixon in Paper 13 were also concerned with nonreversible changes throughout time. It is apropos that lead–zinc deposits are time-restricted: the Australian equivalent of the Karelian, the Carpentarian, is host to Mt. Isa and Broken Hill, the oldest late-stage lead–zinc deposits.

European authors have had the complexities of Alpine orogenies superimposed on Paleozoic mineralization to contend with, and so Pereira's summaries of lead–zinc settings are all the more remarkable. Lacking in expertise in lead–zinc, I wonder how forced is the fit of some deposits into late-stage orogenic and others into Mississippi Valley–type platform deposits. With Dunham (Paper 27) I consider the Permo-Trias to be the prime time of mineralization, related to the terminal stage in the Paleozoic or to the interval between eras. Gilmour has recently built a bridge which includes most varieties of lead–zinc mineralization, from volcanic to sedimentary, and it will be interesting to see the stage–time factor incorporated.

26

Reprinted from *Mining Mag.*, **109**, 265–280 (Nov. 1963)

Further Reflections on Ore Genesis and Exploration

J. Pereira

Introduction

The ideas discussed in the present article are intended to elaborate particular aspects of ore genesis mentioned only briefly in a previous communication.[1] Both contributions are not meant to convey more than a personal viewpoint and comments on other concepts relevant to the author's particular thoughts.

The more important conclusions suggested by the first article were as follows :—

(1) Volcanoes form the first and most important link in the chain of ore genesis.
(2) Volcanic, sedimentary-exhalative, and closely related sedimentary deposits form the most important series of economic orebodies.
(3) Shore lines and volcanoes provide a vital key to exploration problems.

The classification originally adopted is set out again in Fig. 1, although a different series of examples has been chosen to illustrate the particular theme elaborated in these " further reflections." The manner in which various classes of ore deposit fit the pattern of different phases of the orogenic cycle is depicted in Fig. 3. An introductory section deals briefly with the main concepts of the case to be presented, but it must be emphasized that they are highly speculative and not essential to the main conclusions. The ideas are introduced in a very simplified form with the sole object of pointing out that they have an important bearing on problems of ore genesis ; moreover, it may be easier to follow the ensuing detailed picture if some indication of the larger problems is presented first, together with the solutions which are favoured.

Part 1—Orogenic Controls and Fundamental Aspects of Ore Genesis

(a) *Concepts Concerning the Nature of the Earth's Mantle*

L. Egyed, V. Neyman, and others have developed a theory of an expanding earth which is believed to have originated some $4 \cdot 1 \times 10^9$ years ago and has continued to expand to its present size at a rate of $0 \cdot 4$ mm. to $0 \cdot 8$ mm. per annum.[1] It is postulated that there was an initial and almost immediate differentiation into an upper mantle composed of the atmosphere, the hydrosphere, and a crust of acidic rocks underlain by progressively more basic shells. Pressure and equilibrium established in this upper part of the mantle controlled further differentiation of the lower mantle and also of the inner core, the Gutenberg layer representing a transition between the two. Since the lower mantle is assumed to contain all the elements of the differentiated mantle in roughly the same proportions it should be near andesite in composition and rich in volatiles. The only true magmas are thought to be deep-seated volcanic types generated when the build-up of energy due to radioactive decay and initial heat causes expansion, which is expressed at the surface by orogenic phenomena and concomitant volcanic activity.

Expansion is considered to be the natural expression of a decline of total energy, analogous to the effect of surface weathering, which produces new minerals with lower densities and larger volumes than the primary minerals produced at greater depth and temperatures in an environment of higher energy levels.

Alternative suggestions and modifications advanced by Y. U. Sheynmann envisage the possibility that there may be virtually no true magmas in the ordinary sense of the word. However, a version which is largely Egyed's is adopted here because it allows orogenic and volcanic activity to be expressed in the much simplified form shown in Fig. 2, with deep fissures in the crust related to phases of an orogenic cycle.

(b) *Orogenies and the Present-day Geology of the Earth*

A typical orogeny begins with basic and

---

[1] " Reflections on Ore Genesis and Exploration." THE MINING MAGAZINE, Jan., 1963.

[1] A list of references is given at the end of this article.

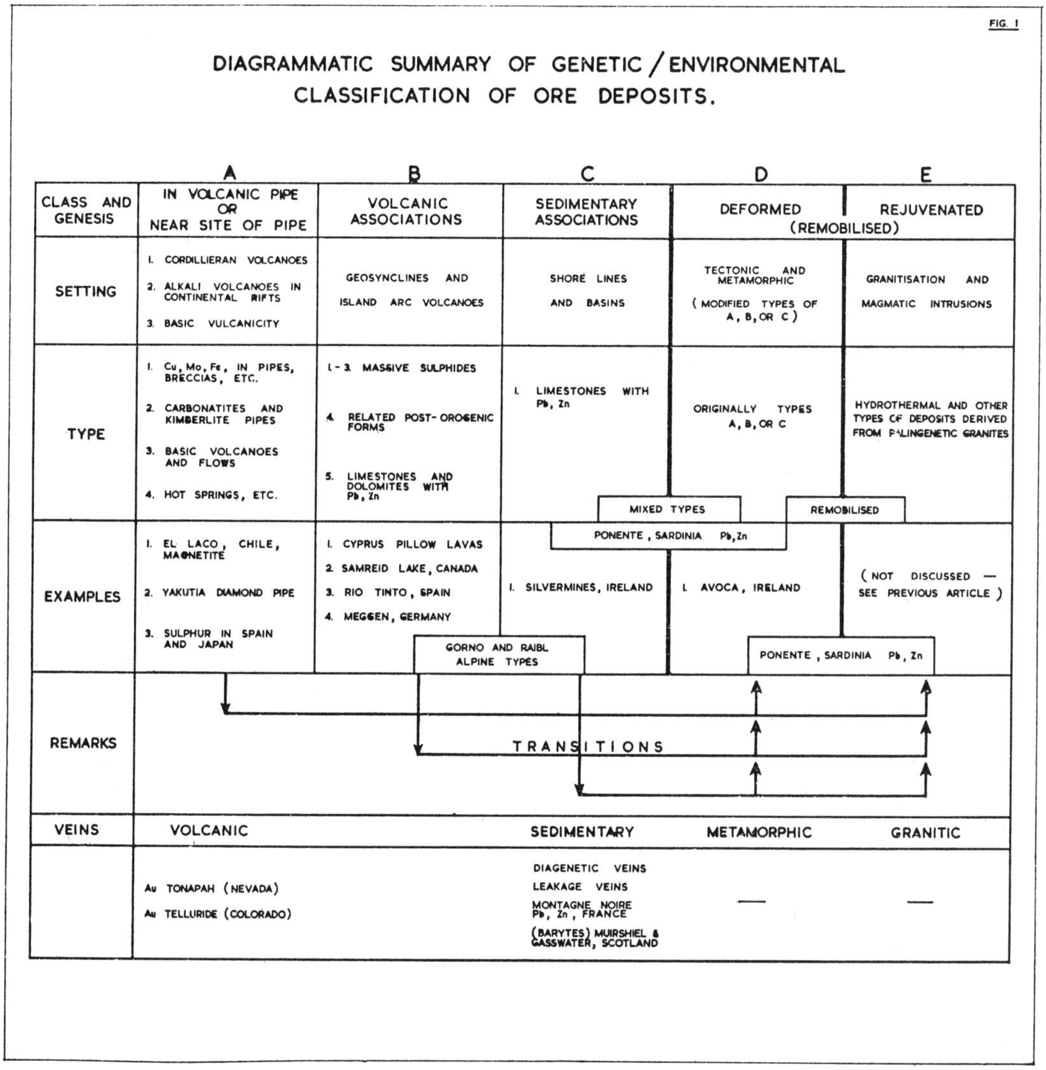

FIG. I

DIAGRAMMATIC SUMMARY OF GENETIC / ENVIRONMENTAL CLASSIFICATION OF ORE DEPOSITS.

	A	**B**	**C**	**D**	**E**
CLASS AND GENESIS	IN VOLCANIC PIPE OR NEAR SITE OF PIPE	VOLCANIC ASSOCIATIONS	SEDIMENTARY ASSOCIATIONS	DEFORMED	REJUVENATED
				(REMOBILISED)	
SETTING	1. CORDILLIERAN VOLCANOES 2. ALKALI VOLCANOES IN CONTINENTAL RIFTS 3. BASIC VULCANICITY	GEOSYNCLINES AND ISLAND ARC VOLCANOES	SHORE LINES AND BASINS	TECTONIC AND METAMORPHIC (MODIFIED TYPES OF A, B, OR C)	GRANITISATION AND MAGMATIC INTRUSIONS
TYPE	1. Cu, Mo, Fe, IN PIPES, BRECCIAS, ETC. 2. CARBONATITES AND KIMBERLITE PIPES 3. BASIC VOLCANOES AND FLOWS 4. HOT SPRINGS, ETC.	1-3 MASSIVE SULPHIDES 4. RELATED POST-OROGENIC FORMS 5. LIMESTONES AND DOLOMITES WITH Pb, Zn	1. LIMESTONES WITH Pb, Zn	ORIGINALLY TYPES A, B, OR C	HYDROTHERMAL AND OTHER TYPES OF DEPOSITS DERIVED FROM PALINGENETIC GRANITES
			MIXED TYPES	REMOBILISED	
			PONENTE , SARDINIA Pb, Zn		
EXAMPLES	1. EL LACO , CHILE, MAGNETITE 2. YAKUTIA DIAMOND PIPE 3. SULPHUR IN SPAIN AND JAPAN	1. CYPRUS PILLOW LAVAS 2. SAMREID LAKE , CANADA 3. RIO TINTO , SPAIN 4. MEGGEN , GERMANY	1. SILVERMINES, IRELAND	1. AVOCA, IRELAND	(NOT DISCUSSED — SEE PREVIOUS ARTICLE)
		GORNO AND RAIBL ALPINE TYPES		PONENTE , SARDINIA Pb , Zn	
REMARKS			TRANSITIONS		
VEINS	VOLCANIC		SEDIMENTARY	METAMORPHIC	GRANITIC
	Au TONAPAH (NEVADA) Au TELLURIDE (COLORADO)		DIAGENETIC VEINS LEAKAGE VEINS MONTAGNE NOIRE Pb , Zn , FRANCE (BARYTES) MUIRSHIEL & GASSWATER, SCOTLAND	—	—

ultrabasic vulcanicity, including the development of extensive serpentine zones which cover appreciable lengths of the deepest parts of geosynclinal troughs. As the trough or troughs are filled with sediments and the geosynclines sink still deeper into the crust, their lower portions are subjected to large-scale metasomatism, migmatization, and granitization. H. H. Read has pointed out that in these mobile belts the early pre-orogenic sediments are usually granitized ; synorogenic sediments are less severely affected and post-orogenic sediments are neither metamorphosed nor migmatized.

During the main phase of orogeny the deeply buried sediments, which are most intensely affected by migmatization, may give

rise to palingenetic cross-cutting granites which, due to their lower specific gravity, rise within the mobile belt to form intrusive granitic plutons. This process takes place at depths of 1 km. to 5 km. while from vastly greater depths fractures starting below the upper mantle may reach the surface and facilitate the rise of andesitic magmas from the undifferentiated lower mantle. Partial equilibrium and stability marks the final phase of orogeny, which is characterized by a return to basic and ultrabasic vulcanicity.

This simple scheme needs to be elaborated in detail to cope with the fact that no particular orogeny is exactly like another and that each is influenced in its timing and position by the history and location of its predecessors. B. B. Brock has pointed out that Precambrian orogenic belts are much less extensive than those of later orogenies. There is a series of types, ranging from the short Precambrian belts described from South Africa to the recent belts which encircle the Pacific and extend from the Alps to the Himalayas. In other words, orogenies represent an evolutionary progression and not the repetition of precisely similar cycles.

Beloussov has compiled a tectonic map of the world which illustrates the way in which to-day's geological pattern has been built up from successive orogenies of the past. He assumes an initial phase during which granitic continental platforms, now represented as vestiges within Precambrian shields were extended by a process of continental accretion. Between the Palaeozoic and the Mesozoic increasing injections of basalt resulted in basification of the continental platforms and an increase in overall specific gravity around the margins of the continents. This trend causes the oceans to increase in size and depth, a process called " oceanification." Island arcs, such as, Japan, are interpreted as the foundering remnants of recent orogenic belts. This view assumes that active belts do not migrate and that continental drift does not take place. In common with other authorities, Beloussov pictures the mechanism of folding and mountain building as being due to the rise and fall of deep blocks which cause the superficial sedimentary cover to move and fold under the force of gravity.

It is hoped that these preliminary speculations on the fundamental nature of orogenies will make it evident that, while the relationship between orogenies and ore deposits is discussed in general terms in the next section, no particular orogeny can be viewed without also considering the special circumstances of its particular epoch.

Part 2—Classes of Orebodies Related to Geosynclines and Orogenies

Fig. 3, showing a re-arrangement of Fig. 1, relates the classes of orebodies in the main sub-divisions to their place in the timing and sequence of a typical orogenic cycle.

Borchert has produced an interesting genetic scheme which relates orebodies to two fundamentally different types of magmatism, one group being sialic and palingenetic in origin and the other related to fundamental basaltic vulcanicity and magmatism. This entails a grouping which at one end covers all types of orebody from dunites with Cr, through norites and gabbros with Cu and Ni, to anorthosites with Fe and Ti. It can be questioned, however, whether such a radical division between palingenetic siliceous magmas and primary basic magmas is valid, for

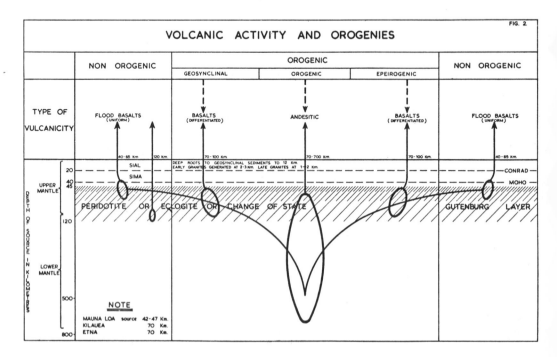

FIG. 2

VOLCANIC ACTIVITY AND OROGENIES

many orebodies are surely due to mixtures of the two.

Basic layered masses and other sub-volcanic intrusives have been included in Fig. 3 but omitted from the discussion that follows, partly because they pose problems that defy immediate answers and partly because they are not germane to the main objects of this article.

One of the major points to be considered is the range and relationship of sedimentary exhalative deposits. Before dealing with particular examples, two points need to be emphasized to meet the criticism levelled by C. F. Davidson against this form of ore genesis on the grounds that it contradicts " actualist principles." From the remarks already made it should be evident that orogenies and ore deposition are essentially evolutionary. It can even be maintained that all large-scale geological processes and the whole history of the earth is an evolutionary sequence, whereby the crust evolves in just the same sense that the biosphere and all living organisms evolve. The present always has many parallels with the past, but individual chapters are never repeated and the sequence is progressive. Precambrian orogenies and ore deposits present features which are unique, just as new features distinguish recent deposits from their predecessors.

Ore deposits formed in pre-Palaeozoic times are notably rich in lithophile elements—such as, Fe, Mn, U, and Th. The giant itabirite deposits of Brazil and other parts of the world have no comparable post-Cambrian successors, nor have the uranium conglomerates; the later deposits of uranium tend to be smaller and associated with palingenetic magmas. In general, with decreasing age there appears to be a diminution in the size of orebodies and an increase in their chalcophile contents, such as might be expected according to the theory of progressive basification.

Petranek has pointed out that one of the important features of Precambrian iron ores is the characteristic lack of oolitic textures, their widespread nature, and their pronounced association with vulcanism of a geosynclinal type. On the other hand, Ordovician iron ores tend to be coastal, shallow water deposits that are very frequently oolitic, limited in extent, and thick or lenticular in habit. The balance of evidence suggests that towards the end of the Precambrian age a vast increase in organic life, more CO_2 in the atmosphere, and lower pH values were responsible for the marked changes in the

style of iron ore deposition, from deep water types to shallow, near-shoreline accumulation. Too little attention has been paid by mining geologists to such evolutionary aspects many of which strongly affect the nature of ore deposits in different epochs.

On the scale on which parallels can be drawn (and many similarities are pointed out in the remarks which follow) it may also be noted that there is yet another reason why Davidson's objections do not appear to be valid. Present-day examples of volcanic ores—such as, the pyrite and sulphur deposits of Japan and the magnetite deposits of El Laco, Chile, are late orogenic manifestations. It would be contrary to all expectations to find present-day volcanoes producing any of the major forms of mineralization associated with early phases of orogeny.

A recent paper by J. Sutton on " Long-Term Cycles in the Evolution of Continents" demonstrates that, in the history of the earth, four major cycles can be discerned behind the shorter successions of orogenies. Each has a length of 750 million to 1,250 million years and they begin in the following periods : (a) 3,600 to (?) (earliest cycle uncertain) ; (b) 2,900 to 2,700 million years ; (c) 1,900 to 1,700 million years, and (d) 1,200 to 1 million years (still in progress). In seeking an evolutionary pattern to changing styles of ore genesis it is probably to these larger cycles that one should look for key periods rather than to the shorter orogenic cycles. It is interesting to note that on this basis the peak period of change in the latest cycle would fall near the Palaeozoic and post-Palaeozoic division and the one before would come in the late Precambrian at a period which has also been suggested for a switch from a pre-actualistic atmosphere to an atmosphere approaching present-day conditions.

(1) Class A

Orebodies Formed in Volcanic Pipes, or Near to the Sites of Volcanic Pipes or Fissures

The first article gave sufficient examples to illustrate the scope and nature of this class and it is not proposed to add to this except with a brief mention of one recent topic of interest.

Russian Diamond Pipes

A long-standing problem that concerns the nature of kimberlite pipes is the absence of any evidence of strong alteration of the rocks surrounding the pipe and of explosive pro-

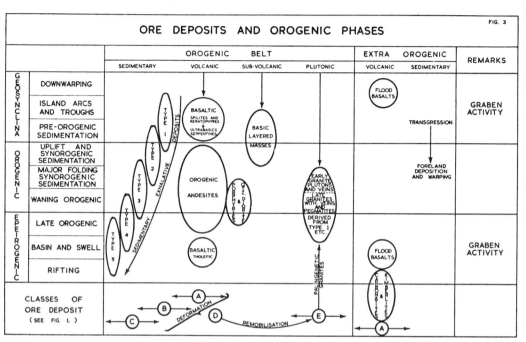

FIG. 3

ORE DEPOSITS AND OROGENIC PHASES

ducts outside it. Recent researches at Yakutia suggest that the diamonds were formed at great depth, long before extrusion, and that the kimberlite emerged at the surface cold, like a crystal mush forced up the pipe and fragmented by this movement and not by explosion. If this is correct it may help to explain the mystery as to how some forms of serpentine were emplaced.

Finally, a small point must be made to elucidate the reason for suggesting that volcanoes constitute the first and most important link in the proposed scheme of ore genesis, in contrast to most schemes which stress the association between granites, intrusive activity, and mineralization. The latter views regard volcanoes as merely ephemeral points of discharge and magma chambers as more fundamental features. Magma chambers, if such exist, are also transitory mixing points, but whereas granite and other magmas form at depths of no more than 1 km. to 5 km. volcanoes associated with mineralization may tap abysmal sources more than ten times as deep. Although a volcano may be no more than a site of discharge, it is at least an established, observable point. Both volcanoes and intrusives are linked by the fact that they are both expressions of very deep-seated fractures in the crust. It is not surprising, therefore, that intrusive rocks often occur in or at the site of volcanic vents, the two phenomena often being complementary.

(2) Class B

Orebodies with Volcanic Associations

Since orebodies are natural phenomena, all schemes of classification face the problem that also bedevils palaeontologists—namely, whether to be a " lumper " or a " splitter " ! This particular section is devoted to further splitting or sub-dividing, in an attempt to relate various forms of sedimentary-exhalative deposits to the orogenic cycle depicted in Fig. 3. It endeavours to fit these deposits into their orogenic setting in time and place, in a way which is of value to the exploration geologist. Five types, corresponding to different orogenic phases, are outlined in the following paragraphs.

Type 1 : Early-orogenic

Pillow Lavas with Cupriferous Pyrite, and Associated Serpentine

The form of the orogenic cycle depicted in Fig. 3 leads one to expect that the earliest manifestations of mineralization should be accompanied by spilitic lavas and serpentine. Moreover, since the deposits may subsequently become deeply buried near the root of the geosynclinal pile, they may be strongly metamorphosed.

The cupriferous pyrite deposits in the pillow lava series of Cyprus, probably of Permo-Carboniferous age, are chosen to

298

exemplify the earliest form of sedimentary-exhalative deposit. Here the Troodos basic complex, including serpentines, forms the bottom of the geosyncline. It is overlain by a lower pillow lava series, now barely recognizable owing to innumerable later intrusions and feeders which ascend to the ore-bearing upper pillow lava series. Mineralization occurs over a vertical range of 2,000 ft., the orebodies being essentially of the massive pyrite type with copper values between 2% and 4% and very minor amounts of As, Zn, Pb, and Ni. It is also worth mentioning that the subaqueous lavas were succeeded by sub-aerial eruptions represented by tuffs within the Jurassic sediments.

The massive sulphide bodies on Balbec Island in the Philippines appear to be younger than the Cyprus ores. Their main sulphide mineral is very fine-grained pyrite, accompanied by chalcopyrite, sphalerite, hematite, and minor amounts of other minerals. The Lorraine orebody averages over 5% Cu. These Philippine deposits form part of a late Cretaceous or early Eocene geosyncline, the sulphide ore being associated with spilitic pillow lavas, cherts, and mudstones at the base of the series. Serpentines, peridotites, gabbros, and other ultrabasic rocks constitute a province rather similar to that of the Troodos basic complex. Moreover, the ore horizons in the Philippines are succeeded by sandstones, shales, and reef limestones, a lithology that also has parallels with the post-mineralization succession in Cyprus.

Some of the iron and pyrite deposits of the Caledonian provinces of Norway, which are typically associated with nearby amphibolites and gabbroic masses, may also fall within this group or be considered as transitional to Type 2.

Type 2 : Synorogenic

Massive Sulphides in Strongly-metamorphosed Country Rocks.

It is well known that degrees of metamorphism are no criteria of age and that some of the most important Precambrian orebodies occur in virtually unaltered environments. It is not unreasonable to suspect that highly-altered geosynclinal rocks are either pre-orogenic or synorogenic and that, broadly speaking, degrees of alteration tend to reflect orogenic relationships.

The Samreid Lake sulphide deposit in Ontario is chosen to illustrate an orebody corresponding to " Type 2 " in Fig. 3. This orebody, which is set in Precambrian terrain,

lies between a northern series in which amphibolite-bearing rocks predominate and tuffs, flows, and quartzites are subsidiary and a southern area consisting mainly of metatuffs, quartz siltites, and quartzites with amphibolite sporadic or absent. The lower part of the succession was invaded during a late stage of the orogeny by numerous dolerite dykes.

The orebody occurs in a pseudo quartzite and it is inferred that most of the original mineral deposition took the form of magnetite that was laid down in a marine environment during a period of intense volcanic activity. Much of the magnetite may have been converted to sulphide at an early diagenetic stage. The minerals which constitute the bulk of the present orebody are pyrite and pyrrhotite, with intergrowths of magnetite, while chalcopyrite and cubanite occur in sub-economic quantities.

The mineralization grades from disseminated to massive, the more important portions being interbedded with the quartzite where it is close to flows. Locally, veinlets of pyrite and calcite cut the massive pyrrhotite ore and alternations of beds rich in magnetite and poor in magnetite are characteristic.

The pre-metamorphic nature of some of the rocks is difficult to determine. However, many of the amphibolites are thought to represent metamorphosed tuffs and the mineralized quartzite may have been a chert.

This type of massive sulphide body is thought to be essentially the same as the Rio Tinto type which follows it, with the exception that it is characteristically pyrrhotite-rich and more strongly affected by metamorphism which followed ore deposition.

Some of the massive sulphides in the Sherridon area of Manitoba and also in New Brunswick are typical pyrrhotite-pyrite bodies with Cu and Zn in settings of high-grade metamorphism and late sub-volcanic intrusions of basic composition.

Type 3 : Late Orogenic

Massive Sulphides in Unmetamorphosed or Mildly Altered Settings

The succession of exhalative types shown in Fig. 3 is not intended to suggest that a particular orogeny produces a corresponding succession of ore deposits. In fact an orogeny commonly evokes only one period of mineralization and, as might be expected, the most productive period tends to coincide with the waning phase of andesitic vulcanicity or the late fumarole stage. Mineralization associated with the earliest and latest episodes of vulcanicity tends to be more specialized and limited in size ; for example, epithermal volcanic-gold veins are a typical tail-end product, rich but small. It may also be true that during the course of geological time the whole nature of the mineralizing phase of orogenies has fundamentally changed. For example, the Mesozoic orogeny which affected the Verkhoyansk-Chukot province in N.E. Siberia produced numerous major mining fields, whereas the ensuing Cenozoic orogeny which involved a shift in vulcanicity to the Pacific or Korzak-Kamchatka province, merely produced occurrences of Hg and Sb. Assuming that most primary minerals are produced from the undifferentiated portion of the mantle, then the scale on which orebodies are produced must be indirectly linked with the extent of deep fissuring and the magnitude of the particular orogeny as a whole.

Massive sulphide bodies within successions whose sediments have not been altered to amphibolitic rocks, or to hornblende gneisses, etc., appear to be the commonest type. Hence it may be deduced that a late orogenic setting for sulphide deposits is more common than an early one. It also appears to be very roughly true that a progression from Type 1 to Type 5 shows tuffs becoming relatively more abundant than flows and a progression from basic to acid vulcanicity until finally in Type 5 there are no close links with vulcanicity and tuffs become only a minor part of the succession.

Another feature which undergoes a progressive change characteristic of different orogenic stages is the form of the associated tuffs and lavas. Spilites, keratophyres, and Na-rich volcanics mark the early stages, while basic sub-volcanic intrusives such as gabbros may be close neighbours. Andesites and trachytes tend to be synorogenic, while orebodies representative of Types 3 and 4 are generally associated with rhyolites and acid tuffs. It must be borne in mind, however, that volcanoes display complex rhythms and changes in types of activity, so that only trends can be discerned but no hard and fast rules. Van Bemmelen, in describing the volcanic history of the Java–Sumatra geosyncline, points out how normal andesitic activity is from time to time interrupted by gigantic eruptions of acid ignimbrites which mark pulses in the orogenic cycle. He has also noted similar phenomena associated with the Alpine orogeny. Hence on a detailed scale, minor phases produce marked changes from the type of eruption characteristic of the main phase.

The pyritic deposits of the Huelva district of Spain and of Alentejo in Portugal are obvious examples of Type 3. Details were given in the previous article, so that here it is only necessary to draw attention to the fact that these deposits are late or epeirogenic. They generally occur at the transition from acid flows to tuffs and the associated sediments are largely altered, but the scale is low-grade.

Types 4 and 5 : Final Orogenic Phases

All the first three types described have many features in common. Volcanic rocks form an important and integral part of their lithology and any differences between them are mainly due to different grades of metamorphism and metasomatic replacement rather than to different types of sedimentation. All are associated with deep geosynclinal troughs and greywacke sediments. They may or may not contain economic quantities of Cu, Pb, or Zn, but either pyrite or pyrrhotite is the predominant mineral.

Types 4 and 5 were deposited after the main events of the orogeny, under altogether less spectacular tectonic and volcanic conditions. While still pyritic, they are less markedly so and in a complete sequence they display a distinctive form of metal zoning with a characteristic suite of minerals. Their setting is predominantly calcareous or dolomitic.

Rammelsberg in W. Germany, Meggen in Westphalia, and Mt. Isa in Australia are typical examples of Type 4. The Meggen deposit, although folded during the Variscan orogeny has not been greatly altered. Detailed investigations by Ehrenberg, Pilger, and Schroder have enabled them to reconstruct in considerable detail the original conditions under which the deposit was formed. During Lower Devonian times a series of keratophyre eruptions took place nearby and the footwall beds of the orebodies contain tuffs which were laid down in fairly shallow water in association with an alternating series of shales, sandy beds, and muddy limestones. A large reef eventually formed a barrier to the north and an important feature of the ore horizons is that they occur within a calcareous series largely of biochemical and volcanic origin, with little debris introduced from outside sources. The basin of deposition was a cut-off arm of the sea some 3 miles in length with a rise towards its centre which marked the locus of ore deposition. The ore

300

was formed by a series of submarine exhalations from minor vents or diatremes, which in the early stages were rich in Fe and Zn while the later stages produced minor Pb and large quantities of Ba. Various types of ore occur, from finely banded to massive varieties. Colloidal and nodular ores are also characteristic and it is suggested that much of the sulphide formed as gels in unconsolidated mud. Fine alternating bands of sphalerite and galena, such as occur at Mt. Isa, are not typical of either Rammelsberg or Meggen but they do occur and at Rammelsberg there is a hanging-wall (= " inverted foot-wall ") orebody known as the " kneist " which is a brecciated, siliceous rock that may be analogous to the silica dolomite at Mt. Isa. Evidence of the original source of the ore fluids is lacking at both Rammelsberg and Mt. Isa but both orebodies appear to be due to deposition in a closed basin receiving alternating supplies of calcareous mud, organic material, and volcanic ash. Meggen is mainly a producer of pyrite, sphalerite (minor galena), and barytes whereas Rammelsberg contains a larger suite of minerals beginning with arsenopyrite, followed by pyrite, chalcopyrite, sphalerite, galena, and finally barytes. Mixed sequences occur and also massive ore which is thought to indicate proximity to the fumarole source, while the fine banded ores tend to thin out towards low-grade areas. Except for coarse sulphides in late cracks and fissures the ore is extremely fine grained, 80% of the massive ore at Rammelsberg having a grain size of less than 60μ.

Biogenetic sulphides are common in all these deposits, the black carbonaceous facies being notably rich in large colonies. Not unnaturally, muddy calcareous environments, rich in organic material, produce complex diagenetic changes, some of which may account for minor textural differences between the various examples.

(3) Class C

Orebodies with Mainly Sedimentary Associations

Type 5 : Post-orogenic (Coastal Deposits)

(a) Ireland

Recent research has highlighted the probability that besides the chemical precipitation of sulphide ores from submarine fumarole sources it is also possible that bacteria play an important role. Hence certain types of stratiform orebodies may be both exhalative and biochemical in origin. Deep stagnant troughs are known to be rich in bacteria and so are cut-off marine basins, lakes, and estuaries. Love has recently commented that biogenetic pyrite is forming at the present day and is far more widespread than is commonly supposed. Deep euxenic conditions for the development of biogenetic sulphides are not essential, provided that there is a sufficiently high level of organic growth. They may be produced by bacteria living on the decaying parts of dead foraminifera in essentially oxygen-rich environments. Algae may also play an important part. Whether bacteria are a direct cause of the formation of base metal sulphides or whether they act indirectly by producing the pyrite that traps base metal ions which are in local excess of equilibrium is not known ; possibly both forms of sulphide precipitation occur.

Type 5 deposits are mainly associated with back-reef facies or lagoonal areas in coral seas along the type of coastline that is protected by a fringing barrier with reef development protecting it from the open water— generally it is a coastline of submergence.

The combination of fumarole activity in or near the type of environment described in Type 4 is necessarily a rare occurrence. On the other hand, volcanic activity is common along many coastlines ; this may explain why there are few deposits like Type 4, whereas the number of bedded or stratigraphic deposits in limestone and dolomite is very large indeed. That there is a close link between a particular type of off-reef deposition which produces a chemically and bacterially favourable environment and regional volcanic sources which produce the initial supply of metallic solutions is suggested in the first instance by the following facts, derived from work on Irish deposits.

The Silvermines deposit in Tipperary occurs in the Lower Limestone Shale division of the Carboniferous. Part of it outcrops and hence has been converted to oxide and supergene forms of mineralization, added to which the deposit has been strongly faulted, resulting in increased superficial alteration and concentration near the fault zones. However, ignoring these later effects, the form of the primary sulphide deposit is quite clear. Sandstones, representing a near-shoreline facies, grade upwards into calcareous shales. The earliest mineral deposition is represented by arsenopyrite which is overlain in turn by dolomites with sphalerite and galena followed by a chert bed which marks the base of a massive bed of barytes. A second chert layer with minor beds of pyrite overlies the barytes and this is capped by reef limestone. Minor copper has been found in the nearby out-cropping sandstones. The sequence displays the full series normal to a sedimentary-exhalative deposit of Type 4 (As, Fe, Cu, Pb, Zn, Ba), together with a change of facies from near-shoreline deposition to the deeper water back-reef facies. The "Shallee Shale" is suspected to be a tuff related to the tholeitic basalts in the Old Pallas complex some 20 miles to the south-east. A notable feature of this region is that in addition to Silvermines there are a large number of less well-known, abandoned mineral deposits which all appear to be grouped around and related to the Old Pallas volcanic centre.

Recent drilling at other sites in Eire shows that characteristically the sulphides occur as disseminations and beds within a back-reef facies. The more carbonaceous or bituminous horizons may include nodular beds of pyrite with sphalerite and galena ; arsenopyrite is characteristic of the earliest deposition and most of the deposits are underlain by a basal tuff. The sulphides, in facies rich in crinoid debris, may occur as fine-grained wisps typical of turbulent deposition ; they are developed in areas which show minor slump breccias such as occur near reef atolls. Coarse sulphides always seem to be post or late dia-genetic and the reef limestone above the sulphide-bearing beds may contain numerous calcite veinlets with coarse galena and sphaler-ite re-distributed by meteoric water. Occa-sionally chalcopyrite is the main sulphide.

Contemporaneous volcanic activity of a late orogenic type in the area of these deposits and the common presence of a basal tuff low in the sequence suggest a tenuous link between vulcanicity and mineralization which becomes stronger when these deposits are compared with similar ones in other localities where the volcanic source of mineralization is more obvious.

The author's former classification placed this type of deposit in Class C (mainly sedi-mentary associations) because, despite the growing evidence of a parent volcanic source, most of the features of these bedded Pb/Zn deposits in limestones are purely sedimentary in origin. The reason for assigning these deposits to Type 5 at the tail-end of the exhalative series is to illustrate a full orogenic

cycle ranging from the earliest form of deposi-tion in a young geosyncline to the final stage of " basin and swell " tectonics at the end of the volcanic sequence before flood basalts, characteristic of non-orogenic periods, mark a return to conditions approaching equilibrium.

(b) Alpine Deposits

More than 20 base-metal deposits have been described in the Trias or post-geosynclinal series of the Alps. They range from the Upper Permian to the Lower Karnian and have been the subject of studies by Taupitz, Hegemann, Maucher, Schneider, and other authorities. The best-known deposits—such as, Bleiberg, Raibl, Gorno, and Monte Calisio, are situated between the Tyrol or Ostalps and the South-ern Alps. They all occur in a limestone-dolomite series with volcanics towards the base.

The diagenetic changes that take place in calcareous base-metal deposits are complex and often destroy much of the more obvious bedded features. Recent papers by Ramdohr, Amstutz, El Baz, and Parks give excellent accounts of such processes.

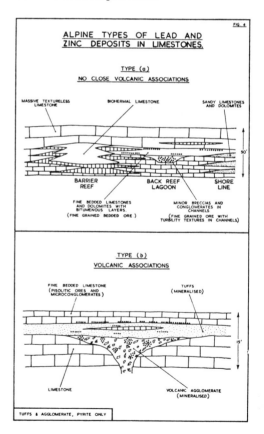

Fig. 4 shows the types of mineral occurrence which may be shown to (a) represent the upper or Karnian type and (b) the lower or Ladinian type. Minor volcanism, of intermediate and basic composition, took place throughout the Ladinian and the type of occurrence depicted in (b)—sketched from an outcrop at Gorno—illustrates mineralization directly related to a small volcanic vent. Here the Zn ore above the tuff has an interesting pisolitic form with minute rings of bitumen within the spherulites of sphalerite. In short, there are two main types of mineralization, with variations between them— namely, deposits directly linked with vulcanicity and deposits which have no such immediate links and may be due to local erosion and re-deposition. In this sense there is a complete transition from truly exhalative types to predominantly sedimentary types. Bleiberg and Raibl suggest a third type, contemporary but vein-like in form.

The non-volcanic types generally show simple mineralogy with predominant pyrite, blende, and galena, while the volcanic types may contain Zn, Pb, Cu, As, and Sb. All of them occur mainly in back-reef facies with a preferential enrichment of Cu in the carbonaceous black shale type of environment and Zn in the dolomitic facies. The near-shoreline facies may be rich in characteristically saline constituents—such as, anhydrite, celestine, and barytes.

No two deposits are ever exactly alike and the foregoing brief remarks are merely intended to demonstrate family relationships and to suggest some close parallels between the Alpine and Irish Pb/Zn fields. Many of the Iranian Pb/Zn deposits also belong to this group or Type 4. They occur, like similar deposits in Turkey, in the Lower Cretaceous or Mesozoic.

Before dealing with the next two classes of ore deposit it may be appropriate to re-consider the evolutionary aspects of ore deposition previously mentioned in connexion with the change in character between early Precambrian and Ordovician iron ores. If Beloussov is correct in supposing that there was a gradual change in world vulcanicity at some time between the Palaeozoic and the Mesozoic, which resulted in a greater outpouring of basic lavas and the onset of oceanification, then there should also be a perceptible change in the character of the ore deposits related to vulcanism. Brief consideration of the range of deposits between Type 1 and Type 5 suffices to make it evident that this does

indeed appear to be true. The vast majority of the classic examples of Types 2 and 3 are either of late Cambrian or Palaeozoic age. Moreover, the author knows of few examples of Type 4 of post-Palaeozoic age and while deposits of Pb/Zn in limestone range from the Precambrian to the Quaternary they change markedly in style between the two extremes in time.

If a better understanding of the nature of orebodies is to be achieved, then the evolutionary nature of all geological processes, as opposed to the more obvious ones which take place in the biosphere, must be given much closer study. In the past the way in which most " Pb/Zn in limestone orebodies " have been indiscriminately described as hydrothermal replacement phenomena has greatly handicapped this type of understanding. Major changes in organic life are not isolated phenomena ; every process, ranging from sedimentation to ore deposition, must be affected and in its own way reflect the sum of evolution. This view is advocated not to decry the method by which geologists gain an understanding of the past by studying the present but to set a limit to its application.

(4) Class D
Deformed Orebodies
(a) Avoca (Ireland)

An important feature of the scheme of classification illustrated in Fig. 1 is that it demonstrates the essentially transitional nature of the relationships between different types of orebodies. Folding, shearing, and metasomatism gradually mask the initial nature of the ore deposit. Class D, or " Deformed " deposits, still show many original features but their later forms illustrated by " Remobilized " and " Rejuvenated " deposits may be too altered to be recognizable as variations of the originals.

The Avoca massive pyrite deposits in Ireland occur in a belt of mixed volcanic and sedimentary rocks of Ordovician age, the sulphides being associated with transitions from flows to tuffs. Since numerous full accounts of these relationships have already been recorded, no detailed comments will be made on their more obvious features.

The Avoca deposits are assigned to Class D because they are heavily sheared bodies. Murphy's account notes that the crests and noses of the orebodies strongly transgress the schistosity, a feature which seems to indicate that the beds of sulphide were strongly folded before being sheared. The deposits display

abundant signs of crushing and movement
and locally the ore simulates the pale, homogeneous, but highly schistose type so common
at Mt. Lyell, Tasmania.

The hanging-wall of the " Pond type " orebodies at Avoca is often rich in Pb and Zn
and in places it may run as high as 4% Pb
and 8% Zn, thereby yielding an ore that is
almost identical to the type known in Spain
as " complejos." Some specimens from this
hanging-wall still reveal original bedding in
the form of thin alternating bands of pyrite,
galena, and sphalerite. Shearing shows
initially as a smearing of the bands and
ultimately coarse areas of re-crystallized sulphide may develop.

Most of the ore (which was mined for
its copper content) was in a form known
as " South Lode type," which shows transitions to the " Pond type " that commonly
occurs on either side of it. In the author's
opinion the South Lode type is the more
heavily sheared portion and originally consisted of sulphides with a fair proportion of
interbedded sediments and volcanics. Much
of its quartz banding and minor boudinage
structures look like relict bedding, though the
" streaky bacon " texture that can often be
seen may also have been derived from quartz
which was sweated out of the original beds
and became distorted and re-distributed
almost beyond recognition. Another interesting and suggestive feature of this ore is
the presence of thin platy layers of graphite,
a further pointer suggesting sedimentary,
bedded ancestry.

The pyritic lodes extend for a total strike
length of 11 miles and are followed at either
end by banded hematite/magnetite bodies.
This distribution of minerals is analogous to
the well-known relationship in Ontario described by Goodwin, with base-metal sulphides
deposited in the deeper troughs and oxides in
shallower water.

Remobilized Orebodies

(a) Sardinian Pb/Zn Deposits (Iglesias Region)

" Remobilized deposits " constitute a class
in which strong tectonic or metamorphic
action has so altered the initial features of
the orebodies that they no longer bear close
resemblance to their original form. Since
describing the Zawar Pb/Zn deposit in India
as an example of this class the author has had
the benefit of visiting numerous deposits in
Sardinia which display some similar features
and has discussed these deposits with

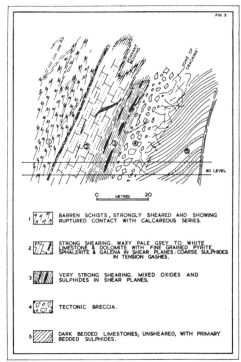

Cross-Section, Iglesiente, Sardinia.

W. Münch who had previously arrived at a
similar interpretation to the one suggested
for Zawar.

The Sardinian deposits are in Cambrian
limestones and dolomites which have been
subjected to intense shearing and the textures of the sheared sulphide ore are almost
identical with those at Zawar. In both areas
fine-grained sphalerite and galena is smeared
along the shear planes and tension gashes are
filled with coarse re-mobilized sulphide. The
calcareous host rocks in Sardinia are tightly
folded although the form of folding has been
largely destroyed except in the less sheared
parts of the orebodies. Schists close to the
orebodies can be shown locally to represent
both sediments and tuffs.

The Sardinian deposits differ from those at
Zawar in two important respects. First, in
some places sizeable blocks of unsheared rocks
demonstrate beyond question the original
sedimentary nature of the deposit and,
secondly, unusually intense oxidation to
depths of more than 700 m. has produced
oxide bodies that in economic terms are more
important than the sulphides. Fig. 5 is a
diagrammatic section through the Iglesias
region, which illustrates the main features of
a typical orebody at a level where oxidation

is still a minor feature. The relationship between grades in unsheared ore, which is a rare and possibly unrepresentative form, and sheared ore suggests that while shearing produced great changes in the limestone and in the appearance of the ore there may have been but little sulphide migration and enrichment except in a few small tension gashes. Oxidation, on the other hand, may upgrade the Pb and Zn by a factor of between 3 and 8.

The importance of the Sardinian deposits from the point of view of ore genesis is largely due to their extensive geographical spread which enables localities to be examined where the original sedimentary form of the deposits and associated tuffs has not been obliterated by later events. It is hoped that in due course W. Münch will publish the wealth of information that he has accumulated about these Sardinian deposits, including evidence bearing on the successive stages which eventually mask their true nature. Economically they are classic examples of the development of oxide orebodies.

A Brief Note on Vein Deposits Classes A–F

While it is broadly true that the commonest types of veins have a granitic parentage, large numbers have no connexion with granite. The following remarks are necessary in order to put the viewpoint of the original article on ore genesis into closer focus.

Amstutz has postulated that veins in sedimentary rocks are usually sedimentary in origin and syngenetic in the broad sense of the term, that veins in volcanic rocks have a volcanic origin, and that veins in or near granites are derived from granitic sources. A few examples will serve to illustrate this dictum which, though subject to inevitable exceptions, expresses a most important truth which is often obscured by needless complications of a psychological rather than a scientific nature.

(a) Volcanic Veins

These occur in or near volcanoes and are directly related to them. They are shown in Fig. 1 as a sub-division of Class A. Typically they are epithermal and theoretically there is a distinction between epithermal veins of direct volcanic origin, veins due to hot meteoric water, and veins due to mixtures of the two. Nevertheless, since no practical distinction can be made between them, it is simpler to consider that any vein close to a volcanic centre is probably volcanic in origin, regardless of whether it is directly related to

extrusive activity, to fumarole activity, or to heated ground water.

The Pacific coast of America has literally hundreds of small vein occurrences of telluride ores and Au/Ag mines. Cinnabar, stibnite, and mercury veins are also typical of the late orogenic, epithermal, volcanic veins and they all have their counterparts on the opposite side of the Pacific. In Fig. 1, Tonapah, Nevada, and Telluride, Colorado, have been noted as typical examples.

It is debatable whether deposits of this nature should be considered as epigenetic or syngenetic, but the author favours Amstutz's contention that, although the veins are cross-cutting features in a local sense and may have formed either during or after the main period of vulcanicity, they are intimately a part of it and occur within the same general unit of time and space. It is thus more logical to consider them as syngenetic rather than epigenetic ; the same argument applies to diagenetic veins and veins derived from sedimentary ores.

D. Williams advocates a hot-spring origin for the native sulphur deposits of Sierra de Gador, S.E. Spain, which are closely related to late Tertiary and early Quaternary volcanism in the Sierra Gata. The sulphur is a metasomatic replacement largely confined to beds of Triassic dolomitic limestone and Tertiary limestones. Although not vein deposits, they conveniently illustrate the point that veins and hot-spring deposits are all intimately linked. In Japan there are both extrusive and hot-spring sulphur deposits and a minor but interesting point is that while almost identical in many ways the sulphur and marcasite of hot-spring origin is rich in phosphorus (the same is true in Spain).

These sulphur deposits and similar orebodies emphasize pitfalls in the use of the terms " epigenetic " and " syngenetic," for clearly the sulphur is syngenetic in relation to the accompanying vulcanicity and epigenetic in relation to its immediate surroundings. On the whole, the use of either term is liable to produce confusion, particularly in view of the fact that both have acquired undertones which are liable to be misleading however employed.

(b) Sedimentary Veins or Veins Derived from Sediments by Diagenetic and other Processes

Metamorphism and metasomatism are potent factors in re-distributing and re-concentrating ores and, since the last events which

305

affect an orebody are often the most obvious, it is not surprising that this suggests that they are fundamental processes. However, it is becoming more and more evident that in practice the earliest factors which cause a re-distribution of ores—namely, those which occur during diagenesis and close to the sur-face—are usually the most important. In some cases, superficial oxidation below old pene-planes is particularly important.

Sandstones and conglomerates are physic-ally comparatively stable, but many of the most important bedded mineral deposits occur in muds, particularly calcareous muds, which are very far from being stable after deposition. Initially they may be thixotropic and conducive to the development of nodular and colloform ores. Fine-grained calcareous deposits re-crystallize and may become dolo-mites and compaction may produce numerous cracks and openings. These zones of cracking are commonly filled with calcite and quartz veinlets carrying coarse sulphides, due to diagenetic or immediately post-diagenetic dispersion.

Ground water often circulates upwards rather than horizontally and in this connexion it is significant that the bedded sulphides in Ireland are frequently overlain by a zone of veins and veinlets, some of which may even reach mineable dimensions. Many veins of this type were the first outcropping signs of ore to attract attention. Indeed, on reading the standard descriptions of known ore occur-rences in Ireland the impression is gained that most of them are minor hydrothermal veins. These Carboniferous veins seem to be a wide-spread diagenetic expression of original bedded ore and since they are by-products of sedimentation they may be referred to as " sedimentary veins."

Diagenetic or post-diagenetic veins of the type previously described by the author as " leakage veins " represent a slightly more complex type. The stockwork veinlets be-neath and alongside the pyrite lodes at Rio Tinto, Spain, are an example.

Barytes veins such as those at Gasswater and Muirshiels in Scotland represent yet another form of sedimentary vein, both of them due to ground water movement. These mines are examples of open fissure fillings with colloform barytes and almost no impurities other than minor late-formed calcite veins and impregnations. The evidence that these veins are due to precipitation from lateral or down-ward-moving meteoric water and not from rising hydrothermal solutions is as follows. The lowest levels are currently at the bottom of the fractured zone and it is possible to trace a downward change from pure barytes, to barytes breccias or autobreccias re-cemented by younger barytes, and then to a breccia large-ly composed of waste rock scaled off the walls and encased in a cement of mud. Finally, the bottoms of the fractures are almost wholly filled with mud. No feeders penetrate the mud, which would clearly have constituted a total barrier to any ascending hydrothermal solutions, and it seems likely that the charac-teristic way in which these barytes deposits " bottom " can only be accounted for by assuming that they are due to underground water circulating in fault fractures. A similar explanation may account for many fluorite veins.

Boyer and Routhier have described lead vein deposits in the Monts de Minervois, France, which although veinlike in form are closely related to sedimentary controls and confined to a particular stratigraphic horizon. It is concluded that in the first instance these deposits must have been of sedimentary origin. Veins of this type, which are com-monly assumed to be of hydrothermal origin, seldom have much economic potential and tend to be abandoned before their bottom limits are reached by mining. However, in the few cases where it is possible to see the full extent of the fissures their form at the bottom is often so different from that of true hydrothermal veins that the author believes a great deal of present-day confusion and uncertainty would be avoided if more veins were fully exposed by mining.

In the absence of conclusive proof con-cerning the nature of stratigraphically con-trolled veins it seems reasonable to adopt Amstutz's suggestion that in a sedimentary environment, unless there is strong evidence to the contrary, it is safest to assume that the veins are of diagenetic or superficial type. Needless to say, the dearth of notable wall-rock alteration alongside such veins is highly characteristic and their gangue is normally limited to minerals prevalent in the sur-rounding country rocks.

Part 3—Exploration

The first article in this series drew atten-tion to the importance and economic potential of exploration specifically directed towards the discovery of bedded or sedimentary deposits. The main points then discussed emphasized the need to study special environ-ments. It is now proposed to apply some of

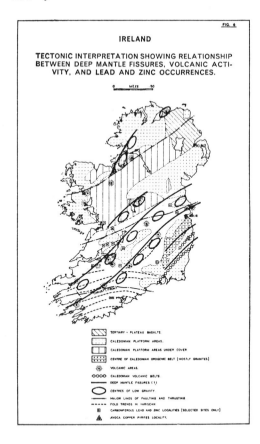

IRELAND

TECTONIC INTERPRETATION SHOWING RELATIONSHIP
BETWEEN DEEP MANTLE FISSURES, VOLCANIC ACTI-
VITY, AND LEAD AND ZINC OCCURRENCES.

the conclusions of Parts 1 and 2 to exploration problems in terms of regional tectonics.

A feature of exploration in the past has been the preference shown by geologists for starting exploration near abandoned mines or sites of known mineralization. However, as time goes on opportunities of this kind decrease and it may be possible to reap greater rewards by exploring areas where deductive reasoning suggests that orebodies might be found, although no direct evidence of their presence may exist. The kind of regional analysis that must lie behind this type of exploration programme is well illustrated by a recent paper by V. T. Matveyenko and E. T. Shatalov on the north-eastern part of Siberia.

Recently the present author has been closely connected with the search for Pb/Zn deposits in Ireland which has led, by a process of rather similar deduction, to a successful discovery in an area where no mineralization was previously suspected, even though Ireland has been carefully prospected for over 100 years. Because this exploration programme is still under way and the territory

is a highly competitive one, only broad assumptions can be mentioned here, for the detailed study necessary before applying these principles to field practice cannot be discussed for the present. For similar reasons, Fig. 6, which shows some of the more important tectonic relationships of certain Irish deposits, is diagrammatic rather than factual.

The opening paragraphs of the present article propounded the theory that major sub-mantle fissures are one of the principal regional controls that govern the distribution of mineral provinces, an aspect dealt with at some length in the above-mentioned Russian paper. It must be pointed out, however, that these initial, primary fissures seldom have any surface expression. They are camouflaged features whose presence must be deduced from the alignment of granites and from the prevalent trend of dykes and hydrothermal veins, also from the distribution of volcanic centres.

In South Africa the diamond pipes on the line Kimberley–New Elands–Theunissen suggest a major sub-mantle fissure and so does the northerly line Koffiefontein–New Elands–Orkney. Pipes occur at intersections while other features point to the intervening presence of deep fissuring.

There is a carbonatite alignment in Canada from Oka, through Lake Nipessing to Wawa. It is also of interest to note that the individual pipes and complexes tend to be aligned with their long axis on the main fissure direction.

In Ireland the alignment of the Leinster granite and its accompanying belt of volcanic rocks suggests the presence of an underlying sub-mantle fissure. "Secondary" fissures and major rifts exposed at the surface also tend to be closely related to sub-mantle breaks. For example, the line of the Linnian fault (Highland Boundary) and the northern edge of the Longford massive (Southern Uplands fault) are obvious major breaks of this nature.

The boundaries between different orogenic provinces, such as the border land between the old Caledonian and Variscan platforms, also tend to be active zones. The relationships shown on the diagram are taken from Coe's book *The Variscan Fold Belt in Ireland*. Four lineaments of volcanic activity, deduced partly from the distribution of known Ordovician and Carboniferous volcanics and partly from Professor Murphy's gravity work, are also depicted. The overall pattern which emerges shows a coincidence between regional gravity trends, vulcanicity, and known mineral occurrences. Particularly at points where volcanic lines meet the old shore lines.

In the earliest days of the search for lead/zinc deposits in the Carboniferous of Ireland it was assumed that they were of sedimentary origin and likely to have been formed along stretches of coast with off-shore barriers due to coral reefs. Since the borders of the Caledonian platforms closely coincide with old shore lines, F. Moss, who initiated the exploration programme, was thus able to narrow down the preliminary areas to be prospected. A later assumption was that the source of the mineralization was closely related to volcanic activity and this led to a further localization of exploration and the type of mineral zone suggested by Fig. 6. Some of the better-known Pb/Zn occurrences in Ireland are shown on the plan in a way which it is hoped will clarify their relationship to shore lines and volcanoes. The Ordovician massive pyrites deposit at Avoca has also been marked since it occurs in a typical tectonic setting.

Regional geological analysis is an aspect of exploration that must assume increasing importance as the number of undiscovered outcropping occurrences becomes fewer and fewer.

Finally, it must be emphasized that the essence of this approach is not to seek favourable structures in which the orebodies themselves may occur, but to analyse the major tectonic features so as to show up sub-mantle fissures which may be the fundamental mineral sources, although their presence can only be deduced from large-scale, regional criteria.

The author wishes to express his thanks to the Rio Tinto Zinc Corporation for permission to publish this article and to Professor D. Williams, Mr. R. F. Lethbridge, Mr. F. Collender, and Dr. W. Münch for helpful criticisms and suggestions. Also to Mr. B. Jordan, who drew the diagrams.

References

AMSTUTZ, G. C., EL BAZ, F., and PARK, W. C. " The Diagenetic Behaviour of Sulphides." [1]

AMSTUTZ, G. C. " L'Origine des Gites Mineraux Concordant dans les Roches Sedimentaires." *Chronique des Mines*, No. 308, 1962.

— — " Some Basic Concepts and Thoughts on the Space-Time Analysis of Rocks and Mineral Deposits in Orogenic Belts." *Geologische Rundschau*, Vol. 50, 1960.

BELOUSSOV, V. V. " Basic Problems in Geotectonics " (McGraw Hill, 1963).

BELOUSSOV, V. V. Tectonic map of the world. *Geologische Rundschau (G.R.)*, Vol. 50, 1960.

VAN BEMMELEN, R. W. " Zur Mechanik der Ostalpine Deckenbildung." *G.R.*, Vol. 50, 1960.

VAN BEMMELEN, R. W. " Volcanology and Geology of Ignimbrites in Indonesia, North Italy, and the U.S.A." *O.G. en M.*, Dec., 1961.

BORCHERT, H. " Zusammenhanger zwischen lagerstattenbildung magmatismus und geotectonik." *G.R.*, Vol. 50, 1960.

BOYER, F., and ROUTHIER, P. " Observations sur Deux Niveaux Minéralisé dans le Palaeozoique Inferieur des Monts Minervois (Montagne Noire, Ande)." *Soc. Geol. de France*, 6e Serie t. VIII, 1958.

BROCK, B. B. " On Orogenic Evolution with Special Reference to South Africa." *Trans. Geol. Soc. S.A.*, Vol. LXII, 1959.

BOWLER, C. Personal discussions on Irish deposits.

EGYED, L. " On the Origin and Constitution of the Upper Part of the Earth's Mantle." *G.R.*, Vol. 50, 1960.

EHRENBORG, H., PILGER, A., and SCHRODER, F. " Das schwefelkies-zinkblend-schwerspatlager von Meggen." *G. Jb.*, H. 12, 1954.

COE, — " Some Aspects of the Variscan Fold Belt in Ireland." Abstract from above, University Press, 1962.

GILMOUR, P. " Notes on Non-Genetic Classification of Copper Deposits." *Econ. Geol.*, Vol. 57, 1962.

GOODWIN, A. M. " Volcanic Complexes and Mineralization in Northern Ontario." *Canad. Min. J.*, Apr., 1962.

HESS, H. " Serpentines, Orogeny and Epeirogeny." *Geol. Soc. America*, Sp. paper, Vol. 62, 1955.

HIRST, T. " Exploration for Manganese in the Guianas and Venezuela." Guiana Conference, 1957.

JOHN, T. U. " Geology and Mineral Deposits of East-Central Balbec Island, Palawan Province, Philippines." *Econ. Geol.*, Vol. 58, 1963.

KOSTELKA, L. " Windisch Bleiberg ", 1960.

LOVE, L. G. " Early Diagenetic Pyrite in Fine Grained Sediments and the Genesis of Sulphides." [1]

— — " Biogenetic Pyrite in Recent Sediments of Christchurch Harbour, England." *Amer. J. Sci.*, May, 1963.

MATVEYENKO, T. V., and SHATALOV, E. T. " Fractures, Magmatism, and Mineralization." *International Geol. Review*, Vol. 5, No. 2 and 3, 1963.

[1] Paper presented at 6th International Sedimentological Congress, Symposium on Sedimentology and Ore Genesis, May, 1963, Delft.

MAUCHER, A., *et al.* " Geologische-Lager-statten Kundlich unter Suchunger im Ost-pontischen Gebirge." Munich, 1962.

MOREL, S. W., and BEAR, L. M. " Memoir No. 7." Geological Survey of Cyprus, 1960.

MÜNCH, W. Personal discussions on Sardinian Pb/Zn deposits.

MURPHY, G. J. " The Avoca Enterprise." *Mine, Quarry Engg.*, July and Aug., 1959.

MURPHY, T. " Some Unusual Low Bouguer Anomalies of Small Extent in Central Ireland and Their Connection with Geological Structure." 22nd meeting, European Assoc. of Expl. Geophysicists, London, 1962.

Also—personal discussions on his regional gravity map of Eire.

PEREIRA, J. " Reflections on Ore Genesis and Exploration." THE MINING MAGAZINE, Jan., 1963.

PETRANEK, J. " Shallow Water Origin of Early Palaeozoic Iron Ores." [1]

RUTTEN, M. G. " The Geological Aspects of the Origin of Life on Earth." Elsevier Press, 1962.

SCHNEIDER, H. J. " Facies Differentiation and Controlling Factors for the Deposition of Pb/Zn Concentration in the Ladinian Geosyncline of the Eastern Alps." [1]

SCHULZ, O. " Lead-Zinc Deposits in the Calcareous Alps as an Example of Submarine Hydrothermal Formation of Mineral Deposits." [1]

SHEYNMANN, YU. " The Mohorovic Discontinuity, Depth of Magma Origin, and Distribution of Ultrabasics." *International Geol. Review*, Vol. 5, No. 2, 1963.

SUTTON, J. " Long-term Cycles in the Evolution of the Continents." *Nature*, May, 1963.

TAUPITZ, K. C. " Erze Sedimentarer Entstehung auf Alpinen Lagerstatten des Typus ' Bleiberg '." *Z. fur E. u M.*, Bd. VII, 1954.

WILLIAMS, D. " Sulphur Deposits of the Sierra Gador Province of Almeira, Spain." *Trans. Inst. Min. Metall.*, 1936.

Editor's Comments
on Papers 27 and 28

The theme of Papers 27 and 28 is the metallogeny of the intervals. Metallogeny of the intervals between cycles is diverse. Mississippi Valley–type lead–zinc mineralization, related by Dunham in Paper 27 to the early Permian, would appear to be related either to the terminal stage or to epeirogenic activity on the platform between eras, yet is most difficult to separate it from the late-stage phenomena described by Pereira in Paper 26. The *Economic Geology* lead–zinc symposium volume (Brown, 1967) is also essential reading.

A continuing theme of this volume is the history of metals. The latest word on the origin of Mississippi Valley–type mineral deposits is by Heyl et al. (1974): "Together, the evidence indicates that the main ore fluids were heated oilfield brines having a largely crustal source for lead and sulfur. Regional distrubances, probably of lower crust–upper mantle origin, initiated large convective systems that produce Mississippi Valley–type mineralization." White's admonition (1974, p. 954) can be applied to the problems of the Mississippi Valley-type deposits: "No simple model of origin is likely to explain all or even most metalliferous deposits—we must be accustomed to diversity."

Dunham (1959, 1960, 1964), whose professional career has shown continuing interest in lead–zinc, drew the following tentative conclusion in his "Neptunist" paper (1964):

> In northern Europe in early Permian Times juvenile waters, perhaps mixed with connate brines, were rising from extreme depth into the fracture systems created by Hercynian earth movements. Much of their load was deposited in or adjacent to the fractures; but my suggestion is that an appreciable amount leaked into the Zechstein lagoon, enabling part of the remaining metal content to be concentrated in the foul bottom that transiently existed when the Marl Slate-Kupferschiefer-Kupfermergel was being formed.

So were resolved twin problems: the date of veining and the relationship of syngenesis and epigenesis. One may go further and suggest that the stage of tectonic cyclicity was more important than physico-chemical-biogenic traps.

Krauskopf has long been concerned with the source of ore metals and in his 1971 paper came to the conclusion that ore accumulation is often a multistage process: in essence, an enriched penultimate source is usual. He gives as one example the Kupferschiefer, following the argument of Wedepohl (1971) that the most likely source for the metals is the red-bed country rocks (the Lower Permian Rotliegendes). Metals accumulated at local reducing centers (logs, etc.), visible as halos and layers of decolorized sandstone. Wedepohl considered that the assumption is confirmed by the absence of extraordinary element accumulations in euxinic sediments from areas without red-bed country rocks. The requisite combination is red-bed country rocks, with soluble trace metal compounds, transgressed by a sea with a reducing environment in its bottom waters.

Paper 28, the benchmark paper on stratabound deposits, is an extract from Plumstead giving the history of ideas on the biogenic deposition of gold and uranium in the Witwatersrand. Garlick must also be accorded due credit for bringing ideas on sedimentary ore deposition to fruition in the Copperbelt, and for the strong support he gave to Haddon King (King and Thomson, 1953; King, 1973) and so to the developments of the Conzinc school.

The relationship of the Sudbury irruptive to the Huronian was noted in the pattern of tectonic development illustrated by Walker (1974). The Bushveld complex is similarly related to the Transvaal System, and here too the inference is that the relationship is genetic as well as geographic. Sudbury was well described in the Geological Association of Canada's symposium volume, Special Paper No. 10, and the Bushveld by Willemse (1969).

One does not readily put the work of C. F. Davidson into a category such as intervals. His 1962 paper is merely eight pages, but as the author said, it ranges far afield. Davidson was an unredeemed epigeneticist at a time when already many leaders were changing to syngenesis. His thoughts were pointed and pertinent. Because he was much concerned with the origin of Witwatersrand and lead–zinc ores, due note is made here.

27

Reproduced from *Econ. Geol.*, **59**(1), 1–21 (1964)

NEPTUNIST CONCEPTS IN ORE GENESIS [1]

K. C. DUNHAM

To be asked to be the first Distinguished Lecturer of the Society of Economic Geologists is a great honor but it is one that confers a great responsibility also. Since the Society has invited a European (and we Britons continue to regard ourselves as Europeans in spite of suggestions to the contrary) I have assumed that some reference to European thinking about ore deposits is expected, with an account not only of aspects of my own researches, but also of those of others.

This is the second occasion upon which I have had the privilege of addressing the Society at large. The first was at its dinner in the Håndvaerker-foreningen in Copenhagen, in September 1960. Then, I referred to the revolt against the hydrothermal hypothesis of ore genesis, and frivolously described as Neoneptunists those who preferred sulfide sedimentation in the seas or oceans. The idea is, of course, no new one in Europe. The orderly succession of strata found in the Thuringian basin in Germany, one of the first stratigraphical sequences ever described (55, 39) was known as the result of mining for copper the Kupferschiefer, manifestly a persistent thin bed in this series. Abraham Gottlob Werner laid great stress upon this orderly sequence in his lectures at Freiberg (1) and there is no doubt that it influenced the Neptunist school to interpret all rocks and ores in terms of a primeval ocean. Werner, in one of his few written works (87), extended his theory to explain cross-cutting veins as due to contraction in the rocks consequent upon expulsion of interstitial water beneath the ocean, followed by filling with ore from sea water.

It appears to me, therefore, that it is proper to describe as Neptunist Concepts those interpretations of ore genesis that postulate syngenetic or diagenetic concentration of sulfides in sediments; and to note that these concepts have in certain cases been extended to embrace vein-like deposits. A brief review will show that Neptunist concepts have now been applied to some of the world's

[1] First Distinguished Research Lecture of the Society of Economic Geologists; delivered before the Society at Dallas, Feb. 22, 1963.

most important and interesting ore deposits. Throughout the history of geology as a science, there has always been a body of opinion that favored a syngenetic origin for the German Kupferschiefer; this I propose to examine in more detail later. The remarkable pyrite-sphalerite-barite deposit at Meggen in Westphalia has been so regarded since the time of Schmidt (71); it is a comformable layer within Upper Middle Devonian strata, averaging about 4 m thick over an area 4 km × 1 km. There is a central zone of sulfides (pyrite 73 percent, sphalerite 12, galena 0.6, little chalcopyrite, tetrahedrite) surrounded by a fringe of high-grade barite. The orebody was certainly in place before the post-Carboniferous Hercynian orogeny folded it into a steep overturned anticline and syncline. Ehrenberg, Pilger and Schröder (34, 26) in an exhaustive study, interpret it as having formed in a stagnant hollow in the bottom of the Massenkalk sea, into which submarine mineral springs poured their metal ions. A similar hypothesis has been proposed by Kraume (53) to explain the two concordant lead-zinc-copper-barite lenses at the famous Rammelsberg mine in the Harz mountains; the deposits show meta-morphic textures and certainly predate the Hercynian orogeny; the hydro-thermal springs are supposed to have been related to Devonian volcanic activity. Here, sulfides and slates are interbanded.

The Copper Belt of Northern Rhodesia and the Congo, covering an area of the order of 77,500 sq km, in which the copper sulfides form extensive concordant layers in argillite (metamorphosed shale), quartzite, and dolomite, is now interpreted by many of the geologists most familiar with it in syn-genetic terms (59). It has been stated (13) that the first suggestion of a sedimentary origin for these important deposits came from Anton Gray in 1927. Schneiderhöhn (73) was an early advocate of such an origin. The syngenetic hypothesis was revived largely as a result of the work of W. G. Garlick and J. J. Brummer; the best statement of it is to be found in the recent volume on the district edited by F. Mendelsohn (59, p. 146–165). It is par-ticularly interesting to note that lateral zones in turn characterized by chalco-cite, bornite, chalcopyrite-carrolite, pyrite are regarded as facies developed successively farther from the original shoreline. This district has recently been discussed by the Society and no further reference to it here is necessary.

Having something in common both with the Kupferschiefer and with the Copper Belt, the White Pine copper deposit in the Nonesuch Shale of the Keweenawan of Michigan contains individual beds 1–3 feet thick carrying up to 3 percent Cu, whose extent is measurable in square miles. The copper oc-curs both as the native metal and as chalcocite, and it is chiefly concentrated in the shale members of a rhythmic series, which also includes sandstones. White and Wright (89) say of this deposit, "It may be too early to deny the possibility that the copper was introduced long after the deposition of the beds, but in general a theory which requires such delicate selective replacement or precipitation over many square miles, and such independence of control by structure or permeability, seems more strained than a theory holding the copper to be essentially sedimentary in origin."

The supposedly neptunian ore deposits so far mentioned all occur in stratal association with shale or slate. According to some, another major example is provided by the galena-sphalerite and chalcopyrite deposits at Mount Isa, Queensland. Here the sulfides of lead and zinc occur concordantly in carbonaceous shale of late Precambrian age, and exhibit what appear to be delicate sedimentary structures. Marker beds of fine acid tuff can be followed through the orebodies. N. H. Fisher (38) has reviewed the conflicting evidence on the origin of these deposits, and concluded in favor of a sedimentary origin; R. L. Stanton (84) on the basis of a statistical examination of their chemistry, also favors a syngenetic/diagenetic origin, with chalcopyrite concentrated in near-shore dolomitic breccias, galena-blende in offshore muds. The latter author postulates a source for the metals in submarine volcanic emanations.

The Broken Hill deposits in New South Wales, area for area probably the greatest known concentration of Pb, Zn, Ag in the earth's crust, are concordant layers in a highly complex structure involving katazone metamorphic rocks. They were interpreted by Gustavson, Burrell and Garretty (41) as replacements of favorable beds, but at the time the second author placed on record a minority view that the manganese minerals, which are a remarkable feature of the paragenesis, were concentrated by sedimentary means. The consistent and striking differences in mineralcgy between the adjacent No. 2 and No. 3 lens deposits, and the fact that mineralization with at least some of the curious characteristics of the Broken Hill assemblage is found over an area of hundreds of square kilometers at what may well be comparable stratigraphic horizons, led H. F. King (50) to electrify the geological world by postulating "a strong stratigraphic influence in the distribution of the original ore constituents." Metamorphism is now accepted by many geologists familiar with the field as having post-dated the ore concentration as Ramdohr (68) first demonstrated, and although the original form of the concentrations is difficult or impossible to establish, Neptunist views are now widely accepted. It should be noted that the association is with dominantly argillaceous rocks, now sillimanite-gneisses, but probably sandy shales at the time the ores first formed. Graphite is known in the district, but it is not now a major constituent.

Geochemical studies have established that bituminous shales the world over show abnormally high metal-contents (54, 30); the stagnant, non-oxygenated environments in which they form favor the precipitation of the metals as sulfides. The same is not generally true of limestones. Nevertheless, the past decade has witnessed persistent attempts especially in Europe to gain recognition for the view that the base-metal deposits (especially those of the 'Alpine' or 'Mississippi Valley' types) were concentrated in the first case by some kind of sedimentary mechanism. The trend may be said to have been foreshadowed by C. L. Knight's idea of source beds (51), which included references to the Tri-State, S. E. Missouri, Upper Silesia and Northern Morocco deposits. Independently, workers interested in the lead-zinc deposits in the Triassic carbonate-rocks of the Southern Alps, for example Bleiberg and the Trento district, K. C. Taupitz (85), A. Maucher (58), O. Schulz (79), W. Siegl (80) were coming to the conclusion that the metals were deposited with the sediments. H. Borchert (7, 8) was willing to see the primary concentration of trace elements into poor ores as a facet of sedimentation in a black bituminous-salinar-dolomite facies. It is nevertheless sufficiently obvious from the epigenetic structures and textures of the sulfide deposits in limestone that if they were originally laid down with the sediments, a great deal of migration, accompanied by further concentration, must have taken place. Here then, neptunism is not enough; it is necessary also to invoke an extension of Sandberger's (70) principle of lateral secretion. This would take place, Borchert suggests, in deep groundwater stockworks, the hydatogene zone of Maucher. But at this stage, as Charles Behre (6) has pointed out, solutions hardly distinguishable from the modern conception of hydrothermal fluids become involved, though they are not of igneous origin. As to the mechanism of sedimentary concentration of metal, D. L. Kendall (49), who has shown that zinc deposits of Jefferson City, Tennessee, occur in algal reef breccias, thinks that the algae may have played a significant role since some algae are known to contain 400–1,400 times as much Zn as the normal figure for sea water.

To all the Neptunist views briefly indicated, and to the many not mentioned, there have been and still are numerous objections. I imagine that not many members of my present audience take sedimentary syngenesis very seriously. At this point, I must confess that I am myself by training and inclination a Neoplutonist. Beginning under Arthur Holmes at Durham, well known for his advocacy of a deep source for lead deposits (42), I spent part of my graduate time with L. C. Graton and D. H. Maclaughlin at Harvard; I was also a member of Waldemar Lindgren's last Seminar at M.I.T. I am, incidentally, especially honored to see Mr. Graton here today. With this background I could hardly be expected to be prejudiced on the side of syngenesis or of lateral secretion. But I feel, as I am sure we all do, that since all great orebodies demand a high degree of unusual coincidence to promote their formation, no hypothesis must be rejected without careful examination.

There is no doubt that sulfides can and do form as part of a normal process of sedimentation. The bottom sediments of the Black Sea provide actualistic evidence of present day precipitation of black iron sulfide and of its conversion, below the sediment/water interface, into pyrite (84). Pyritic shales throughout the sedimentary column testify to the existence of comparable conditions at intervals throughout the geological past. Sedimentary iron orefields both of the hematite-greenalite-siderite type and the oolitic chamosite-siderite type show sulfidic facies (44, 29). In small amount, other sulfides such as sphalerite accompany the latter type in conditions that cannot have been other than diagenetic. But may major ore concentrations be formed in this way?

I propose to select two European orefields for more critical evaluation. The Kupferschiefer is an obvious choice, for among the deposits in shale, it is uncomplicated by severe tectonic disturbance or metamorphism; and the time is ripe for a new review in English. It has a production of over 2.5 million tons of copper metal to its credit. For the other I shall deal with the English Pennines, the eastern part of which lies close to the outcrop of the English equivalent of the Kupferschiefer, the Marl Slate of Durham. The Pennine orefields, credited with a production (in round figures) of 7 million tons lead, 1 million zinc, a few thousand tons of copper, and at least 4 million tons of fluorite, witherite and barite, have often been regarded as examples of the 'Mississippi Valley' type of base metal field in limestone.

THE KUPFERSCHIEFER-MARL SLATE

Description.—The formation is a bituminous-calcareous or dolomitic shale lying near the base of the Zechstein (Middle Permian) of Northern Europe. The Kupferschiefer or its equivalents can be traced from northern England eastward through the Low Countries, across Germany into Poland (Fig. 1). The southern and to some extent the western margins of the Zechstein basin or lagoon in which it was deposited are in part well defined; the northern

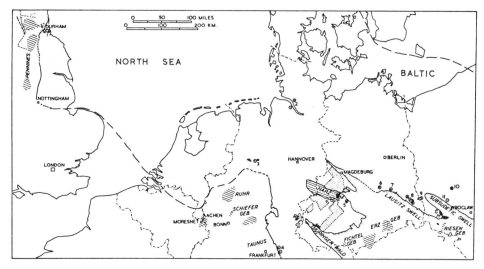

FIG. 1. Geography of the Kupferschiefer-Marl Slate. Heavy black lines, present margins of the Lower Zechstein deposits. Diagonal shading, important mineralized districts in pre-Permian rocks. Solid black, copper ore in Kupferschiefer; stipple, Pb + Zn over 1 percent. M = Mansfeld; S = Sangerhausen; E = Edderitz; R = Richelsdorf Hills; H = Haasel; G = Groditz. Borings: 1 = S.E. Durham; 2 = Elbe estuary; 3 = Osnabruck; 4 = Spessart; 5 = Halle; 6 = Schönewald; 7 = Bornsdorf; 8 = Spremberg; 9 = Zibelle; 10 = Wschowa; 11 = Sieroszwice, Lubin.

margin is concealed, probably below the Baltic. In England the Marl Slate (Lower Permian Marl, Hilton Plant Bed, Kirksanton Grey Beds) rest variously on a thin basal conglomerate, on aeolian dune sands (the Yellow Sands) or on red sandstone (Penrith Sandstone). In Germany a basal Zechstein conglomerate may underlie it, but over considerable areas it rests on a red sandstone, shale, and conglomerate formation, the Rotliegende. This may be bleached for a short distance below the shale contact (Weissliegende). Ancient dunes, elongated ESE have been recognized in the Weissliegende

Sands. Everywhere, dolomitic limestone, the Magnesian Limestone or Zechsteinkalk overlies the Marl Slate-Kupferschiefer and this in turn succeeded
by the great evaporite series now proved in N. E. England, Denmark, Han-
nover, Werra, around the Harz Mountains and at Strassfurt. The deposition
of the Kupferschiefer-Marl Slate was, in short, merely a brief prelude to a
prolonged episode of cyclic lagoonal evaporation.

The copper workings, continuous since the 13th century, began in the
Mansfeld basin, east of the end of the Harz Mountains, now in East Germany.
The workable shale averages 22 cm thick over an area of 140 sq km and
carries a little less than 3% Cu. A lateral eastward passage into a barren, red
facies known as the "rotfäule" occurs in the basin. South of the Harz,
a smaller worked area of 9 sq km occurs around Sangerhansen. The formation at Mansfeld is divided into a number of recognizable thin beds (in upward succession, Feine Lette, Grobe Lette, Kammschale, Schwarzer Kopf,
Grauer Kopf, Kopfchen, Schwarze Berge) above which transitions (Klotz,
fäule) link it with the limestone. The sulfides in the shale occur mainly in
two forms; as "speise," minute spheres or groups of spheres generally in the
size-range 5–15 microns; and as "erzlineale," lenses parallel to the bedding
about 1–4 mm long and up to 50 microns thick. The minerals are chalcocite,
bornite, chalcopyrite, pyrite, argentite, galena, sphalerite. In places these
minerals occur also in the top part of the underlying sands, forming "sanderz."
Adjacent to the red facies, they rise into higher subdivisions of the shale.
In the Mansfeld basin, a series of minor tension faults, the "rücken," trending
N 70–90° W cut through the ore bed and adjacent strata. These tend to
occur in particular belts; they carry chalcocite, digenite, bornite, chalcopyrite,
niccolite, covellite, maucherite, chloanthite, skutterudite, argentite, molybdenite,
galena, blende, pyrite, uraninite, bismuth and in their vicinity the copper values
in the shale may become impoverished though the combined yield of shale
plus vein is above the average for the shale alone. Unmineralized compression faults probably contemporary with the gentle late Mesozoic or Tertiary
folding which produced the basins, also occur. This brief summary of the
established facts is based upon the excellent handbook published by Eisenhuth
and Kautsch in 1954 (35), which contains the fullest bibliography so far
assembled for this deposit.

A 9-inch shale carrying workable copper values over more than 50 square
miles is sufficiently remarkable but the activities of an exploration company
shortly before World War II showed that this is only a small fraction of the
area over which this bed shows metal concentration abnormal even for the most
bituminous of black shales. Boring in the basins N and S of the Harz revealed that although the total extent of the bed with copper values over 1
percent is probably not much over 100 square miles, there are at least 2,200
square miles with over 1 percent zinc, and almost equally large areas enriched
in lead. Another small area formerly worked for copper lies in the Richelsdorf
(W. Germany), where the thickness over 50 square km is 24 cm, but the other
base metals are too lean for exploitation at present. Nevertheless, their wide
extent is of the greatest geological interest. The facts about this exploration
were published by Kautzsch (45), and Richter-Bernburg (69), during the

war, but they did not reach an English language publication until the important paper of T. Deans (16) appeared in 1950. Although by no means fully explored by boring, it could at this time be concluded that the Kuperschiefer, in its black bituminous marl facies as at Mansfeld, covered 20,000 sq km "between Gelnhansen and Bieber as the southernmost point in the neighborhood of Frankfurt (Main), and the eastern margin of the Rheinischer Schiefergebirge, past Thalitter and Korbach to near Osnabrück in the west; and in the east, over the Thuringer Wald to Gera and northwards to the Flectinger Swell" near Magdeburg (32, p. 61). The basin of deposition was believed to be open to the north. Deans showed that virtually the same facies is represented by the Marl Slate in N. E. England, so that towards the west the basin is certainly far more extensive than the quotation above indicates. The Marl Slate varies from a few centimeters to 3.9 m in thickness, averaging 1.3 m; as will be shown below, it also carries abnormal lead-zinc values.

To the east, exploration in Lower Silesia had by 1939 established the occurrence, in marl and marly limestone of a similar stratigraphic horizon, of workable copper deposits in the Haasel and Groditz basins, 300 km E. of Mansfeld (Fig. 1). The Kupfermergel in these two basins is respectively 180 cm and 120 cm thick (36), the greater workable thickness offsetting a lower copper content as compared with Mansfeld. Marl enriched in lead overlies the Kupfermergel. There was thus an indication that base-metal concentration in early Zechstein sediments continued over a far wider area than the immediate neighborhood of the Harz Mountains; but in the region between Mansfeld and Haasel, the formation is largely concealed beneath alluvium. Since 1950, however, the results of a series of deep borings in East Germany and in Poland [2] have appeared, strikingly confirming the continuity of high metal contents in that direction (Fig. 1). The principal localities are Schönewald, Bornsdorf, Spremberg, and Zibelle along the N. side of the Lausitz massif of older rocks, and Sieroszowice and Wschowa N. of the Subsudetic swell (52). At Bornsdorf the facies resembles that of Lower Silesia; the Copper Marl carries 0.43% Cu, the Lead Marl up to 0.14 Cu, 6.2 Zn, 0.14 Pb; the combined thickness of the two beds is 4.5 m. Sphalerite in the overlying Magnesian Limestone accounts for a zinc content of 1.5 percent. Among the Spremberg borings (52, 77) No. 13E showed dark bituminous shale 74 cm thick carrying 0.03–1.09% Cu; 0.13–3.6% Pb; 0.13–3.99% Zn; the average combined base metals is between 3 and 4 percent. In No. 3 boring, Zn reaches 4 percent; and in No. 29 the Kuperschiefer contains 3% Zn and 0.5% Pb. Copper predominates in No. 15, ranging from 0.3 to 2.8 percent. Borings at Sieroszowice and Lubin (67), show, respectively, 225 cm with Cu 1.4 percent and 230 cm with Cu 1.69. Here a potentially workable area of 30 sq km is indicated, containing 6 million tons copper metal, while it has also been claimed that major reserves probably exist N of the Lausitz massif. At Wschowa, the Kupferschiefer, penetrated at a depth of 285 m, is 38 cm thick and is still in a typical Mansfeld facies. Surveying these results, Ekiert (37), maintains that if Richter's estimate of 20,000 sq km for the

---

[2] I am very grateful to the late Professor O. W. Oelsner of Freiberg for his kindness in providing references to publications in East Germany.

area around the Harz is correct, the figure for Poland is 200,000 sq km, with 0.5 percent combined base metals. The total area of the Zechstein basin is, of course, much greater than this (Fig. 1), but over the northern part, the Kupferschiefer is, for the most part, unproved. We may accept a figure of 8,000 sq mi of black shale greatly enriched in Cu, Pb, Zn, as compared with normal shale as the minimum basis for the present discussion.

Certain aspects of the mineralogy deserve attention before the genetic problems are examined. First, typical vertical sections through the shale show copper predominant in the lower part of the bed, commonly to a very marked extent; lead above this, giving place upwards to zinc (35, 16). There is some suggestion of a similar zonal arrangement in a lateral sense with respect to the Zechstein sea shore-lines, with copper closest to the Variscan Swells and to the "rotfäule" regions, though not, according to Dette (18), in the littoral and sublittoral facies. Second, there is evidence in many places of small amounts of galena and sphalerite and less commonly of copper minerals in the Zechstein Limestone above the Kuperschiefer. Third, the microscopical evidence shows that replacements of various kinds have occurred among the sulfides. Schouten (76), objected to Schneiderhöhn's contention (72), that the microspheres were fossil bacteria, and it may well be that minerals other than pyrite are not normally precipitated in this way. But in view of the experiments of Baas-Becking and Moore, which showed that *Desulfovibrio desulfuricans* can precipitate covellite, digenite, argentite, sphalerite and galena (4), one hesitates to be too sure about this. The whole subject has been given a new impetus by the work of Love, who has shown that an organic sac or cortex surrounds each sphere or framboidal group (56). If the spheres were originally deposited as biogenic pyrite, then substitution has been extensive, either during diagenesis or at some late stage.

In the case of the English Marl Slate, my colleague D. M. Hirst and I were fortunate enough to have the cores from seven recent National Coal Board borings through the Middle Permian, placed at our disposal. On these we have made a geochemical and petrographical study, in some respects more detailed than any yet made for the Kuperschiefer of N. Europe. Dean's pioneer demonstration (16) that in Durham, the formation is in a zinc-lead facies was confirmed; the microspheres, which are abundant, are pyrite; the "erzlineale" carry the sphalerite and galena. A survey of Mo, Ni, Co, Rb, Sr, Cu, Pb, Sn, Mn-contents showed that although the first four and most of the copper could be related to rates of sedimentation, with steady or slowly changing additions of metals, a very fluctuating supply for the zinc, lead and part of the copper must be postulated (33). The zinc content ranges up to 3.1 percent (200 × average shale), the lead to 2,590 ppm (1,250 × average shale). The amounts of trace metals show a close correlation with organic carbon. The only sulfides actually observed under the microscope were pyrite, sphalerite, galena and chalcopyrite. In N. E. England a large number of coal-mine shafts and borehole intersections, over an area of 390 sq km, have shown that the thin Marl Slate is widely present; the frequency with which sphalerite and galena can be seen in it, noticed by Sedgwick as long ago as 1829, testifies to the regional character of its abnormal base-metal content.

Genesis.—The essential problem is to account for the base-metal concentration at levels far above any other known black shale, in an impermeable bed measuring only tens of centimeters thick over an area certainly amounting to tens, possibly hundreds of thousands, of square kilometers; and with economic concentrations of copper over a total of at least 230 sq km. The almost two-dimensional nature of the deposit makes a pure Neptunian hypothesis very attractive and only a few investigators have regarded it as other than syngenetic.

A "hydrothermal" origin was, however, postulated by Beyschlag and Krusch before the real dimensions of the problem were known. G. H. White (88), supposed that warm solutions travelled in the upper part of the permeable standstones beneath the shale, which trapped the metal ions they were carrying. An attempt has been made to suggest a relationship between the mineralization and the nature of the rock underlying the Kuperschiefer in the Edderitz basin, N. of Mansfeld (82), but the argument is not very convincing and is outweighed by the difficulty of introducing juvenile solutions or groundwaters over the huge area involved.

An interesting variant of the hydrothermal hypothesis has recently been proposed by C. F. Davidson (14), whose uncompromising opposition to syngenesis is well known. He suggests a lean epigenetic regional enrichment in metals connected with deep granites, followed by concentration in chemically receptive strata such as the Kupferschiefer. The concentration process is held to be the result of intra-stratal solution and redeposition of metals connected with tectonic events; an analogy is drawn with chertification and dolomitization. Undoubtedly the base-metal sulfides do participate in segregation and cementation processes in shaly sediments; sphalerite, for example, is quite common in siderite nodules in the Lias black shales in England, and galena and chalcopyrite occur; but can this process produce concentrations of sulfides of consistent tenor over tens or hundreds of square kilometers? The suggestion deserves further investigation, especially from the geochemical side.

The "rücken" at Mansfeld are unquestionably epigenetic as Kautzsch (46) and others have maintained but their presence does not prove the epigenetic origin of the Kupferschiefer mineralization any more than the "ore-pegmatites" indicate such an origin for the Copper Belt ores. The fillings, localized in or near the black shale, were probably derived, in part or whole, from selective bleaching of the shale by warm acid waters released from the Basement by tectonic pressures. According to Kautzsch this occurred in mid-Zechstein times, but Oelsner (64), considers the rücken are post-Triassic and thus probably Saxonian. The latter author points out that there is no evidence of magmatic activity hereabouts in late Cretaceous times.

To me, the difficulty of emplacing the metals in an impermeable bed subsequent to consolidation over the areas involved appears insuperable.

Turning now to a Neptunian interpretation, we may note first that the existence of stagnant conditions throughout the period of deposition of the bed is conceded by all; there is no bottom fauna other than of anaerobic bacteria; the fossils are remains of nekton, especially fish, of plants and of light floating shells. This is true of the Marl Slate, the Kupferschiefer and the Kupfermergel; the Bleimergel of Poland contains bottom-dwellers, but it could well have received its lead from below, as the Zechsteinskalk has almost certainly done (77). As far as the deposition of pyrite is concerned, the analogy with the Black Sea, pointed out as long ago as 1930 by Gregory (40), provides an actualistic instance of the deposition of FeS as a result of the activity of sulfur bacteria and its conversion during diagenesis into FeS_2—in the form of microspheres. The Black Sea, however, shows no concentration of base metals during the process; indeed, the Cu content of the bottom sediments is only about half that in average sediment (14, 84). Thus it is not easy to agree with Dean's tentative suggestion that the deposition of sulfides of heavy metals is to be regarded as "an initial minor phase of the established succession of chemical precipitates which characterize the Zechstein" (16), if normal sea water was being evaporated. The interesting recent study by J. L. Mero (60) of manganese nodules from the ocean floor, shows that heavy metals may become concentrated to an appreciable extent during very slow chemical sedimentation.; but the metal ratios of Cu, Pb, Zn, relative to Ni, Co, are very different from those in the Kupferschiefer-Marl Slate as Table 1 shows.

TABLE 1

COMPARISON OF OCEANIC MANGANESE NODULES WITH KUPFERSCHIEFER

| | Average of 54 nodules from Pacific (Mero, 53) | Mansfeld Kupferschiefer ore (Goldschmidt; 54) | |
	Percent	Percent	Recalc:
Mn	24.2	0.3	24
Fe	14.0	2.5	200
Co	0.35	0.004	0.32
Ni	0.99	0.01	0.8
Cu	0.53	2.9	272
Zn	0.047	0.9	72
Pb	0.09	0.5	40
Mo	0.052	0.03	2.4
Ag	0.0003	0.016	0.33

Relative to manganese, copper, lead and zinc are very much higher in Kupferschiefer ore than in the oceanic nodules.

If a syngenetic or diagenetic origin is to be accepted, a special influx of Cu, Zn, Pb and other heavy metals into the Zechstein lagoon must be sought. Three main hypotheses have been proposed:—(i) intermediate concentration of the metals in conglomerates and red sandstones of the Rotliegende, followed by upward migration during acid weathering and further concentration in superficial "rasenkupferstein" soils, which ultimately contributed sediment to the Kupferschiefer. With this thesis the distinguished names of Friesleben, Stelzner, Pompecki, Fulda, Goldschmidt and Schneiderhöhn are associated. It is pointed out that "Red Bed" copper deposits occur in the Rotliegende and recently Schüler (78), has noted that the pebbles in the conglomerates contain sulfides. In a variant of this "hypothesis," metal-enriched groundwater from the Rothliegende reached the lagoon; (ii) rivers draining the land surface, transported the heavy metals to the lagoon. Richter-Bernberg's calculations (69), demonstrate the insufficiency of the veins that might have been exposed on the Variscan land-surface to supply the metals, and our calculations for the Marl Slate (33), make it clear that the exposed veins in the English Pennines could not have supplied the lead and zinc. However, if the trace metal content of the exposed rocks is taken into account quite a different picture emerges. Oelsner (64), has calculated that the weathering of 160,000 cubic km from the Variscan Mountains in an area 400 × 100 km would yield sufficient Cu, Zn, Pb, to explain the metal content of the Kupferschiefer, if 0.9 percent of the available Cu, 4.4 percent of the Zn and 12.8 percent of the Pb was retained in the stagnant bottom of the lagoon. It is supposed that (35), the acid streams produced oxidizing "rote faule" conditions, with the fresh water floating on the salt water of the lagoon as in the case of the Congo or Amazon. The metals began to be precipitated adjacent to these streams. The most serious argument against this hypothesis is the absence of heavy metal enrichment in the Black Sea, already mentioned; (iii) the exhalative-sedimentary hypothesis by which the heavy metal enrichment was due to submarine springs. Perhaps because of an English disposition to compromise, I regard this as the most plausible explanation. As long ago as 1936, a National Research Council Committee, under the Chairmanship of E. S. Bastin (5), posed the question: "what becomes of ore-depositing solutions after they have accomplished their mineralization?" I had already suggested that the (epigenetic) mineralization of the English Pennines was still in progress when the Zechstein lagoon advanced over the area, and that the fluids found their way into the lagoon (21). If an influx of hydrothermal fluids occurred during the prevalence of foul-bottom conditions, metal concentrations could be built up; if it continued into Magnesian Limestone (Zechsteinkalk) times, the foreign metals would be widely scattered, as indeed they are, in England and Germany, in the dolomitic limestone.

It is pertinent to draw attention to the fact that numerous districts in which the mineralization is considered to be of Permo-Carboniferous age occur round the eastern and southern margins of the Zechstein. This was, in fact, the great period of European mineralization as Schneiderhöhn (74) has often stressed. Some of the districts shown on Figure 1 were almost certainly partly or wholly covered by the lagoon, for example the Pennines and the Harz Mountains. If it is accepted that these major orefields were receiving hydrothermal mineralization in Zechstein times, there seems no reason why the fluids should not have discharged into the lagoon. Petrascheck (66, Fig. 77), has pointed out that the Kupferschiefer both north and south of the Riesengebirge in Bohemia is richer in arsenic than is normal, while the epigenetic veins in the older rocks of that district are likewise arsenical. There may well be other districts containing Hercynian mineralization concealed beneath the Saxonian basins in which the Zechstein Series now occurs. It would be desirable to have many more absolute ages for the Hercynian and Zechstein leads; such evidence as is now available suggests, though inconclusively, that the Hercynian mineralization could in part be of Zechstein age. Thus Moorbath (61) gives four model ages for galena from the Northern Pennines ranging from 260 to 310 million years, and three for galenas from the Durham Permian ranging from 280 to 310 m.y.

The objection by Davidson (14), that advocates of the exhalative sedimentary hypothesis accept the passage of hydrothermal waters from depth through thousands of feet of strata into the sedimentary basin and yet argue against the introduction of similar emanations into the rocks of the stratigraphic succession, does not apply to the case here presented; both processes are contemplated, and the stagnant bottom of the lagoon merely becomes another environment where metals may become fixed. Its advantage over lateral secretion in an impermeable shale is the great mobility conferred upon the metals, so that they could spread widely over a large area. Here, of course, not merely electrolytic solutions are likely to be involved; adsorption on clay and bitumen would quickly take place; but that these materials were even more widely distributed as the Kupferschiefer lagoon received sediment is obvious enough.

The most eloquent cases for an exhalative—sedimentary origin for the Kupferschiefer is to be found in the writings of Richter-Bernburg (69) and Ekiert (37). My own position is that it appears to me that the balance of evidence—especially the areal scale of the metal enrichment—favors a sedimentary syngenetic origin, with a biogenic source for much of the sulfur; but hydrothermal springs augmented the metals supplied by weathering solutions, to give the Kupferschiefer its unique character. Essentially, the geological coincidences required here were (i) stagnant bottom conditions, (ii) very slow sedimentation, (iii) hydrothermal mineralization in progress at many places below or within the drainage of the lagoon. A connection with submarine volcanic activity in the general sense proposed by Oftedahl (65), can hardly be justified, even though there was some local activity (the Melaphyres—Kautzsch, 47), shortly before Zechstein times over a small area in Germany.

THE NORTHERN PENNINES

Description.—I now propose to select one of the districts of Hercynian mineralization marginal to the Zechstein lagoon for review, and of these the northern part of the English Pennines is the one with which I am most familiar. The Pennine ridge, running 150 miles from Derbyshire to the Tyne, contains three separate lead-zinc orefields; (i) North Derbyshire, where the veins and pipes occur in massive Carboniferous (mainly Viséan) limestone containing thin intercalated basaltic lavas ('toadstones'), mainly rotted by contemporaneous weathering; (ii) the Askrigg field, occupying the southern half of the sigma-shaped Northern Pennine fault block (Fig. 1), and separated from the Derbyshire field by a region of thick grits and shales, almost unmineralized; (iii) the Alston field, occupying the northern half of the fault block. We have much more complete information for the third of these, and I shall confine my remarks mainly to this. Since, however, the mineral suite galena-sphalerite-fluorite-barite-quartz appears throughout the Pennines, the basic genetic problems are regarded as the same in all three districts.

The Alston field occupies 550 square miles in the north of England, and the deposits occur in fissure veins of which about 600 have so far been discovered. These occur, as do the corresponding veins in the Askrigg district to the south, in a cyclothemic series in which the rhythmic unit, in upward succession, is limestone-shale-sandstone-(coal), and in which the unit thickness averages about 33 m (23). The age is Carboniferous (Viséan and Namurian), but some of the deposits extend up into the overlying Westphalian coal-measures. The Westphalian in east Durham is overlain with moderate unconformity by Zechstein, comprising dune sands and/or conglomerate not over 60 m thick (and generally much less), at the base; the Marl Slate (as already described) is succeeded upward by Magnesian Limestone and thick evaporites. The Northern Pennines may have been, in part, land in Zechstein times, but this is by no means certain, for Permian deposits reappear immediately to the west of the fault block, including possible correlatives of the Marl Slate. It is suggested that there must have been places where the Marl Slate Kupferschiefer lagoon overlay the veins, which could have discharged directly into the lagoon water.

At this point it is perhaps necessary to dismiss the truly Wernerian idea that the veins could have been filled by downward percolation of water from the stagnant bottom of the lagoon. The fact that the mineralization is now known to extend at least 1,600 m below the former base of the Marl Slate, the nature of the veins, and their relationship to basement structure all make such a suggestion fantastic and it is only made to be shot down. It has no serious advocates.

The Northern Pennine fault block is bounded to the north, west, and south by major faults with displacements up to about 3,000 m, the main movements of which took place in Tertiary times. The block had, however, been defined by Permo-Carboniferous faulting before this, and since it is an area of thin sedimentation relative to adjacent regions, hinges are believed to have been operative from early Carboniferous times onward. At the end of the Carboniferous important events connected with the Hercynian earth movements occurred. A series of linked tholeiitic sills (the Whin Sill) were intruded at stratigraphic horizons ranging from lowest Viséan in the center of the Alston Block to Westphalian at its margins; the maximum thickness of an individual sill is about 80 m. The marginal western faults became active as compression-features at this time and ENE tholeiite dikes were injected. Soon afterwards gentle doming in the orefield and basining in the coalfield to the east occurred; with these movements the beautifully geometric pattern of ENE, NNW and WNW fractures that were later mineralized, is believed to be associated (20, 22 and Fig. 3A). These cut through sediments and Whin Sill alike and they have the peculiarity that whereas they stand nearly vertical in the hard, brittle limestones and sandstones, they assume a considerable dip in the shale-members of the cyclothems. This applies especially to faults with displacements up to 10 m, which include a great majority of the veins; the few with larger displacement are less sensitive to the physical properties of

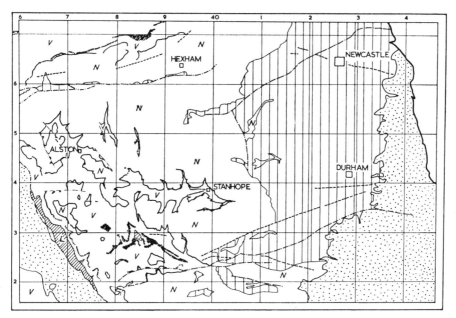

SURFACE GEOLOGY

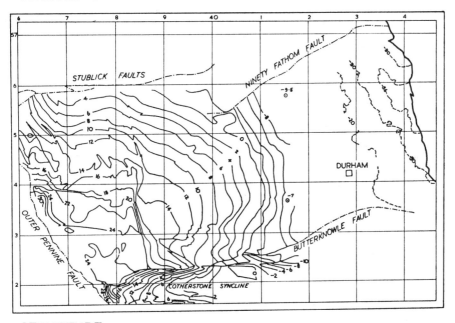

STRUCTURE

Fig. 2. Northern Pennines-Alston Orefield. *Surface Geology:* L = Lower Paleozoic slates and lavas; V = Viséan; N = Namurian; Vertical shading = Westphalian (Carboniferous); Stipple = Permian and Trias, with Marl Slate near base. Diagonal shading, Whin Sill; dashed lines, dikes; dot-dash, faults. *Structure:* Structure contours on the base of the Great Limestone (base of Namurian) in hundreds of feet relative to sea-level.

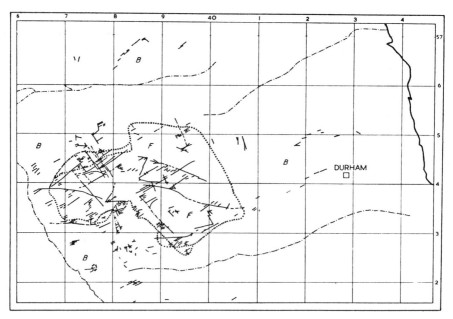

MINERAL ZONES

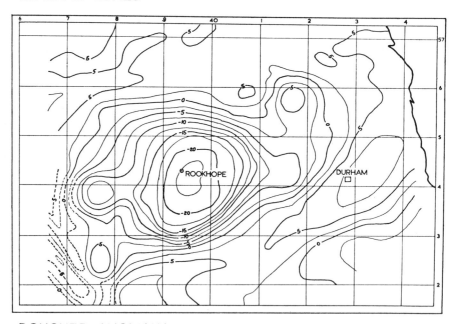

BOUGUER ANOMALY

FIG. 3. Northern Pennines-Alston Orefield. *Mineral Zones:* Lines = mineral veins; dot-dash = unmineralized faults; dots = zone boundaries. F = Fluorine zones (I) to (III). B = Barium zones (IV) and (V). *Bouguer Anomaly:* Contours at 2.5 milligals interval, based on the observations of M. H. P. Bott and D. Masson-Smith (10). The grid squares throughout Figs. 2 and 3 are 10 km.

the rocks they traverse. The mineralization may have begun soon after these movements. It was no longer in progress when a Tertiary dike-system related to the volcano of Mull cut through the veins. The final stage in the structural history of the field was the tilting of the block by uplift along the western faults, converting the dome into a half dome closing eastward, and producing the horizontally-grooved slickensides that cut through the mineral fillings in the veins.

The control by the physical nature of the wallrocks on the shape of the fissures imparts a characteristic ribbon-shape to individual oreshoots, since these are normally confined to the thin limestones and sandstones, where clean openings have formed in the fissures; but in shale, the faults are generally plugged up with gouge. Since the thickest limestone of the cyclothemic sequence is only about 20–25 m, the length of the oreshoots, commonly 1,000–1,500 m, greatly exceeds their vertical dimension. A series may, however, occur one above the other in successive hard beds, separated by barren shale. There is a tendency for oreshoots to terminate laterally at rectangular intersections, for example of a ENE vein by a NNW "cross" vein, and in this may be traced the lateral movements of fault-bounded blocks under tension. Replacement "flats" in the limestones may or may not accompany the veins, according to local permeability conditions; but they never occur away from feeding fissures.

It is now thirty years since I demonstrated that a systematic zonal disposition of mineral species occurs in the Alston field (22, 25), (Fig. 3). Regionally this is expressed as concentric zones characterized by (I) quartz-pyrrhotite-chalcopyrite-fluorite; (II) fluorite-galena-quartz (sphalerite); (III) fluorite-sphalerite-galena-barite-(witherite); (IV) barite-witherite-galena-(sphalerite); (V) barite-witherite with little or no sulfide. The main lead producers have been found in the outer part of zone (II), in (III) and the inner part of zone (IV); nowadays the district is only important for its fluorspar output (from zone (II)), its barite (zone (V)) and for possessing the only witherite mines in the world; but in the past, especially in the 18th and 19th centuries, it was among the world's leading lead districts.

Since the main movement of mineralizing fluids was lateral, along the ribbon or tube-like openings produced by the concidence of fissure and thin hard bed, it is natural that the mineral zoning is most conspicuous in the lateral sense; but there are clear cases of upward passage, for example from fluorite to barite, or from pyrrhotite-chalcopyrite to galena-sphalerite. It must be emphasized that the zoning is completely independent of the nature of the wallrocks, and it cannot by any stretch of the imagination be interpreted in terms of the syngenetic zoning which Amstutz (2) has recently discussed in principle, and which probably occurs in the Kupferschiefer.

In passing through the Whin Sill dolerite, the mineralizing solutions have converted this, to distances up to 5 m from the vein-walls, into "white whin," in which the labradorite has gone to an illitic clay mineral, the pyroxenes to ankeritic carbonates. The chemical balance shows losses in Si, Fe, Mg, but notable gains in K and CO_2. The evidence would be consistent with potash-rich brines as the mineralizing fluids. Ankerite and siderite are abundant in the veins above the Whin Sill and the source of the Fe, Mg is almost certainly the sill; thus these can be explained by Sandberger's hypothesis; but this appears not to be the case for the other metals, or for fluorine.

Genesis.—The zoning invites comparison with the upper zones in the well-known Cornwall-Devon field (19), and it was natural to suggest a concealed granite as the source of the fluids, with mainly lateral movement from a small number of ascendent centers (22). I very little thought when this suggestion was first made that an opportunity to test the hypothesis would ever arise.

However, in 1952, M. H. P. Bott and D. Masson-Smith, following up a preliminary traverse by J. L. Hospers and P. L. Willmore (43), which had shown a gravity low about the center of the Alston district, worked out the gravitational field for N. E. England (10), (Fig. 3). This shows two marked gravity lows with a maximum Bouguer anomaly of 30 milligals corresponding precisely with the centers of the mineral zoning (Fig. 3). Other geophysical features also combined to reinforce the suggestion of a buried granite, which most satisfactorily explained the gravity picture. It now became a matter of considerable interest to prove the granite by boring and in 1960 the Department of Scientific and Industrial Research sanctioned an expenditure sufficient for 5,000 feet of drilling. The hole was sited at Rookhope, Co. Durham, close to the minimum of the larger gravity anomaly, and in such a position that the two principal veins of the district, the Red Vein (WNW) and Boltsburn Vein (ENE), were dipping towards it and could be expected to be penetrated in depth. The hole was cored throughout, except in a few short stretches where particular difficulty was encountered; the starting diameter was 9 inches, the final diameter 6 inches, and overall core-recovery was 81 percent.

About the time boring commenced, Bott (9), using a method devised in collaboration with R. A. Smith (11), revised the previous estimate of depth to the top of the granite (within 5,500 feet of the surface), and presented mathematical arguments indicating 4,000 feet as the maximum depth; and favoring 2,000 feet. In the event, the hole passed from cyclothemic sediments into granite at 1,281 feet below surface, fully justifying the geophysical prediction (31); it was continued in granite to a final depth of 2,650 feet. There is now direct and indirect evidence that two major cupolas of granite underlie the Alston Block, with their apices corresponding with the centers of the mineral zones; a smaller apex corresponds with an "inlier" of the fluorite zone at Scordale, two others with strong mineralization in stratigraphical high positions. Nevertheless, the cupolas are *not* the sources of the hydrothermal solutions that mineralized the Pennines, for the geological evidence clearly shows that the top of the granite is weathered, and that fragments of it are incorporated in the overlying unmetamorphosed Carboniferous sediments. Radiometric dating carried out immediately at Oxford (62), gave a Devonian age of 362 ± 6 million years, at least 80 m.y. older than the lead-zinc ores. It appears that the influence of the granitic stocks can be interpreted in no more intimate terms than those of control of structural channelways. Perhaps a comparison might be drawn with S. E. Missouri.

The drilling showed, however, that the mineralization continues to far greater depths than had previously been proved in the district. The present local depth of mining is 360 feet below the collar of the borehole; in the hole not merely the two hoped-for vein intersections were obtained, but a total of 17, the deepest being 2,350 feet, 1,070 feet inside the granite, but it is clear that zone (I) mentioned above has been penetrated. An interesting feature of the boring is that the present temperature gradient is substantially higher than in the Durham Coalfield and other comparable areas.

Viewed stratigraphically, the mineralization in the Alston field thus occurs through at least 1,000 feet of "basement" granite, and in individual sandstone and limestone beds through a cyclothemic Carboniferous sequence not less than 3,500 feet thick; few hard beds are known which do not somewhere in the field carry an oreshoot but certain groups are more favorable, on present evidence, than others; the Great Limestone, at the base of the Namurian, being most favorable. Even allowing for some redistribution (and not very much is possible because of the numerous impermeable shale barriers), it seems impossible to contemplate a syngenetic or diagenetic origin for these orefields, and though, as Amstutz (3) has pointed out, some of the features are similar to those shown by supposedly syngenetic lead-zinc districts in limestone, these features are in my view better explained by epigenesis and the great weight of evidence is hostile to an origin for the metal deposits contemporaneous with Carboniferous sediments or even during diagenesis. Nevertheless, we are now engaged in a program of geochemical work designed to test, among other things, the primary distribution of base metals in the cyclothemic sediments.

The mineralization of the Northern Pennines is very similar to that of the Illinois-Kentucky fluorite field; the rocks have many similarities, there are major veins that cannot be other than epigenetic; the mineralization is known to continue in depth, though in a volcanic pipe (12) here rather than a basement granite. The veins in this field that I have been privileged to see underground are indistinguishable from those of zone (II) in the Pennines, "Sheet ground"—supposed by most workers to represent metasomatism of limestone is at present more important in Illinois than in the Pennines, but perhaps we have not searched sufficiently. But in view of the invariable association with through-going veins, a syngenetic or diagenetic origin for the sheet ground or flats can hardly be acceptable.

A source at depth is, then, an inescapable postulate, but in neither district is there any good evidence as to its nature. Deep circulating, chemically-active brines, deriving their metal solutes by lateral secretion, cannot yet be ruled out, though quantitatively it appears inadequate (27); but true juvenile water from great depth and in some way associated with movement or birth of sub-continental magmas is to my mind a better alternative.

My tentative conclusion is, then, that in northern Europe in early Permian times juvenile waters, perhaps mixed with connate brines, were rising from extreme depth into the fracture systems created by Hercynian earth movements. Much of their load was deposited in or adjacent to the fractures; but my suggestion is that an appreciable amount leaked into the Zechstein lagoon, enabling part of the remaining metal content to be concentrated in the foul bottom that transiently existed when the Marl Slate-Kupferschiefer-Kupfer-mergel was being formed.

DEPARTMENT OF GEOLOGY,
UNIVERSITY OF DURHAM, ENGLAND,
August 10, 1963

REFERENCES

1. Adams, F. D., 1938, The birth and development of the geological sciences, New York.
2. Amstutz, G. C., 1959, Syngenetic zoning in ore deposits: Geol. Assoc. Canada Proc., v. 11, p. 95–113.
3. ——, 1959, Syngenese und Epigenese in Petrographie und Lagerstättenkunde: Schweiz. Min. Petr. Mitt., v. 39, p. 1–84.
4. Baas Beckling, L. G. M., and Moore, D., 1961, Biogenic sulfides: ECON. GEOL., v. 56, p. 259–272.
5. Bastin, E. S., 1936, Suggestions concerning desirable lines of research for the fields of geology and geography: Nat. Res. Council, p. 32–38.
6. Behre, C., 1962, Types of evidence for genesis of ore deposits in the East Tennessee and other lead-zinc deposits: ECON. GEOL., v. 57, p. 114–119.
7. Borchert, H. 1958, Neuere geochemische Untersuchungen von Blei und Zink in lager-stättenkundlicher Sicht.: Z. deutsch Geol. Ges., v. 110, p. 450–473.
8. ——, 1959, Geosynklinale Lagerstätten, was dazu gehört und was nicht dazu gehört, sowie deren Besiehungen zu Geotektonik und Magmatismus: Freiberger Forschungs-hefte, C. 79, p. 8–49.
9. Bott, M. H. P., 1960, Depth to top of postulated Weardale granite: Geol. Mag., v. 97, p. 511–514.
10. Bott, M. H. P., and Masson-Smith, D., 1957, The geological interpretation of a gravity survey of the Alston Block and the Durham Coalfield: Geol. Soc. London Quart. Jour., v. 113, p. 93–117.
11. Bott, M. H. P., and Smith, R. A., 1958, The estimation of the limiting depth of gravitating bodies: Geophys. Prospecting, v. 6, p. 1–10.
12. Brown, J. S., Emery, J. A., and Meyer, P. A., 1954, Explosion pipe in test well on Hicks Dome, Hardin County, Illinois: ECON. GEOL., v. 49, p. 891–902.
13. Brummer, J. J., 1955, The geology of the Roan Antelope Orebody: Inst. Min. Met. Trans., v. 64, p. 257–318.
14. Davidson, C. F., 1962, The origin of some strata-bound sulfide ore deposits: ECON. GEOL., v. 57, p. 265–275.
15. ——, 1962, On the cobalt-nickel ratio in ore deposits: Mining Mag., v. 106, p. 78–85.
16. Deans, T., 1950, The Kupferschiefer and associated mineralization in the Permian of Silesia, Germany and England: Rpt. 18th Int. Geol. Congress, v. 7, p. 340–351.
17. Dette, K., 1956, Der Kupferschieferklözkern: Zeit. für angewandte Geologie, v. 2, p. 252–257.
18. ——, 1960, Erkundungsarbeiten auf Kupferschiefer: Zeit. für angewandte Geol., v. 6, p. 619–623.
19. Dewey, H., 1925, The mineral zones of Cornwall: Geol. Assoc. Proc., v. 26, p. 107.
20. Dunham, K. C., 1933, Structural features of the Alston Block: Geol. Mag., v. 70, p. 241–254.
21. ——, 1934, Genesis of the North Pennine Ore Deposits: Geol. Soc. London Quart. Jour., v. 90, p. 689–720.
22. ——, 1948, Geology of the Northern Pennine Orefield, v. 1: Mem. Geol. Survey G. B., London.
23. ——, 1950, Lower Carboniferous sedimentation in the Northern Pennines (England): Rpt. 18th Int. Geol. Cong., v. 4, p. 46–63.
24. ——, 1952, Age-relations of the epigenetic mineral deposits of Britain: Geol. Soc. Glasgow Trans., v. 21, p. 396–429.
25. ——, 1952, Fluorspar: Spec. Rpts. Min. Res. G. B., v. 4, 4th Ed.
26. ——, 1955, Das Schwefelkies-Zinkblende-Schwerspatlager von Meggen (Westfalen); Re-view: Inst. Min. Met. Bull., v. 64, p. 8–10.
27. ——, 1959, Epigenetic mineralisation in Yorkshire: Yorks. Geol. Soc. Proc., v. 32, p. 2–29.
28. ——, 1959, Non-ferrous mining potentialities of the Northern Pennines: *in* Future of non-ferrous mining in Great Britain and Ireland, London, p. 115–147.
29. ——, 1960, Syngenetic and diagenetic mineralisation in Yorkshire: Yorks. Geol. Soc. Proc., v. 32, p. 229–284.

30. ——, 1961, Black shale, oil and sulphide ore: Adv. Science, v. 18, p. 284–299.
31. Dunham, K. C., Bott, M. H. P., Johnson, G. A. L., and Hodge, B. L., 1961, Granite beneath the Northern Pennines: Nature, v. 190, p. 899–900.
32. Dunham, K. C., and Bott, M. H. P., 1961, The foundations of the Northern Pennines: Times Science Review, no. 41, p. 13–14.
33. Dunham, K. C., and Hirst, D. M., 1963, Chemistry and petrography of the Marl Slate of S. E. Durham, England: ECON. GEOL., v. 58, p. 912–940.
34. Ehrenberg, H., Pilger, A., and Schröder, F., 1954, Das Schwefelkies-Zinkblende-Schwerspat lager von Meggen (Westfalen): Monographien der Deuts. Blei-Zink-Erzlagerstätten, n. 12.
35. Eisenhuth, K. H., and Kautzsch, E., 1954, Handbuch für der Kupferschieferbergbau, Leipzig.
36. Eisentraut, O., 1939, Der niederschlesische Zechstein und seine Kupferlagerstätte: Arch. f. Lagerst. Forsch., v. 71, Berlin.
37. Ekiert, F., 196–, Neue Anschauungen über die Herkunft des in der Sedimenten des unteren Zechstein auftretenden Kupfer: Freiberger Forschungsheft, C. 79, p. 180–201.
38. Fisher, N. H., 1960, Review of evidence of genesis of Mount Isa orebodies: Rpt. 21st Int. Geol. Congress, Copenhagen, v. 16, p. 99–111.
39. Fuchsel, G. C., 1771, Historia terrae et Maris ex historia Thuringiae per montium descriptionem erecta, Erfurt.
40. Gregory, J. W., 1930, The copper shale (Kupferschiefer) of Mansfeld: Inst. Min. Met. Trans., v. 40, p. 1–55.
41. Gustavson, J. K., Burrell, H. C., and Garretty, M. N., 1950, Geology of the Broken Hill Deposit, Geol. Soc. America Bull., v. 61, p. 1369–1437.
42. Holmes, A., 1937, The origin of primary lead ores: ECON. GEOL., v. 32, p. 764–781.
    ——, 1938, The origin of primary lead ores: paper II, ECON. GEOL., v. 33, p. 830–865.
43. Hospers, J., and Willmore, P. L., 1953, Gravity measurements in Durham and Northumberland: Geol. Mag., v. 90, p. 117–126.
44. James, H. L., 1954, Sedimentary facies of iron-formation: ECON. GEOL., v. 49, p. 236–281.
45. Kautzsch, E., 1942, Untersuchungsergebnisse Über die Metallverteilung im Kupferschiefer: Arch. f. Lagerst-Forsch., v. 74, Berlin.
46. ——, 1593, Tektonik und Paragenese der Rücken im Mansfelder und Sangerhänser Kupferschiefer: Geologie, v. 2, p. 4–24.
47. ——, 1957, Die Metallführung des Melaphyres von GroBörner und ihre Beziehung zur Metallführung im Kupferschuper: Neues Jahrb. f. Min. Abh. 91.
48. ——, 1958, Die sedimentären Erzlagerstätten des Unteren Zechstein: Freiberger Forschungshefte, C. 44, p. 14–21.
49. Kendall, D. L., 1960, Ore deposits and sedimentary features, Jefferson City Mine, Tennessee: ECON. GEOL., v. 55, p. 985–1003.
50. King, H. F., and Thomson, B. P., 1953, The geology of the Broken Hill district: in Geology of Australian Ore Deposits, Melbourne-Australasian Inst. Min. Metall., p. 533–577.
51. Knight, C. L., 1957, Ore genesis—the source bed concept: ECON. GEOL., v. 52, p. 808–817.
52. Kolbel, P., 1958, Zur Stratigraphie und Erzführung des Zechstein 1 (Werra Serie) in SüdBrandenberg und in der Subsudetischen Zone: Zeit. für angewandte Geol., v. 4, p. 504–508.
53. Kraume, E., 1955, Die Erzlager des Rammelsberges bei Goslar: Monographica der Deuts. Blei-Zink-Erzlagerstätten, v. 4, p. 4.
54. Krauskopf, K., 1955, Sedimentary deposits of rare metals: ECON. GEOL. 50TH ANN. VOL., v. 1, p. 411–463.
55. Lehmann, J. G., 1756, Versuch einer Geschichte von Flötzbirges, Berlin.
56. Love, L. G., 1962, Biogenic primary sulfide of the Permian Kupferschiefer and Marl Slate: ECON. GEOL., v. 57, p. 350–366.
57. Maucher, A., 1957, Erzmikrokopische Untersuchungen an Blei-Zink-Lagerstätten in Ramme von Trento (Norditalien): R. v. Klebelsberg-Fetschr. Geol. Ges. Wien, v. 48, p. 139–153.
58. ——, 1957, Die Deutung des primaren Stoffbestandes der kalkalpinen Pb-Zn Lagerstätten als syngenetisch-sedimentäre Bildung: Berg-u. Hüttenm. Monatsch., v. 102, p. 226–229.
59. Mendelsohn, F., 1961, The geology of the Northern Rhodesian copper belt, London.
60. Mero, J. L., 1963, Ocean-floor manganese nodules: ECON. GEOL., v. 57, p. 747–767.
61. Moorbath, S., 1962, Lead isotope abundance studies on mineral occurrences in the British Isles and their geological significance: Phil. Trans. Royal Soc., Series A., v. 254–360.
62. Moorbath, S., and M. H. Dodson, 1961, Isotopic ages of the Weardale granite: Nature, v. 190, p. 900–901.

63. Oelsner, O., 1958, Zur Methodik der geologischen Erkundung in Abhängig von der Lagerstättentypen : Zeit. für angewandte Geol., v. 4, p. 322–332.
64. ——, 1959, Bewerkungen sur Herkunft der Metall im Kupferschiefer : Freibergen Forschungshefte, C. 58, p. 106–113.
65. Oftedahl, C., 1958, A theory of exhalative-sedimentary ores: Geol. Foren. i. Stockholm Forh., v. 80, p. 1–19.
66. Petrascheck, W. E., 1961, Lagerstättenlehre, 2nd Ed., Vienna, p. 96.
67. Plodowski, T., 1958, Kupfererzbau : Przegl. Geol., v. 6, p. 96.
68. Ramdohr, P., 1951, Die Lagerstatte von Broken Hill in New South Wales im Lichte der neuer Geologischen Erkenntnisse und Erzmikroskopischer Untersuchungen : Heidelberger Beit. z. Min. u. Pet., v. 2, p. 4–.
69. Richter-Bernburg, G., 1941, Geologische Gesetzmä Bigkeiten im Metallgehalt des Kupferschiefers : Arch. f. Lagerst. Forsch. v. 73, Berlin.
70. Sandberger, F., 1882, Untersuchungen über Erzgänge, Wiesbaden.
71. Schmidt, W. E., 1918, Über die Entstehung und über die Tektonic des Schwefelkies- und Schwerspatlagers von Meggen a.d. Lenne nach neueren Aufschlüssen : Jb. preuss. Geol. L. A., v. 39, p. 23–72.
72. Schneiderhöhn, H., 1921, Chalkographische Untersuchungen der Mansfelder Kupferschiefers : Neues. Jahrb. f. Min., BB 47, p. 1–38.
73. ——, 1931, Mineralische Bodenschätze im sudlichen Africa : Berlin, Nein-verlag, p. 84–102.
74. ——, 1952, Genetische Lagerstätten-gliederung auf geotektonischer Grundlage : Neues. Jahrb. Min. Monabs.
75. ——, 1955, Erzlagerstätten : Kurzvorlesungen zur Einführung zur Wiederholung. 3rd Ed., Stuttgart.
76. Schouten, C., 1946, The role of sulfur bacteria in the formation of the so-called sedimentary copper ores and pyritic ore bodies : ECON. GEOL., v. 41, p. 517–538.
77. Schüller, A., 1958, Die Metallisation im Kupferschiefer und Dolomit des Unteren Zechstein in der Bohrungen Spremberg 13E/57 und 3/54 : Geologie, v. 8, p. 651–660.
78. ——, 1958, Metallisation und Genese des Kupferschiefers von Mansfeld : Abh. d. D. Wiss. du Berlin, Ki. Chem. Biol. Geol., v. 6.
79. Schulz, O., 1957, *In* Entstehung von Blei-Zinkerzlagerstätten in Karbonatgesteinen : Berg.-u. Hüttenm. Monatsch., v. 102, p. 241–242.
80. Siegl, W., 1957, *In* Entstehung von Blei-Zinkerzlagerstätten in Karbonatgesteinen : Berg.-u. Hüttenm. Monatsch., v. 102, p. 237–238.
81. Stanton, R. L., 1962, Elemental Composition of the Black Star Orebodies, Mount Isa, Queensland, and its interpretation : Inst. Min. Metall. London, Trans., v. 72, p. 69–124.
82. Steinbrecher, B., 1959, Die petro- und erzfazielle Differenzierung der Kupferschieferzone in der Edderitzer Mulde : Zeit. für angewandte Geol., v. 5, p. 201–204.
83. ——, 1959, Die Sedimentation in Saaletrog im Bereich des östlichen Harzvorlandes während des Zechsteins 1 und 2 : Zeit. für angewandte Geol., v. 5, p. 381–385.
84. Strakhov, N. M., 1960, Osnovy Teorii Litogeneza, v. i, Gosgeoltekhizdat, Moscow.
85. Taupitz, K. C., 1954, Über Sedimentation, Diagenese, Metamorphose, Magmatisimus und die Entstehung der Erzlagerstätten : Chemie der Erde, v. 17, p. 104–164.
86. ——, 1954, Erze sedimentären Entstehung auf alpiner Lagerstätten des Typus Bleiberg : Erzmetall., p. 7.
87. Werner, A. G., 1791, Neue Theorie von der Entstehung der Gänge, mit Anwendung auf den Bergbau besonders den freibergischen Freiberg.
88. White, C. H., 1942, Notes on the origin of the Mansfeld Copper Deposits : ECON. GEOL., v. 37, p. 64–68.
89. White, W. S., and Wright, J. C., 1954, The White Pine copper deposit, Ontonagon County, Michigan : ECON. GEOL., v. 49, p. 676–716.

28

Reprinted from *Alex L. du Toit Memorial Lecture 11, Geol. Soc. South Africa,*
72, 11-18, 68-72 (1969)

THREE THOUSAND MILLION YEARS OF PLANT LIFE IN AFRICA

Edna P. Plumstead

[*Editor's Note:* In the original, material precedes this excerpt.]

C. THE WITWATERSRAND TRIAD

(see Plate III)

1. General

The oldest South African formation which is not intruded by the Archaen, or Older Granites is now regarded as a super-system which has been termed the Witwatersrand Triad. Alternatively it is known as three systems; the Dominion Reef, the Witwatersrand and the Ventersdorp Systems in order of age, but the unconformities between them are often no greater than those which occur at intervals in different areas within each member of the group and since they are known to follow one another fairly closely in time and each has yielded gold and uranium in economic·quantities, it is more convenient to consider them all together. Each contains both sediments and lavas in thick successions but only portions of each are found in any single area and in particular the oldest, or Dominion Reef Group is often absent.

2. Age of the Witwatersrand Triad

A recent age determination of upper Dominion Reef lava from the Western Transvaal, is $2\,800 \pm 60$ million years and the underlying sediments must be older (Van Niekerk and Burger, 1969b). Ventersdorp lavas have been dated as $2\,300$ million years (Van Niekerk and Burger, 1964) and the Witwatersrand Formation, approximately $7\,600$ metres thick, lies between the two.

3. Biological Importance

Because of the enormous economic importance and potential of these rocks, their extent, petrology, origin, structure and especially the origin of ores have been subjects for study and discussion and often of heated arguments.

Professor R. B. Young (1917) in writing about the banket, as the gold-bearing conglomerate is called, drew attention to the presence of carbon in these "reefs". He described it as occurring in the form of small black, opaque, spheroidal or nodular grains with a dull, warty outer surface and believed it to be of inorganic origin. According to current belief at the time, the possibility of any biological interest or of finding fossils was dismissed at a very early stage for it was believed that the rocks were far too old for any type of life to have been in existence.

The pros and cons of the various theories of origin of the Witwatersrand gold formed the subject-matter of a special issue by the Geological Society of South Africa in the form of an Annexure (1930) in which a modified placer theory was finally accepted by all the geologists most closely concerned with the gold-mines, but the possibility of biological assistance in enriching the ore was not even contemplated.

In recent years a biological origin of the carbon began to be reconsidered even before the realization that every one of the rock types associated with the early plant life of the Primitive Systems occurs in the Witwatersrand Triad also, for example — (a) Beds containing a high percentage of carbon, these include the

PLATE III

The Witwatersrand Triad

Early Precambrian

1. Basal Reef, St. Helena Gold Mine, Orange Free State.
N.B. The 'Carbon' band is now crystallised perpendicular to the bedding. It is scattered with visible gold and pyrite and contains uraninite. A large vein-quartz pebble is sunken in the carbon which is believed to have been originally a mat of algal.growth. The gold value of the narrow carbon band far exceeds that of the banket above it. (X2)
Specimen presented to the B.P.I.(Pal) by Union Corporation Ltd.

2. Carbon Leader from the Far West Rand.
A variable layer of pebbles, lying and sometimes partly embedded in a footwall of 'carbon' with a high sulphide content. Values are concentrated especially in the 'carbon' layer. (X$\frac{1}{2}$)
(After de Kock, 1964, Pl. X, fig. 1, with permission of the Geol. Soc. S.A.)

3. 'Carbon'. Polished specimen photographed by reflected light showing distribution of secondary uraninite in carbon and the resemblance of the whole to a Palaeozoic torbanite or rich oil shale, both in appearance and composition. (X60)
(After Snyman, 1965, Pl. II, fig. 4, with the permission of the Geol. Soc. S.A.)

4. and 5. Algal filaments separated by maceration from the carbon of the Basal Reef greatly magnified (scale provided).
(Reproduced from Oberlies and Prashnowsky, with permission, 1968.)

6. Road cutting through Contorted Bed of Banded ironstone, Lower Witwatersrand Series. De Beer street, Braamfontein, Johannesburg. Grass blades on the left give the scale.
N.B. This type of rock only occurs in geological formations older than 1 800 m.y. Bacteria are believed to have caused the precipitation of the ferric iron.
Photo: J. Anderson.

7. Core of small stromatolites — calcareous algae — upper Ventersdorp System, O.F.S. (X1)
(After Winter, 1963, with permission of the Geol. Soc. S.A.)

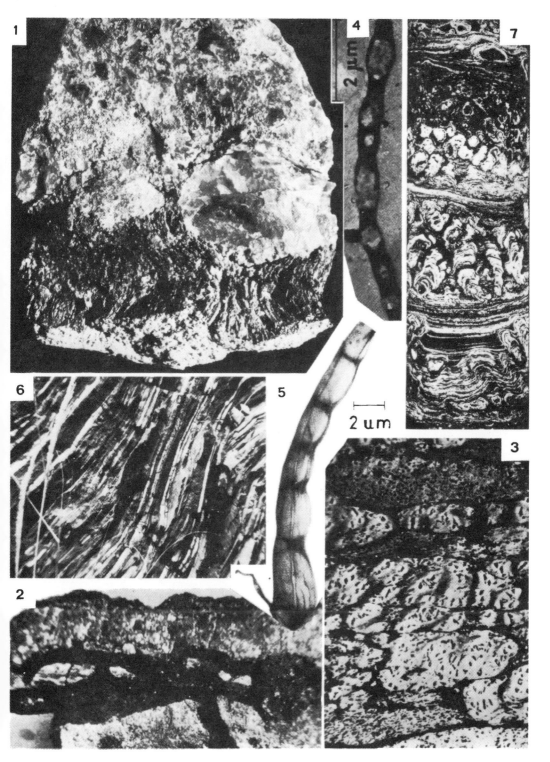

PLATE III
The Witwatersrand Triad
Early Pre-Cambrian

335

"carbon" reefs, the gold-bearing banket and pyritic quartzites. (b) Banded iron-stones. (c) Stromatolitic calcareous deposits. Group "a" has received far greater attention because it includes the "host"-rocks of immensely valuable gold and uranium concentrates. However each of the rock types will be considered briefly because the probability now exists that primitive plants and a high concentration of metals in sedimentary rocks are often closely connected. It is obvious that much more work remains to be done before the close association of the two can be universally accepted not only as factual but as largely causal.

4. A high percentage of carbon with gold and uranium — Is there a biological connection?

The new interest in carbon was the direct result of two unrelated investigations. Firstly, the prospecting and subsequent opening up of mines on the Far West Rand in the 1930s (where the Upper Witwatersrand lies beneath a thick cover of younger formations) revealed that two new, highly payable reefs in this area, the Carbon Leader and the Ventersdorp Contact Reef were intimately associated with large amounts of carbon. This area now includes the richest gold-mines in the world. The second factor was that during the war the importance of atomic energy soared and the realization that many of the Witwatersrand Reefs contained highly radioactive ingredients, encouraged detailed investigations. In these ventures two men played especially active parts. To Dr. Liebenberg of the Government Metallurgical Laboratories, the task of investigating the source, chemistry and mineralogy of the radioactive substances was allocated. The results of his work over ten years were permitted to be published in 1955. Meanwhile Dr. W. P. de Kock, senior geologist of Gold Fields of South Africa Ltd. was studying the geology of the new gold-field and in particular of the new carbon-rich reefs. The economic importance of this work, which revitalised the gold industry of South Africa, subsidised research-work of academic, geological and biological value, which might not otherwise have been done. A number of other geologists have contributed valuable papers and observations but De Kock's (1964) summary of the geology and economic significance of the "West Wits Line" after more than twenty-five years of work in the area (see De Kock, 1940) can be applied to the whole Witwatersrand.

Previously little notice had been taken of the fact that there was a close association between gold and carbon and even more so, between uranium and carbon, for this had been less apparent in the Main Reef Group, which had been the chief producers of gold up to that time. It was only when very high gold values were found on the Far West Rand, at several new horizons, that the close association with carbon became patently obvious and new theories began to be promulgated. The biological origin of the carbon however remained obscure for many more years.

Liebenberg's investigation of the radioactive elements in the Witwatersrand banket soon proved that uranium was the chief source and that it was present in the form of the oxide, uraninite. It occurs as small, round, detrital grains and also in large amounts of secondary uraninite, both of which are often, and the secondary products always, closely associated with hydrocarbon. The detrital grains are frequently concentrated around a grain of carbon although they are entirely free from microscopically visible specks of carbon. The secondary uraninite however occurs as small enclosures in the carbon and these are often concentrically arranged. For this reason Liebenberg proposed the adoption of the name "Thucholite" for the uranium-rich carbon, but the term, although still used sometimes, is not favoured because it was originally proposed for thorium-rich carbon and this element is present in the South African radioactive carbon in very small amounts. The terms uraniferous carbon or just carbon are more commonly used but both imply a radioactive carbon. Liebenberg noticed also such sympathetic relationship between the amounts of gold and uranium in various reefs that he believed both must have had the same detrital origin and have been deposited simultaneously (Liebenberg, 1948, but this paper had a very restricted circulation). In his later paper (1955) he analysed every gold-bearing reef zone in the Witwatersrand Triad and found that the triple association of uranium, gold and carbon was invariably true. Even in the areas where the normal conglomerate of quartz pebbles was missing and the zone was represented by a mere pencil line of carbon, it often carried payable amounts of gold and uranium and was worth mining e.g. the Vaal Reef of the Klerksdorp area and the Main Reef Leader in some parts of the Central Rand. His investigations included also the new Orange Free State goldfields where the Basal Reef and the Leader Reef were both carbon-rich and highly payable. (See especially Pl. III, fig. 1 showing a quartz pebble partially sunken into the carbon of the Basal Reef, St. Helena Gold-mine, O.F.S.)

Liebenberg states that 60-70 per cent of all uraninite grains are altered by partial solution. They may show frayed margins, or even be completely replaced by metallic sulphides e.g. pyrite, pyrrhotite, galena, pentlandite, chalcopyrite, sphalerite etc. In qualitative spectrographic analyses of uraninite concentrates from amalgam-barrel residues, thirty-nine different elements were found although a number of these were present only as traces.

He studied, too, the banded pyritic quartzites (see Pl. II, fig. 10) which are formed in many areas and at a number of different horizons. In these no pebbles are present but a number of parallel gold- and uranium-bearing pyritic and carbon bands, from a few mm to approximately 6 cm in width, occur in a dense quartzite. Liebenberg considered that they were originally stratified "black sands" with a concentration of heavy detrital minerals, but that the carbon, like the heavy minerals, was concentrated and confined to the pyritic bands and he found it difficult to explain the seemingly anomolous association of the light carbon granules with the heavy detrital minerals but suggested that "thucholite" had been formed by the replacement of detrital grains of uraninite by hydrocarbon.

Liebenberg's long and detailed discussion of all the aspects of the gold and radioactive mineral association in the Witwatersrand Triad and the vast number of illustrations he supplied left no doubt about the placer (sedimentary detrital) origin of both gold and uraninite, nor of their intimate association together with smaller amounts of metallic sulphides with the hydrocarbon. In order to explain why only a small percentage of the metallic minerals remained in their original detrital form while the majority were recrystallised he resorted to the assumption that hydrocarbons in gaseous or liquid form had been introduced into the reefs and had been irradiated by the detrital uraninite and so produced the hydrocarbon which always includes grains of secondary uraninite and fragments of many other metallic minerals. He apparently did not at any time consider that the carbon might have represented living plant matter.

Nine years later De Kock's paper (1964) on the "West Wits Line" was published. His chief concern was gold and in particular the rich new reefs — the Carbon Leader and the Ventersdorp Contact Reef. Both of these led him inevitably to a study of the carbon but he reached quite different conclusions about its origin. He compared it with low volatile bituminous coal in which rounded and angular inclusions of uraninite occur. A brief synopsis of his description of the two reefs is as follows:—

(a) The Carbon Leader (see Pl. III, fig. 2 based on De Kock, 1964, Pl. X, fig. 1) is worked particularly on the Far West Rand, especially on the rich West Driefontein, Blyvooruitzicht and Doornfontein Gold Mines. De Kock described it as having a thin foot-wall seam of soft friable carbon-rich quartzite, heavily mineralized with pyrrhotite and pyrite, on which rests a variable layer of pebbles often embedded in the carbon-rich material. The carbon band varies from a streak up to 7 cm in thickness. In the latter case columnar structure perpendicular to the bedding plane is usually developed. In this grains of visible gold are common and often in considerable quantities. Very high values are contained almost exclusively in the carbon band.

(b) The Ventersdorp Contact Reef, or V.C.R. (De Kock, 1940) is a coarse, partly alluvial, scree lying unconformably on truncated Witwatersrand rocks. It is concentrated especially in erosionchannels and is obviously enriched on the lower stream side of suboutcrops of Witwatersrand reefs. c.f. De Kock, 1940, Pl. XIX which is a block-diagram showing the pre-Ventersdorp erosion surface and the relationship of the V.C.R. to the dipping outcrops of reefs of the Witwatersrand System.

The reef varies considerably in thickness but carries visible gold in particularly coarse particles ranging up to 2 mm in diameter and is especially rich where the carbon percentage is high. It forms the base of the Ventersdorp System and is overlain by thick flows of andesitic lava which sealed this surface-deposit in its original position.

(c) Origin of Carbon, De Kock (1964, p. 370) in emphasising the immense importance of the gold-carbon association listed all the important gold-bearing reefs in recent years, the Carbon Leader and Ventersdorp Contact Reef of the "West Wits Line", the Kimberley Reef of the East Rand, the Vaal Reef of Klerksdorp and the Basal Reef of the Orange Free State and states that it is true of all these, as well as of the older well-known reefs of the Main Reef Group, and the Bird Reefs, that it *can be regarded as a general rule that the higher the carbon content, the higher the gold value will be.* De Kock (1964, p. 371-373) gave chemical analyses of a number of specimens of carbon from different reefs on the West, Central and East Rand which showed extraordinarily consistent comparisons with South African coals ranging from low volatile bituminous to semi-anthracite (with the exception of the mineral inclusions which gave a very variable ash content). He reiterated also, as a number of other geologists have done, that the carbon was of aliphatic origin i.e. probably directly from plant matter and not aromatic or distilled from an oil, a bitumen or gas as Liebenberg had suggested, since it is completely insoluble when tested. Finally he offered his own solution, p. 373, "that the carbon is the organic remains of a primitive form of algae which may have existed in the waters of the basin in which these reefs were deposited is a suggestion which deserves further investigation."

The following year Snyman (1965) carried the argument a little further by comparing the "carbon" with a sapropelic coal or oil shale whose composition had been changed by reducing the hydrogen content "as a result of alpha particle bombardment from the uraninite". He provided illustrations, one of which is reproduced on Pl. III, fig. 3 showing that the uraninite is often distributed concentrically inside many of the oval shaped bodies of "thucholite" (carbon) and that the size and arrangement of these are comparable with the small algal colonies in Palaeozoic torbanite (i.e. a rich oil-shale). His explanation of polished blocks of "carbon" from three different mines revealed that it consisted in each case of a concentration of rounded and oval bodies having a spongy appearance due in some cases to numerous small cavities and in others to the minute concentrically arranged inclusions.

Snyman's comparative chemical analyses of Witwatersrand "carbon" and Transvaal torbanites, on a dry ash-free basis, showed them to be extremely close save for lower hydrogen in the former.

As a student of coal, De Kock's and Snyman's suggestions interest me very much. Torbanite has been studied for many years and is known from the Devonian Period onwards but its algal origin has only been understood since Blackburn and Temperley (1936) showed it to be so remarkably like the small waxy algae *Botryococcus braunii*, living today in a number of different parts of the world, that they could be regarded as co-specific. Up to that time living waxy algae were unknown and the efforts of coal-petrologists and geologists to introduce a form of wax, in liquid or gaseous form, into what used to be called "the little yellow bodies" in oil shales are reminiscent of the efforts of some of our own geologists to explain the Witwatersrand carbon. Blackburn (1936) provided a diagram of the living *Botryococcus* showing it to be made up of colonies of cells. Each cell was a small open cup which exuded a waxy waste product to form what she called a thimble. Each cell divided again and again until a spherical lump, several cells deep made up of crushed waxy thimbles was formed of which only those on the outside of the sphere were still living. At an optimum size of a few mm the sphere broke up and the process was repeated. Fossil torbanites are common in Transvaal coalmeasures. They are deep-water coals in the sense that the algae float on the surface of deep water and only sink when they die. (See Pl. IX, fig. 4, a thin section of a Transvaal torbanite, and also the coloured frontispiece where the algae are shown floating on the lake on the right hand side). They are green in the spring and summer, but turn yellow as they die and either sink, or are piled up on the shore as a waxy, rubbery mass. Any Precambrian counterpart would of course have had to live below a protective cover of water. The living algae of today float on the surface and have been recorded from Portuguese East Africa although I have been unable to obtain a sample. Barghoorn and Tyler,

Pflug and others have emphasized that many of the Precambrian algae are remarkably like living ones. It is therefore conceivable that a morphologically simple form like *Botryococcus*, or something similar, might have existed in Witwatersrand times. The waxy waste-product would undoubtedly have been carbonised with time but the spongy nature, the open cups and the protective coating are all sufficiently relevant to the ancient fossil product to promote some speculation.

In 1967 Prashnowsky and Schidlowski had published a detailed biochemical investigation of Witwatersrand "Thucholite" from the Basal Reef of the Orange Free State in which they found not only a number of amino acids but also carbohydrates of which they analysed seven monosaccharides. These results convinced them of the biogenic derivation of the carbon and encouraged them to proceed with an electron-microscopic examination, the results of which were published the following year. The small septate algal threads isolated from the Basal Reef by Oberlies and Prashnowsky (1968) and illustrated here on Pl. III, figs. 4 and 5, had been separated by acid maceration and belonged to different genera in what was obviously a mixed assemblage of primitive plants.

That a large proportion of the gold and uraninite occurs in placer deposits and that a number of the grains were originally detrital, has been accepted for many years by most of the South African economic geologists familiar with them. They did not question that the vast hinterland of very ancient igneous rock and enriched Primitive System "rafts" could have provided every mineral and element which has been found in the Witwatersrand Triad but the concentration of these into a limited number of zones of economic significance and the fact that a large proportion of the metallic minerals occur in crystallised form and many as sulphides has been the subject of arguments which lent support to the various hydrothermal theories proposed from time to time. Our newly acquired knowledge of the reducing and toxic nature of the early atmosphere, coupled with the proved existence of primitive plant life, provides a reasonable excuse for adding a few more pages to the already voluminous literature on the origin of the payable Witwatersrand gold reefs but it will be presented very briefly.

5. Living Plant Matter in the Role of a Natural Reduction Works

(a) This satisfies a number of geological observations, conditions and problems.

I believe that the direct intervention by primitive living algae and bacteria was the only agent able both to trap the heavy detrital metalliferous minerals and also to precipitate the metals formerly in solution in the fast-flowing streams — which, normally, would have been carried much further, and to deposit them together in single narrow zones. The zones would have been controlled by the limited areas and depths at which the plants could grow at that time, i.e. platforms beneath a cover of about ten metres of water. Here the velocity of the mineral-carrying currents would have been checked and the mat of water-weed would have acted much as a corduroy table does today, trapping heavier detrital grains of gold, uraninite and pyrite etc. which would sink down together with any pebbles which may have been present. The oxygen released by the plants would have begun to attack some of the detrital uraninite accounting for the many "frayed" grains. Sulphuretted hydrogen, which always accumulates where rotting plants are present in anaerobic conditions, would have been available to precipitate secondary uraninite as well as other metals in solution, as sulphides.

(b) Experimental Evidence

(i) Sulphides: Cheney and Jensen (1966) confirmed that bacterially produced hydrogen sulphide precipitated uranium minerals and sulphides and that the pyrite cement associated with the secondary uranium differed, from that magmatically produced and from hydrothermal pyrite, being significantly lighter. There are many other reports of the importance of biogenic activity in precipitating metallic sulphides e.g. Pflug's (1966) evidence of the metallic sulphide coating of globular organisms from the Fig Tree Group (confirmed by laboratory experiments). Also in the Onverwacht Group, Viljoen *et al* (1969) state that flakes of graphite are abundant in some chert horizons with chalcopyrite, pyrite and pyrrhotite.

(ii) Uranium: Harrison *et al* (1966) record a most interesting experiment, using bacteria to extract U_3O_8 cheaply, from the acidic mine-water of Elliot Lake mine and also by washing broken rock in stopes with water containing bacteria. In eleven weeks 53,5 per cent of the U_3O_8 was recovered while in a control experiment with non-bacterial water only 1 per cent U_3O_8 was obtained in the same period. They reported that, under the microscope, bacteria could be seen clustering around the ore particles which they did not attack directly but set up a chemical condition which caused the dissolution of U_3O_8. The treatment could be continued at intervals.

(iii) Gold: Williams (1918) showed that certain fungi extract gold from colloidal gold solutions containing some tannin. The metal was stored in the cell walls and in spores but could be recovered completely by soaking the plants in chlorine water. (I am indebted to R. Melville of Kew for tracing this reference for me.)

(c) Plants and Coal: Experiments prove that the ash from plants living in any particular area contains a much higher percentage of many metals than can be found in the surrounding soil. Similarly the organic ash in coal, as distinct from any mineral matter introduced into the swamps, contains a very large number of elements. The plants appear to select rare metals particularly and cases are known where wood ash has as much as 1 000 times more of a particular element than can be found in the surrounding soil. The ash from Spruce trees in Norway has been recorded as having 20 per cent MnO_4. Ash from deciduous trees averages 2 per cent of their original weight but may be much more. An average of 54 per cent of the total sulphur in coal is of organic origin. In higher plants the mineral matter is stored, in different proportions, in various parts of the plant e.g. Moore (1940, p. 52) quotes 5,8 per cent ash in birch leaves and only 4 per cent in birch stems. The leaf-ash included seventeen different elements. Of course the environment in which the plants grew is the controlling factor of the number of elements present but the plants themselves determine the quantities absorbed. It is significant that Liebenberg (1955) recorded 39 trace-elements in Witwatersrand carbon. In fact all plants have a strong affinity for metals.

(d) Plants in the Witwatersrand Triad provide answers to a number of geological problems

e.g. They could explain why

(i) in the lowest members of the Witwatersrand Triad, the Dominion Reef System, which has very limited amounts of carbon, the distribution of detrital gold is so patchy that the reefs are no longer worked.

(ii) Liebenberg's (1955) problem of light carbon and heavy gold being deposited together is overcome because the plants were there first.

(iii) There are conglomerates in both the Lower and Upper Witwatersrand which do not contain payable quantities of gold while alternatively there are fine-grained pyritic quartzites and zones, where only a parting plane of carbon exists, which are rich in gold. In fact the presence of carbon appears to be more important than any other factor as an agent of enrichment.

(iv) Both Davidson (1957) and Hiemstra (1968) found it difficult to explain why the alteration products of metals were deposited so close to the originals (within 1 cm) if migrating solutions had been responsible.

(v) Sharpe's (1949) observation that payable reefs often followed a disconformity, would provide the time interval long enough for the growth of a mat of water weeds.

I think the few examples given above serve to indicate that biological assistance in increasing payability should be considered more seriously than in the past.

Plate III of the Witwatersrand Triad includes pictures of banded ironstone (Fig. 6) from the lower Witwatersrand in which iron bacteria probably played a major role. While fig. 7 is stromatolitic limestone from the Upper Ventersdorp System in which calcareous algae were instrumental (see Winter, 1963). Details of these plant-mineral associations will be discussed in the next section on the Transvaal System where both types play even more important roles.

[*Editor's Note:* Material has been omitted at this point.]

REFERENCES

BARGHOORN, E. S. and TYLER, S. A., 1965. Micro-organisms from the Gunflint Chert. *Science, 147* (3658): 563-577.

BLACKBURN, K. B. and TEMPERLEY, B. N., 1936. *Botryococcus* and the Algal Coals. *Trans. roy. Soc. Edinb.*, 58(3): 841-868.

CHENEY, E. S. and JENSEN, M. L., 1966. Origin of the Wyoming Uranium Deposits. *Econ. Geol., 61*(1): 44-71.

DAVIDSON, C. F., 1957. On the occurrence of Uranium in ancient conglomerates. *Econ. Geol., 52*: 668-693.

DE KOCK, W. P., 1940. The Ventersdorp Contact Reef. *Trans. geol. Soc. S. Afr., 43*: 85-107.

————, 1964. The Geology and Economic Significance of the West Wits Line, *in* S. H. Haughton, Ed. "*The Geology of some Ore Deposits of Southern Africa.*" Vol. 1. 1964. Publ. Geol. Soc. S. Afr., p. 323-391.

HARRISON, V. F., GOW, W. A. and IVARSON, K. C., 1966. Leaching of Uranium from Elliot Lake Ore in the presence of bacteria. *Can. Min. J., 87*(5): 64-67.

HIEMSTRA, S. A., 1968a. The Mineralogy and Petrology of the Uraniferous Conglomerate of the Dominion Reefs Mine, Klerksdorp Area. *Trans. geol. Soc. S. Afr., 71*(1): 1-65.

————, 1968b. The Geochemistry of the Uraniferous Conglomerate of the Dominion Reefs Mine, Klerksdorp Area. *Trans. geol. Soc. S. Afr., 71*(1): 67-100.

LIEBENBERG, W. R., 1955. The Occurence and Origin of Gold and Radioactive Minerals in the Witwatersrand System, the Dominion Reef, the Ventersdorp Contact Reef and the Black Reef. *Trans. geol. Soc. S. Afr., 58*: 101-227.

MELVILLE, R., 1962. A new theory of the Angiosperm Flower. *Kew Bull., 16*(1): 1-50.

————, 1963. A new theory of the Angiosperm Flower. *Kew Bull., 17*(1): 1-65.

————, 1969. Leaf Venation Patterns and the Origin of the Angiosperm. *Nature, 224*(5215): 121-125.

MOORE, E. S., 1940. *Coal.* Second Edition. John Wiley and Sons, Inc. New York. pp. 473.

OBERLIES, F. and PRASHNOWSKY, A. A., 1968. Biochemische und electronenmikroskopische Untesuchung prakanbrischer Gesteine. *Die Naturwissenschaften*, Heft 1, 5. 25-28. 55 Jahrgang.

PFLUG, H. D., 1966. Structural Organic Remains from the Fig Tree Series of the Barberton Mountain Land. *Econ. Geol. Research Unit, Univ. Wits, Inform. Circ.* No. 28, 14 pp. and 4 pls.

PRASHNOWSKY, A. A. and SCHIDLOWSKI, M., 1967. Investigation of Pre-Cambrian Thucolite. *Nature, 216*(5115): 506-563.

SHARPE, J. W. N., 1949. The Economic Auriferous Bankets of the Upper Witwatersrand Beds and their relationship to Sedimentation Features. *Trans. geol. Soc. S. Afr., 52*: 265-300.

SNYMAN, C. P., 1965. Possible Biogenetic Structures in Witwatersrand Thucolite. *Trans. geol. Soc. S. Afr., 68*: 225-236.

VAN NIEKERK, C. B. and BURGER, A. J., 1964. The Age of the Ventersdorp System. *Ann. geol. Surv. S. Afr., 3*: 75-86.

VAN NIEKERK, C. B. and BURGER, A. J., 1969b. Lead Isotopic Data Relating to the Age of the Dominion Reef Lava. *Trans. geol. Soc. S. Afr., 72*(2): 37-45.

VILJOEN, R. P., SAAGER, R. and VILJOEN, M. J., 1969. Metallogenesis and Ore Control in the Steynsdorp Goldfield, Barberton Mountain Land, South Africa. *Upper Mantle Congress*, Pretoria, July 1969. Econ. Geol., 64(7): 778-797.

WINTER, H. de la R., 1963. Algal Structures in Sediments of the Ventersdorp System. *Trans. geol. Soc. S. Afr., 66*: 115-121.

WILLIAMS Maud, 1918. Absorption of Gold from Colloidal Solutions by Fungi. *Ann. Bot., 32*: 531-534.

YOUNG, R. B., 1917. *The Banket — A Study of the Auriferous Conglomerates of the Witwatersrand and the Associated Rocks.* Gurney and Jackson, 125 pp.

Editor's Comments
on Paper 29

29   BILLINGSLEY and LOCKE
      Excerpt from *Structure of Ore Districts in the Continental Framework*

Cross structures, those major features which trend with or across mobile belts and cross platforms, are undeniably important in guiding ore emplacement. In Paper 29, Billingsley and Locke use the figure of $1 billion gross as the hallmark of a major deposit. Even at today's values, $1 billion is still "a major." Perhaps 90 percent of the major ore deposits of North America are on major structures. All the porphyry coppers of Arizona, New Mexico, and Sonora are on the 150-mile-wide Texas lineament; Bingham on the Utah rifts; Coeur d'Alène and Butte on the Lewis and Clark lineament; Thompson on the Thompson lineament; Pine Point on the McDonald fault; Sudbury at the intersection of the Murray, Temiskaming and Ottawa Valley systems, and so on.

Not surprisingly the theme has continuing interest. King (1964) showed the major structures transecting the cordillera. Mayo (1958) described the lineament tectonics and some ore districts of the southwest. Robert (1968) made similar studies of Spain, Tunisia, Corsica, Bulgaria, Yugoslavia, and Central America. Kutina (1972) compared "regularities in the distribution of hypogene mineralization along rift structures" in the Canadian Shield, southwest United States, and Madagascar, and Scheibner and Stevens (1974) plotted the distribution of deposits in the Lachlan geosyncline of New South Wales, as they are enhanced by a major crossfault system. As yet, nevertheless, the origin of the major linear features is unknown. Commonly, facies changes indicate an early origin and subsequent reuse is prevalent.

29

Reprinted from *Trans. Amer. Inst. Mining Metallurg. Eng.,* **144,** 39-51 (1941)

STRUCTURE OF ORE DISTRICTS IN THE CONTINENTAL FRAMEWORK

Paul Billingsley and Augustus Locke

[*Editor's Note:* In the original, material precedes this excerpt.]

CONTINENTAL FRAMEWORK (FIG. 12)

The foregoing analyses serve to illustrate the manner in which, as facts become available, district structures fit into the larger continental framework. The motions represented in the districts are local phases of the crustal movements of the continent.

These continental movements are not haphazard; they are *systematic* in time and place. They occur predominantly in definite *places,* and recur at intermittent *epochs of orogeny.* The full pattern of folds, faults, uplifts and associated erosional dumps of coarse detritus made by each successive epoch will, we believe, be successfully deciphered in the future by detailed regional mapping, supplemented by drill data and the interpretation of geophysical studies.   At this time the general patterns only can be vaguely seen, with sparse bright spots here and there marking the localities where tectonic study is more advanced.   But, dim as it still appears, the pattern of North American orogenies forms the essential framework for the ore deposits.

1. *Archean Orogenic* belts are found on the Canadian shield, which they traverse in southwesterly lines of deeply eroded mountain roots. They abound in unsolved problems of correlation and structure.

2. *The Late pre-Cambrian Orogeny* can be identified in southern Quebec.   It makes an arc to the northward around Lake Huron, with Sudbury at the point; then passes westward through the Michigan iron ranges and under the Great Plains to the Black Hills, where it seems to turn sharply southward.   It reappears again in the iron districts of eastern Wyoming, traverses the Front Ranges of Colorado, and shows intermittently in New Mexico and Arizona (Mazatzal Range) as far as Jerome. From Sudbury to Jerome the trends in these pre-Cambrian mountain roots are southwesterly or north-south.   In western Arizona and throughout the Pacific States, however, they are northwesterly, so th t this orogenic system grew as a V, enveloping the then Continental shie d from Labrador to Arizona to Alaska.   Inside this V are found its dumps of conglomerates and coarse sandstones, as well as foreland uplifts such as the Uintas and Yellowstone Park.

3. *An Early Paleozoic Orogeny* can be recognized in Quebec and New England, and again along the Pacific Coast.   In the latter region it was *superimposed* on the late pre-Cambrian, but in the east the action had *retreated* southeastward, off the shield.

4. *The Final Paleozoic Orogeny* was again continent wide.   It formed the Appalachians, the Arbuckles and Ouachitas; traversed southern New Mexico and Arizona, with a spur into Colorado; raised again the early Paleozoic Sierra Nevadas; swung northeasterly across Oregon (Blue Mountains) and finally renewed the northern Coast Ranges in Idaho, Washington and British Columbia.   The clastic fans of this system make the Appalachian and Middle Western coal measures, the earlier red beds of Colorado and the Southwest, and the upper Pennsylvanian quartzites and conglomerates of the Great Basin.   In this orogeny, in the west, the great *Pacific Echelon* first becomes noticeable, with the Oregon arc linking the complex masses of the Sierras and northern Coast Ranges.   In these masses formed by superimposed orogenies there is a great development of *granite*.   In the east, the Final Paleozoic orogeny again shows a retreat off the shield.

5. *The Early Cretaceous Orogeny* has in the east retreated to the West Indies, extending through Cuba into Honduras.   Turning up the Pacific line, it traverses Sonora and southwest Arizona, and is superimposed on the Sierra Nevada complex.   The Oregon arc is advanced southward

into northwestern Nevada, linking up through Idaho with the northern Coast Ranges. This orogeny greatly advanced the granitization of the Sierra Nevadas, Idaho, and the north coast. The advance of the Oregon arc into the Great Basin must have been accompanied by tear faulting on both Walker and Lewis and Clark lines. The Early Cretaceous clastic fans are found in Cuba, throughout the Southwest and the eastern Great Basin, and the eastern Rockies of Montana and Canada.

6. *The Early Tertiary Orogeny* is found only in western North America, for the eastern arm has retreated to the Caribbean. In the west it is superimposed on the Sonoran and Arizona belts of earlier orogenies, jumps the Walker line to form the Cordilleran thrust zone across Nevada, Utah and Southern Idaho; and, broken again by the Lewis and Clark line, makes a new belt of mountains (Canadian Rockies) east of the old north Coast Ranges. The expansion of the *Pacific Echelon* is marked by the advance of the Cordilleran belt of thrusts to the eastern margin of the Great Basin, accompanied by strong tear faulting on both Walker and Lewis and Clark lines and by stretching in the western Great Basin. In addition to these main orogenic elements, this early Tertiary orogeny enlivened the entire western part of the continent. Sectors of earlier mountain belts were rejuvenated (north Coast Ranges), old foreland domes were sharpened into horsts wedged up between thrusts (Colorado Front Ranges, Uinta Mountains), old roots were re-sheared. A new structure element, undoubtedly active much earlier, can also be recognized; a strong northwesterly tear line on the south Pacific Coast, showing motion of the continent southeastward with respect to the ocean. This we call the Anza line. The clastic dumps include the great lower Tertiary conglomerates in Colorado, Utah, Wyoming and Montana and the Tertiary clastics in troughs along the tear faults.

7. *The Parade of Orogenies* may be briefly expressed in terms of motion as shown in Table 1. To simplify still further, we can note that the two

TABLE 1.—*Parade of Orogenies*

Period	Pacific					Atlantic
	North		Great Basin[a]		South	
Late pre-Cambrian...	Coast range	Lewis and Clark line	?	Walker line	Sierra Nevada	S. margin of Archean shield
Early Paleozoic......	Superimposed		W. Oregon link		Superimposed	Retreat to N. New England
Final Paleozoic......	Superimposed		E. Oregon		Superimposed	Retreat to Appalachians
Early Cretaceous....	Superimposed		Nev.-Id. link		Superimposed	Retreat to Cuba
Early Tertiary.......	Superimposed and adv. to east		Cordilleran		Superimposed	Retreat to Caribbean

[a] See Fig. 14.

masses of superimposed orogenies; i.e., the north coast and Sierra-Nevada, lie in echelon along the Pacific coast, like drag folds made by the southeastward movement of the continent. And that this same movement is marked by a minor southeasterly advance of orogenic arcs into and across the Great Basin, which lies *between* these masses, and by a major southeasterly retreat of the contemporary Atlantic arcs. In all cases orogenic activity has spread *equatorward*.

Fig. 12 shows in a very generalized way the composite pattern made by the successive orogenic belts in the United States.

The belts are readily differentiated east of the Rocky Mountains, occurring there in a well separated series which retreats southeastward off the Canadian shield into the Caribbean region. In and west of the Rocky Mountains, however, they are partly superimposed and difficult to disentangle. The earliest, late pre-Cambrian, turned north from a vertex in Arizona and seems to have extended up the coast. It cannot, however, now be distinguished there from the Early Paleozoic orogeny, which has left a trail of metamorphics of Ordovician and earlier age from southern California to Alaska.

By Final Paleozoic times the Pacific coastal region showed a three-part structure, made up of the Sierra-Sonora complex on the south, the north coast complex on the north, and the Oregon arc spanning the gap between these two echeloned massives. Subsequent orogenic activity, in Early Cretaceous and Tertiary times, increased the complexity of the massives, but shows on the map mainly because of the advance of the Oregon arc across Nevada and Utah, in a bulge bounded on north and south by northwesterly tear lines (Lewis and Clark, and Walker lines). Minor advanced Tertiary arcs are apparent in Wyoming, Colorado and New Mexico, where they both sharpen older tectonic structures and bulge between old high masses. A major tear line (Anza) parallels the California coast and bounds the Sierra-Sonoran block, which has moved southeastward relative to the ocean.

Sources of Information, Continental Framework

It is impossible to list in full the sources from which the stratigraphic-tectonic picture of the continent has been built up.

Billingsley has been assembling such data since 1922, when he found it desirable, in connection with oil exploration for the Anaconda Copper Mining Co., to break down the Rocky Mountain Mesozoic formations to small units of sedimentation and to trace the principal clastic deposits to their sources in various rising mountain masses along the Cordilleran thrust belt. A series of thickness contour maps of these fans, and of the intervening shale-basin deposits, was made by Billingsley and Lyon in 1922. These maps were based on: (1) U. S. Geological Survey publications; (2) an almost complete collection of oil-well logs as of 1922; and

(3) seven months spent in continuous field mapping during which the Mesozoic outcrops were followed continuously from central Montana across Wyoming into northwest Colorado.

The line of thought thus initiated led to the collection of similar data for the later Paleozoic formations, which include thick clastic deposits, obviously of tectonic derivation, in Colorado, Utah, Nevada, etc. The bulletins of the American Association of Petroleum Geologists were found rich in material on the Paleozoic mountain belts in Arkansas and Oklahoma, and their buried extension in Texas and New Mexico. U. S. Geological Survey publications supplied many stratigraphic data. Keith's and Schuchert's papers in bulletins of the Geological Society of America were particularly valuable for Appalachian structure. Field work involving these Paleozoic formations covered the years 1922–1933, when mapping of Utah, Colorado and Arizona mining camps was frequently expanded into the adjoining regions. A paper by Billingsley and Locke on Tectonic Position of Ore Districts in the Rocky Mountain Region [*Trans.* A.I.M.E. (1935) **115**, 59], summarized conclusions as of that date. (Also Geologic Structure of Salt Lake Region, by Billingsley, *Guidebook* 17, Int. Geol. Cong. 1933).

The late pre-Cambrian (Huronian, etc.) scheme of mountain belts was studied after the others. Interest in these early orogenic periods came when work in deepening Rocky Mountain mines began to lead again and again to the basement formation as the place of fundamental motion and structure. More mapping and analysis of the pre-Cambrian is necessary than has been usually given in publications on western mining districts. Too often the basement is lumped as pre-Cambrian schist or gneiss or granite, with the stratigraphy and structure unknown. To secure the necessary knowledge of the basement will be a work of years; but sound progress has been made recently in several areas. Much valuable material has appeared in the publications of the Colorado Scientific Society, for example. With more detailed study has come a new outlook on the processes that have operated in the basement. "Granitization" by replacement, suggested by Quirke and Collins in 1930 to explain the disappearance of the Huronian sediments east of Sudbury, Ontario, seems to be the best explanation of certain features in the West also. (Colorado Front Ranges, Sierra Nevada, Inyo Range, Idaho batholith, Coast Ranges in British Columbia.) Field work by the writers in these areas in 1923–1937 was fortunately supplemented by some months in 1934 and 1937 in the Michigan copper range, where structures, intrusions, and clastic sediments can be correlated in sequences recurring intermittently from Huronian into Paleozoic time.

In 1933 was published, as *Guidebook* 28, Int. Geol. Congress, Philip B. King's Outline of the Structural Geology of the U. S. The writers found in this concurrent study a welcome reinforcement of their own

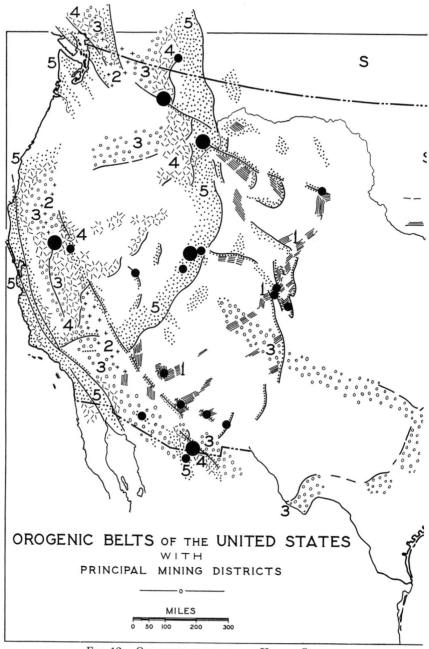

 OROGENIC BELTS OF THE UNITED STATES
WITH
PRINCIPAL MINING DISTRICTS

MILES

0   50 100   200   300

FIG. 12.—OROGENIC BELTS OF THE UNITED STATES.

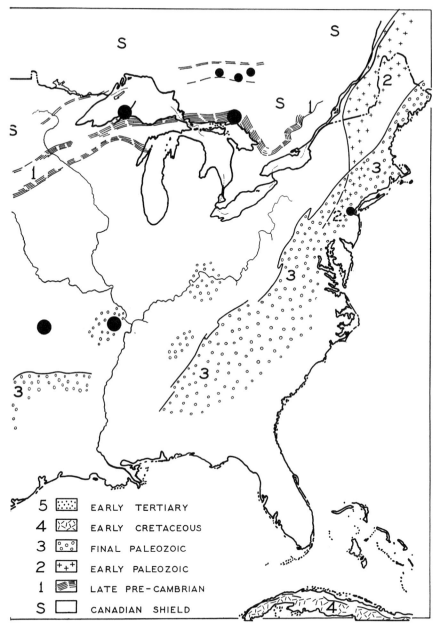

S

S

S

S

S

S

1

2

1

3

3

3

3

3

5 ⬚ EARLY TERTIARY

4 ⬚ EARLY CRETACEOUS

3 ⬚ FINAL PALEOZOIC

2 ⬚ EARLY PALEOZOIC

1 ⬚ LATE PRE-CAMBRIAN

S ⬚ CANADIAN SHIELD

4

FIG. 12.—(CONTINUED).

attack on the problem. Believing, however, that the maintenance of their independent synthesis, based on slightly d fferent selections from the literature and on independent field work, will have a certain value, they have in no way utilized King's work in their assembled maps or conclusions. Their principal differences from King's map are three; (1) Billingsley and Locke emphasize more the Huronian and late pre-Cambrian structures; (2) Billingsley and Locke in the Great Basin region use the fold and thrust alignments within the basin ranges rather than the Quaternary normal faults along their edges; (3) Billingsley and Locke have recently begun to emphasize certain tear-fault lines (Lewis and Clark in Montana-Idaho Walker in Nevada-California; and Anza on the California coast).

The authors wish freely to acknowledge their indebtedness to all workers whose maps, stratigraphic measurements and structural data they have incorporated in their tectonic syntheses. They must also acknowledge the stimulation and broadening of outlook that has come from their too brief readings in European tectonic studies.

POSITION OF DISTRICTS*

When the mining districts are placed on a structure map on which the above superimposed and intersecting orogenic elements are emphasized, it appears at once that they are grouped in *clusters of districts* at particular places (Fig. 12). The chief clusters are as follows:

A. In the *roots* of old orogenies, schists, gneisses, drag folds:
 1. Canadian gold districts—earliest orogenies.
 2. Sudbury copper-nickel—late pre-Cambrian, chonolith.
 3. Black Hills gold—late pre-Cambrian, drag folds.
 4. Jerome copper—late pre-Cambrian, drag fold, intrusives.
B. In the *complex blocks* of superimposed orogenies. These, like the roots, are characterized by the widespread development of schistosity, gneiss and granite. Movement has become adjusted into shear zones in these formations. Parallel structures are dominant:
 1. *North Coast Block*
     Juneau gold belt—shear zone on western margin.
     Britannia copper—shear zone within complex.
     Bridge River gold—late intrusives within complex.
     Hedley gold—cross thrusts, intrusives, in cover.
 2. *Sierra Nevada Block*
     Shasta copper—
     Plumas copper—shear zones within complex.

---

* Although in the preceding pages we have described only 10 districts, we here place more than 50 on the structure map. Our justification lies in the fact that we are familiar with the additional districts, and have in fact written detailed descriptions of them in a manuscript which is now being prepared for publication. We do not wish to restrict the scope of the grouping by using only the districts that space has permitted us to describe in this paper.

Sierra gold—shear zone on western margin.

Silver Peak gold—shears, thrusts, on eastern margin.

C. In sectors of *early orogenies*, eroded to roots and *rejuvenated* or intersected by later elements:

1. *Michigan Copper*—a "living" structure from Huronian into Cambrian.
2. *Colorado Front Ranges*—late pre-Cambrian plus Tertiary.
    Central City, etc., gold in basement.
    Climax molybdenum in basement.
    Cripple Creek gold in basement.
    Leadville—lead, zinc in cover.
    Aspen—lead, silver in cover.
    Red Cliff—zinc-lead-copper-silver-gold in cover.
3. *Yellowstone Horst*—late pre-Cambrian plus Lewis and Clark tears.
    Pony—gold in basement.
    Jardine—gold, tungsten in basement.
    Chromite belt—chromium in basement.
    Cooke City—copper in cover.
4. *Central Arizona*—late pre-Cambrian plus Tertiary and Walker tears.
    Miami—copper in basement.
    Ray—copper in basement.
    Morenci—copper in basement and cover.
    Globe—copper in cover.
    Superior—copper in cover.
5. *Southern Arizona-Sonora*—Final Paleozoic and Early Cretaceous plus Tertiary.
    Bisbee—copper in cover.
    Cananea—copper in cover.
    Nacozari—copper in cover.
6. *San Juan Mountains*—Final Paleozoic horst plus Tertiary.
    Rico—silver, lead zinc in cover.
    Ouray—silver lead in cover.
    Telluride—gold in cover.

D. At *nodes* on the *later orogenic belts*. The nodes are usually found where the belts are cut by tear faults. Folds and thrusts are sliced by steep faults:

1. *Salt Lake Area*—Cordilleran thrusts plus Uinta tears.
    Park City—lead silver in cover.
    Bingham—lead zinc, in sediments, copper in intrusion.
    Tintic—silver lead in cover, gold in basement.
2. *Boulder Dam Area*—Cordilleran thrusts plus Walker tears.
    Goodsprings—zinc lead in cover.
    Pioche—zinc lead in cover, gold silver in basement.
    Oatman—gold in basement.
3. *Western Nevada*—Great Basin stretching plus Walker tears.
    Comstock—silver gold in volcanic cover.
    Tonopah—silver gold in volcanic cover.
    Goldfield—gold in volcanic cover.
4. *Eastern Nevada*—Early Cretaceous (?) thrusts rejuvenated.
    Eureka—lead silver in upthrust prong.
    Ely—copper in crumpled trough.
5. *Montana Gap*—Cordilleran thrusts, Lewis and Clark tears.
    Butte—copper in intrusion.
    Phillipsburg—silver in intrusion.
    Corbin-Wickes—silver in intrusion.

Georgetown—gold in sediments.
Hecla—silver in sediments.
Garnet—gold in sediments.
6. *Coeur d'Alene*—Early Cretaceous (?) thrusts, Lewis and Clark tears.
Mullan—copper, zinc, lead.
Wallace—silver, lead.
Kellogg—silver, lead.
Pine Creek—zinc lead.
Big Creek—silver.
E. In *foreland uplifts* of early orogenies, mildly rejuvenated.
1. *Tri-State* lead, zinc—Final Paleozoic plus Early Cretaceous.
2. *Southeast Missouri*, lead—Final Paleozoic plus Early Cretaceous.

The importance of the great tear-fault lines is obvious, with Coeur d'Alene, Butte and the Yellowstone districts on the Lewis and Clark line, the Comstock, Tonopah, Goldfield, and most of the Arizona copper districts on Walker-line elements, and Park City, Bingham and Tintic on tear faults marginal to the Uinta Mountains. Since the occurrence of great ore bodies requires the existence of an open channel at precisely the right time, this superiority of the tear faults suggests that of all types of crustal breaks they are the most frequently renewed and reopened. The distribution of historic earthquakes corroborates this conclusion, for Montana earthquakes are on the Lewis and Clark line, Nevada earthquakes on the Walker line, and California earthquakes on the Anza line. Occasional deep-focus earthquakes suggest that these tear lines may be deeply penetrating breaks extending to depths such as 500 km. We suspect that the pattern of tear faults controls the method whereby the earth's rigid crust adapts itself to the equatorial bulge of the spinning globe.

IDEAS ON ORE GENESIS

We find, then, that the ore districts of upper magnitudes are clustered at nodes within the Continental framework, and that these nodes are determined either by the presence of superimposed orogenic movements or of intersecting lines of successive motion, or of persistent deep-seated breaks. The effect of these factors has invariably been to stiffen the region, to rebreak it and to culminate in widespread heat effects near the surface. The least common denominator is *heat*, moving to the surface through reopened channels in a basement made strong by earlier metamorphism or intrusion, and so competent to carry channels to the depths. Above the basement, the action of the heat is modified by expanding channels in shallow structures and by approach to the surface and to ground water (Fig. 13).

The effects of the heat may cover a wide range. At one extreme is the *melt*, an igneous fluid moving by intrusion and injection. A milder

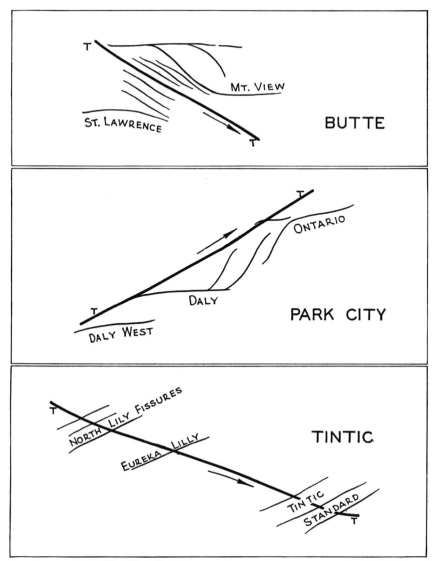

FIG. 13.—TEAR FAULT AND FISSURE PATTERNS.

but more widespread effect is produced by *vapors*, gases, etc. This may take the form of granitization, of pegmatitic replacement, of quartz-orthoclase or quartz-sericite or garnet alteration, or merely of recrystal-lization. Still cooler, below the boiling point, is *hydrothermal* action, "disseminated" ore bodies, vein fillings, ore clusters. At this extreme is the mineral-bearing artesian ground water; and no one can as yet tell to what extent ground water has contributed, with the addition of heat, to the more intense phenomena.

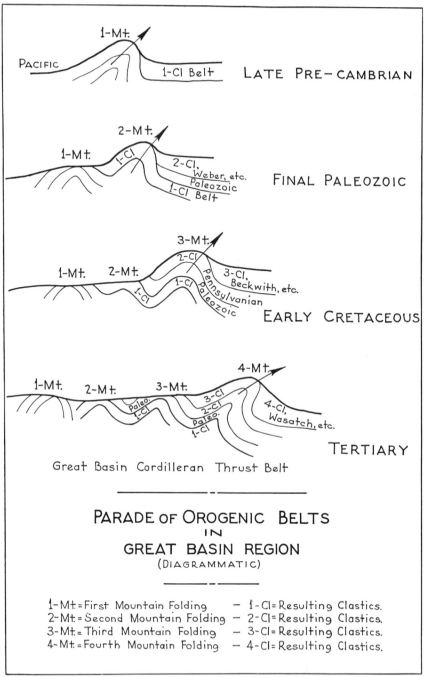

FIG. 14.—PARADE OF OROGENIES ACROSS GREAT BASIN.

The economic minerals occur in the cooler range of this series. The hot end produces "regional metamorphism" with granite "batholiths" growing at nodes out of schists via gneisses, and with local true melts cutting upward into less altered regions, and possibly reaching the surface as volcanoes. The intermediate stages produce alteration funnels, with vein and ore minerals in their later channels. The cold end ranges from mineral springs depositing metals to cold artesian water. The conventional dogma of igneous hearths seems increasingly artificial and unnecessary in this scheme. The deeper the development of a district the farther such hearths seem to retreat from view. A rush of heat through a channel producing in turn melts, gases and solutions that move toward the surface and there freeze and rebreak, seems for the present an adequate concept of ore genesis, and orogenic motion with breaking seems an adequate cause for the channel and the escape of heat.

The sketches in Fig. 13 are diagrammatic only. They illustrate, however, the patterns formed by the failure of rock under the stresses produced by the motions shown by the arrows.

In Butte the rock is quartz monzonite, which, intruded among Cretaceous shales and volcanics, behaves as a "perched" basement.

In Park City the rock is the Weber formation, a massive Pennsylvanian quartzite 2000 ft. thick, which also forms a perched basement.

In Tintic the rock is Tintic quartzite, over 6000 ft. thick, forming the true pre-Cambrian basement.

Fig 14 is intended to show what is meant by the successive advances of an orogenic belt; in this case, the Oregon arc of the western United States, which has advanced from Oregon to eastern Utah, between the limits of the Lewis and Clark tear fault on the north (Spokane to Yellowstone Park), and the Walker tear fault on the south (Reno to Boulder Dam region).

As can be seen, the crust itself has not moved the entire distance. It buckled and thrust eastward to some extent in the first folding (late pre-Cambrian) and then froze into the rigidity of a completed mountain belt. Its clastic deposits spread eastward over the present Great Basin area. These clastics in turn were involved in the second and more easterly folding, which in turn froze. And so on through the series.

Editor's Comments
on Paper 30

30   SPURR
Excerpt from *The Ore Magmas*

Spurr (Paper 30) is perhaps the one author in this volume who had a true flair for a name. The Great Silver Channel conjures up streets paved with gold, the Malemute Saloon, and dreams that became a reality at Comstock and Virginia City—an honest-to-god metallogenic province. Yet we still have no idea why, of two identical rock sequences, one is host to riches and the other is barren. Even defining the boundaries of a metallogenic province is not easy once one moves away from productive deposits. Burnham (1959) is one of the few workers who have attempted to do so.

30

Reprinted from J. E. Spurr, *The Ore Magmas*, McGraw-Hill Book Co., New York, 1923, pp. 457–466

THE ORE MAGMAS

J. E. Spurr

[*Editor's Note:* In the original, material precedes this excerpt.]

Since writing the above, I have found what appears to me to be a very remarkable piece of additional evidence bearing on these problems. I have received and studied the "World Atlas of Commercial Geology," published by the United States Geological Survey (Washington, 1921). On these maps the ore-bearing localities in the world are mapped separately, accurately, and quantitatively. With these maps in hand, let us take up again the distribution of silver in the Western Hemisphere, adding this new evidence to the discussion above. I have previously remarked on the relatively wide distribution of silver, as compared with its

[*Editor's Note:* Certain plates and figures have been omitted owing to limitations of space.]

relative abundance in the earth's crust, but that it was most concentrated in that part of the peri-Pacific province which lies in North and South America. In these maps under discussion the quantitative basis brings the concentration of silver into a clearer symbolic representation. They are, to be sure, production charts, based on the production of the different districts for the year 1913; and so only partly suit our purposes. A chart showing total production to date is what we should have for this purpose. Let us take, however, Plate 42, showing the production of silver in North America for 1913. We perceive a very remarkable string of silver-producing camps in Mexico, separated each from the next by a hundred miles more or less, from the Arizona border to near Mexico City, a distance of about a thousand miles. These camps are famous and productive —Cananea, Dolores, Santa Eulalia, Parral, Mapimi, Velardeña, Concepcion del Oro (Mazapil). Charcas, Zacatecas, Guanajuato, Pachuca, and El Oro. These are all in a *straight line,* or a narrow straight zone; the extension of this line a little farther south—about 300 miles—runs through Oaxaca and so into the Pacific. Extending this same straight line to the northwest, it runs through Tonopah, in Nevada, and so, in Southern Oregon, into the Pacific. I have copied the chart, adding only the straight line which I describe, in Fig. 74.

Now if we turn to Plate 48 (Fig. 75), where the 1918 silver production of the United States is given in more detail, we find further evidence that this line or narrow northwest zone is indeed continuous between Mexico and Tonopah. The map shows, on this line, the districts of Cochise County, Pima County, Globe, Prescott, and San Francisco, in Arizona; and Clark County, in Nevada. Continuing the line northwest of Tonopah, it runs through Mineral County, the Comstock Lode, and the district of the Northern Sierras. Between these "high spots" thus mapped, there are, as is well known, many other camps which contain silver ores. There is thus aligned a very remarkable string of silver deposits, including many renowned for their richness and production, lying in an almost

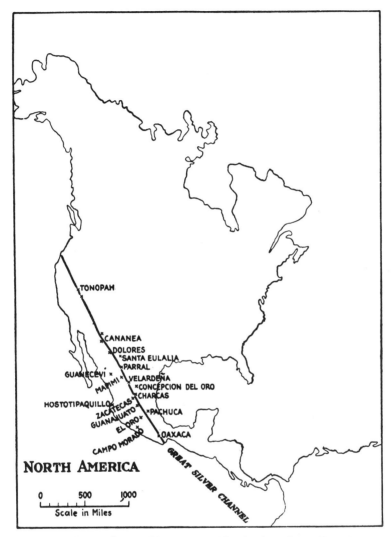

Fig. 74.—The Great Silver Channel in North America. Based on Plate 42, "World Atlas of Commercial Geology," U. S. Geol. Surv., Part I; Washington, 1921. The continuous line indicating the chain or channel, and the designation Great Silver Channel, are my own; also the recognition of this line.

perfectly straight line or narrow zone nearly 2,500 miles in length—a chord subtending the swelling arc of the Pacific Coast line of Mexico and California.

All this is remarkable enough, but if we extend this line straight southeast till it again strikes land in South America, we find that it strikes quite accurately and coincides

Fig. 75.—Principal silver camps of the Western United States. Based on Plate 48, "World Atlas of Commercial Geology," Part I., U. S. Geol. Surv., 1921. The crosses indicate the principal productive camps in 1918. The lines indicating ore belts or ore channels, and the recognition of the alignment into two systems, northeast and northwest, are my own.

with the great silver belt of Peru and Bolivia! This is indicated clearly on the world map (Plate 41), but is better proven in the map of South America (Plate 43), where the line may be followed from the coast at the Zaruma district, through the districts of Hualgayoc, Cerro de Pasco, and

Fig. 76.—The Great Silver Channel, or Silver Belt, in South America. Based on Plate 43 of the "World Atlas of Commercial Geology," Part I, 1921. Crosses show principal silver-producing camps in 1913. The drawing of the lines indicating belts or channels of ore, and the designation of the Great Silver Channel, are my own.

Santa Lucia, in Peru, into Bolivia, where it passes through Potosi and other silver-bearing camps on this belt. The

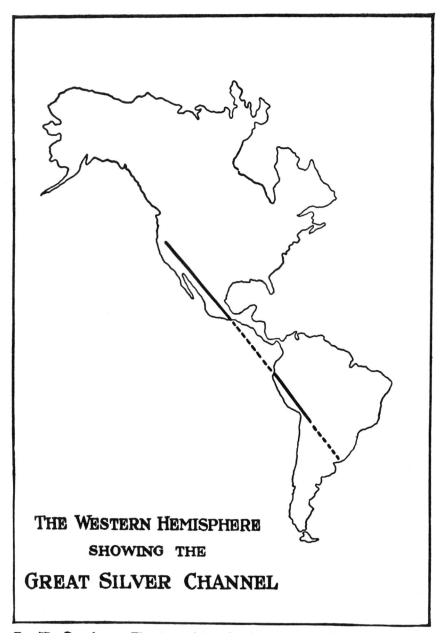

FIG. 77.—Based upon Figs. 74 and 76: showing the Great Silver Channel of the Western Hemisphere. See text.

total distance of this northwest-trending silver-bearing belt in Peru and Bolivia is some 1,600 miles, and there is no question as to the continuity of the belt. I have shown this in Fig. 76. Nor does there seem to be much question as to the identity and continuity of· the North American line (above described) with this South American line, as is shown in Plate 41, a portion of which is reproduced herewith, on which I have drawn the silver belt in question, in North and in South America, and indicated the connection (Fig. 77). The total length of these two belts, and the connecting gap which lies in the Pacific, is about 6,000 miles, the connecting segment submerged in the Pacific being nearly two thousand miles long.

Along this line in North and South America occur most of the celebrated silver mines in the world, like the Comstock, Tonopah, Santa Eulalia, Guanajuato, Pachuca, Cerro de Pasco, and Potosi. This wonderful straight line slashes clean across mountain ranges and other geologic structures, and continues its course independent of them. This is shown in Nevada, where it cuts at an angle of nearly 45° across the trend of the north-south-trending desert ranges. In Mexico, it shows an utter disregard of the main geologic features and mountain chains—starting in northern Sonora west of the main range, or Sierra Madre, it cuts diagonally across this on to the central Mexican Plateau, with its short desert ranges; and in the south of Mexico, its unswerving course carries it back across the Sierra Madre again, for this range curves much like the west coast line. In Peru and Bolivia, it coincides with the general trend of the Andes. It is in the gap between Mexico and Peru, where its extension is assumed, that the most utter lack of any relation between it and the mountain chains of the crust is shown, for here it crosses the depths of the Pacific, and part of it runs where the ocean is 2,000 to 3,000 fathoms (12,000 to 18,000 feet) deep; while in South America, where it coincides with the Andes, it lies mainly in country more than 12,000 feet above sea level; and in Mexico and in the

United States it lies in a plateau country several thousand feet above the sea.

Not only does the straight course of this Great Silver Belt disregard geography, but it disregards geology as well. Tonopah and the Comstock, in Nevada, represent deposits of the later Tertiary. In Arizona, the silver camps on this line correspond with the copper camps, and vary accordingly in geologic age from pre-Cambrian through post-Jurassic to post-Cretaceous. In Mexico, most of the camps in the northern part of the country are probably immediately post-Cretaceous, those in the southern part middle to late Tertiary. In Peru and Bolivia, the deposits are partly Tertiary, partly perhaps older. Therefore, the same conclusions are indicated as were deduced above for the copper deposits of Arizona—that this silver belt is very ancient and stable, and has manifested itself at any geologic period when igneous activity along this belt has furnished the necessary excitant and means. Indeed, in Arizona, the silver belt coincides in general with the belt of maximum copper deposition.

In this marvelously straight and persistent Great Silver Belt we have a striking symbol which we may decipher if we can. It is not at all, as manifested in the visible rocks, a fissure or a fissure system. The veins in the districts mentioned have their own courses and systems. At Tonopah, for example, they run east-and-west; at El Oro (in Mexico) they run northwest; at Asientos northeast, at Charcas nearly east-and-west; at Santa Eulalia mainly north-south; and so forth. It is not a mountain-making structure, a fault or a fold zone: as noted above, it cuts across all minor and major structural features without paying any attention to them. In what then does it consist? Its persistence, definition, and great length tempts us to classify it vaguely, gropingly, as a structural fracture, but it is an invisible one, disregarding and underlying all known rock structures, mountain chains, and lines of igneous intrusion. Here we have a striking supporting proof of our conclusions reached in considering the Arizona copper province: that this metallographic province means a stable highly copper-bearing substance in depth, not only far below the crust but below the unstable magma zone as well. Here, in this narrow metallographic province or ribbon which I have called the Great Silver Belt or Channel, is a Structural Feature which plainly underlies the wrinkles of the crust, with all its faults and folds, great and small— even such faults of semi-continental scale as the fault which bounds the Sierra Nevada (and probably the Cascades) on the east, and blocks out these great ranges; it underlies the effect of the loftiest mountain chains, and the magma reservoirs beneath them. We may conceive of it as a Structure; but it is a structure of geometric simplicity such as we do not have in the crust. We perceive here dimly a symbol defining to some extent a peculiar and simple framework or arrangement of the heterogeneity of the under-earth. We may perhaps interpret it as a belt along which—in that period which precedes all recognized geologic history, when the elements migrated freely in the fluid earth, to achieve the heterogeneous distribution which is indicated by the phenomena of metallographic provinces—silver notably was concentrated and segregated. We naturally picture such a straight line of segregation as a rift or fissure in the cooling globe, at a depth untouched by the subsequent flowage and adjustments of the solid or fluid outer zones; and thus we are led to set up as a trial hypothesis the formation of the heterogeneity of the stable under-earth as aided by regular fissuring. There presents itself to my mind at present at least one other trial hypothesis, which also postulates deep and regular globe-fissuring, but does not necessarily link up this fissuring with the period of origin of globe-heterogeneity. Rather it pictures the grand fissure systems as perhaps later, and that along them the stable highly metallic under-earth has bulged up in geometrically regular ridges above the normal deep zone; and that the magma tides in the overlying unstable magma zone became specially metal-liferous by flowing against and over these long reefs. Such an alternative trial hypothesis, which I think I favor over the first, would still, I repeat, postulate globe-fissuring, and still assign an immensely remote geologically prehistoric period for the fissuring; but it would nevertheless contemplate a possible immense lapse of time between the globe-fissuring and the earlier origin of globe-heterogeneity.

[*Editor's Note:* Material has been omitted at this point.]

Editor's Comments
on Paper 31

A compendium on metallogeny such as this would be incomplete if no mention were made of the manner in which the science is used in practice to find ore. Semenov and Labazin (Paper 31) tell us how this is done in the USSR. Unfortunately, not only are no examples given, but no metallogenic maps at the working scale (about 1/200,000) are known by the writer to have found their way to the Western world. I expect to see the New South Wales products at a comparable scale in the near future, but it is unlikely that the authorities in New South Wales will publish enough copies to satisfy the needs of the geological fraternity, which their leadership merits.

Many national geological survey organizations are now participating in the compilation of geologic, tectonic, and metallogenic maps of the world at a scale of 1 : 5 million. As increasingly more of the basic compilation is done at 1 : 250,000, improved correlation and analyses for ore forecasting are envisaged.

I find the study of the history of a particular metal or suite of metals to be most illuminating, starting with the magmatic origin and going on through concentration by metamorphism or intrusion, erosion, and reconcentration in placer. Gold, for example, may be introduced in basic vulcanism, concentrated with exhalite iron carbonate and sulfide, mobilized during orogeny (Sawkins and Rye, 1974) into anticlinal or shear zones, and ingested by intrusives to reappear as veins or disseminated in gold porphyries, eroded, and reconcentrated in placers. Uranium was expelled during the major granulite event at 3 b.y. along with water, thorium, potassium, and rubidium (Watson 1973b; Heier, 1973), and so has its first concentration in gneisses flanking the granulites; from then on one can follow a history similar to that of gold.

369

Domarev (1962), Itsikson (1960), Ivanov (1962), Radkevich (1961), and Shatalov (1958) provide summaries of thought in the USSR on some details of metallogenic patterns of copper, tin, pyrite, and on types of and the metallogeny of ore districts. I find that, as I come to know more about metallogenesis, these leaders in the USSR repeatedly provide further insights.

Geology is dynamic. Metallogeny in exploration takes cognizance of that fact.

Reprinted from the *Intern. Geol. Rev.*, 4, 139–150 (1962)

BASIC PROBLEMS OF INVESTIGATION
IN THE FIELD OF METALLOGENY[1]

A. I. Semenov and G. S. Labazin

• *translated by Royer and Roger, Inc.* •

ABSTRACT

Geologists of the metallogeny department of VSEGEI [All-Union Research Institute of Geology of the Ministry of Geology and Conservation of Natural Resources] have taken a leading role in perfecting methods of mapping ore-mineral resources. Numerous other geologic groups of the U. S. S. R. have joined in the amassing of data and the mapping of various regions. VSEGEI mappers stress the need for differentiating type, age and mode of occurrence in mapping. Four scales from 1:2,500,000 to 1:200,000 are employed, each with its own function. With the growing complexity of metallogenic mapping, the authors suggest multiple overlays to show the many factors involved in a given region. The article reflects the economic urgency of the Soviet plan to survey its vast mineral reserves. --A. Eustus.

* * *

Regional metallogeny is a new branch of the study of economic minerals, originated and developed in the Soviet Union in connection with the constantly growing needs of the planned development of Socialist industry for a great variety of mineral raw materials.

The purpose of regional metallogeny is to study the geologic conditions of the occurrence and the laws governing the time-space distribution of deposits of economic (chiefly metallic) minerals of various kinds, in connection with the geologic history and structure of the individual structural elements of the earth's crust (folded zones, platforms, etc.). A knowledge of these laws is necessary to provide a scientific basis and to increase the effectiveness of exploration and prospecting operations, as well as to further develop the ore formation theory. Investigations of many scientists in the various branches of the geology, as well as the experience of workers in applied geology, who have been occupied in the location and study of economic minerals, have gradually established a new scientific trend.

The appearance and the successful development of regional metallogeny, particularly in the Soviet Union, has been made possible by a Socialist economic system of economy creating unlimited possibilities for the development of science. One of the basic factors has been the

successes achieved by Soviet geologists in the various branches of geologic knowledge, particularly study of geologic structure and economic minerals of the enormous Soviet territory. Only with this broad development of geologic science has it been possible to study regional metallogeny, synthesizing and supplementing the complex of investigations with practical application in mind.

Metallogic generalizations and examinations of individual problems involved in the discovery of the laws governing the distribution of economic minerals in the earth's crust may be found in papers by many Soviet investigators published even before the Great Patriotic War [World War II]. Among these, special mention should be made of the papers by Academician A. Ye. Fersman, especially, "Geochemical Problems of the Soviet Union" (1931) and "The Prospective Distribution of Economic Minerals within the Territory of the Soviet Union" (1932). His identification of the Mongolian-Okhotsk ore zone was of great importance.

Problems of the laws governing the spatial distribution of economic minerals in the earth's crust have received much attention in the papers of Academician V. A. Obruchev. His papers, "Metallogenic Epochs and Regions of Siberia" (1926), "Geologic Conceptions behind the Distribution of Important Economic Minerals" (1932), "Laws Governing the Distribution of Economic Minerals in the Principal Regions of the U.S.S.R." (1933) and "Geologic Laws and Forecasts" (1940) have not lost their timely significance even today.

Obruchev stated, in particular (1940), that "the great variety to be found in the composition and structures of the earth's crust and the economic minerals contained in it is not accidental, but is fully governed by laws, and has been determined in the case of each part of the earth's crust by its geologic history".

---

[1]Translated from Osnovnyye problemy issledovaniy v oblasti regional'noy metallogenii. In Zakonomernosti razmeshcheniya poleznykh iskopayemikh [Regularities in the distribution of minerals], ed. by N.S. Shatskiy, v. 2, Akademiya nauk SSSR, Otdelenie geologo-geograficheskikh nauk, Moscow, 1959, pp. 60-77.

[2]All-Union Scientific Research Institute of Geology of the Ministry of Geology and Conservation of Natural Resources.

Many valuable studies of the geologic conditions of formation and the laws governing the distribution of economic minerals were made in the years before the World War II in particular regions of the Soviet Union.   These researches have been extremely valuable in geologic prospecting. There is room to mention only a few of these. In the Urals such studies were made by: A. N. Zavaritskiy, A. G. Betekhtin, P. M. Tatarinov, V. S. Domarev, V. M. Sergiyevskiy, I. I. Malyshev, S. N. Ivanov and Ye. Ye. Zakharov; in Kazakhstan and its subdivisions by: N. G. Kassin, M. P. Rusakov, K. I. Satpayev, I. S. Yagovkin, N. I. Nakovnik, Ye. D. Shlygin, G. S. Labazin, G. I. Vodorezov, R. A. Borukayev and I. I. Knyazev; in the Altay by: V. K. Kotul'skiy, I. F. Grigor'yev, V. P. Nekhoroshev, N. A. Yeliseyev, N. N. Kurek and P. P. Burov; in Central Asia by: D. I. Shcherbakov, S. F. Mashkovtsev, A. V. Korolev, V. M. Kreyter, V. I. Smirnov and F. I. Vol'fson; in the Caucasus by: L. A. Vardan'yants, V. G. Grushev and I. G. Magak'yan; in Western Siberia by: M. A. Usov, F. I. Shakhov, I. V. Derbikov, V. A. Kuznetsov and Yu. A. Kuznetsov; and in the Transbaykal by: S. S. Smirnov, G. L. Padalka, O. D. Levitskiy and Yu. P. Den'gin.

Academician S. S. Smirnov's works on metallogeny are extremely important. As is well known, he has devoted much attention to general metallogenic theory and to laws governing the emplacement of economic minerals in particular regions, such as the Eastern Transbaikal, the Verkhoyansk region, the Pacific Ocean metallogenic belt. He has produced some some brilliant studies of the metallogeny of individual minerals (tin, polymetallic ores) and the application of resulting scientific data. The establishment of the Soviet Union's tin resources is intricately associated with the name of S. S. Smirnov, universally recognized as one of the founders of regional metallogeny as an independent branch of the science.

During and after World War II, as a result of the sharply increasing demands for mineral raw materials, problems of laws governing the economic mineral emplacement have received much attention. In this brief article it will not be possible to enumerate all the pertinent works from this period.

Smirnov's ideas on regional metallogeny have been developed by Yu. A. Bilibin, who shares VSEGEI credit for developing the scientific basis of regional metallogenic analysis. In VSEGEI, Bilibin has established a staff of geologists which, under his direction, has analyzed and drawn generalizations from a large amount of factual data on U. S. S. R. regional metallogeny to develop general principles of regional metallogenic analysis. His staff of geologists has included: Yu. A. Arapov, T. V. Bilibin, D. V. Voznesenskiy, V. G. Grushevoy, Ye. N. Goretskaya, K. I. Dvortsova,

V. S. Domarev, M. I. Itsikson, Ye. D. Karpova, G. S. Labazin, N. K. Morozenko, A. P. Nikol'skiy, G. L. Padalka, Yu. I. Polovinkina, L. I. Salop, A. I. Semenov, V. M. Sergiyevskiy, V. I. Serpukhov and P. M. Tatarinov. This staff was later augmented by other geologists, including young workers, is still at work. Bilibin and his VSEGEI staff have formulated the basis for regional metallogeny as an independent and important branch of the science of economic minerals.

In the last five to seven years papers on regional metallogeny have become especially common, and work in this branch is being carried out by various scientific research organizations, in cooperation with the main and territorial geologic administrations of the Ministry of Geology and Conservation of Mineral Resources and other government branches.

The bulk of field work is being done by: VSEGEI (Department of Metallogeny and Ore Deposits and regional departments), IGEM of the Academy of Sciences of the U. S. S. R. (O. D. Levitskiy, G. A. Sokolov, Ye. A. Radkevich, Ye. M. Shatalov and others), Institute of Geology of the Academy of Sciences, Kazakh S. S. R. (under Academician K. I. Satpayev), the Siberian branch of the Academy of Sciences of the U. S. S. R. (V. A. Kuznetsov and others), the Armenian S. S. R. Academy of Sciences (I. G. Magak'yan, S. S. Mkrtchyan and others), the Uzbek S. S. R. Academy of Sciences (under Kh. M. Abdullayev), the Academy of Sciences of the Azerbaydzhan S. S. R. (M. A. Kashkay and others), the Caucasus Institute of Mineral Raw Materials (G. A. Tvalchrelidze and others), the VNII-1 in Magadan (V. T. Matveyenko and others), Academy of Sciences of the Ukrainian S. S. R. (under N. P. Semenenko), and Lomonosov State University of Moscow (V. I. Smirnov and others).

To achieve constant coordination and direction of the work done on the problem of "The Laws Governing the Emplacement of the Principal Economic Minerals in the Earth's Crust", there was established in 1955 at the Academy of Sciences of the U. S. S. R. a permanent interdepartmental committee under the direction of Academician Shatskiy.

Recent investigations in metallogeny in the U. S. S. R. have been carried out in three chief areas: 1) development of the general principles of regional metallogenic analysis; 2) the construction of metallogenic and forecasting maps; 3) metallogenic investigations of individual species of economic minerals.

Let us look in general terms at the status of the work being done in each of these branches of regional metallogeny so as to understand the successive tasks involved in their development in the near future.

ELABORATION OF THE GENERAL
PRINCIPLES OF REGIONAL
METALLOGENIC ANALYSIS

In recent years geologists have devoted much
attention to the general principles of regional
metallogenic analysis, elaborated on the basis
of S. S. Smirnov's ideas by the staff of VSEGEI,
under the direction of Yu. A. Bilibin. The importance and fruitfulness of this new trend in
the study of economic minerals has been noted
by numerous investigators in a large number
of published papers (Shcherbakov, 1957; Betekhtin, 1955; Smirnov, 1957; Abdullayev, 1957;
Belyayevskiy, 1957; Kharkevich, 1957). The
essence of these principles has been repeatedly
set forth at various conferences, including All-
Union conferences. They are also known to a
wide community of geologists from Bilibin's
paper, "Metallogenic Provinces and Metallogenic
Epochs" (1955). In their most complete form
these principals have been set forth in "General
Principles of Regional Metallogenic Analysis
and Methods of Constructing Metallogenic Maps
for Folded Regions" (1957), VSEGEI metallogeny staff.

As the basis for a regional metallogenic
analysis, whose purpose is to discover the laws
governing the disposition of economic mineral
deposits in a given region, this staff has established the historical principle, which takes
into account the close interrelationship of the
processes of ore formation, the regime of
tectonic movements, the specific features of
sedimentation, and the nature, form and intensity of the igneous activity and metamorphism at particular stages in the development
of the structures in the earth's crust.

At the first stage of investigation, the chief
attention was devoted to finding the laws governing the distribution of endogenic mineralization
during the geologic development of the mobile
belts of the earth's crust. Later the scope of
such investigations began to include problems
regarding the laws governing the distribution
of exogenic mineralization during the development of these zones. Less attention has been
paid to elaborating the principles and methods
of metallogenic analysis of platform regions.
VSEGEI has begun to work in this field only
recently.

Without going into detail of the basic principles, as they are already widely known, we
may nevertheless set down some of the most
essential aspects. First, it should be said
that analysis of an enormous amount of factual
data on many regions of the Soviet Union not
only confirms Smirnov's statement (1955) that
"specific complexes of mineral deposits correspond to specific complexes of igneous rocks"
but further develops it. The various mineral
and intrusive complexes usually replace each
other, in a definite and more or less uniform

succession, during the geologic development of
the mobile zones of the earth's crust, which
have existed at various geologic times. It has
also been discovered that this regular succession is closely associated with all the other
aspects of the geologic development of mobile
zones and, in particular, with the development
and alterations of the forms of tectonic movements and sedimentation — that is, it is in complete agreement with the existing general concept of the development of geosynclinal systems
of folded regions on their sites.

On the basis of data, it has been possible to
subdivide the processes of geologic and metallogenic development of the mobile zones of the
earth's crust into a number of natural historical stages, each of which is characterized by
a definite regime of tectonic movements, by
specific features of sedimentation, igneous
activity and, especially, of ore formation (both
endogenic and exogenic). Mobile zones of
various ages, according to their tectonic re-
gimes, very clearly pass through three stages
of development: a) a period of predominance of
overall downward movements within the geosynclinal system and of the first structural
redispositions within this system (in the initial
and early stages of development of a mobile
zone); b) a period of predominance of overall
upward movements and of the main phases of
folding (middle stages of development of a
mobile zone); c) a period of development and
consolidation of the mobile folded zone and of
its transition into a young platform (late and
terminal stages of development of a mobile
zone).

In the case of each of these stages of development of mobile zones, it has been possible
to distinguish characteristic intrusive and
corresponding mineral or ore complexes, and
later, using Shatskiy's formational method
also to note the principal sedimentary and sedimentary-volcanogenic formations, especially
those with which the most important industrial
deposits of economic minerals are paragenet-
ically associated. The individual stages of
development have been characterized from the
standpoint of sedimentation, igneous activity
and metallogeny in the above-mentioned paper
by the staff of VSEGEI (1957).

As the work in this direction has developed
in recent years, the ore or mineral complexes
have been subdivided into genetic types of deposits. This, in turn, has created the basis for
a new and considerably improved classification
of ore deposits, which as Smirnov has indicated
takes into account the geologic conditions of
formation of the deposits, their localization in
existing structures and their association with
tectonic-igneous complexes.

The conception of the distribution of mineral
deposits during the geologic development of the

mobile zones of the earth's crust, as elaborated by the staff of VSEGEI, is, as already pointed out, the result of analysis and generalization from a large amount of factual material on the territory of the U. S. R., and is not derived from some general abstract ideas, as certain critics would have us believe. This scheme merely reflects the general tendencies in the metallogenic development of mobile zones, and is both a summation and a generalization. It must, of course, be used with intelligence and imagination in studying the metallogeny of specific regions. It cannot be regarded dogmatically as a system reflecting all the possible peculiarities of each particular region.

At the present time it is important to establish clearly the concept of mobile belts, since various misunderstandings have arisen on this problem, as will be indicated below. Each mobile zone exists in time from the beginning of the active development of a geosyncline to the end of the formation of folded structures on the site of the geosynclinal system and their transition into a platform or a platform-like state. The length of development of the existing mobile zones (the known intervals of time) are frequently different. Geographically a given mobile zone is bordered by relatively stable structures of the earth's crust, which might be either ancient (Precambrian) platforms or regions of folding that have terminated earlier.

This concept of a mobile zone, as adopted by the metallogeny staff of the VSEGEI, agrees in general with the principles underlying the tectonic map of the U. S. R. on a scale of 1:5,000,000 (1956), constructed under the direction and editorship of Academician Shatskiy. The folded regions shown on this map have at various times undergone the development of a mobile zone. They are distinguished on the map according to the time when their formation came to an end (as folded regions) by the names Cenozoic, Mesozoic, Hercynian, Caledonian and more ancient folded regions.

The distribution of mineral deposits during the geologic development of the mobile zones of the earth's crust, as conceived by the staff of the department of metallogeny of VSEGEI, applies specifically to mobile zones as defined above. It should be noted that inasmuch as this scheme of distribution is a general one, it does not specifically reflect the particular features of the metallogenic development of concrete types of mobile zones, or the provincial differences between the developments of mobile zones of a single type. It reflects only the general tendency in the metallogenic development of the majority of mobile zones that have existed in the past or exist in the present.

The scheme of metallogenic development of each mobile zone, taken separately, will

almost always be less complete than the general scheme, since the individual links in the chain which are clearly reflected in one zone may be only slightly manifested or even altogether absent in another. This will be readily seen if one compares, for example, the development of igneous activity and metallogeny during the formation of the Urals and of the Verkhoyansk folded zones with the general scheme of development. The first of these contains extremely widespread igneous formations and deposits of economic minerals originating in the initial and early stages of development of a mobile zone. Within the Verkhoyansk folded zone, on the other hand, the igneous rocks and endogenic deposits of economic minerals characteristic of the middle and late stages of development of mobile zones are clearly represented, whereas the igneous formations and endogenic deposits of economic minerals of the initial and early stages of development are very slight or practically absent.

Taking this circumstance into account, Yu. A. Bilibin (1948) posed the question of the types of igneous activity and endogenic mineralization in individual folded regions, as determined by the peculiar features of their geologic and metallogenic development (such as a Urals, a Far East and other types of mineralization). Bilibin (1948) observed that "such sharp differences in igneous activity and metallogeny between similar regions indicates that each of them had an entirely different course of tectonic and igneous development and that they are, in essence, representatives of different types of geosynclinal orogenic zones".

In view of what has been said, it becomes extremely important to work out additional schemes more closely reflecting the principal features of the distribution of mineral deposits during the development of mobile zones of different types, along with more accurate schemes of development of the individual mobile zones that reflect their provincial metallogenic features.

Some investigators, in studying some small part of a mobile zone, frequently only one of its subordinate elements, attempt to find all the links in the general chain within it, and when they do not find them, conclude that this scheme cannot be applied. It is quite obvious that in such cases, within the particular territory under consideration one may find only one or a few links in the general chain of development of mobile zones. For example, if one considers the Kalba-Narym rare-metal zone, part of the folded region on the site of the Zaysan geosynclinal system, which has been described in detail in papers by V. P. Nekhoroshev, one must remember that the sedimentary and intrusive rocks, and also the deposits of rare metals associated genetically with the intrusives located within this zone, were formed primarily in the middle stages of development of the Zaysan mobile zone. The geologic formations characteristic of the early and late stages of development of this zone are most completely represented in the zones adjacent to the Kalba-Narym zone (the Rudnyy Altay and the Charskaya zones).

Other investigators, such as Sheynmann (1958), define a mobile zone as a combination of all the folded zones of various ages located between the ancient platforms, such as all the folded zones between the Siberian and Chinese platforms. On this basis, Sheynmann declares that the scheme elaborated in VSEGEI is incorrect in principle. It is naturally quite useless to attempt to apply the VSEGEI scheme to a mobile zone as defined by Sheynmann, or to criticize it from the standpoint of this definition, since its elaboration was based on another definition of mobile zones, as explained above. It appears that Sheynmann's opinion is also clearly contradicted by the opinion of the large staff of geologists headed by Shatskiy, as expressed in the tectonic map of the U. S. S. R. and adjoining countries.

In recent years, as is known, much work has been done in studying the laws governing the formation of sediments and the ore formations associated with them during the development of the individual major structures of the earth's crust (mobile zones and platforms). Special note should be taken of the valuable investigations carried out in the geological institutes of the Academy of Sciences of the U. S. S. R. (by Shatskiy, N. M. Strakhov, V. V. Beloussov, B. N. Krotov, G. I. Bushinskiy and others), as well as by V. Ye. Khain, V. L. Popov, L. B. Rukhin, Yu. K. Goretskiy and other structural geologists and lithologists. The VSEGEI metallogeny geologists have been able to make use of the results of these investigations in developing the general principles of regional metallogenic analysis. Of especial value has been the method of distinguishing between sedimentary and sedimentary-volcanogenic formations, as defined by Shatskiy. He remarks (1955) that "one of the methods by which one may discover the laws governing the distribution of sedimentary mineral concentrations consists in recognizing those formations in the earth's crust with which are associated deposits of the particular economic mineral under investigation. The determination of the laws governing the distribution of sedimentary formations and in particular of ore-bearing formations, encounters considerable difficulties, since the formation of one type or another is determined not only by the geotectonic, but also by the physical and geographic conditions. There are formations and series of formations characteristic of only one stage of development of mobile zones (for example, spilitic-keratophyric, jasper-siliceous, flysch and certain others, in the purely geosynclinal

stages of development of mobile zones). On the other hand, there are formations that are repeated in various stages of development (for example, different types of carbonate formations). In spite of this fact, in the process of sedimentation and its associated ore formation one may observe (although not always with sufficient clarity) a general tendency in the course of development of particular mobile zones, which from this standpoint confirms the theoretical correctness of the above-mentioned general scheme of distribution of mineral deposits during the development of the mobile zones of the earth's crust.

In the future more attention will obviously have to be devoted to the problems of exogenic metallogeny, especially its relationship to endogenic metallogeny. The study of exogenic metallogeny will, of course, require more specific investigations directed at discovering more concrete laws of the time-space distribution of exogenic ore deposits and at the application of the resulting conclusions in prospecting.

Of great value to improving regional metallogeny has been recent Soviet work in elaborating aspects of the relationship of ore formation processes and igneous activity (A. G. Betekhtin, D. N. Korzhinskiy, G. D. Afanas'yev, Kh. M. Abdullayev, I. G. Magak'yan, V. S. Koptev-Dvornikov, V. N. Kotlyar, G. A. Tvalchrelidze, V. A. Kuznetsov, M. B. Borodayevskaya, N. L. Borodayevskiy, G. N. Shcherba and others).

Among these significant investigations should be mentioned the work devoted to the relationship between endogenic mineralization and igneous dikes. Most prominent among these are papers by Borodayevskaya and Borodayevskiy (1945-1956) and Abdullayev (1957), which contain the results of some very fruitful investigations and provide an historical and many-sided approach to the relationship of ore mineralization and minor intrusives.

One of the achievements of the staff of the department of metallogeny of VSEGEI in the treatment of the general problems of regional metallogeny is the identification of structural-metallogenic zones and the clarification of the concept of metallogenic provinces. Metallogenic provinces and zones are elongated linear zones whose position coincides with those of mobile zones and platforms (or parts of the latter) existing in various periods of geologic history. They are characterized by a predominance of particular ore complexes or combinations, as determined on the one hand by different types of geologic and metallogenic development of mobile zones and platforms, and on the other hand by provincial differences in the development of these basic structural elements of the earth's crust. Metallogenic provinces and zones are thus the major units in the metallogenic classification of ore-bearing territories (for example, the Verkhoyansk and the Altay-Sayan metallogenic provinces, the Urals and the Pacific Ocean metallogenic zones).

Within metallogenic provinces and zones are structural-metallogenic zones with a predominant distribution of a specific type of mineralization, as determined by the geologic development of the given zone (the tectonic regime, the nature of the sedimentation, igneous activity, metamorphism and other factors). Examples of structural-metallogenic zones are the Rudnyy Altay polymetallic zone and the Southern Fergana antimony-mercury zone.

Studies and generalizations from the data on various metallogenic provinces of the U. S. S. R. have resulted in the identification of the principal types of structural-metallogenic zones and their descriptions, which will considerably facilitate the evaluation of the ore-bearing prospects of regions that have been little studied.

The individual types of structural-metallogenic zones have arisen successively in specific stages of development of the mobile belts of the earth's crust, so that they are disposed more or less regularly within metallogenic provinces relative both to the general contours of the province and to each other. The formation of certain structural-metallogenic zones embraces long periods of time (several stages), with the result that such zones have a complex metallogenic appearance. Within the structural-metallogenic zones are subzones, and within the latter are ore zones, ore nodes and individual deposits. A study of these problems has been necessary in order to make practical use of the principles of regional metallogenic analysis. The practical suitability of distinguishing structural-metallogenic zones on the basis of such principles has come to be widely recognized, and is used in metallogenic analysis by many scientific staffs and individual investigators.

In conclusion, it may be remarked that principles of regional metallogenic analysis similar to those developed by VSEGEI have been employed in recent years by Chinese geologists in determining the laws governing the distribution of economic minerals within the territory of China, as may be seen, for example, in papers by Sè-Tszya-Zhun (1953, 1956).

THE CONSTRUCTION OF METALLOGENIC AND ORE-FORE-CASTING MAPS

Systematic work has long been carried out in the Soviet Union in the mapping of economic minerals to various scales on a geologic base. Such maps are required in producing the state geological maps on the scales of 1:1,000,000 and 1:200,000. Maps of economic minerals are sometimes drawn on a tectonic base for individual regions. One of the first small-scale schematic maps of this type was the "Geochemical Map of the U. S. S. R.", published by A. Ye. Fersman in 1931. One of the more detailed maps of economic minerals on a tectonic base is Nekhoroshev's [not in references] map of the Altay, showing not only the disposition but type of mineralization of deposits. Similar maps have been constructed for certain other regions.

The great importance of such maps, both on a geological and tectonic base, is quite obvious, since they are frequently used to determine certain regularities in the spatial distribution of various types of deposits in the individual regions. Nevertheless they usually fail to reflect many important factors of the interrelationship between deposits and geological conditions, and also fail to show the successive formation of various types of deposits during the geologic development of a specific part of the earth's crust. For this reason maps of economic minerals which do not reflect all the combinations of causes determining the regularities of localization of the individual types of deposits are of limited value in forecasting ore occurrences on a scientific basis.

From the elaborated general principles, and from the vast experience and data accumulated on Soviet ore regions, the department of metallogeny of VSEGEI has been able to set forth a method of metallogenic mapping and to publish, beginning with 1949, the first of such regional maps on scales of 1:2,500,000, of 1:1,000,000, of 1:500,000 and of 1:200,000. Metallogenic maps by other staffs and geologic organizations have introduced certain changes into the VSEGEI methods, or else have worked out other methods.

By now the construction of metallogenic and forecasting maps has been greatly expanded and systematized. This is quite natural, since the volume of prospecting work for various economic minerals has increased immeasurably over previous years. Moreover it has become exceedingly important to provide scientific and methodological guidance for metallogenic and forecasting mapping required for scientifically-based prospecting. On the other hand, up to now there has been no complete agreement on the requirements that should be met by metallogenic maps and no generally accepted mapping method. Thus, it seems suitable to use the accumulated existing experience in constructing metallogenic maps in determining and agreeing to agree upon the basic requirements of such maps and to consider their types, scales and designations.

It is presupposed that these maps will sum up a complex of geological investigations for practical ends and for further ore mineral studies. From this standpoint, any metallogenic map should be based on factual data, including: a) a geologic map of the appropriate or more detailed scale (especially for ore districts); b) a well-founded system of tectonic regionalization reflecting main geologic features; c) petrological, geochemical, lithological-facies and geophysical data of these areas; and d) concentrate or metallometric assays.

What are the general requirements that can be made of metallogenic maps? The VSEGEI geologists believe a metallogenic map should:

1) Reflect the general laws on the spatial distribution of deposits of economic minerals of various types — that is, it should reflect the metallogenic regionalization of the territory of the U. S. S. R. based on analysis of all the existing data. Chief result of such regionalization should be the identification on the map of the structural-metallogenic zones or districts, with the actual or possible distribution of deposits of various types of economic minerals; this is especially pertinent to the rational planning of prospecting.

2) Give some idea of the distribution of deposits, and also of the other geologic formations, at the various stages in the geological history of the main structures of the territory under consideration, to provide a more accurate estimation of ore-bearing prospects since the individual stages usually are characterized by their own special features of mineralization.

3) Clearly reflect existing relationships between mineral deposits geological factors (intrusive complexes and their individual parts, sedimentary and volcanogenic-sedimentary formations, to the greatest possible extent showing the lithology of the surrounding rocks and the tectonic structures, zones of metamorphism, etc.).

4) Reflect the genetic types of deposits shown, since this is most important in judging the conditions of their localization.

Thus, there arises the very important problem of the types of metallogenic maps. Experience in the construction and practical use of such maps has led to the conclusion that metallogenic maps should be divided into two principal types. The first of these includes complex metallogenic maps reflecting the laws of distribution, and the relative time and general geologic conditions of formation of deposits of various economic minerals (endogenic and exogenic) corresponding to the above-mentioned requirements. Maps of the second type are special forecasting metallogenic maps drawn for a single economic mineral (such as iron, copper, etc.) or a group of related minerals (such as polymetallic ores, rare metals, etc.). These maps, in contrast to those of the first type, should reflect not only the laws governing

the distribution of deposits (in association with the various geologic factors) and the favorable geological conditions for the occurrence of a given economic mineral, but should also show both the direct and indirect prospecting criteria and indications of this economic mineral, as well as the degree to which it has been studied from the standpoint of general geology and special prospecting, and should also distinguish in detail the districts and areas in which it may be located in accordance with the prospects for finding the given mineral.

The use of special metallogenic forecasting maps hastens the process of locating deposits of economic minerals, and in particular may facilitate the discovery of "blind" ore bodies, currently particularly important.

Special metallogenic forecasting maps (especially on small scale) may usually be constructed on the basis of composite large-scale metallogenic maps. Without complex metallogenic maps it is usually impossible to take into account all the geologic conditions for the occurrence of deposits in new areas, and it becomes necessary to distinguish prospective areas for the occurrence of any given economic mineral on the sole basis of the direct prospecting criteria for this mineral, thus, lowering the reliability and scientific basis of these forecasts.

The next important problem associated with the construction of metallogenic maps is that of their scale. In the Soviet Union metallogenic maps are constructed on various scales, depending on the size of the territory to be studied and the nature of the problems to be solved.

Metallogenic Map of the U. S. S. R. on the Scale of 1:2,500,000

This map, which embraces the entire territory of the Soviet Union, can reflect the principal general and particular metallogenic features of major tectonic structures of the earth's crust, such as: a) platforms, b) folded regions distinguished by type of development and by age formed on the sites of the mobile zones of the earth's crust, and c) areas linking ancient platforms with Paleozoic or younger folded zones.

Within these major structures of the first order it is convenient to make a more detailed tectonic regionalization reflecting the specific development and metallogeny of their individual parts. Such regionalization must follow the historical principle.

The analysis of sedimentation, igneous activity, metamorphism and their associated processes of ore formation will be the basis for distinguishing the various types of metallogenic provinces, and within them of the main established and potential structural-metallogenic zones. Such a map will provide a comparative estimate of the prospects of the known ore districts, and will also probably reveal new districts that show prospects for the occurrence of particular mineral raw materials. This may be the basis for long-range planning of statewide geologic prospecting operations and surveys. A metallogenic map of the U. S. S. R. is now being produced at VSEGEI with the participation of various geological organizations. Tectonic regionalization of the U. S. S. R., as reflected on a map drawn to 1:5,000,000 edited by Academician Shatskiy (1956), is being used as the base for this map.

Metallogenic Maps on the Scales of 1:1,000,000 and 1:500,000

These maps are constructed for relatively large folded zones and platforms or parts of platforms, usually including one, or more rarely two or three, metallogenic provinces (such as the Urals or the Mongolian-Okhotsk Province) in order to distinguish within them all the structural-metallogenic zones, as a basis for evaluating the ore mineral prospects and for the rational organization of prospecting operations. These maps more clearly show the interrelationship of the ore deposits and the igneous and sedimentary rocks, the various types of tectonic structures and the zones of metamorphism.

The sedimentary and volcanogenic rocks on maps of this scale are usually subdivided into formations (as defined by Shatskiy and his successors). Special distinction is made of those formations with which some ore-mineral deposits are or may be associated, such as: spilitic-keratophyric rocks, which usually contain pyrites; siliceous-jasper rocks, which often contain manganese and iron-ore deposits; and terrestrial volcanogenic rocks, which often are the locations of molybdenum-copper, polymetallic and silicate deposits.

Intrusive rocks are divided into complexes showing the relationship between these and the ore deposits. The individual series of sedimentary and volcanogenic formations, as well as groups of intrusive complexes characterizing specific stages of the geologic development of the main structural elements of the earth's crust (folded regions or zones and platforms), are divided into structural stages and substages. The stages are usually separated by major regional gaps and unconformities. Such maps also reflect the directions and general nature of the folded structures, as well as the types of disjunctive dislocations. The maps show not only the deposits of industrial value, but also many or all of the known ore occurrences, reflecting their relationship to given sedimentary and volcanogenic formations and intrusive complexes, which have been formed

at various stages of development of the mobile zones and platforms. Maps drawn to the scale of 1:500,000 usually also reflect the results of concentrate and metallometric assaying. These maps are used as the basis for selecting areas with good prospects for finding specific ore types, and for establishing geologic surveys and special prospecting on scales from 1:50,000 to 1:200,000. Metallogenic maps on these scales have already been constructed or are being constructed for many regions of the Soviet Union.

Metallogenic Maps on the Scale of 1:200,000 and More Detailed Maps

When used to establish prospecting operations, such maps are usually drawn to a scale of 1:10,000 or even larger, so as to determine the laws of ore location and local structures within zones and subzones. Large-scale metallogenic maps are especially important in prospecting in ore districts. Large-scale mapping may differ somewhat from small-scale mapping. Such methods have not yet been completely worked out. The first such maps were produced by VSEGEI. Such work has also been carried out successfully by IGEM of the Academy of Sciences of the U. S. S. R. (Ye. A. Radkevich, Ye. T. Shatalov), by the Academy of Sciences of the Uzbek S. S. R. (under direction of Kh. M. Abdullayev), and the Altay Mining and Metallurgical Institute of the Academy of Sciences of the Kazakh S. S. R. On large-scale metallogenic maps, it is especially important to reflect the following geologic elements which effect minerals distribution: lithology and facies of sedimentary and sedimentary-volcanogenic formations; intrusive rocks, divided not only into complexes but also according to the composition of the particular petrographic varieties formed at specific intrusives phases, as well as those separated by the processes of interaction with the surrounding rocks and of crystallizational differentiation; dikes of igneous rocks, subdivided according to composition and relative time of formation; types and forms of folded structures and disjunctive dislocations, also subdivided into types; zones and areas of hydrothermally altered rocks and manifestations of contact metamorphism; and generalized conclusions drawn from geochemical and geophysical investigations, if these indicate the laws governing the disposition of deposits or their locations. These maps will show the deposits and all the ore occurrences, dividing them as much as possible into their ore-mineral formations within the individual genetic groups. They will also show the results of concentrate and metallometric assaying.

In view of the enormous amount of detail to be shown on metallogenic maps, experience to date indicates that some of these details may perhaps be more conveniently shown on separate overlays to be placed over the map.

The drawing of metallogenic maps, especially those showing much detail, requires not only analysis and generalization of all the available data, but also special preliminary field investigations. Thus, in recent years, in ore-bearing and prospective ore-bearing districts, various specific investigations have been carried out (studies of the ore content of igneous rocks from determination of the role of the surrounding rocks in the processes of endogenic ore formation, studies of the lithology and facies of sedimentary formations from the standpoint of their ore content, studies of the role of disjunctive dislocations in the localization of deposits, and studies of individual types of deposits, etc.).

In considering the problem of metallogenic maps it may be observed that up to the most recent years these maps have been constructed primarily for folded regions. Thus, the methods of metallogenic mapping for platforms have not yet been fully worked out. This is now being done in VSEGEI in connection with the production of metallogenic maps for the western part of the Siberian platform and of a metallogenic map of the U. S. S. R. on the scale of 1:2, 500, 000. It should also be said that the determination of the metallogeny of the Siberian platform at the present time has been greatly facilitated by the publication in VSEGEI by T. N. Spizharskiy of a tectonic map of the Siberian platform and the adjoining parts of folded regions on the scale of 1:2,500,000.

METALLOGENIC STUDIES OF INDIVIDUAL SPECIES OF ORE MINERALS

This branch of regional metallogeny has existed for a comparatively long time, but has only recently acquired importance and direction. Chief aspects of the study of individual types of ore minerals that are now being developed are: a) analysis and generalization from all the existing data on the deposits of a given mineral within the U. S. S. R. or its separate regions, with the production of appropriate monographs; b) study of the deposits of individual ore districts; c) determination of the laws governing the emplacement of deposits of a particular ore type (iron, copper, titanium, etc.) or groups of related minerals (rare metals, lead-zinc and others), both in individual regions and throughout the U. S. S. R. as a whole, with the drawing of appropriate special metallogenic forecasting maps.

At present extensive work is being done on the distribution laws governing deposits of iron, bauxites and aluminum ores, copper, polymetallic ores, molybdenum, mercury, nickel and cobalt. It should be noted that, despite the great successes achieved in this area, the volume and the rates of the investigations being carried out here are still insufficient. Considerably less effort is being devoted to metallogeny investigations of tungsten, manganese, chromium, vanadium, and other economic minerals.

FUNDAMENTAL TASKS OF FURTHER METALLOGENIC INVESTIGATIONS

Results so far show that regional metallogenic study has great bearing on increasing industrial mineral resources and warrants further investigations. The unprecedented volume of ore prospecting points up the need for making it more effective, particularly since chance discoveries are becoming less likely.

The improvement of the principles of regional metallogenic analysis, on the basis of the achievements in various geologic fields and their greater practical use in combination with the newest methods of exploration and prospecting, have made it possible, within a relatively short time and with the smallest expenditure of resources, to discover new mineral deposits and to establish reserves of ore minerals in the amounts required. This assurance is based on the fact that the potential for discovering new deposits of various types and even new ore districts in the enormous area of the U. S. S. R. are immeasurable. This has been indicated by the preliminary data from regional metallogenic analysis of the U. S. S. R. , as well as year-in, year-out results of exploration and prospecting.

Since regional metallogeny must take account of the entire gamut and interrelationship of geologic factors affecting ore-deposit formation and distribution, the problem of its further development is quite complicated and varied. This problem is a complex one, requiring the collective and cooperative efforts of metallogenic specialists and ore geologists, along with petrologists, structural geologists, stratigraphers, geochemists and specialists in the field of lithologic-facies and paleogeographic analysis. Thus, the future tasks of regional metallogeny will require the active participation of specialists in the various branches of geologic knowledge. The chief tasks of further research in the field of regional metallogeny at the present stage of its development are the following.

I. Further elaboration of the general principles of regional metallogenic analysis:

1) Development of the principles of metallogenic analysis and the methods of constructing metallogenic maps of platform regions.

2) Comparative study of the metallogeny of folded regions of different ages and different types, in the light of the history of their formation on the sites of mobile zones (also taking into account available data on territories outside of Soviet Union). This work should first of all result in a classification of the types of mobile zones of the

earth's crust and of the corresponding metallogenic provinces. For each type distinguished,
and initially for each mobile zone (folded region of a given age, there should be established
an outline of distribution of ore deposits
throughout their geologic history, based on
latest data. From this it will be possible to
discover and express the differences between
types and provinces in the metallogenic development of mobile zones, and on this basis
to improve the existing data, and in individual
areas to change the general outline of development of mobile zones, if necessary. It may be
assumed that such outlines will be used with
great success in determining the laws governing
the physical distribution of various types of
mineral deposits in specific regions and to con-
struct complex and special metallogenic forecasting maps.

3) Continuation of the researches to determine the laws governing the physical distribution of ore deposits of various types within metallogenic provinces, distinguishing
the structural-metallogenic zones and characterizing and classifying them.

4) A more thorough study of the processes
of metamorphism, especially in regard to
their role in the formation of economic mineral deposits at various stages of development in the formation of folded systems.

5) Study of the influence of deep-seated
structures in the earth's crust, and particularly of systems of deep faults which are
active over a prolonged period, on the development of the mobile zones of the earth's
crust and on the distribution of endogenic
mineralization within them.

6) Establishment of the generalizations
characterizing the localization of ore deposits
within structural-metallogenic zones, taking
account of all the factors that control their
distribution (local structure, lithology of the
enclosing rocks, intrusions, depth of erosion, distribution of ore concentrations
relative to the various types of hydrothermally altered rocks, etc.).

7) Further elaboration of the problem of
establishing more distinct and objective
criteria of the connection between igneous
and ore formations, and of subdividing igneous rocks and endogenic deposits into complexes. The establishment of additional
criteria for distinguishing ore complexes
must also include a discovery of the distinguishing features of deposits which have a
similar general composition but occupy a
different position in the course of development of the mobile zones in the earth's crust.

8) For a more reliable evaluation of the
prospects for endogenic ore mineralization

in the individual ore districts and their parts,
and to establish additional criteria for distinguishing related igneous and endogenic
mineral complexes, it is very urgent to
determine ways to find the depth of formation of intrusive rocks and endogenic deposits
of different types. In particular, this must
also include a study of vertical zonality in
the distribution of endogenic ore mineralizations of various types.

9) The introduction of greater precision
into the existing criteria for distinguishing
between formations of sedimentary, volcanogenic-sedimentary and volcanogenic rocks;
the discovery of their relationship to economic minerals; the determination of the
general laws governing the distribution and
conditions of formation of sedimentary and
volcanogenic-sedimentary formations during
the development of mobile crustal zones of
different ages and types and on platforms,
as well as the determination of the general
laws governing the spatial disposition of
various types of formations in relation to
the overall geotectonic, paleogeographic and
especially paleoclimatic conditions. Here
arises the need for expanding the construction of facies and paleoclimatic maps.

10) Study of the concentration and dissemination of chemical elements under natural conditions as a function of their basic
properties, in order to obtain additional
geochemical data on the causes leading to
the formation of deposits.

II. Tasks involved in the further construction of metallogenic maps:

1) Perfection of methods of drawing
metallogenic maps to different scales, on
the basis of general conclusions derived
from cumulative experience, and also the
specific requirements and the purpose of
such construction.

2) Continuation of work on combined
metallogenic maps of the U. S. S. R. on the
scale of 1:2,500,000, whose great importance in planning geologic prospecting operations and in the further development of the
theory of ore formation will be readily
seen.

3) Continuation of the construction of
metallogenic maps, on the scales of
1:500,000 and 1:1,000,000, for the major
regions and especially for the Siberian
platform, the Soviet Far East and Eastern
Siberia.

4) In connection with the vast scope of
detailed prospecting, it will be especially
important in the near future to produce a
sharp expansion of the work done in scien-

tific research organizations and geological administrations in the construction of metallogenic maps on the scale of 1:200,000, and in even greater detail for the principal ore districts and areas likely to contain new deposits of economic minerals.

III. Tasks involved in studying the metallogeny of the individual economic minerals:

1) The organization of more thoroughgoing investigations of the metallogeny of individual metals or closely interrelated natural groups of metals, with the publication of generalizing monographs and special metallogenic forecasting maps both for the individual districts and for the entire territory of the Soviet Union.

2) A detailed study of the individual and still insufficiently known types of deposits.

It must be stressed, in conclusion, that the problems in the field of regional metallogeny are enormous and complicated, but nevertheless all of them are very urgent. Their solution should considerably strengthen the theoretical foundations of this new branch in the study of economic minerals and should have a large and favorable effect on expanding the mineral raw material resources of our country.

REFERENCES

Abdullayev, Kh. M., 1957, Dayki i orudeneniye [DIKES AND ORE MINERALIZATION]: Gosgeoltekhizdat.

Anonymous (VSEGEI), 1957, Obshchiye printsipy regional'nogo metallogenicheskogo analiza i methodika sostavleniya metallogenicheskikh kart dlya skladchatykh oblastey [GENERAL PRINCIPLES OF REGIONAL METALLOGENIC ANALYSIS AND METHODS OF CONSTRUCTING METALLOGENIC MAPS OF FOLDED REGIONS]: Gosgeoltekhizdat.

Belyayevskiy, N. A., 1957, XX Sessiya Mezhdunarodnogo Geologicheskogo Kongressa [THE XX INTERNATIONAL GEOLOGICAL CONGRESS]: Sov. geologiya, sb. 57.

Betekhtin, A. G., 1955, Sovremennoye sostoyaniye i ocherednyye zadachi izucheniya rudnykh mestorozhdeniy [THE PRESENT STATUS AND FUTURE PROBLEMS OF THE STUDY OF ORE DEPOSITS]: Sov. geologiya, sb. 43.

Bilibin, Yu. A., 1948, Voprosy metallogenicheskoy evolyutsii geosinklinal'nykh zon [PROBLEMS OF THE METALLOGENIC EVOLUTION OF GEOSYNCLINAL ZONES]: Izv. Akad. nauk SSSR. ser. geol., no. 4.

_____ 1955, Metallogenicheskiye pro-

vintsii i metallogenicheskiye epokhi [METALLOGENIC PROVINCES AND METALLOGENIC EPOCHS]: Gosgeoltekhizdat.

Borodayevskaya, M. B., 1955, Nekotoryye voprosy geologii, petrogenezisa i metallogenii malykh intruziy pozdnykh etapov razvitiya tektonomagmaticheskogo tsikla. v kn.; Magmatizm i svyazi s nim poleznykh iskopayemykh [SOME PROBLEMS OF THE GEOLOGY, PETROGENESIS AND METALLOGENY OF THE MINOR INTRUSIVES ASSOCIATED WITH THE LATER STAGES IN THE DEVELOPMENT OF THE TECTONIC-MAGMATIC CYCLE. In the book: IGNEOUS ACTIVITY AND ITS ASSOCIATED MINERAL DEPOSITS]: Izd. Akad. nauk SSSR.

Borodayevskiy, N. I., 1945, Geologicheskiye nablyudeniya nad daykami, soprovodayushchimi rudnyye polya [GEOLOGIC OBSERVATIONS OF DIKES ACCOMPANYING ORE FIELDS]: Izv. Akad. nauk SSSR, ser. geol., no. 2.

_____, 1946, O znachenii nablyudeniy nad daykami izverzhennykh i zhilnykh porod pri izuchenii rudnykh mestorozhdeniy. Sb. "Rudnichnaya geologiya" [ON THE IMPORTANCE OF OBSERVATIONS OF DIKES OF IGNEOUS AND VEIN ROCKS IN STUDYING ORE DEPOSITS. In symposium volume, "ORE GEOLOGY"]: Gosgeolizdat.

_____, 1931, Geokhimicheskiye problemy Soyuza [GEOCHEMICAL PROBLEMS IN THE SOVIET UNION]: Izd. Akad. nauk SSSR.

Fersman, A. Ye., 1932, Perspektivy rasprostraneniya poleznykh iskopayemykh na territorii Soyuza [PROSPECTS FOR THE DISTRIBUTION OF MINERAL DEPOSITS WITHIN THE SOVIET UNION]: Izd. Akad. nauk SSSR.

Kharkevich, D. S., 1957, O soderzhanii i postroyeniya kursa "Metallogenicheskiye provintsii" [ON THE CONTENT AND ORGANIZATION OF A COURSE IN "METALLOGENIC PROVINCES"]: Uchen. Zap. (Kishinevskiy Univ.), v. 25.

Obruchev, V. A., 1926, Metallogenicheskiye epokhi i oblasti Sibiri [METALLOGENIC EPOCHS AND REGIONS OF SIBERIA]:.

_____, 1933, Zakonomernosti raspredeleniya poleznykh iskopayemykh v glavnykh rayonakh SSSR, [LAWS GOVERNING THE DISTRIBUTION OF MINERAL DEPOSITS IN THE PRINCIPAL AREAS OF THE U.S.S.R.]: Sotsial. rekonstr. i nauka, no. 6.

_____, 1940, Geologicheskiye zakonomernosti i prognozy [GEOLOGIC LAWS AND FORECASTING]: Sovetskaya nauka, no. 5.

Satpayev, K. M., 1953, O prognoznykh metallogenicheskikh kartakh Tsentral'nogo Kazakhstana [METALLOGENIC FORECASTING MAPS OF CENTRAL KAZAKHSTAN]: Izv. Akad. nauk Kazakh. SSR, ser. geol. no. 6.

_____, 1955, O metodologii, fakticheskoy baze i osnovnykh vyvodakh metallogenicheskikh prognoznykh kart Tsentral'nogo kazakhstana [ON THE METHODOLOGY, THE FACTUAL BASIS AND THE CHIEF CONCLUSIONS DRAWN FROM METALLOGENIC FORECASTING MAPS OF CENTRAL KAZAKHSTAN]: Izv. Akad. nauk Kazakh. SSR, ser. geol., no. 20.

Se Tszya-zhun, 1953, O napravlenii poiskovykh rabot na osnove nekotorykh zakonomernostey razmeshcheniya mestorozhdeniy poleznykh iskopayemykh v Kitaye [THE DIRECTION OF PROSPECTING OPERATIONS ON THE BASIS OF SOME LAWS GOVERNING THE EMPLACEMENT OF MINERAL DEPOSITS IN CHINA].

_____, 1956, Zakonomernosti raspredeleniya poleznykh iskopayemykh Kitaya i ikh perspektivy. Zhurn. "Geologicheskiye znaniya" [LAWS GOVERNING THE DISTRIBUTION OF MINERAL DEPOSITS IN CHINA AND THEIR POTENTIAL RESOURCES. In the journal, "GEOLOGIC SCIENCES"]: no. 8, [in Chinese?].

Shatalov, Ye. T., 1958a, Metallogenicheskiye issledovaniya rudnykh rayonov [METALLOGENIC INVESTIGATIONS OF ORE DISTRICTS]: Vestn. Akad. nauk. SSSR, no. 9.

_____, 1958b, O metallogenii rudnykh rayonov [ON THE METALLOGENY OF ORE DISTRICTS]: Izv. Akad. nauk SSSR, ser. geol., no. 9.

Shatskiy, N. S., 1955, Fosforitonosnyye formatsii i klassifikatsiya fosforitovykh zalezhey [PHOSPHORITE-BEARING FORMATIONS AND THE CLASSIFICATION OF PHOSPHORITIC DEPOSITS]: Soveshch. po osad. porod., no. 2, Dokl. Akad. nauk SSSR.

Shatskiy, N. S. et al., 1951, K voprosu o periodichnosti osadkoobrazovaniya i o metode aktualizma v geologii. Sb., "K voprosu o sostoyanii nauki ob osadochnykh porodakh" [ON THE PERIODICITY OF SEDIMENTATION AND ON THE METHOD OF ACTUALISM IN GEOLOGY. In the collection, "ON THE STATUS OF THE STUDY OF SEDIMENTARY ROCKS"]: Izd. Akad. nauk SSSR.

_____, 1956, Tektonicheskaya karta SSSR i sopredel'nykh stran v masshtabe 1:5,000,000 [A TECTONIC MAP OF THE U. S. S. R. AND ADJOINING COUNTRIES ON THE SCALE OF 1:5,000,000]: Gosgeolizdat, Moscow.

Shcherbakov, D. I., 1957, Puti razvitiya geologii v SSSR [THE PATHS OF DEVELOPMENT OF GEOLOGY IN THE U. S. S. R.]: Priroda, no. 1.

Sheynmann, Yu. M., 1958, Nekotoryye cherty evolyutsii magmatizma skladchatykh poyasov [SOME FEATURES OF THE EVOLUTION OF IGNEOUS ACTIVITY IN FOLDED ZONES]: Mat. k II Vses. petrograf. soveskhch., Tashkent.

Smirnov, S. S., 1955, Izbrannyye trudy [SELECTED PAPERS]: Izd. Akad. nauk SSSR.

Smirnov, V. I., 1957, Uspekhi i nekotoryye ocherednyye zadachi teorii rudoobrazovaniya [ACHIEVEMENTS AND FUTURE PROBLEMS OF THE THEORY OF ORE FORMATION]: Sov. geologiya, sb. 60.

BIBLIOGRAPHY

Agricola, G., 1556. De re metallica, trans. H. H. Hoover and L. H. Hoover. Dover, New York, 1950.

Aubouin, J., 1965. Geosynclines. Elsevier, Amsterdam, 335 p.

Bailes, A. H., 1971. Geology of the Snow Lake-Flin Flon-Sherridon area. Manitoba Mines Branch Geol. Paper 1/71.

Bateman, A. M. 1951. Economic mineral deposits Wiley, New York.

Bilibin, Yu. A., 1968. Metallogenic provinces and metallogenic epochs, Trans. E. A. Alexandrov. Geol. Bull., Dept. Geology, Queens College Press, Flushing, N.Y.

Billingsley, P., and A. Locke, 1941. Structure of ore districts in the continental framework. AIME Trans., v. 144, p. 9-64.

Boldy, J., 1968. Geological observations on the Delbridge massive sulphide deposit. CIM Bull., p. 1045-1054.

Borrello, A. V., 1972(a). La estructura assyntica de la Argentina. Rev. Brasileira, v. 2, no. 2, p. 71-84.

——, 1972(b). The Precordillera as a type of geosyncline in Argentina. 24th Intern. Geol. Congr., Sec. 3, p. 293-299.

Brock, B. B., 1972. A global approach to geology. A. A. Balkema, Capetown.

Brown, J. S. (ed.), 1967. Genesis of stratiform lead-zinc-barite-fluorite deposits (Mississippi Valley-type deposits). Econ. Geol. Monogr. 3, 443 p.

Buddington, A. F., 1927. Coincident variations of types of mineralization and of Coast Range intrusions. Econ. Geol., v. 22.

Burke, K. C., and J. F. Dewey, 1972. Orogeny in Africa, in African geology, ed. A. J. Whiteman. University of Ibadan.

Burnham, C. W., 1959. Metallogenic provinces of the southwestern United States and northern Mexico. New Mexico Bur. Mines Bull. 65, 76p.

Burton, C. K., 1970. The geological environment of the mineralization in the Malaya-Thailand peninsula, in A Second Technical Conference on Tin, ed. W. Fox. International Tin Council, London, p. 105-122.

Cahen, L., 1970. Igneous activity and mineralization episodes in the evolution of the Kibaride and Katangide orogenic belts of Central Africa, in African Magmatism and Tectonics, eds. T. N. Clifford and I. G. Gass. Oliver & Boyd, Edinburgh.

383

——, and N. J. Snelling, 1966. The geochronology of equatorial Africa. North-Holland, Amsterdam.

Clark, A. H., and M. Zentilli, 1972. The evolution of a metallogenic province at a consuming plate margin. The Andes between latitudes 26° and 29° south. Abs. CIM Bull., v. 65, no. 719, p. 37.

Clifford, T. N., 1966. Tectono-metallogenic units and metallogenic provinces of Africa. Earth Planetary Sci. Letters, v. 1, p. 421–434.

——, 1970. The structural framework of Africa, *in* African Magmatism and Tectonics, eds. T. N. Clifford and I. G. Gass. Oliver & Boyd, Edinburgh.

Corliss, J. B., 1971. The origin of metal-bearing submarine hydrothermal solutions. Jour. Geophys. Res., v. 76, no. 33, p. 8128–8138.

Daly, R., 1933. Igneous rocks and the depths of the earth. McGraw-Hill, New York.

Dana, J. D., 1873. On some results of the earth's contraction from cooling, including a discussion on the origin of mountains and the nature of the earth's interior. Amer. Jour. Sci., v. 5, p. 423–443, v. 6, p. 6–14, 104–115, 161–176.

Davidson, C. F., 1962. The origin of some strata-bound sulfide ore deposits. Econ. Geol., v. 57, no. 2, p. 265–274.

Dewey, J. F., and J. M. Bird, 1970. Mountain belts and the new global tectonics. Jour. Geophys. Res., v. 75, no. 14, p. 2625–2467.

Dietz, R. S., 1961. Continent and ocean basin evolution by spreading of the sea floor. Nature, v. 190, no. 4779, p. 854–857.

Domarev, V. S., 1962. Basic features of the metallogeny of copper. Intern. Geol. Rev., v. 4, p. 263–270.

Dunham, K. C., 1959. Epigenetic mineralization in Yorkshire. Yorkshire Geol. Soc. Proc., v. 32, p. 2–29.

——. 1960. Syngenetic and diagenetic mineralisation in Yorkshire. Yorkshire Geol. Soc. Proc., v. 32, p. 229–284.

——, 1964. Neptunist concepts in ore genesis. Econ. Geol., v. 59, no. 1, p. 1–21.

Emslie, R. F., and J. M. Moore, 1961. Geological studies of the area between Lynn Lake and Fraser Lake. Manitoba Mines Branch Publ. 59-4.

Fersman, A., 1926. Mongolo-Okhotsky metallichesky pojas. Poverkhnost y Nedra, No. 3. Moscow.

Finlayson, A. M., 1910. The metallogeny of the British Isles. Quart. Jour. Geol. Soc. London, v. 66, pt. 2, no. 262, p. 281–298.

Fyfe, W. S., 1974. The ocean ridge environment: heat and mass transfer (abs). GAC/MAC '74, St. John's Newfoundland.

Garlick, W. G., 1953. Reflections on prospecting and ore genesis in Northern Rhodesia. Trans. IMM, v. 63(I), no. 563, p. 9–20.

Gastil, R. G., 1960. The distribution of mineral dates in time and space. Amer. Jour. Sci., v. 258, p. 1–35.

Gilmour, P., 1971. Strata-bound massive pyritic sulphide deposits—a review. Econ. Geol., v. 66, no. 8, p. 1239–1244.

Goodwin, A. M., 1961. Genetic aspects of Michipicoten iron formations. CIM Bull., Jan., p. 38–42.

——, and R. H. Ridler, 1970. The Abitibi orogenic belt, *in* Symposium on Basins and Geosynclines. Geol. Surv. Canada Paper 70-40.

Gorzhevskii, D. I., and V. N. Kozerenko, 1970. On the depth problem of postmagmatic deposits (abs.), *in* Problems of Hydrothermal Ore Deposition, eds. Z. Pouba and M. Stemprok. IUGS, Ser. A, no. 2, p. 161–165.

Griggs, D., 1939. A theory of mountain building. Amer. Jour. Sci., v. 237, p. 611–650.

Guild, P. W., 1971. Metallogeny: a key to exploration. Mining Eng., v. 23, p. 69-72.

Haile, N. S., 1969 (for 1968). Geosynclinal theory and organizational pattern of the north-west Borneo geosyncline. Quart. Jour. Geol. Soc. London, v. 124, p. 171-195.

Hall, J., 1859. Description and figures of the organic remains of the Lower Helderberg Group and the Oriskany Sandstone. Nat. Hist. N.Y.; Palaeontology. Geol. Surv., Albany, N.Y., 3, 544 p.

Harland, W. B., 1969. Interpretation of stratigraphical ages in orogenic belts, *in* Time and Place in Orogeny, eds. P. E. Kent et al. Geol. Soc. London, Spec. Publ. 3, p. 115-135.

——, A. G. Smith, and B. Wilcock, 1964. The Phanerozoic time-scale. Geological Society, London.

Haug, E., 1900. Les géosynclinaux et les aires continentales. Contribution à l'étude des régressions et des transgressions marines. Bull. Soc. Geol. France, v. 28, no. 3, p. 617-711.

Heier, K. S., 1973. Geochemistry of granulite facies rocks and problems of their origin. Phil. Trans. Roy. Soc. London, Ser. A, v. 273, p. 429-442.

Hess, H. H., 1962. History of ocean basins, *in* Petrologic Studies: A Volume to Honor A. F. Buddington. Geological Society of America, p. 599-620.

Heyl, A. V., G. P. Landis, and R. E. Zartman, 1974. Isotopic evidence for the origin of Mississippi Valley-type mineral deposits: a review. Econ. Geol., v. 69, no. 6, p. 992-1006.

Holmes, A., 1931. Radioactivity and earth movements. Trans. Geol. Soc. Glasgow, v. 18, pt. 3, p. 559-606.

——, and L. Cahen, 1955. African geochronology. Colonial Geol. Min. Resources, v. 5, no. 1, p. 3-38.

Hosking, K. F. G., 1970. Aspects of the geology of the tin-fields of south-east Asia, *in* A Second Technical Conference on Tin, ed. W. Fox, v. 1, p. 39-80. Intern. Tin Council and Dept. Min. Resources, Thailand.

Hutchinson, R. W., and D. L. Searle, 1971. Stratabound pyrite deposits in Cyprus and relations to other sulphide ores. Soc. Mining Geol. Japan, Spec. Issue 3 (Proc. IMA-IAGOD Meetings, 1970), p. 198-205.

Huttenlocher, H. F., 1926. Metallogenese und metallprovinzen der Alpen. Metall und Erz.

Irving, E., and J. K. Park, 1972. Hairpins and superintervals. Canadian Jour. Earth Sci., v. 9, no. 10, p. 1318-1324.

Isacks, B., J. Oliver, and L. R. Sykes, 1968. Seismology and the new global tectonics, Jour. Geophys. Res., v. 73, no. 18, p. 5855-5899.

Itsikson, M. I., 1960. The distribution of tin-ore deposits within folded zones. Intern. Geol. Rev., v. 2, no. 5, p. 397-417.

Ivanov, S. N., 1962. Some basic problems of the distribution of sulfide mineralization in pyrite ore provinces. Intern. Geol. Rev., v. 4, p. 151-160.

Kay, M., 1951. North American geosynclines. Geol. Soc. America Mem. 48, 143 p.

King, H. F., 1953. Environment of orebodies in relation to search for ore deposits. Proc. 15th Empire Mining Metallurg. Congr. Melbourne, v. 7, p. 151-156.

——, 1973. Some Antipodean thoughts about ore. Econ. Geol., v. 68, no. 8, p. 1369-1374.

——, and B. P. Thomson, 1953. The geology of Broken Hill district. Proc 5th Empire Min. Metallurg. Congr., v. 1, p. 533.

King, P. B., 1964. The North American cordillera, *in* Tectonic History and Mineral Deposits of the Western Cordillera. Canadian Inst. Min. Met. Spec. Vol. 8, p. 1-25.

385

Knight, C. L., 1957. Ore genesis—the source bed concept. Econ. Geol., v. 52, no. 7, p. 808–817.

Kober, L., 1921–1928. Der Bau der Erde I. Aufl. Borntraeger, Berlin, 500 p.

Krauskopf, K. B., 1967. Source rocks for metal bearing fluids, *in* Geochemistry of Hydrothermal Ore Deposits, ed. H. L. Barnes. Holt, Rinehart and Winston, New York.

——, 1971. The source of ore metals. Geochim. Cosmochim. Acta, v. 35, p. 643–659.

Kutina, J., 1972. Regularities in the distribution of hypogene mineralization along rift structures. 24th Intern. Geol. Congr., Sec. 4, p. 65–73.

Lambert, I. B., and T. Sato, 1974. The Kuroko and associated ore deposits of Japan: a review of their features and metallogenesis. Econ. Geol., v. 69, no. 8, p. 1215–1236.

de Launay, L., 1900. Sur les types régionaux de gîtes métallifères. Compt. Rend., v. 130, p. 743–746.

Lindgren, W., 1933. Mineral deposits. McGraw-Hill, New York.

Lowell, J. D., 1968. Geology of the Kalamazoo orebody. San Manuel district, Arizona. Econ. Geol., v. 63, no. 6, p. 645–654.

——, and J. M. Guilbert, 1970. Lateral and vertical alteration—mineralization zoning in porphyry ore deposits. Econ. Geol., v. 65, no. 4, p. 373–408.

Maucher, A., 1972. Time- and stratabound ore deposits and the evolution of the earth. 24th Intern. Geol. Cong., Sec. 4, p. 83–87.

Mayo, E. B., 1958. Lineament tectonics and some ore districts of the Southwest. AIME Trans., v. 211, p. 1169–1175.

McCartney, W. D., and R. R. Potter, 1962. Mineralization as related to structural deformation, igneous activity, and sedimentation in folded geosynclines. Canadian Mining Jour., v. 83, no. 4, p. 83–87.

McGregor, V. R., 1973. The early Precambrian gneisses of the Godthaab district, West Greenland. Phil. Trans. Roy. Soc. London, Ser. A, vol. 273, no. 1235, p. 343–358.

Meyerhoff, A. A., 1968. Arthur Holmes: Originator of the spreading ocean floor hypothesis. Jour. Geoph. Res., v. 73, no. 20, p. 6563–6565.

Miller, W., and C. Knight, 1915. Metallogenic epochs in the Pre-Cambrian of Ontario. 24 A. R. Ontario Bur. Mines, p. 243–248.

Mitchell, A. H., and J. D. Bell, 1973. Island-arc evolution and related mineral deposits. Jour. Geol., v. 81, no. 4, p. 381–405.

——, and M. S. Garson, 1972. Relationships of porphyry copper and circum-Pacific tin deposits to palaeo-Benioff zones. Trans IMM, Sec. B, v. 81, p. B10–B25.

——, and H. G. Reading, 1971. Evolution of island arcs Jour. Geol., v. 79, p. 253–284.

——, 1975. Cenozoic arc systems and collision belts as models for ancient orogens and related metallogeny, *in* Metallogeny and Plate Tectonics, ed. D. F. Strong. Geol. Assoc. Canada Spec. Paper 14.

Naldrett, A. J., 1973. Nickel sulphide deposits—their classification and genesis, with special emphasis on deposits of volcanic association. CIM Bull., v. 66, no. 739, p. 45–63.

Niggli, P., 1923. Gesteins- und mineralprovinzen. Berlin, Gebruder Bornträger.

Obruchev, V., 1926. Die metallgenetischen epochen und gebiete von Siberien. Halle.

Oftedahl, C., 1958. A theory of exhalative-sedimentary ores. Geol. Foren. Stockholm Forh., v. 80, pt. 1, no. 492, p. 1–19.

Pereira, J., 1963. Further reflections on ore genesis and exploration. Mining Mag., v. 109, Nov., p. 265–280.

——, and C. J. Dixon, 1965. Evolutionary trends in ore deposition. Trans. IMM, Sec. B, v. 74, p. 505–527.

Petrascheck, W., 1942. Gebirgsbildung, vulcanismus und metallogenese in den Balkaniden und Sudkarpathen. Forschr. Geol. Paleont., v. 14, pt. 47, p. i-viii, 131-181.

Peyve, A. V., and V. M. Sinitzyn, 1950. Certains problèmes fondamentaux de la doctrine des géosynclinaux. Izv. Akad. Nauk SSSR, Ser. Geol., 4, p. 28-52.

Plumstead, E. P., 1969. Three thousand million years of plant life in Africa. Geol. Soc. South Africa, Annex to v. 72. A. L. du Toit Memorial Lecture 11.

Radkevich, Y. A., 1961. On the types of metallogenic provinces and ore districts. Intern. Geol. Rev., v. 3, no. 9, p. 759-783.

Ramović, M., 1968. Principles of metallogeny. University of Sarajevo, 271 p.

Ridler, R. H., 1970. Relationship of mineralization to volcanic stratigraphy in the Kirkland-Larder lakes area, Ontario. Geol. Assoc. Canada Proc., v. 21, p. 33-43.

Robert, D., 1968. Géologie linéamentaire et richesses minérales. Rev. Ind. Min., p. 743-757.

Sawkins, F. J., 1972. Sulfide ore deposits in relation to plate tectonics. Jour. Geol., v. 80, p. 377-397.

——, and D. M. Rye, 1974. Relationship of Homestake-type gold deposits to iron-rich Precambrian sedimentary rocks. Trans. IMM, Sec. B, May, p. B56-B59.

Scheibner, E., 1973. A plate tectonic model of the Paleozoic tectonic history of New South Wales. Jour. Geol. Soc. Australia, v. 20, pt. 4, p. 405-426.

——, E., and B. P. J. Stevens, 1974. The Lachlan River lineament and its relationship to metallic deposits. Quart. Notes Geol. Surv. New South Wales, no. 14, p. 8-18.

Schneiderhöhn, H., 1923. Chalkographische Untersuchung des Mansfelder Kupferschiefers. Neues Jb. Miner, v. 47, p. 1-38.

——, 1941(a). Lehrbuch der Erzlagerstättenkunde, Erster Bard, Jena, p. 857.

——, 1941(b). Erzlagerstätten, Kurzvorlesungen zur Einführung und zur Wiederholung, Jena, p. 371.

Schuchert, C., 1923. Sites and natures of North-American geosynclines. Bull. Geol. Soc. America, v. 34, p. 151-260.

Semenov, A. I., and G. S. Labazin, 1962. Basic problems of investigation in the field of metallogeny. Intern. Geol. Rev., vol. 4, p. 139-150.

Shatalov, Y. T., 1958. The metallurgy of ore districts. Izv. Acad. Sci. USSR, Geol. Ser., no. 9, p. 28-40.

Sillitoe, R. H., 1972. Formation of certain massive sulphide deposits at sites of seafloor spreading. Trans. IMM, v. 81, Sec. B., p. 141-148.

——, 1973. Environment of formation of volcanogenic massive sulfide deposits. Econ. Geol., v. 68, p. 1321-1326.

Smirnov, V. I., 1971. Essays on metallogeny. Geol. Bull. 4, Dept. Earth Environ. Sci., Queens College, Flushing, N.Y.

Snelgrove, A. K., 1971. Metallogeny and the new global tectonics. Bull. Min. Res. Expl. Inst. Turkey, p. 130-149.

Spence, C., 1966. Volcanogenetic setting of the Vauze base metal deposit, Noranda district, Quebec. Preprint, CIM Annual Convention, 1966.

Spurr, J. E., 1923. The ore magmas. McGraw-Hill, New York, p. 457-466.

Stanton, R. L., 1955. Genetic relationship between limestone, volcanic rocks and certain ore deposits. Australian Jour. Sci., 17.

Stille, H., 1913. Evolutionen und revolutionen in der Erdgeschichte. Borntraeger, Berlin, 32p.

Stoll, W. C., 1965. Metallogenic provinces of South America. Mining Mag., v. 112, no. 1, p. 22-33, no. 2, p. 90-99.

Suess, E., 1883. Das Antlitz der Erde. Prag. F. Tempsky.

387

du Toit, A. L., 1937. Our wandering continents. Oliver & Boyd, Edinburgh.

Upadhyay, H. D., and D. F. Strong, 1973. Geological setting of the Betts Cove copper deposits, Newfoundland: an example of ophiolite sulfide mineralization. Econ. Geol., v. 68, no. 2, p. 161–167.

Wager, L. R., and W. A. Deer, 1939. Geological investigations in east Greenland, Pt. 3. Meddelelser om Grønland, bd. 105, no. 4, p. 64–67 and plt. 8.

Walker, W., 1971. Time and place in orogeny: the Precambrian of Manitoba. Geol. Assoc. Canada Spec. Paper 9, p. 61–68.

——, 1972. Mantle cells and mineralization. SME of AIME Trans., v. 252, p. 314–327.

——, 1973. The evolution of mobile belts: a comparison. The Hellenides (Alpine), La Ronge (Hudsonian), and Rice Lake (Kenoran). Geol. Assoc. Canada Proc., v. 25, p. 19–26.

——, 1975. Eras, mobile belts, and metallogeny, *in* Metallogeny and Plate Tectonics, ed. D. F. Strong. Geol. Assoc. Canada Spec. Paper 14.

Watanabe, M., 1923. The geological distribution of ore deposits in Japan. Econ. Geol., v. 18, p. 173–183.

Watson, J. V., 1973(a). Influence of crustal evolution on ore deposition. Trans. IMM, Sec. B, vol. 82, p. B107–B113.

——, 1973(b). Effects of reworking on high-grade gneiss complexes. Phil. Trans. Roy. Soc. London, Ser. A, v. 273, p. 443–455.

Weber, W., 1971. The evolution of the Rice Lake–Gem Lake greenstone belt, southeastern Manitoba. Geol. Assoc. Canada Spec. Paper 9, 97–103.

Wedepohl, K. H., 1971. "Kupferschiefer" as a prototype of syngenetic sedimentary ore deposits. Soc. Mining Geol. Japan, Spec. Issue 3, p. 268–273 (Proc. IMA-IAGOD Meetings 1970, IAGOD Vol.)

Wegener, A., 1967. The origins of continents and oceans. translated by J. Biram from 4th ed. Reproduced from Dover Publications 1966 edition by Methuen, London.

White, D. E., 1974. Diverse origins of hydrothermal ore fluids. Econ. Geol., v. 69, p. 954–973.

Willemse, J., 1969. The geology of the Bushveld igneous complex, the largest repository of magmatic ore deposits in the world, *in* Magmatic Ore Deposits, ed. H. D. B. Wilson, Econ. Geol. Monor. 4.

Wong, W., 1920. Les provinces métallogéniques de Chine. Geol. Surv. Bull. China, v. 2.

Wynne-Edwards, H. R., 1972. The Grenville Province, in Variations in Tectonic Styles in Canada, eds. R. A. Price and R. J. W. Douglas. Geol. Assoc. Canada Spec. Paper 11, p. 263–334.

Zonenshain, L. P., M. I. Kuzmin, V. I. Kovalenko, and A. J. Saltykovsky, 1974. Mesozoic structural-magmatic pattern and metallogeny of the western part of the Pacific belt. Earth Planetary Sci. Letters, v. 22, p. 96–109.

BIOGRAPHICAL NOTES ON
REPRINT AUTHORS

Biographical sketches for some authors appear with their respective articles; a number of others are reproduced here. Owing to various geographical dislocations, we did not receive biographies for all authors by the time this volume went to press; we hope nonetheless that those which do appear will be of interest and help to the reader.

J. D. BELL was born in Yorkshire, England, and studied geology at the University of Oxford (B.A., 1956). After completing his D. Phil. research on the Tertiary granites and hybrid rocks of the Isle of Skye under L. R. Wager, he joined the Overseas Geological Survey and worked in the British Solomon Islands protectorate. He returned to Oxford in 1962 and is now University Lecturer in Petrology there.

PAUL BILLINGSLEY and AUGUSTUS LOCKE: For their first twenty years as mining geologists, Billingsley's interest was *intricate three-dimensional form correlating with ore*; Locke's was the *processes of metal supply and entrapment*. Thereafter, the two joined forces and, in the 1930s, published papers with small-scale diagrams.

In 1940, Billingsley compiled, from sources including his own notes, a $1 : 2\frac{1}{2}$ x 10^6 tectonic sketch of western United States and southern British Columbia forms relating to ore district regional position; in 1952, a $1 : 5$ x 10^5 sketch $32°$ to $52°$ north latitude; and in 1960, a sketch from $52°$ north latitude into Alaska. None of these have been published; the last is in the possession of the Bear Creek Mining Company of Salt Lake City.

From time to time in the 1940s and 1950s, they worked on a manuscript on forty western ore districts; diagrams and texts passed between them in attempts to fit ore bodies into unique patterns of inlets, channel systems, districts, megadistricts, megalodes or belts, and regions or provinces, by specification, on various scales, of loci of supply and entrapment.

Billingsley died 1962. Locke continues work on the manuscript.

389

JOHN MALCOLM BIRD was born December 27, 1931, in Newark, New Jersey. He received his B.S. from Union College in New York, his M.S. from Rensselaer Polytechnic Institute in 1959, and his Ph.D. from Rensselaer in 1962. He has served as instructor, assistant professor, associate professor, and since 1969 as chairman of the Department of Geological Sciences at the State University at Albany, New York. Bird has received numerous research grants. He is a member of the Geological Society of America and the American Geophysical Union. His interests include geologic mapping of the Taconic region, sedimentary structures, petrography of sediments and metasediments, lithosphere plate tectonics, and the evolution of mountain belts.

LUCIEN CAHEN was born in Brussels, Belgium, and took degrees in mining engineering (1936) and geology (1947) at the University of Brussels. From 1938 to 1941 he was connected with the Comité Spécial du Katanga and was active in geodetic work and topographical and geological mapping of parts of Shaba (formerly Katanga), Zaire (formerly the Belgian Congo). He then worked for several years in Bas Zaire (formerly Bas-Congo), performing geological mapping for the geological survey and later as leader of a prospecting party for the Bamoco Syndicate for mining research. He has also been and still is actively engaged in geological work in other parts of Central Africa. In 1947, Cahen entered the Geological Department of what is now the Musée Royal de l'Afrique Centrale at Tervuren, Belgium, and became director of this institution in 1958. He is also professor of the Geology of Africa at the University of Brussels.

ALAN H. CLARK was born July 14, 1938, in London, England. He is currently Associate Professor of Economic Geology at Queen's University, Kingston, Ontario. He was educated at Christ's Hospital, the Royal School of Mines (1957-1966), and received his Ph.D. from the University of Manchester in 1964 and N.R.C.P.D.I. from McGill University (1964-1965). From 1965 to 1967 he was Lecturer at University College, London. In 1967, he came to Queen's University. His special interests include the geology and mineralogy of ore deposits and the regional metallogeny of the central Andes and southern Yukon.

JOHN FREDERICK DEWEY was born May 22, 1937, in London, England. In 1958, he received a B.Sc. from the University of London, and, in 1960, a Ph.D. in structural geology from Imperial College. In 1964, he received his M.A. from the University of Cambridge. He has served as Assistant Lecturer at the University of Manchester and as Lecturer at the University of Cambridge. Since 1970, he has been professor at the State University at Albany, New York. Dewey has received numerous research grants. He is a member of the Geological Society, London, the Geological Association of Canada, and the American Geophysical Union. His interests include the stratigraphic–structural evolution of the Appalachian–Caledonian orogen and the Alpine orogen of Europe, and the significance of plate tectonics for the evolution of continental margins and orogenic belts.

ROBERT S. DIETZ (Ph.D., University of Illinois, 1941) is a marine geologist with NOAA, Atlantic Oceanographic and Meteorological Labs, in Miami, Florida. His extensive scientific publications include papers on the evolution of continental margins, marine sedimentation, geotectonics of ocean basins, continental drift, actualistic geosynclines, and sea-floor spreading (a concept of which he is the cooriginator with the late Harry Hess). Additional oceanographic interests are marine resources and underwater sound, diving with deep research vehicles (he coauthored with J. Piccard the documentary account of the bathyscaph *Trieste, Seven Miles Down)*. He also has led several deep-sea oceanographic expeditions and traveled widely overseas for various geologic studies. A contrasting specialty has been meteoritics, selenography, and astroblemes (impact structures).

COLIN J. DIXON was born in London, England, and graduated in mining geology at Imperial College in 1957. He has made his career teaching mining geology at Imperial College, becoming Lecturer in 1959 and Senior Lecturer in 1973, although he has worked extensively for mining companies and foreign governments in Europe, Africa, and the Middle East as a consultant. In 1968–1969 he was Research Fellow at the École des Mines de Paris, working on problems of data processing in geology, and has worked in collaboration with J. Pereira on regional aspects of mineral distribution. He is now engaged in research on ore reserve and grade control problems and is supervisor of the one-year advanced course in mineral exploration.

SIR KINGSLEY DUNHAM is Director of the Institute of Geological Sciences, set up in 1965 to incorporate the Geological Survey of Great Britain, Overseas Geological Surveys, and the Geological Museum. A member of Hatfield College, he took his first degree and Ph.D. in geology at the University of Durham under Arthur Holmes. He was a Commonwealth Fund Fellow at Harvard University, during the tenure of which he acquired the M.S. degree in mining engineering and the S.D. for a thesis on a metalliferous area of New Mexico. From 1935 to 1950 he served on the Geological Survey of Great Britain, of which he had become Chief Petrographer before he resigned to return to Durham as Professor of Geology. During his time there he acted as geological adviser to a number of leading industrial companies. In 1967 he returned to South Kensington to direct the new Institute. He was elected a Fellow of the Royal Society in 1955 and has been Foreign Secretary of the Society since 1971. Recently he became Chairman of the newly formed Council for Environmental Science and Engineering. At present he serves as British member of the council of the International Institute for Applied Systems Analysis in Vienna and has represented the Royal Society in the international discussions leading to the setting up of the European Science Foundation linked with EEC.

Sir Kingsley is best known for his researches on the ores of lead, zinc, and iron and on the evaporite deposits, but his other interests include Carboniferous stratigraphy and the petrology of basalts.

R. GORDON GASTIL was born in San Diego, California, and studied geology at San Diego State University and the University of California at Berkeley (Ph.D., 1953). His dissertation was on Precambrian rocks in central Arizona. He has subsequently worked in southern Alaska, the Ungava Peninsula, southern and Baja California, and the states of Sonora and Nayarit, Mexico. He is currently a professor at San Diego

391

State University. His recent work has been concerned with the reconstruction of Mesozoic California and western Mexico.

ALAN MURRAY GOODWIN was born June 20, 1924, in Quebec, Canada. He received his B.Sc. from Queen's University, Ontario, in 1949, his M.Sc. from the University of Wisconsin in 1951, and his Ph.D. from the University of Wisconsin in 1953. From 1953 to 1961 he was research geologist for Algoma Steel Corp. Ltd. Since then he has worked as field geologist for the Ontario Department of Mines (1961–1965) and as research geologist for the Geological Survey of Canada (1965–1969). Since 1969 he has been Professor of Geology at the University of Toronto. In addition, Goodwin is a member of the Geological Society of America, where he has served as Associate Editor. He is also a member of Geological Association of Canada, which he served as president from 1967 to 1968, the Canadian Institute of Mining and Metallurgy, and the Society of Economic Geologists. He has contributed articles to numerous professional journals.

HARRY HAMMOND HESS was born in New York City on May 24, 1906, and died on August 25, 1969, at Woods Hole, Massachusetts, literally with his boots on, while chairing a Space Science Board conference on the scientific objectives of lunar exploration.

After receiving a B.S. degree from Yale in 1927, he spent eighteen months as an exploration geologist in Northern Rhodesia. He received a Ph.D. degree from Princeton in 1932 with a thesis on a hydrothermally altered peridotite at Schuyler, Virginia. While completing his thesis, he had the opportunity of accompanying the famous Netherlands scientist, Vening Meinesz, on a gravity-measuring cruise in an old World War I submarine in the Caribbean region. Thus, under these circumstances, in this year, were initiated the major interests of his whole career: ultrabasic rocks, gravity, oceanic crustal studies, the geology of the Caribbean region, and the U.S. Navy.

In 1932–1933, Hess taught at Rutgers University and in 1933–1934 was a Research Associate at the Geophysical Laboratory in Washington. In 1934, he returned to Princeton. He had attained the rank of full professor by 1948 and was made Chairman of the Department of Geology in 1950. He had been made a reserve officer in the navy in 1935, and in 1941, immediately after Pearl Harbor, was assigned to active duty in the navy. He was made commanding officer of the U.S.S. *Cape Johnson* in 1945; in 1961, he received the rank of Rear Admiral in the U.S. Naval Reserve.

Ultrabasic rocks were one of Harry Hess's first and most enduring interests. His work on serpentinites and peridotites, their mineral components, and their role in crustal history is classic. Among major contributions in this field are his memoir on *The Stillwater Complex in Montana* (1960) and his contribution to *The Crust of the Earth* (1955). His numerous detailed published studies of the pyroxenes are fundamental to the understanding of this mineral group.

From 1957 until its demise in 1966, Hess was one of the recognized leaders of the famous Mohole Project. The Caribbean region provided Hess with an alpine-type mountain system in Venezuela–Trinidad, an island arc in the West Indies, and extensive occurrences of his beloved ultrabasic rocks in both. Thus came into being the Princeton Carribean Research Program. He became deeply involved in the Moon Project, and in 1962 was a leader in the National Academy's review of the NASA program. This led to his appointment as Chairman of the academy's prestigious Space Science Board (1962-1969), from which position he influenced many of the decisions on the lunar and planetary program.

KENNETH F. G. HOSKING, a Cornishman from the tin-mining district of Camborne, has B.Sc., M.Sc., Ph.D., and D.Sc. degrees in geology from the University of London. He is also a Fellow of the Institution of Mining and Metallurgy. After twenty years (1948-1968) at the School of Mines, Camborne, Cornwall, where he was head of the Department of Geology and Applied Geochemistry, he left to take up an appointment as Visiting Professor in Applied Geology in the Department of Geology, University of Malaya. On leaving the School of Mines he was made an A.C.S.M. (Honoris causa). After three years in the University of Malaya he was appointed to the Chair in Applied Geology and still holds this position.

Hosking has worked as a consultant in Great Britain, France, Spain, Portugal, Greece, Nigeria, Rhodesia, Canada, Malaysia, Thailand, and Indonesia. The award of an Anglo-American traveling scholarship enabled him to visit South Africa and Zambia. His major fields of research are tin and associated deposits, applied geochemistry, aids to mineral identification, and the genesis of minerals in recent placers. He has written about 130 papers.

V. I. KOVALENKO was born in the Tula district, USSR, and studied geology at the Moscow State Geological Prospecting Institute. From 1960 until the present he has worked at the Institute of Geochemistry of the Siberian Branch of the USSR Academy of Science in Irkutsk. In 1965, Kovalenko received his candidate degree (Ph.D) for work dealing with the petrology and geochemistry of the alkaline rocks of the eastern Sayan. In 1967 he began investigations into the petrology, geochemistry, and metallogeny of Mesozoic granites of Mongolia, taking part in the Joint Soviet-Mongolian Geological Expedition.

M. I. KUZMIN was born in Moscow, USSR. He graduated from the Geological Department, Lomonosov Moscow State University, in 1960, and began work at the Institute of Geochemistry of the Siberian Branch of the USSR Academy of Science, organized in Irkutsk in the end of 1950s. In 1966, Kuzmin received his candidate degree (Ph.D.) for work dealing with the geochemistry of Transbaikalian granitoids. Since 1967 he has investigated the petrology, geochemistry, and metallogeny of Mesozoic granites of Mongolia, participating in the Joint Soviet-Mongolian Geological Expedition.

AUGUSTUS LOCKE: *see* PAUL BILLINGSLEY and AUGUSTUS LOCKE

J. DAVID LOWELL was born in Tucson, Arizona, and studied mining engineering and geology at the University of Arizona (B.S. Min. Engr., 1949; Engr. Geol., 1959) and geology at Stanford University (M.S., 1957). He worked as engineer, mine foreman, and mine manager, and as Geologist, Senior Geologist, District Geologist, and Chief Geologist for Asarco, U.S.A.E.C., subsidiaries of Ventures Ltd. of Canada and

Utah International. Since 1962 he has been a consulting geologist specializing in porphyry copper exploration and has done consulting work for fifty companies and government organizations in fifteen countries.

ALBERT MAUCHER was born in Freiberg, Germany, in 1907 and studied mining and metallurgy at the technical university of Aix-la-Chapelle (Dipl. Ing., 1930). During this time, Paul Ramdohr stimulated his interest in ore deposits and ore microscopy. He graduated at the Technical University of Munich (Dr. Ing., 1932), spent half a year with V. M. Goldschmidt at Göttingen (1934), and became Assistant Professor at the Technical University of Berlin in the Institut of F. K. Drescher-Kaden (1934). In 1936 he spent a year as mineralogist at the mining research institute of the Turkish government (MTA-Enstitüsü) in Ankara, visiting most of the ore deposits in Turkey. In 1937 he became Assistant Professor at the University of Göttingen, where he worked on geochemical problems, especially of ore deposits in metamorphic rocks (massive pyrite, graphite, and copper).

In 1947 he became Professor and Director of the Institute of General and Applied Geology at the University of Munich, where he organized a special branch for geochemistry and ore deposits, with a particular interest in stratabound deposits (Alpine lead–zinc, stibnite–scheelite–cinnabar, uranium, massive sulfides, and Kuroko ores). From 1956 to 1965 he spent two months every year in Turkey consulting for the Department of Economic Geology at the MTA-Enstitüsü, Ankara. In 1973 he retired from his professorship but is still devoted to the problems of time-bound ore deposits and the evolution of the Earth's crust and mantle. Since 1972 he has been Managing Editor of *Mineralium Deposita*.

ANDREW MITCHELL was born in England in 1939. Two years articled to a Chartered Accountant in London were followed by a year at Sir John Cass College, London; he obtained a B.A. in Geology at Trinity College Dublin in 1963. Three years with Overseas Geological Surveys in the New Hebrides Condominium and work at Oxford under Harold Reading on sedimentary facies in volcanic arcs led to a D. Phil. in 1968.

Field seasons in Thailand (1969), Burma (1970 to 1973), and briefly in Oman, working with the Institute of Geological Sciences for the U.K. Overseas Development Ministry, were interspersed with six months with Magellan Petroleum (Australia) in Fiji, the Trobriand Islands, and Australia, and with periods at Oxford financed by Rio Tinto Zinc working on plate tectonics related to mineralization in Southeast Asia. He joined the United Nations Development Program as Chief Geologist on a mapping and exploration project in Burma in 1974.

EDNA PAULINE PLUMSTEAD (née Janisch) was born in 1903 and studied at the University of the Witwatersrand, Johannesburg, where she graduated M.Sc. with distinction in 1925 and was awarded a postgraduate research scholarship. Subsequently, she studied coal petrology and the palaeobotany of South African coals at Cambridge University. In 1959, she was awarded the degree of D.Sc. from Witwatersrand.

After leaving Cambridge University, she served on the geology staff of the University of Witwatersrand until her marriage in 1934 to Edric R. A. Plumstead, a mining engineer. After an interval of twelve years spent raising a family, she was recalled by the university to assist in the teaching of ex-service men. She remained as Senior Lecturer until her retirement, but continues her research work today.

She has published 49 scientific papers on geological and palaeobotanical subjects, especially on southern hemisphere fossil floras, including those of Antarctica. In addition, she has been a delegate to a number of overseas scientific congresses and from 1964 to 1973 served as Chairman of the Gondwana Subcommission of the Commission of Stratigraphy of the International Union of Geological Sciences, presiding at Gondwana Congresses in India, South America, South Africa, and Australia. As a result, she has acquired a keen interest in continental drift and plate tectonics.

She has been awarded all three medals of the South African Geological Society and in 1959 the Crestian Mica Gondwana Gold Medal of the Mining, Geology, and Metallurgy Institute of India. In 1973 she received the South African Science Medal and Grant of the South African Association for the Advancement of Science. Dr. Plumstead is a Fellow of the Royal Society of South Africa and, in addition, has held office in a number of professional organizations.

ROLAND H. RIDLER, a Torontonian, received an Honors B.A. Sc. in 1964 and an M.A. Sc. in 1966 from the University of Toronto. A Ph.D. program in economic geology at the University of Wisconsin, Madison, with fieldwork supported by the Geological Survey of Canada, led to a doctorate in 1969 on the relation of mineralization to volcanic stratigraphy in the Kirkland Lake area, Ontario. In 1968, Ridler assumed an N.R.C. postdoctoral fellowship at the University of Western Ontario, and with field support from the Canadian Survey studied the role of exhalites in the metallogeny of the Kirkland Lake–Noranda area. Since 1970, Ridler has been a research scientist with the Geological Survey of Canada, Ottawa, pursuing volcanic stratigraphic and metallogenic studies in the Rankin Inlet–Ennadai Belt of the District of Keewatin and the Abitibi Belt of Ontario and Quebec.

ARTHUR I. SALTYKOWSKY was born in Moscow, USSR, and studied geology at Moscow State University (B.A., 1957). He works at the Institute of the Physics of the Earth in Moscow. Saltykowsky is studing problems associated with volcanism and its relations with different structures of the earth's crust. For 12 years he worked with J. M. Scheinmann in the Institute of the Physics of the Earth. In recent years he has worked in Transbaikalia and Mongolia.

ERWIN SCHEIBNER was born in Jasina, Czechoslovakia, and studied geology at the J. A. Comenius University in Bratislava (Dipl. Geologist, 1958), where he later became lecturer and senior lecturer in the Department of Geology and Palaeontology. In 1964, he obtained the degrees of Candidate of Geological Sciences and Rerum Natur. Doc. (equivalent to a Ph.D.). In 1967 he qualified as Docent.

Scheibner's main research interest has been regional geology (tectonics) of West Carpathians, especially of the Pieniny Klippen Belt, where he continued the work of his teacher D. Andrusov. In 1964 he took part in the Czechoslovak Geophysical Expedition to Northern Afghanistan (geological interpretation of seismic measurements). In 1967–1968 he obtained a senior fellowship with the Alexander von Humboldt Stiftung and was with the university in Munich. After August 1968 he left Czechoslovakia. He spent three months working with R. Trümpy at the Technical University in Zürich and since December 1968 he has been with the Geological Survey of New South Wales. At present, he is a senior research scientist studying the structure and tectonics of New South Wales and eastern Australia.

JOSIAH EDWARD SPURR was born in 1870 in Gloucester, Massachusetts, and died in 1950. He received his A.B. from Harvard in 1893 and his A.M. in 1894. Spurr's mining interests led to a varied career in industry, academe, and government. From 1901 to 1902 he served as a mining engineer and geologist to the Sultan of Turkey; from 1902 to 1906 as geologist with the U.S. Geological Survey; from 1906 to 1908 as chief of the geology department of American Smelting and Refining Company; from 1908 to 1911 as mining specialist for Spurr & Cox, Inc.; and from 1911 to 1917 as vice-president in charge of mining for the Tonopah Mining Company of Nevada. Spurr was very active in government during World War I, serving as chief engineer of the War Mineral Relief in 1918–1919. From 1919 to 1927 he was editor of the *Engineering and Mining Journal* and from 1930 to 1932 Professor of Geology at Rollins College.

His associations include the Mining and Metallurgical Society of America, of which he served as president in 1921, the American Institute of Mining and Metallurgical Engineers, the Society of Economic Geologists, of which he was president in 1923, the Geological Society of America, and the American Geological Society.

Spurr authored a number of books, including *The Iron-Bearing Rocks of the Mesabi Range in Minnesota* (1894), *Through the Yukon Gold Diggings* (1900), *Geology Applied to Mining* (1904), *Political and Commercial Geology* (editor; 1921), *The Ore Magmas* (1923), *Geology Applied to Selenology* (1944), *Features of the Moon* (1945), *Lunar Catastrophic History* (1947), and *The Shrunken Moon* (1949), as well as various monographs and reports on economic geology.

Mt. Spurr peak in Alaska was named after Spurr by the U.S. Geological Survey in honor of his exploration in Alaska in 1896 and 1898.

DAVID STRONG was born in Botwood, Newfoundland, in 1944 and received his B.Sc. from Memorial University in 1965. He has an M.Sc. from Lehigh University and a Ph.D. from the University of Edinburgh (petrology, 1970). Since 1970, Strong has been with the Geology Department of Memorial University and has been appointed acting head of the department for 1974–1975. Strong is an Associate Editor of *Geoscience Canada* and was Chairman of the NATO Advanced Studies Institute on Metallogeny and Plate Tectonics held in St. John's in May 1974. He is editing the proceedings of this symposium, which will be published as a special paper of the Geological Association of Canada.

HANSA UPADHYAY was born in Lucknow, India, in 1944 and received his B.Sc. in 1964 and M.Sc. in 1966 from the University of Lucknow. He received an M.Sc. from Memorial University in 1970 and his Ph.D. in 1973. He is presently teaching economic geology at Northeastern University in Chicago, Illinois.

LAWRENCE RICKARD WAGER was born February 5, 1904, and died November 20, 1965. He was a born natural scientist; his interest in geology was already well established while still a schoolboy, and it is not surprising that his earliest work was concerned with Carboniferous Limestone tectonics in the Craven area and with meta-somatism in the Whin Sill. He soon widened his horizon, however, and after a short interlude in Connemara, in particular in the Roundstone district of Galway, he be-gan in 1930 his work in Greenland—a work that was to remain his great passion un-til his death. In 1935-1936, Wager organized his own expedition, the British East Greenland Expedition, which spent 15 months in the Kangerdlugssuaq region. Dur-ing this period the detailed field investigation of the Skaergaard intrusion, the pre-liminary examination of the Kangerdlugssuaq alkaline complex and the banded gab-bro complex of Kap Edvard Holm, together with the reconnaissance mapping of some 30,000 square kilometers was accomplished.

Wager achieved a remarkable synthesis in presenting the bold lines of the geology between Scoresby Sound and Angmagssalik. In parallel with this broad geological canvas and deriving particularly from his discovery of the Tertiary igneous rocks in the Kangerdlugssuaq region, he made what many believe to be a greater contribution by executing a small part of the canvas in fine detail. This was the work on the Skaergaard intrusion, especially that part of it relating to the mechanics of fractional crystallization, convective circulation, and sedimentation processes in basic magmas. Much of the terminology now in common use to describe these processes and the mineralogical textures to which they give rise are the direct result of Wager's work in this field.

After World War II, Wager, although continuing with his Greenland work, widen-ed his horizon to carry out investigations in islands as far apart as St. Kilda and St. Vincent, and made the time and found the energy to visit and to work on the Bush-veld Complex and in the British Tertiary Igneous Province, particularly in Skye.

When Wager arrived in Oxford in 1950, the department was hardly equipped to take advantage of the rapidly broadening fields of geological research. It was not long, however, before the department took on a new look, and became the first laboratory in the United Kingdom for isotope geochemistry and radiometric dating. It is not surprising that Wager pressed for more accommodation, and the present enlarged building is a lasting memorial to his personal scientific reputation and the doggedness of his persistance to obtain the best conditions for the department's work. (This biography is taken from *Proceedings of the Geological Society of Lon-don, 1636*, Feb. 1967, pp. 211–212.)

LEV P. ZONENSHAIN was born in Moscow, USSR, and studied geology at Moscow State University (B.A., 1952). From 1953 to 1964 he worked in the Aerogeological Trust, Ministry of Geology of the USSR, being engaged in mapping the south Siberian regions. In 1965 he joined the Laboratory of Geology of Foreign Countries. He studied the tectonics and metallogeny of Mongolia and adjacent parts of central and eastern Asia. He received his candidate degree (Ph.D) in 1962 and his full doc-torate degree in 1970. He is involved now in research problems associated with the new global tectonics and metallogeny. He now works at the Institute of Oceanology, Academy of Science of the USSR.

AUTHOR CITATION INDEX

SUBJECT INDEX

About the Editor

WILFRED WALKER is an exploration geologist whose primary interest is in base and precious metals. Following graduation from Durham in 1950, he moved to Canada, where he was deeply involved in airborne geophysics and photogeology in his work on the worldwide projects of the Hunting Group. As a consultant to his prime client of recent years, Barringer Research Ltd., he aided in the evaluation of regions proposed for exploration and of the large volume of geological, geophysical, and geochemical data resulting from regional exploration. The results of this work, as well as the concepts underlying it, have been published in a series of papers. In an effort to develop a broader understanding of patterns of global geology and metallogeny, Mr. Walker recently began a series of extended trips to mobile belts and metallogenic provinces in all parts of the world.

Date Due

MAY 3 1983			
			UML 735